AGRICULTURE ISSUES AND POLICIES

WHEAT: GENETICS, CROPS AND FOOD PRODUCTION

AGRICULTURE ISSUES AND POLICIES

Additional books in this series can be found on Nova's website
under the Series tab.

Additional E-books in this series can be found on Nova's website
under the E-books tab.

AGRICULTURE ISSUES AND POLICIES

WHEAT: GENETICS, CROPS AND FOOD PRODUCTION

MARIA T. ALMEIDA
EDITOR

Nova Science Publishers, Inc.
New York

For permission to use material from this book please contact us:
Telephone 631-231-7269; Fax 631-231-8175
Web Site: http://www.novapublishers.com

Library of Congress Cataloging-in-Publication Data

Wheat : genetics, crops and food production / editor: Maria T. Almeida.
p. cm.
Includes index.
ISBN 978-1-61209-307-9 (hardcover)
1. Wheat. I. Almeida, Maria T.
SB191.W5W356 2011
633.1'1--dc22

2010048367

Published by Nova Science Publishers, Inc. † New York

CONTENTS

PREFACE

Globally, wheat is the leading source of vegetable protein in human food. This book presents current research in the genetics, crop and food production of wheat. Topics discussed include weed management in wheat production; aquaporins and their potential in wheat improvement; improving the drought and salinity tolerance in wheat; isolation of starch from wheat and endophytic fungi in biocontrol and biofertilizer applications. (Imprint: Nova Press)

Chapter 1 - This chapter deals with negative effects of weeds on wheat crop, the impact of weed competition on quantity and quality of wheat yield, allelopathic effects of certain weeds on crop growth and development and the poison effect of other weed species to humans and animals. Weeds usually found in wheat fields and associate with crop plants were identified and grouped based on their morphology. Studies on the determination of the critical weed-free period in wheat were reviewed and the importance of this period in weed management was discussed. Chemical and none-chemical weed control methods were reviewed. Certain agricultural and management practices aimed at promoting early canopy crop development and reduce weed growth are mentioned. Importance of plant factors, timing of farm operations, and selection of proper herbicide are discussed. Most recent research findings on weed management in wheat fields are incorporated. Herbicides recommended and practiced in wheat, their type, method of action, method and time of application, physiological effects on crop plants and weeds, and recent research findings on this aspect were included. Some suggestions and recommendations that sustain weed management and aimed at increasing crop productivity and minimizing weed effects are discussed.

Chapter 2 - Wheat, a noble cereal of high global consumption is, produced in quantity and quality by a few countries and imported by developing countries, which in most cases do not have enough production to supply the internal market. In this chapter, we will present the efforts of countries as Brazil and Mexico, which require large volumes of wheat import for bread making products, in production and evaluation of cultivars adapted to environmental conditions, as in development and evaluation of products with the use of additives and ingredients to improve the quality of bread obtained. Initially, we will focus on research conducted to improve technological and commercial quality of Mexican wheat cultivars aimed at improving the production processes for the producer and milling industries, which increasingly require good quality wheat for bread. In this chapter, we will detail relevant work done in Mexico. These studies have been done considering that technological quality of wheat depends on genetic fitness of the variety, climatic conditions during the crop cycle, soil resources to be cultivated, technological resources applied in cultivation, post-harvest management of field production, and industrial flour processing. Finally, the chapter will detail the work of Research Institutes and Universities in the area of Cereal Technology being directed at improving nutritional and technological quality of wheat products, with major studies in recent years involving the use of gamma radiation and infrared radiation on whole wheat flour as a method to extend the shelf-life, the effect of enzymes in whole grain breads using the enzyme transglutaminase to strengthen the gluten network in bread making products, the use of mixed flours, and application of thermoplastic extrusion, among other works that will be described in this chapter. With this chapter, it is expected that the results and trends pointed out can contribute to other countries that also need large volumes of wheat import for bread.

Chapter 3 - Wheat is the largest food crop that humans rely on, and as such, the world wheat production is expected to keep pace with demands of the growing population and changing climate. There are also new pressures on cereal grains from the bioethanol industries. However, wheat as well as other cereals are severely affected by climatic factors such as salinity, drought, and nutrient deficiency or toxicity in a number of areas around the world. These environmental stresses variously impact plant growth, health, nutritional value and crop yields in various geographic regions and climates. A number of genetic factors are known to participate in the responses of plants to abiotic stresses, and involve interacting networks of specific genes and proteins with the ultimate goal of stress tolerance. Aquaporins, members of the major intrinsic protein (MIP) family, are channel proteins located in cellular membranes, and their roles in water uptake in plants are now well established. In addition, recent years have seen an explosion in the number of substrates found to be transported by these channels, many of the substrates being of physiological importance. Further, aquaporins are also shown to contribute to processes such as germination, plant development and photosynthesis. In this work, the key biochemical properties of aquaporins will be summarised, followed by an overview of the main subfamilies of MIPs, the types of substrates that permeate through these channels and their effects on plant physiology, and perspectives on applications of these proteins in cereal biotechnology.

Chapter 4 - Wheat provides about 19% of the food calories and 21% of the protein to the world population. In 2008, the total wheat area grownwas about 224 million ha and with a harvest of 690 million t of grain.The Green Revolution in wheat production, spawned by the development of short-stature, photoperiod-insensitive, high yielding wheat cultivars with host plant resistance to rusts and other pathogens, was able to increase wheat productivity at a 3.6% annual rate in the 1960s and 1970s. However, yield growth in wheat slip to 2.8% from the mid-1980s to the mid-1990s, and to 1.1% since the mid-1990s. In the period leading up to 2020, demand for wheat for human consumption in developing countries is expected to grow at 1.6 % per annum, and for feed at 2.6 % per annum. The global average of wheat yield will have therefore to increase in the coming years at least to 3.5 t ha^{-1}. This yield increase, essential to maintain global food security, requires a continuing supply of improved germplasm and appropriate agronomy in order to sustain enhanced productivity and preserve the natural resource base of the wheat farming. Global warming, as a result of climate change, may negatively affect wheat grain yields —potentially increasing food insecurity and poverty. Wheat breeders will depend on the crop´s genetic diversity and modern genetic enhancement methods (both genomics and transgenics) for improving wheat yield and its stability. They should conservewheat germplasm, study it, add value to it, and share it in enhanced form with clients worldwide. While changes in plant architecture (dwarfism and harvest index), photoperiod response and breeding for host plant resistance (especially for rusts) led to the productivity gains of the Green Revolution, adaptation to climate change (water and heat stresses, emerging pathogens and pests), efficient input use, food safety and quality, are additional global challenges to be tackled to fit newly bred-genotypes into sustainable and improved wheat farming in this 21st Century.

Chapter 5 - Wheat is the world's most popular crop. It is grown over a large area and under a wide range of conditions and provides more nutrition to humans than any other species. Wheat is the leading source of vegetable protein in human food, having higher protein content than either maize (corn) or rice, the other major starchy crops. Wheat is the primary food staple in North Africa, south Asia and the Middle East, and is growing in popularity in Asia. In world different types of species are grown according to their geographical conditions. Some species are Einkorn (*T. monococcum*), Emmer (*T. dicoccon*), Durum (*T. durum*), Kamut (*T. polonicum), Bread* wheat (*T. aestivum*) and Spelt (*T. spelta*). These species are used for production of different cultural foods based on religion or customs including different breads. Various types of breads are made in different cultures that are consumed as staple or at special festivals. In addition breeding wheat for quality was discussed and possibilities of increasing yield through breeding products such as synthetics, hybrids and biotechnology product i.e. transgenics wheats was reviewed.

Chapter 6 - Wheat has a broad genetic potential for the expression of various gluten proteins and is strongly influenced by the environment, e. g. with respect to the fertilization of nitrogen (N) and sulfur (S) during wheat plant development. Nitrogen and S availability markedly change wheat flour and baking quality. However, without adequate management of N and S fertilization, the genetic potential of wheat cannot be exploited. Nitrogen fertilization primarily affects the concentration of gliadines and glutenines as it is source-regulated. Storage protein concentration, particularly of gliadine and glutenine, is highly correlated with loaf volume. The use of three instead of one dressing of fertilizer, especially as a late dressing at ear emergence, the gliadine and glutenine concentration of the grain and, furthermore, loaf volume can be enhanced. However, not only quantity, but also the quality of gluten proteins is important for their composition and characteristics. Despite protein amount being only slightly influenced by increasing S fertilization levels, the composition of gluten proteins is highly affected. Increasing S availability for plants enhances S-containing amino acids, namely cysteine and methionine, which are foremost in S-rich gluten proteins. For an improvement of baking quality, a high concentration of S-rich proteins is of particular importance in order to form an appropriate gluten network within the dough. Additionally, a late S dressing can further improve baking quality, since such a dressing increases the synthesis of high molecular weight (HMW) glutenins that considerably affect baking quality in wheat. S-rich components such as sulfate and glutathione are, moreover, important molecules for the transport and supply of S to the grain. Both these molecules represent important S transport forms from source to sink organs. The potential to synthesize large amounts of storage protein depends on the ability of the plant to take up and to transport S into sink organs such as ears and kernels. Sulfur and glutathione are therefore important molecules functioning as actuators during plant growth and grain development. Furthermore, glutathione can function as internal plant signal for the S status of the plant. In terms of baking quality, glutathione is able to interact with gluten proteins responding in a change of existing disulfide bonds between gluten proteins and resulting in a different rheological property and baking quality. Concerning the interaction of N and S in wheat, a well-balanced

N/S ratio is exceedingly important for suitable gluten and baking quality. If high N fertilization is accompanied by inadequate S fertilization, S deficiency is provoked, resulting in changed gluten protein composition and a loss of nutritional quality of the grains.

Chapter 7 - Drought and salinity are major constraints on crop production and food security, and adversely affect entire countries over several years and result in serious social, economic, and environmental cost. Wheat production in the Mediterranean region is limited mainly by the availability of water resources. Water-deficit stress caused by drought and soil salinization adversely affects plant growth and crop productivity. This situation is compounded firstly by the predicted change in climate with increased temperatures and decreased precipitation as a result of global warming and secondly by the narrowness of genetic diversity of modern wheat cultivars due to domestication, selection and conventional breeding programs that minimizes plant tolerance to such stresses. The challenge of the majority of breeding programs, which are in continuous profit from the development of biotechnology, genetic engineering and genomic–based approaches, is the release of improved wheat cultivars able to withstand these adverse conditions. The identification of genes that control discrete agronomic traits, the mapping of QTLs (Quantitative Trait Loci) affecting tolerance to drought and salinity in model species and crops are powerful support for the development of functional markers, which will assist selection in the breeding process.

Chapter 8 - The main component of wheat is starch, which constitutes 63 – 72 % of wheat grain (d.m.). Wheat flour, produced by wheat grain milling, is major ingredient in most European breads and pastries. Since part of the grain is removed during production of flour, starch content in flour rises to app. 80 % d.m., and in bread it averages around 50 % of total bread mass. Starch is main source of energy in human diet. Therefore, it is very important to understand its properties and changes it undergoes during processing, as well as its physiological effects and influence of botanical source on these characteristics. It is well known that starch constitutes of two polysaccharides: linear amylose and branched amylopectin, organised in granules of different sizes and shapes, depending on botanical source. The cluster model of organisation of amylose and amylopectin is widely accepted. However, its chemistry, biochemistry, genetics, processing and modifications, as well as behaviour of starch and starchy materials in different thermodynamic conditions, although revealed to some extent, are still object of many researches and are not yet fully understood. This chapter will deal with isolation of starch from wheat and its characterisation. The first part of book chapter will give literature review of different processes of starch isolation from wheat and its properties, while in the second part of chapter laboratory isolation of starch from two Croatian wheat varieties (*V1* and *V2*) and results gained by its characterisation will be presented. Properties of starch: gelatinisation, retrogradation, pasting properties, swelling capacity, gel texture, resistant starch content etc. will be discussed.

Chapter 9 - Wheat (Triticum aestivum L.) is an important source of food for people of the world. It is widely planted and is subjected to different stresses. Therefore, it is pertinent to suggest and test methods and/or techniques, which may contribute to enhanced wheat production under stress. For example, a large part of cultivating lands are saline, acidic, and compacted, as a result of using agricultural machinery in the field, or are subjected to drought stress. Methods or techniques, which are tested, must be environmentally and economically friendly and recommendable. There have been biological and molecular methods, which have shown to be promising for wheat cultivation under different stresses. For example, using the symbiotic soil fungi, arbuscular mycorrhiza (AM) or identification and insertion of stress controlling genes in the planted seeds are among such methods. Using AM fungi under compaction stress has been effective on alleviating the stress, or molecular techniques have made possible the use of stress controlling genes for higher production under stress. The most important related findings are reviewed in this article.

Chapter 10 - Environmental hazards and health problems due to massive application of chemical pesticides in agriculture incite huge public concerns. A solution is introduction of biological control means to manage outbreaks of plant diseases. To date, only small numbers of beneficial microorganisms, belonging to the category of hyperparasitic or mycoparasitic fungi, have proven to protect crops from plant pathogen infection. Wheat crops, in all phases of growth, are subjected to infections by several groups of pathogens or pests. *Fusarium* pathogenic fungi are responsible for many different severe diseases, such as Fusarium head blight (FHB), Fusarium damaged kernels (FDK), Fusarium wilt, Fusarium crown rot, as well as Fusarium root rot. *Fusarium* mycotoxins produced are also important contaminants on food and agricultural products. Control of Fusarium pathogens is, therefore, crucial. *Fusarium*-associated biotrophic mycoparasitic fungi represent important source of novel biological control agents to manage *Fusarium* diseases and promote plant growth. The specificity of biotrophic mycoparasites distinguishes them as the most promising, eco-friendly solutions for suppressing *Fusarium* pathogens on wheat and preventing mycotoxins accumulations in food and feed. Compared to necrotrophic generalists, biotrophic mycoparasites preserve other beneficial micro-organisms in wheat ecosystems.

Chapter 11 - Wheat is the second most important cereal crop in Bangladesh and its production is expanding rapidly. However, wheat production is very sensitive to environmental production conditions and managerial practices. The present chapter first provides a detailed account of the profitability and production practices of wheat cultivation in Bangladesh using a sample survey of 293 households from three regions. The chapter then examines the effect of environmental production conditions and managerial factors on wheat yield using a reduced form regression analysis. Results reveal that wheat cultivation is profitable as indicated by Benefit-Cost Ratio of 1.41 with significant variation among regions. There are significant variations in production practices of wheat cultivation across regions as well. A host of environmental and managerial factors adversely affect wheat yield explaining a substantial 77% of variation in its yield performance. Wheat yield is significantly lower in low lying areas and poor soils. Also, yield is significantly affected by pest infestation and weather variations. Among the managerial factors, farmers' education and access to various sources of agricultural information significantly improves wheat yield, whereas a delay in sowing and fertilization significantly reduces yield. Farm size category as well as geography does matter. Productivity of wheat is significantly higher for large farmers whereas it is significantly lower in Jamalpur region. Policy implications include soil fertility improvement; development of wheat varieties that are suitable for low lying areas and are resistant to insect/pest attacks as well as can withstand weather variations; improvement in managerial practices through extension services; and investment in farmers' education.

Chapter 12 - Fungal endophytes spend at least part of their lifecycle living within healthy plant tissues and form associations with the majority of terrestrial plant species. Endophytic fungi of grasses include arbuscular mycorrhizal fungi (AMF) and non-mycorrhizal fungal endophytes. While these two distinct groups of mycoendosymbionts can co-occur in plants, the functions and diversity of non-AMF endophytes is less well understood than that of AMF. Although the diversity of fungal endophytes of wheat is incompletely characterized, its potential can be appreciated based on the diversity of these organisms in other grass and cereal species. These mycobionts can benefit host plants in a variety of ways, including enhanced plant biomass formation and tolerance for biotic and/or abiotic stress. These benefits can be conferred during seed germination (mycovitality) or later in the lifecycle of the host (mycoheterotrophy). The mechanisms by which fungal endophytes and host plants interact are poorly understood, but may include fungal metabolite production, scavenging of reactive oxygen species by the fungal partner, phytohormone production or changes in plant gene expression. A more complete appreciation of the biology of endophytic fungi of grasses in general, and wheat in particular, is likely to be invaluable in developing strategies to utilise these organisms in sustainable agricultural production. In light of the agricultural importance of wheat, a further study on the use of endophytic fungi in biocontrol and biofertilizer applications is merited.

Chapter 13 - *Fusarium* species are worldwide causal agents of *Fusarium* Head Blight (FHB) in cereals. Head Blight can not only reduce yield but also produce fungal toxins that seriously damage grain quality for human and animal consumption. Several studies have shown that *F. graminerarum sensu stricto* and deoxynivalenol (DON) are the main contaminants detected in wheat in Argentina. Chemotype characterization has been extensively used to characterize *F. graminearum* for its toxigenic potential. Many reports have shown differences over space and time of its chemotype distribution, suggesting that this approach is valuable to map a population and identify population changes in the field. The analysis of the trichothecene chemotype distribution of the *F. graminearum sensu stricto* strains across the Argentinean wheat cropping area has revealed that the 3-DON chemotype is always present. Differences in chemotype distribution may result in the shift in toxin accumulation in grains. However, no definitive proof on the correlation between chemotype detection and grain contamination has been found to date. This is mainly due to the variability of toxin production found among strains and to the fact that the samples analyzed have been taken from different origins and environmental conditions. Thus, the combination of genetic chemotyping with chemical analysis of toxin production is desirable. Many authors who have studied this important pathogen on wheat suggest the presence of only one population with high genetic variation in Argentina. Whether or not this variation implies differences in chemotype is an issue that needs further studies. Another point of remarkable interest is the increase in the presence of other casual agents of FHB, such as *F. poae*, a relatively weak pathogen compared with *F. graminearum*, but capable of producing a large number of mycotoxins, including type A and type B trichothecenes, beauvericin and enniatins. All these aspects show the need for a better knowledge of the population dynamics and distribution of the pathogens associated to wheat in the large production area of Argentina. The objectives of this study were to identify the chemotype distribution of *F. graminearum* isolates from wheat obtained in different regions of the main wheat production area in Argentina and to determine whether there are variations in this distribution among crop seasons or geographical locations.

Chapter 14 - Sourdough is a mixture of cereal flour and water which is fermented by a heterogeneous population of lactic acid bacteria (LAB) and yeasts and is traditionally used as inoculum for the production of sourdough bread and other baked products. Yeasts are primarily responsible for leavening, while LAB metabolism is the main cause of acidification, flavour formation and is partially responsible for leavening. Sourdough fermentations enhance dough properties, improve the volume, texture, flavour and nutritional value of the final baked products, retard the staling process and protect the baked goods from mould and bacterial spoilage. The sourdough microbiota usually originate from flour, dough ingredients, the environment and a stable microbial consortium is reached in a short time. The stability of a ripe sourdough depends on several factors mainly including the type and the quality of the flour and the technological process parameters applied to the fermentation. As a consequence of the heterogeneity of these ecological determinants and because of the artisan and region-dependent handling of sourdoughs, mature sourdough microbiota differ greatly in the number and type of species found. The prevalence of certain species and strains confers typical characteristics to the final products resulting in a wide regional variety of baked products. In Italy sourdough technology is used in more than 30% of bakeries for the production of more than 200 different types of sourdough breads, snacks, pizza and various sweet baked products such as Panettone, Pandoro and Colomba. Some Italian sourdoughs have been extensively characterized for their microflora. This chapter deals with the microbial ecology of Italian sourdoughs examined both with traditional methods and using a novel molecular approach; the importance of using a combined analytical approach to explore the microbial communities of sourdoughs has often been demonstrated.

Chapter 15 - Agricultural residues, mainly cereal straws, have gradually been recognized as very important renewable biomass. Wheat straw is a relatively less heterogeneity material compared to maize and other crop straws, facilitates the subsequent harvesting, baling, transport, and the final fractionation and degradation. However, at present wheat straw hasn't been effectively utilized, which is otherwise an important and abundant biorefinery feedstock for production of bioenergy, biochemicals and biomaterials. The utilization of wheat straw has great advantages such as reducing dependence on fossil resources, alleviating food shortage caused by producing corn-based ethanol and avoiding air pollution from straw combustion. Therefore, to study and develop an array of integrated utilization methods of wheat straw is of significant importance. Our group has devoted more than two decades' work in primary biorefinery of wheat straw; formed several key platform techniques and preliminarily constructed an ecological industry chain system. Steam explosion pretreatment of wheat straw is an ideal method to cope with the complex and recalcitrant structure with the merits of high efficiency, simple operation and clean processing. After steam explosion, wheat straw can be processed to various products generally by three pathways viz. mechanical pathway, biochemical pathway and physicochemical pathway. The present chapter summarizes these achievements in detail.

In: Wheat: Genetics, Crops and Food Production
Editor: Maria T. Almeida, pp. 1-79
ISBN 978-1-61209-307-9

Chapter 1

WEED MANAGEMENT IN WHEAT: A REVIEW

Jamal R. Qasem *
Department of Plant Protection
Faculty of Agriculture, University of Jordan, Amman 11942, Jordan

ABSTRACT

This chapter deals with negative effects of weeds on wheat crop, the impact of weed competition on quantity and quality of wheat yield, allelopathic effects of certain weeds on crop growth and development and the poison effect of other weed species to humans and animals. Weeds usually found in wheat fields and associate with crop plants were identified and grouped based on their morphology. Studies on the determination of the critical weed-free period in wheat were reviewed and the importance of this period in weed management was discussed. Chemical and none-chemical weed control methods were reviewed. Certain agricultural and management practices aimed at promoting early canopy crop development and reduce weed growth are mentioned. Importance of plant factors, timing of farm operations, and selection of proper herbicide are discussed. Most recent research findings on weed management in wheat fields are incorporated. Herbicides recommended and practiced in wheat, their type, method of action, method and time of application, physiological effects on crop plants and weeds, and recent research findings on this aspect were included. Some suggestions and recommendations that sustain weed management and aimed at increasing crop productivity and minimizing weed effects are discussed.

Keywords: Wheat, weeds, negative effects, competition, allelopathy, management, control methods, herbicides.

* Professor of Weed Science, University of Jordan, Phone (96265355000 22515) Fax (96265300806) E-mail: jrqasem@ju.edu.jo

INTRODUCTION

Cereal crops are members of the Gramineae (Poaceae) family, produce edible grains for food and feed. Wheat (*Triticum aestivum* L.) is foremost among cereals and indeed among all crops, as a direct source of food for human beings. Wheat grains provide about one-half of man's food calories and a major portion of his nutrient requirements. Researchers estimated that weeds cause 17-25% annual yield loss in wheat. However, in modern agriculture, and in order to overcome factors cause crops yield reduction and for high productivity, intensive and complex crop production practices including application of fertilizers, insecticides, herbicides and fungicides are common. Herbicides however, represent 44% of the world's pesticide markets out of which 57% is for the USA (Kiely *et al.*, 2004). Many products are specifically manufactured for weed control in cereal crops and mainly in wheat although at present, the total number of resistant weeds may exceeds 668 species of more than 346 biotypes (htpp://www.weedscience.org/summary/MOASummary.asp) and many are usually infesting wheat fields. The worldwide expansion in wheat cultivation and food production facilitated the use of agro-chemicals, particularly chemical fertilizers and herbicides. In wheat cultivation a considerable portion of production cost is involved in weed control. Hand weeding and other traditional control methods are time consuming and involve high labor cost. Chemical weed control method offers an advantage to save time, labor and money. This method has become popular in different parts of the developed world mainly due to scarcity or expensive labor during peak of growing season, and relatively the lower weeding cost. Continuous use of a single herbicide or herbicides of the same family or mechanism of action eventually resulted in weed resistance, herbicide persistence as well as the buildup of toxic residues in food chain and toxic effects to humans and animals. In this way herbicides have become international burden and raised much public concern. So, appropriate measures should be taken regarding safe handling and use of low toxic herbicides for sustaining farm productivity as well as protecting environment. Information under third world countries context, however, is limited regarding the performance of herbicides and their effects on surrounding environments.

As alternatives, current weed control strategies concentrate on integrated weed control programs and include crop rotation, mechanical practices, cultivation; hand weeding and herbicide use (Soufizadeh et al, 2007). The development of an integrated weed management system however, requires detailed information on weed-crop interactions, including the relative competitive ability of the crop and weeds during various phases of developments (Tollennar *et al.*, 1994). Cultural, mechanical, chemical and biological practices exert a selective influence on weed population composition and biology (Moss, 1985; Wilson et al, 1988, Van Acker *et al.*, 1997). In this review study, wheat crop growth and importance and weed problems and their control methods are reviewed, evaluated and discussed.

1. WHEAT CLASSIFICATION AND DESCRIPTION

Botanically, wheat is a member of the grass family (Poaceae) to which rice, barley, corn, and several other cereal grain crops also belong. The *Triticum* genus includes a wide range of wheat forms. Taxonomic studies place the goat grasses (*Aegilops*) and wheat (*Triticum*) in one genus, *Triticum*. Wheat has been crossed with rye (*Secale*) and with *Agropyron* (a grass). New forms, called *Triticale*, have been derived from crossing rye and wheat followed by doubling the chromosomes in the hybrid.

Classification: Triticum aestivum L.

Kingdom	*Plantae* – Plants
Subkingdom	*Tracheobionta* – Vascular plants
Super division	*Spermatophyta* – Seed plants
Division	*Magnoliophyta* – Flowering plants
Class	*Liliopsida* – Monocotyledons
Subclass	*Commelinidae*
Order	*Cyperales*
Family	*Poaceae* – Grass family
Genus	*Triticum* L. – wheat
Species	*Triticum aestivum* L. – common soft wheat *Triticum durum* –hard wheat

The wheat inflorescence is a spike bearing sessile spikelets arranged alternately on a zigzag rachis. Two, three, or more florets may develop in each spikelet and bear grains. The grain may be white, red (brown), or purple, and it may be hard or soft in texture. Size of the grain or caryopsis may be large, as in *durum*, or very small, as in *Triticum sphaerococcum*. Wheat varies in plant height and in the ability to produce tillers. The stems are usually hollow. The wheat grain is composed of the endosperm and embryo enclosed by bran layers. The endosperm portion is principally starch and is therefore used as energy food. Wheat is also an important protein source, especially for vegetarians.

2. Wheat Importance

Wheat (*Triticum aestivum* L) ranks first in the world in terms of cultivated area, yield and economic return. In addition, it provides humans by almost 25% of the total calories and proteins. Therefore, it is considered as a main food staple for more than one and a half billion people in 43 countries representing more than one-third of the world's population (CimmyT, 1996). The principal food use of wheat is as bread, either leavened or unleavened. The United States is second to Russia in total production, but the average yield per acre is about twice that of Russia. Other major wheat-producing countries are Canada, China, India, France, European Union, Argentina, and Australia. Wheat production however, was doubled by the end of the 20th century due to tremendous developments and expanding in production systems and technology including the development of new bred cultivars, application of pesticides (mainly herbicides), chemical fertilizers and mechanization. In this regard, wheat producing countries usually have wheat breeding programs aiming at development of more productive and stable cultivars. Methods are integrated in nearly all and selected parent types are crossbred followed by pure-line selection among the progeny to develop new combinations. Varieties and genetic types from all over the world become candidate parents to provide the desired recombination of a good quality, winter and drought hardiness, straw strength, yield, and salinity and disease resistance. Wheat must be also bred for specific milling processes to provide quality end-use products.

In spite of all these production developments and increases in wheat yield obtained food gape remains widening due to the misuse of natural resources and huge rapid increase in the world population in a linear fashion but exponential increases in food production (Al-Sahooki *et al.*, 2009).

However, statistical data indicates that world population was 6 billion in 2002 and if increases remained in the same trend the number reaches 14, 4 billion in 2050. Rajaram (2002) pointed out that world population needs is a one billion ton of wheat in 2020 to fulfill the worlds' need from food compared with the present world production which is not exceeding 600 million ton. Based on these data, there is an immediate need for a hard extensive work aimed at vertical increase in wheat yield (Mason et al, 2007).

Wheat is best adapted to cool dry climate conditions, but is also grown in a wide range of soils and climates.

There is a high possibility for horizontal expansion in the area cultivated by this crop worldwide and possible introduction to certain areas of limited rainfall or soils of certain degree of salinity. Much of the world's wheat is seeded in the fall and, after being dormant or growing very slowly during winter, it makes rapid growth in the spring and develops grains for harvest in early summer. Wheat for milling is classified according to hardness, color, and best use. Among important classes are the hard red winter, hard red spring, soft red winter, white and durum.

3. Weed Species in Wheat Fields

Many weed species differ in morphology and life cycles are usually invading wheat fields (Table 1). Most weeds are annuals reproduce by seeds, associate with wheat and became very well adapted to narrow rows and dense planting system.

Many of these have developed weed races that became highly specialized in wheat microhabitat. However, narrow-leaved weeds and grasses in particular are troublesome, cause serious yield loss and difficult to control since resemble crop in morphology, growth habit and requirements.

Under certain conditions such as drought and low seeding rates perennial weeds may find their way through, establish and forming a net-like structural layer below the soil surface (rhizomatous species) or deeply buried vegetative parts (bulbs, tubers, corms) enable them to survive and make their control far reaching even by use of Tran located herbicides. Certain weed species tend to grow in patches or colonies chocking out crop plants or other species, heavily exhaust the below ground resources and many exhibit alleloapthic activities (Figure 1).

Table 1. Reported weed species that usually infest wheat fields in different parts of the world

Common Name	Scientific Name	Family Name	Remarks
Broad-leaved weeds			
Camel thorn	*Alhaji Camelorum* Fisch.	Leguminosae	Perennial, difficult to control
Bishop's weed	*Ammi majus* L.	Umbelliferae	Competitor, tall growing weed in different crops
Shepherd's clock	*Anagallis arvensis* L.	Primulaceae	Short annual weed, in vegetables and cereals. Abundant under irrigation
Stinking Mayweed	*Anthemis cotula* L.	Compositae	strong competitor annual weed of allelopathic activity
Dutchman's Pipe	*Aristolochia maurorum* L.	Aristolochiaceae	Found in diverse climates worldwide, seeds of contaminating wheat grains may be the cause of endemic nephropathy and a high rate of endemic renal failure
Blue woodruff	*Asperula arvensis* L.	Rubiaceae	Competitor weed at high density, lowers crop yield quality
Astoma	*Astoma seselifolium* L.	Umbelliferae	Perennial weed with edible tuber
Beet	*Beta vulgaris* L	Chenopodiaceae	Edible, cultivated as a crop
field marigold	*Calendula arvensis* L.	Compositae	Wide spread, found in patches in the field where wheat is absent
Shepherdspure	*Capsella bursa-pastoris* (L.) Medik. L.	Cruciferae	Weed of dry, open areas, showing some adaptability to moderate droughts. Listed as noxious weed widely distributed throughout different parts of the world.
White top	*Cardaria draba* L.	Crucifereae	Strong competitor, allelopathic and invading perennial weed of high ecological tolerance
Musk thistle	*Carduus nutans* L.	Compositae	Noxious spiny weed interferes with harvesting

Table 1. (Continued)

Common Name	Scientific Name	Family Name	Remarks
Wild Safflower	*Carthamus nitidus* Boiss	Compositae	Noxious spiny weed interferes with harvesting
Wild safflower	*Carthamus oxyacantha* Bieb.	Compositae	Noxious spiny weed interferes with harvesting
Iberian Star-thistle	*Centaurea iberica* Speng.	Compositae	Spiny weed interferes with harvesting
Syrian scaious	*Cephalaria syriaca* L.	Dipsacaceae	Noisnious weed in different crops including wheat. Its seeds lower flour quality
Forked chickweed.	*Cerastium dichotomum* L.	Caryophyllaceae	Short statured competitive weed
Lamb's quarter	*Chenopodium album* L.	Chenopodiaceae	Worst weed worldwide, nutrient accumulator and allelopathic weed
Nettle-leaf Goosefoot	*Chenopodium murale* L.	Chenopodiaceae	Nutrient accumulator and allelopathic weed
Blue mustard	*Chorispora tenella* (Willd) DC	Cruciferae	Leaves are edible but when fed to cattle it can cause distasteful milk with a disagreeable odor.
Mallow bindweed.	*Convolvulus althaeoides* L.	Convolvulaceae	Perennial binding weed, difficult to control, spread in semi-arid regions
Field bindweed	*Convolvulus arvensis* L.	Convolulaceae	Perennial binding weed, difficult to control, and of allelopathic activity
Hawk's beard	*Crepis aspera* L.	Compositae	Annual weed of high seed production
Flixweed	*Descurainia sophia* L.	Cruciferae	One of the most Infestive weeds in winter wheat fields in different parts of the world. It is tribenuron-methyl resistant.
White wall rocket	*Diplotaxis erucoides* (L.) DC.	Crucifereae	Wide spread in wheat, competitor and allelopathic weed
Garden rocket	*Eruca sativa* Mill	Cruciferae	Edible, annual widely spread in different crops. High infestation totally smother wheat plants
Spanish pink mustard	*Erucaria hispanica* (L.) Druce.	Cruciferae	Annual competing weed
Treacle mustard	*Erysimum repandum* L.	Cruciferae	Resistant weed to the ALS inhibitor herbicides
Annual spurge	*Euphorbia aleppica* L.	Euphorbiaceae	Host for plant diseases

Common Name	Scientific Name	Family Name	Remarks
Sun spurge	*Euphorbia helioscopia* L.	Euphorbiaceae	Host for plant diseases
Sickleweed	*Falcaria vulgaris* Bernh	Umbelliferae	Competitor in wheat fields
Goosegrass	*Gallium aparine* L.	Rubiaceae	Seeds contamination deteriorate flour quality and give better taste of the bread
Rough bedstraw	*Gallium tricorne* Stockes	Rubiaceae	Seeds contamination deteriorate flour quality and give better taste of the bread
Tumble thistle	*Gundelia tournefortii* L.	Compositae	Edible, biennial or perennial spiny weed, interferes with harvesting
Sunflower	*Helianthus annuus* L.	Compositae	Volunteer, or farmers escaper weed
Hoary mustard	*Hirschfeldia incana* L.	Cruciferae	Very common in wheat fields, strong competitor.
Henbane	*Hyoscyamus reticulates* L.	Solanaceae	Highly poisonous weed in wheat
Aromatic Inula	*Inula viscosa* (L.) Ait.	Compositae	Widely spread in uncultivated land, perennial with seed reproduction
Aleppo weed	*Isatis lusitanica* L.	Cruciferae	Small annual in semi-arid regions
Kochia	*Kochia scoparia* Schrad.	Chenopodiaceae	Difficult to control, herbicide resistant weed
Prickly lettuce	*Lactuca serriola* L	Compositae	Summer weed, competitor with hard stem at maturity that make hay unpalatable
Henbit	*Lamium amplexicaule* L.	Labiatae	Weak competitor but at high density
Rakaf	*Leontice leontopetalum* L.	Berberidaceae	Perennial with deep buried tuber, spread in semi-arid regions
Common mallow	*Malva sylvestris* L.	Malvaceae	Edible but highly competitor for water and nutrients, spread under different climatic conditions
Wild medic	Medicago dentatus	Leguminosae	-
Wild medic	*Medicago dentatus* Pers.	Leguminosae	-
White sweet clover	*Melilotus alba* Medikus.	Leguminosae	Reduce grain yield quality, and gives better taste to bread
????	*Neslia apiculata* Fisch.	Cruciferae	Annual weed in semi-arid and arid regions

Table 1. (Continued)

Common Name	Scientific Name	Family Name	Remarks
Syrian Thistle	*Notobasis syriaca* (L.) Cass	Compositae	Spiny biennial weed
Cottonthistle	*Onopordum jordanicolum* Eig.	Compositae	Spiny biennial weed
Corn poppy	*Papaver rhoeas* L.	Papaveraceae	Annual, may serve as host for yellow rust disease
Common peganum	*Peganum harmala* L.	Zygophyllaceae	Poisnous weed in arid and semi-arid regions
Ribwort Plantain	*Plantago lanceolata* L.	Plantaginaceae	Perennial weed grow in a variety of habitats, and exhibit phenotypic plasticity It is genetically variable for a broad variety of physiological, morphological, and life history traits.
Prostrate knotweed	*Polygonum aviculare* L.	Polyginaceae	Summer annual weed in different crops, lower yield quantity and quality
Wild buckwheat	*Polygonum convolvulus* L.	Polyginaceae	One of the most common weeds in cereals and root crops. Owing to its winding growing habit, it seriously reduces yields and interferes with combine harvesting. May cause skin irritation and photosensitivity in cattle.
Pennsylvania smartweed	*Polygonum pensylvanicum* L	Polyginaceae	-
Corn Buttercup	*Ranunculus arvensis* L.	Ranunculaceae	Competitor, poisonous, and allelopathic weed
Curled Dock	*Rumex crispus* L.	Polyginaceae	Edible, biennial or perennial weed
Sheep sorrel	*Rumex dentatus* L.	Polyginaceae	Edible, biennial or perennial weed
Russian thistle	*Salsola kali* L.	Chenopodiaceae	Difficult to control, resistant to certain herbicides
Syrian sage	*Salvia syriaca* L.	Labiatae	Perennial rhizomatous competitor and allelopathic weed under moist and drought conditions
Shepherd's Needle	*Scandix pecten-veneris* L.	Umbelliferae	-
Common Name	Scientific Name	Family Name	Remarks
Groundsel	*Senecio vulgaris* L.	Compositae	One of the worldwide spread annual weed
Egyptian Campion	*Silene aegyptiaca* L.	Caryophyllaceae	Winter annual weed

Milk thistle	*Silybum marianum* L.	Crucifereae	Spiny noxious weed, interferes in harvesting
Wild mustard	*Sinapis arvensis* L.	Crucifereae	One of the worldwide spread annual weed in wheat, competitor and allelopathic.
Tumble mustard	*Sisymbrium altissimum* L.	Cruciferae	Competitor weed
London rocket	*Sisymbrium irio* L.	Crucifereae	Edible weed, spread in different crops
Sow Thistle	*Sonchus oleraceous* L	Compositae	Winter annual weed found also in vegetables
Common chickweed	*Stellaria media* (L.) Vill	Caryophyllaceae	Annual weed in vegetables and cereal crops
Globe mustard	*Texiera glastifolia* DC	Cruciferae	Annual in different regions but concentrated under semi-arid conditions
Field pennycress	*Thlaspi arvense* L.	Cruciferae	Lower yield and quality.
Vetch weed	*Vicia hirta* (Balb.) Loisel.	Leguminosae	Lower yield quality
Violet, Field	*Viola arvensis* Murr	Violaceae	-
Narrow-leaved weeds			
Jointed goatgrass	*Aegilops cylindrica* Host	Graminae	Lower yield quality, and difficult to control
Black-grass	*Alopecurus myosuroides* Huds.	Graminae	Strong competitor, difficult to control in cereals
Wild onion	*Asphodelus tenuifolius* Cav.	Liliaceae	Lower wheat yield quality and bad tasting.
Wild oat	*Avena fatua* L.	Graminae	One of the worst weeds in the world
Wild oat	*Avena sterilis* L.	Graminae	One of the worst weeds in cereals, competitor and allelopathic, difficult to control
Japanese brome	*Bromus japonicus* Thunb.	Graminae	Herbicide resistant annual weed, difficult to control
Downy brome	*Bromus tectorum* L.	Graminae	Herbicide resistant annual, difficult to control
Bermudagrass	*Cynodon dactylon* (L.) Pers.	Graminae	Perennial rhizomateous weed with strong competition for water and of allelopathic influence
Sedge nut grass	*Cyperus rotundus* L.	Cyperaceae	Perennial weed, competitor and allelopathic, mainly in humid, moist or irrigated soils, difficult to control
Deep root grass	*Desmostachya bipinnata*	Graminae	-
Gladiolus	*Gladiolus aleppicus* Boiss.	Iridaceae	Perennial bulbous weed widely spread in different climatic conditions from semi-arid to moist soils
Bulbous barley	*Hordeum bulbosum* L.	Graminae	Difficult to control in cereals, strong competitor for soil moisture, lower yield quality

Table 1. (Continued)

Common Name	Scientific Name	Family Name	Remarks
Wild barley	*Hordeum glaucum* Steud	Graminae	Weed of deteriorated soils, unpalatable at flowering, with high cyanide contents
Italian ryegrass	*Lolium perenne* L	Graminae	Difficult to control, host ergot disease, cause death when contaminating wheat flour
Annual ryegrass	*Lolium rigidum* L.	Graminae	Difficult to control weed, host ergot disease, cause death when contaminating wheat flour
Darnel	*Lolium temulentum* L.	Graminae	Difficult to control weed, host ergot disease, cause death when contaminating wheat flour
Dwarf Canary grass	*Phalaris minor* Retz.	Graminae	Worldwide spread in wheat fields, difficult to control
Canary grass	*Phalaris* spp.	Graminae	Annuals worldwide spread weeds in wheat fields and difficult to control

Figure 1. Colonies of white top (*Cardaria draba*) scattered in wheat field and repelling out crop plants through strong competition and allelopathy effects.

The relationship between weeds and wheat plants is determined by prevailing environmental condition and agricultural practices employed in addition to weed species and wheat cultivars densities and growth habits. It is well established that different cultivars were different in their competitive abilities with weeds and in responses to agricultural practices followed including the application of fertilizers and herbicides (Duwayri and Saghir, 1983; Qasem and Tell, 1993).

4. Yield Losses and Cost of Weed Control

Wheat is attacked by different agricultural pests, but weeds remained the major problem play the main role. Weed losses in wheat may occur from initial stages to the last stage of maturity, harvest, threshing, winnowing and storing of wheat grains. Weed plants are more resistant, hardy and making faster growth than wheat and cause great growth and yield loss due to competition before crop harvest.

Generally weeds reduce wheat yield by 30-50%. Losses, may reach 70% depending on weed species and density (Jennie *et al.*, 2005, Chaudary *et al.*, 2007 and Heather *et al.*, 2007). However, different workers reported different estimations of wheat yield losses depending on the infesting weed species, crop cultivar and their densities, and the agricultural practices employed. McCuen and Silver (1943) reported that weeds in cereal fields increased crop loss by 13% since prevent separation of the grains from the straw. One to three Canada thistle plants per square-meter reduced wheat yield by 12% while 13-20 plants reduced yield by 36% (Infanger, 1956). Blackgrass (*Alopecurus mysuroides*) at density of 170 to 1170 plants m^{-2} caused wheat yield reduction equal to 10 and 30%, respectively (Vizantinopoulos and Kotrens, 1998). Yield reduction due to *Chenopodium album* competition for 20, 40, 60, 80 and 110 days after emergence were 13, 28, 29, 32 and 33%, respectively, while weed competition for the entire growing season caused 33% reduction in grain yield and 34% straw, 31% in total tillers/plant, 43% in effective tillers/plant, 15% grains/spike and 35% in total dry weight (Mondol *et al.*, 2007). Results of two years experiment showed that weed competition (dominated by *A. sterilis*) caused reductions up to 19.4% in wheat grain yield and 23.4% in straw yield (Qasem, 2007). *A. sterilis* emerges with wheat, resembles crop plants in morphology, physiology and has a great demand for water and nutrients (Johri *et al.*, 1992; Dhima and Eleftherohorinos, 2001; Gonzalez-Ponce and Santin, 2001). Chopra *et al.*, (1999) reported that uncontrolled weeds caused 30.7% grain yield loss, *Avena fatua* and *Phalaris minor* are troublesome grassy weeds in wheat fields and cause more than 30% grain yield reduction. (Hobbs et al, 1998). However, grain yield was not significantly affected by the presence of 76 plants m^{-2} of two *Phalaris* species (*minor* and *brachystachis*), but reduction was 36 to 39% by the presence of 304 canary grass plants m^{-2} (Afentouli and Eleftherohorinos, 1996). Full season weed infestation caused 48% in grain yield (Chaudhary *et al.*, 2008).

However, weeds are widely different in their effects on wheat. High infestation of *Cirsium arvense, Lactuca tartarica, Sonchus arvensis*, and *Convolvulus arvensis* decreased wheat yield by 36, 26, 27 and 9%, respectively (Shashkov, 1977). When wild oats exceeded 200 plants m^{-2}, tillering was reduced by more than 50% (http://www1.agric.gov.ab/ca/$department/deptdocs.nsf/all/crop1280). Wheat (*Triticum durum*) infested by Syrian sage (*Salvia syriaca* L.) until harvest reduced grain yield by 58 and 78% of the crop infested by the same weed for only 2 weeks after crop emergence in moderate and low rainfall locations, respectively (Qasem and Abu-Irmaileh, 1983).

Weed infestation may also reduce wheat yield quality. Wild onion and wild garlic bulblets have drastically affected the color, flavor and texture of bread (Qasem, 2003). Weed seeds, buds and bulblets in grains, lowered the price of small grain, flour, breakfast food and other products (Timons, 1970). Some weeds (e.g. wild onion and white sweet clover) produce odourful seeds, that spoil flour quality The strong competition effects of weeds with wheat and their negative impacts on yield quantity and quality necessitated intensive research on their control through which satisfactory results were obtained (Tessema *et al.*, 2001, Andrew *et al.*, 2005; Ashiq *et al.*, 2007). However, the adverse effects of weeds in wheat field can be generally summarized as:

- Reduce growth through competition for nutrients and light at early growth stages and for water at grain filling stage.
- Reduce tiller production and survival, and thus reduce wheat growth and yield.
- Weeds grow fast and shade over wheat plants, which may result in weak crop stand. Because, until when, wheat plants establish before that weeds bind over the plants and wheat crop receive less time for growth, which result in crop yield reduction.
- Weeds reduce quality of wheat grains when mixed with at harvest time.
- Weeds make hindrances and reduce water flow in irrigated wheat
- Some weeds induce wheat lodging by binding and climbing over crop plants. Bind weed is such a species that binds over the wheat reducing crop growth and yield.
- Spiny weeds interfere with harvesting and many bear thorns (such as camel thorn, prickly poppy etc.) and create difficulties in harvest and cause losses in grain yield and troubles to labors. .

- Weeds increase dockage, harvest, transportation and storage cost.
- Immature weeds (green material) at harvest impose swathed crop
- Weed seeds in the soil can result in increased future tillage and weed control cost.
- Increase cost of farm operation and crop production cost in general.
- Increase problems of insects and plant diseases. Some weed species, nourishes insect and diseases and provide shelter to these pests.
- Many weed species exert allelopathic harmful influence on germination, growth or development of wheat.

5. WEED COMPETITION

Research on the effect of weed competition on wheat has been an integral part of weed science research throughout the world for many years and there is considerable information available on the nature and extent of weed interference in this crop. Growth and yield reduction result from weed competition with wheat for water, light and nutrients. At intense weed infestation the crop can be a complete failure and yield may be almost denied (Figure 2). In addition, certain weed species exert an allelopathic influence on wheat, extenuate crop suffering and negatively affect wheat germination, growth and productivity. However, to benefit from fertilizer application weeds must be controlled (Qasem and Tell, 1993) and for maximum yield of spring wheat, weeds must be removed early (Blackman and Teplman, 1938, and Bowden and Friesen, 1967).

Figure 2. White wall rocket (*Diplotaxis erucoides*) (upper) and Garden rocket (*Eruca sativa*) (lower) covering wheat field, crop plants were almost absent.

Wheat cultivars differ widely in their competitive abilities with wheat; short-stature varieties did not compete with weeds as well as tall growing varieties (Qasem and Tell, 1993,

Korres and Willams, 2002). In addition to plant height canopy development can influence the competitive ability (Musick, 1969). The leaf area of a single plant of wild mustard (*Sinapis arvensis* L.) (Figure 3) at blooming stage was 52 times higher than that of wheat compete with (Pavlychenko, 1949).

Figure 3. *Sinapis arvensis* invading wheat field, smothering crop plants through competition and allelopathy.

However, in weed/crop competition the following must be taken into account:

- Weeds must be controlled early in the growing season to avoid serious weed competition effects and yield loss.
- Weeds usually emerge after one-third of the growing season mostly causes no yield reduction.
- Weed density effect on crop yield is curvilinear while weed weight is linearly increases.
- Most annual weeds do not relatively tolerate competition.
- Crop species and cultivars are quite different in their competitive abilities.
- Annual weeds tend to cause higher losses in annual crops while perennial weeds are more effective on perennial crops.
- Leaves are the site of above ground competition, and competition is determined by the factors affect quantity and quality of absorbed light.
- Roots are the site for below ground competition which is determined by the soil volume they occupied.
- The life cycle of many annual weeds is usually shorter than that for crops associate with.
- In most cases competition for a single growth factor leads to competition over other factors.

- Competition may also affect production and activity of plant hormones in crops and weeds.

Other direct factors include the following:

- Complete reliance on one method of weed control leads to emergence of a weed population very well adapted to these conditions with time.
- Single cropping system mostly tends to decrease number of weed species but increased species richness.
- It is possible to apply herbicides to the soil to prevent weed emergence with crop plants.
- Reduced tillage favors growth of perennial weeds.
- Light tillage is more appropriate to control weeds emerged from the soil surface but not for weeds emerged from deep soil layers.
- Herbicides could be easily used on weeds growing with crop plants to prevent their competition effects.
- Tillage type and depth determine weed seed distribution in soil profile
- Plant residues left on the soil surface may affect survival of weed seeds, species germination and composition and effectiveness of soil applied herbicides.
- Crops and their cultivars are different in competitive abilities and this may affect weed population composition.
- High crop seeding rate and narrow rows may favor growth of tall growing weeds.
- Crop sequences in rotation affect weed and species numbers.
- Herbicides may be used to prevent production of weed seeds.
- Weed emergence and number of emerged plants may be predicted with available information on weed seed bank and environmental conditions.
- It may be better dealing with perennial weeds difficult to be controlled in certain crops before or after the growing season or in other crops
- Weed control program may be planned to effectively reduce weed numbers and cost with time.

Critical Period of Weed Competition

Critical period of weed competition (CPC) represents the time interval between two separately measured components; the maximum weed-infested period or the length of time that weeds which have emerged with the crop can remain before they begin to interfere with crop growth, and the minimum weed-free period or length of time a crop must be wee-free after planting in order to prevent yield loss (Weaver *et al.*, 1992, Qasem, 2009). In other words, it is the period after crop emergence or planting at which weeds must be removed for a non-significant yield loss (Nieto *et al.*, 1968). It is essential in reducing weed control expenses. Weeds at this period must be removed by any mean or in economic language it is the shortest time span in the ontogeny of crop growth when weeding result in the highest economic returns. It is determined by crop species and cultivars, weed species and density, agricultural practices employed and prevailing environmental conditions.

CPC in wheat has been determined and reported by different workers in different parts of the world. Shebeski and Frieson, (1955) reported that maximum competition of wild mustard with wheat did not take place until after the 5-leaf stage; but six days later, competition was intense. Blue mustard was competitive mostly at early stages of wheat growth and continued competing throughout the growing season (Musick, 1969). It appears that properly timed hand-weeding may be effective in listening the competitive effect of *Avena sterilis* on wheat. Singh and Bajpai (1992) reported that 4 hand-weeding at 25 to 70 days after sowing gave the best weed control and the lowest weed dry weight and the highest wheat yield. The longer that wheat and *A. sterilis* competed together the shorter is the time that the crop remains each stage of its tillers. Little competition effect of the weed if removed before crop tillering after that wheat growth and yield are significantly affected (Gonzalez and Correa, 1979). None of the canary grass species (*minor* and *brachystachis*) at any date had any effect on grain yield of wheat when cold and dry conditions prevailed during the early stages of growth. Emergence of *Phalaris minor* was completed by the middle of February. Grain yield of wheat grown with 150 *Phalaris minor* plants m^{-2} was not affected even when these plants were removed early on April. However, the presence of the weed until harvest reduced grain yield of wheat by 23 to 28% (Afentouli and Eleftherohorinos, 1996). Critical period of *Chenopodium album* competition in wheat lay between 20 and 40 days after emergence (Mondol *et al.*, 2007). It occurs between 32 to 40 days after sowing (Chopra *et al.*, 1999), and between 30-50 days after sowing (Chaudhary *et al.*, 2008). Agostinetto *et al*, (2008) found that effective weed control should be carried out at 12 to 24 days after wheat emergence. The critical period of blackgrass (*Alopecurus mysuroides*) infestation occurred at 4 to 5.5 weeks after wheat emergence and its economic threshold level varied between 100 and 125 plants m^{-2} (Vizantinopoulos and Kotranis, 1998). However, the existence of the critical period of weed competition depends on the imposition of an acceptable yield loss. If a 5% yield loss gives a marginal benefit compared with cost of weed control, the critical period of *Alopecurus myosuroides* and *Tripleurospermum inodorum* begins at 506°C days after sowing (November) and end at 1023°C days after sowing (February) (Welsh *et al.*, 1999). However, studies on the critical weed competition period in wheat in general are somehow limited.

Wildoats as Problematic Weeds in Cereals: A Case Study

1. Avena Fatua

Wild oat (*A. fatua*) is a weed of 20 crops (Gounnay, 1964). It has a long history of association with cereal crops. It is primarily a troublesome weed in spring-seeded small grains (Allard, 1965). Wild oat is considered to be nearly a cosmopolitan weed wherever cereal crops occur. It is one of those plants that were cultivated as food cereal and became weed due to changes in cultural practices. It is adapted to a wide range of geographical religions and climatic conditions. It is one of the 12 most successful colonizing weed species in the world (Allard, 1965). Its adaptation areas range from Europe to Asia, Canada and USA and considered as one of the trouble weed species distributed in 55 countries (Gounnay, 1964). The weed is common in arable and waste land, infesting almost all cultivated fields and became widely spread in response to intensive cereal production methods and herbicide application that removed other weed species in cereals (Thurston, 1976).

A. *fatua* is an annual grass that reproduces by seed. The reproduction is bisexual, which produces flowers and seed from June to August. Mature seeds are completely dormant when freshly harvested. At maturity the shattered seeds are exposed or buried, germinated or died within 12 months of shattering. Studies indicated that possibility of about 46.5% of the seeds to lose their dormancy storage within 33 weeks at room temperature. However, seeds can remain dormant at least for about 10 years depending on conditions under which they are stored and the composition of seed. One study indicated that an initial infestations of 2000 seeds gave 248, 28 and 1 plants m^{-2} in the second, third and fourth year, respectively and less than 1% viable seeds remained at the end of 4.7 years (Conn, 1990). Depth at which seeds are buried in the soil has a significant effect on weed seed viability, remaining higher at 15 cm than at 2 cm. Seeds on the surface are more viable when buried deep into the soil, however, viability was decreased by 80% shortly after burial, but 1-7 % of seeds were still viable after 9 years burial. Burying by deep tillage induces dormancy while warm and dry conditions can break dormancy after the seeds ripen.

Gibberilic acid (GA) was very effective in overcoming dormancy in freshly harvested or stored seeds (Sharma, *et al.*, 1976). The magnitude of either delay in germination or loss of seed viability depends upon the duration of seed storage under aerobic condition because *A. fatua* is susceptible to anaerobic decomposition (Parasher and Singa, 1984).

Wild oat reduces crop growth, yield, and quality when allowed to interfere with normal crop development (Wit and Bergh, 1965, Bell and Nalewaja, 1968, Bowden and Friensen, 1967, Chancellor and peters. 1974, Cussans and Wilson, 1975, Peters and Wilson, 1983, Carlson and Hill, 1985). Based on relative height and canopy architecture, rooting system (Regnier,1986), rate of expansion and ability to obtain limiting resources in the environment, *A. fatua* and crop could compete for resources contained in their shared growing space (Spitters and Bergh, 1982).

The relative time of emergence between weed and crop may determine their relative competitiveness. Wild oat emerged before the crop, caused greater yield losses than when the crop emerged first. Wild oat had the heaviest and most stems per plant (tillers) when emerged first (Peters, 1983, Peters and Wilson, 1983). The loss in crop yield caused by the earlier-emerging weeds was 16 % compared with 2 % for the later emerging. A difference of fortnight in the time of emergence of the wild oats and wheat made a 14 % difference in the yield loss. The majority of yield loss is accounted for by a reduction in the number of fertile stems per unit area and the 1000-grain weight (Peters, 1983). Grain size is only reduced if wild oats were competition during the period of grain filling.

Relative leaf cover (Regnier *et al.,* 1989) and plant height are the most important factors determining competition effects (Kropff, 1988) of weed and crop plants. Relative leaf covers accounts for both the effect of density and the period between crop and weed emergence. Higher densities and growth rates of wild oats than crop are expected to cause more crop yield losses (Carlson and Hill, 1985) because more resources are shared or portioned by weed (Harper, 1977). Yield of wheat grain was inversely and linearly proportional to relative density of wild oats (Cudney *et al.*, 1991).

When *A. fatua* is taller than the crop, there may be suboptimal distribution, availability, and quality of light in the lower canopy. This affects the photosynthetic capacity and growth rate of the crop (Felton, 1976, Stivers and Swearingin, 1980; Beatty *et al.*, 1982). The shortest stature of the new wheat cultivars give wild oat a height and light interception advantage (Stivers and Swearingin, 1980) which may account for greater yield reductions. Wild oat grew taller than wheat, had a greater portion of its canopy above 60 cm at maturity and reduced light penetration in a mixed canopy than in monoculture wheat (Cudney *et al.*, 1991). Wild oat produced a greater proportion of its leaf area near the top of the canopy where light intensity was maximal. Higher wild oat's leaf area and its placement relative to available light had significant effect in reducing light penetration and wheat yield.

Wild oats can compete with the crop at various stages of crop growth. Critical duration of competition in small-grain cereals range from before crop emergence to the two-to three tiller stage of the crop growth stage (Bowden and Friensen, 1967, Bell and Nalewaja, 1968, Chancellor and Peters. 1974, Peters and Wilson, 1983). Removing the early emerging weeds in good crops does not lead to large-scale compensatory growth by the later-emerging weeds. The important time in which weed should be removed is during the critical weed competition period. Any control beyond this time is not economical.

However, the severity of wild oat interference in small-grain cereals varies with geographical locations and climatic conditions (Bell, and Nalewaja, 1968; Carlson and Hill, 1985). In wheat, Bowden and Friesen (1967) found that competition from wild oats began earlier with greater wild oat densities while high nitrogen and low moisture stress enhanced wild oat interference (Henson, and Jordon, 1982; Carlson and Hill, 1985).

2. Avena Sterilis

This species is a dominant weed in cereals in different parts of the world and widely abundant (Figure 4) in the Mediterranean region (Fernandez-Quintanilla *et al.*, 1997; Barroso *et al.*, 2004). It competes with wheat for soil nutrients, water and light (Gonzalez, 1987), cause great yield losses (Barroso *et al.*, 2004, Fernandez-Quintanilla *et al.*, 1997), ranged between 17-62% in grain yield at a density of 146-162 plants m^{-2} depending on wheat cultivar (Balyan *et al.*, 1991) and up to 75% in certain parts of the world (Wilson, 1979).

Figure 4. Wild oat (*Avena sterilis*) heavily infesting wheat field and smothered crop plants.

Wheat cultivars have been reported to be different in their competitive abilities with this weed species (Gonzalez, 1982) at which semi-dwarf cultivars were found more sensitive than tall grown cultivars (Ponce *et al.*, 1990). However, the economic threshold for *Avena sterilis* subsp. *ludoviciana* was found between 7 and 12 plants m^{-2} (Zanin *et al.*, 1993) and high significant linear regression was obtained between crop yield and weed density (Hoesle, 1981). While Gonzalez and Hernando (1977) mentioned that the greatest wheat yield reduction was at 2 plants of *A. sterilis* per 6 plants of wheat, Dordio *et al.*, (1981) reported that significant wheat yield reduction occurred only with infestation of 30 to 60 plants m^{-2}. *A. sterilis* however, is a strong competitor on moisture and nutrients mainly nitrogen and phosphorus (Conzalez, 1987; Johri *et al.*, 1992; Dhima and Eleftherohorinos, 2001; Gonzaliz-Ponce and Santin, 2001) and has been reported to affect growth, grain yield and total accumulation of nutrients of wheat without having any effect on K levels (Gonzalez-Ponce, 1989).

Stronger competition of *A. sterilis* was found when germination of wheat and wild oat coincided, while increased weed density affected weed adversely, particularly if the weed germinated before crop plants, while the weed affected grain weight rather than straw weight and wheat is affected in its development, productivity and total accumulation of N and P, whereas in the weed, competition from the crop affected the straw rather than the seed weight (Gonzalez and Fernandez, 1978). The earlier the emergence of wheat relative to that of wild oat, the greater was the suppression of the weed, with a lower total accumulation of N and lower P content in the plant (Gonzalez, 1987). In another study Gonzalez (1980) reported that *A. sterilis* emerges simultaneously with wheat or even one day earlier, the growth and production of the crop are seriously impaired, but when the crop emerges before *A. sterilis*, the weed growth was significantly reduced, the number of panicles shedding seed to re-infest the soil, however, is only reduced when *A. sterilis* emerges as the crop is tillering, and then only where infestation is low, and increasing density of *A. sterilis* populations lead to inhibition of the development of both crop and weed. Gonzalez and Fernandez (1978) reported that both inter- and intra-specific competition of *A. sterilis* was much greater than that of wheat, since it is predominantly growth and development of wheat rather than its nutrient content that is affected by competition. They expected that *A. sterilis* competition affected nutrients in the straw and N in the grain of wheat and suggested that the large absorption of cations by the weed may explain its competitive ability. It has been suggested that the correct level of potassium fertilization might be used to favor wheat over *A. sterilis* (Gonzalez, 1978) and as the rate of N fertilizer increased, the straw and grain yield of wheat and the straw yield and spiklete numbers of *A. sterilis* were increased (Gonzalez, 1979). However, at the highest rate of ammonium sulphate and calcium nitrate wheat no longer responded to but *A. sterilis* did, and thereby gaining a competitive advantage over wheat. *A. sterilis* reduced wheat grain yields through the reduction in number of tillers and its competition effect could be reduced by increasing wheat density without totally overcome its effect even beyond the weed-free optimum (Martin *et al.*, 1987).

However, wheat is grown in poor marginal lands in different parts of the world, inadequately planted and fertilized and is hampered by erratic rainfall, periodic drought as well as weed competition in many regions (Qasem, 1982). The lack of suitable weed control methods early after crop and/or weed emergence worsen the effect of weed competition especially in the absence of selective grass killers. Wild oats (*A. ftua* and *A. sterilis*) are the greatest nuisance, dominating wheat cultivated area in the world and became abundant in regions receive relatively high rainfall. They are difficult to control because of morphological and physiological similarities to crop plants and long period of emergence. While both are strong competitors for water and nutrients (Johri *et al.*, 1992; Dhima and Eleftherohorinos, 2001; Gonzaliz-Ponce and Santin, 2001) and their effects on crop plants is severe under certain conditions. This however, depends on weed density, time of germination and emergence, development and duration in relation to prevailing weather and soil conditions.

6. Weed Allelopathic Effects

The concept of plant interference was first proposed by Harper (1964) to include both competition for growth factors and allelopathy. The same concept was also applied by Muller *et al.* (1969) to the adverse effect of neighboring plants in an association and including the two components. Allelopathy is a chemical interaction between plants in nature and is probably of wide spread significance in the functioning of natural plant communities (Rice, 1974, Qasem and Foy, 2001).

It depends on chemicals being added to the environment (Rice, 1974). It affects germination, growth, health, development, and population biology or behavior of inflected plant species. Allelochemicals are secondary metabolites influence the growth and development of agricultural and biological systems (excluding animals) (IAS, 1994).

Allelopathic effect of weeds on wheat crop has been noted at prehistoric time (Willis, 1996) and there is a large number of weed species reported to exhibit such an effect on wheat (Table 2) and include annual, biennial and perennial weed species. Inhibitory effects of weeds on wheat are widely varied but may reach 100% germination or growth inhibition (Qasem and Foy, 2001).

Allelochemicals may be released through volatilization, root exudates, leachates from foliage parts, or through residues on soil surface or soil incorporated. These chemicals may be toxic or non toxic upon their release into the soil but may be also modified by soil microorganisms. They affect wheat seed germination, emergence, seedling growth, development and yield. (Figure 5). However, allelopathy interaction under field condition may be exchangeable at which crop plants may chemically fight back against weeds invading wheat fields (Kohli *et al.*, 2001).

On the other hand, wheat has been reported to exert an allelopathy effects on different weed species (Wu, 2005) while different wheat varieties/cultivars/ accessions have been also reported to have different toxicity effects through this mechanism on different weed species (Ma & Han, 1995; Pereira *et al.*, 1996; Weston, 1996; Jordan *et al.*, 1999; Li-Xiang *et al.*, 2000; Blum *et al.*, 2002; Narwal *et al.*, 2002; Ni & Zhang, 2005; Kong , 2005; Li *et al.*, 2005, Blez, 2007). This phenomenon opened a wide venue of research on the development of wheat allolopathic cultivars able to suppress weed growth associate with this important crop (Wu, 2005) and attracted different researchers' world wide to exploit this mechanism in developing engineered wheat cultivars able to smother weed growth through allelopathy.

Table 2. Weed species with allelopathic inhibitory activity on wheat

Weed species	Source of toxin	Reference
Amaranthus gracilis	Extract	Qasem, 1994b, 1995b
Amaranthus retroflexus	Extract, exudate, leachate, residue	Bhowmik & Doll, 1983; Menges, 1987, 1988; Qasem, 1994b,1995b,c
Amaranthus palmeri	Extract	Altieri and Doll, 1978
Ambrosia trifida	Extract	Le Tourneau *et a*l., 1956; Rassmussen & Einhellig, 1979
Ammi majus	Extract, residue	Qasem, 1995a
Anagallis foemina	Extract	Qasem, 1995a
Anaphalis araneosa	Extract	Tripathi *et al.*, 1981
Antennaria microphylla	Extract	Selleck, 1972; Manners & Galitz, 1986
Antennaria neglecta	Extract	Selleck, 1972
Anthemis cotula	Residue	Qasem, 1995a
Argemone mexicana	Living plants	Sharma & Nathawat, 1987
Ageratum conzoides	Extract	Sugha, 1979
Asclepias syriaca	Extract	Le Tourneau, *et al.* 1956; Rassmussen & Einhellig, 1975
Avena fatua	Exudate	Schumacher *et al,* 1983
Brachiaria mutica	-	Chou, 1989
Brassica campestris	Extract	Jimenez-Osornio & Gliessman, 1987
Brassica napus	Extract, volatile	Muller, 1969, Mason-Sedun *et al.*, 1986
Brassica nigra	Extract	Muller, 1969, Mason-Sedun *et al.*, 1986
Calendula officinalis	Residue	Qasem,1995a
Cradaria draba	Extract, residue	Qasem, 2001
Carthamus nitidus	Extract	Qasem, 1995a
Cassia sophora	Leachate	Datta & Dasmahapatra, 1984
Centaurea maculosa	Residue	Fletcher & Renney, 1963; Locken & Kalsey, 1987
Centaurea nigra	Residue	Vezina & Doyon, 1983
Centaurea repens	Residue	Fletcher & Renney, 1963
Cerastium decotomum	Extract	Qasem, 1995a
Chenopodium album	Extract, exudate, residue	Caussanel, 1979; Porwal & Gupta, 1986; Qasem & Hill, 1989
Chenopodium murale	Extract, leachate, residue	Qasem, 1990, 1993, 1995a
Circium arvense	Extract, leachate, residue	Stachon & Zimdahl, 1980; Hussain *et al.*, 1987
Circium discolor	Extract	Le Tourneau *et al.*, 1956
Convolvulus arvensis	Extract	Qasem, 1995a
Crotolaria pallida	Leachate	Datta & Dasmahapatra, 1984
Descurainia Sophia	Extract	Yang & Mu, 2006
Diplotaxis erucoides	Extract, residue	Qasem, 1995a
Elytrigia repens	Extract, leachate, residue	Kommedahl *et al.*, 1959; Weston & Putnam, 1986

Table 2. (Continued)

Weed species	Source of toxin	Reference
Eruca sativa	Extract	Qasem, 1995a
Erucaria hespanica	Extract	Qasem, 1995a
Eupatorium adenophorum	Extract, residue	Tripathi *et al.*, 1981; Song *et al.*, 1999
Eupatorum riparium	Extract	Tripathi, Singh, & Rai, 1981
Euphorbia esula	Extract, residue	Le Tourneau & Heggeness, 1957; Manners, 1987
Euphorbia geniculata	Extract	Sugha, 1979
Euphorbia helioscopia	Residue	Qasem, 1995a
Falcaria vulgaris	Extract	Qasem, 1995a
Fumaria densiflora	Extract	Qasem, 1995a
Galinsoga ciliate	Extract	Tripathi et al.1981
Galium mollugo	Extract	Kohmuenzer, 1965
Gendelia tournefortii	Residue	Qasem, 1995a
Helianthus annuus	Toxin, extract	Spring & Hager, 1982; Leather, 1987
Helianthus mollis	-	Anderson *et al.*, 1978
Hieracium umbellatum	Extract, leachate	Becker & Guyot, 1951
Hieracium vulgatum	Extract, leachate	Becker & Guyot, 1951
Iva xanthifolia	Extract	Le Tourneau *et al.*, 1956
Lantana camara	Volatile	Das & Pal, 1970; Tripathi *et al.*, 1981; Sahid & Sugua, 1993
Lepidium draba	Extract, exudate, leachate, residue	Qasem, 1994a, Qasem, 1999b
Lychnis alba	Extract	Bhowmik & Doll, 1979
Melilotus alba	Extract	Oudhia & Tripathi, 1999
Mercurialis annua	Residue	Qasem, 1995a
Nepeta cataria	-	Le Tourneau et al., 1956
Notobasis syriaca	Residue	Qasem, 1995a
Onopordon jordanicolum	Residue	Qasem, 1995a
Parthenium hysterophorus	Extract, leachate, residue	Sarma *et al.*, 1976; Mersie & Singh, 1988
Phalaris minor	Extract, exudate	Porwal & Gupta, 1986; Oudhia & Tripathi, 1999
Plantago lanceolata	Extract, residue	Sugha, 1979; Qasem, 1995a
Poa pratensis	Residue	Cochran *et al.*, 1977
Polygonum pensylvanicum	Extract	Le Tourneau *et al*, 1956
Portulaca oleracea	Extract	Le Tourneau *et al,* 1956
Prosopis glandulosa	Extract, leachate	Sankhla *et al*, 1965; Alam & Azmi, 1989
Ranunculus arvensis	Extract, leachate, residue, volatile	Qasem, 1999a

Ranunculus asiaticus	Extract	Qasem, 1995a
Saccharum spontaneum	Leachate	Amritphale & Mall, 1978
Salvia reflexa	Leachate, volatile	Lovett & Lynch, 1979
Weed species	Source of toxin	Reference
Salvia syriaca	Extract,exudate, leachate, residue, volatile	Qasem & Abu-Irmaileh, 1985; Qasem, 1999b, Qasem, 2001
Sasa cernua	Soil	Li *et al.*, 1992
Scandix pecten-veneris	Extract	Qasem, 1995a
Schinus molle	Extract	Anaya & Comez-Pompa, 1971
Solanum nigrum	Extract	Badhani *et al.*, 2008
Solidago virgaurea	-	Becker et al., 1951
Sonchus arvensis	Toxin	Bukolova, 1971
Sphaeranthus indicus	Extract	Oudhia & Tripathi, 1999
Urtica urens	Residue	Qasem, 1995a

Figure 5. White top (*Cardaria draba*) foliage leachates inhibited germination and growth of different wheat and barley cultivars grown in pots in the glasshouse (upper), and *C. draba* infestation of wheat field which nullified wheat germination and growth (lower).

7. Other Negative Effects of Weeds

Poison weeds cause direct toxicity to humans and animals when fed on these weeds or their parts by mistake especially from small children who may wrongly believe that these or certain parts of weeds are edible. Contamination or mixing weed parts with wheat hay may poison animals, cause severe illness or even death to cattle. Examples are *Hyoscyamus* spp., *Leontice leontapetalum*, *Hypericum* spp., *Lolium temulentum* and *Peganum harmala*. Other weeds in wheat fields serve as hosts for plant disease and insects enable spread of these pests and cause sever growth damage to crop plants and yield loss. Some of these disease may cause toxic effects or death to humans and grazing animals. Examples are many including darnel (*Lolium* spp.) host the fungus cause Ergot disease resulted death to any who fed on bread produced from contaminated wheat grains with darnel seed infected by the fungus. Rust and smut fungal diseases which are mainly spread and hosted by *Barberis vulgaris*, grassy weeds (*Sorghum halepense*) and *Euphorbia* spp., respectively. Many weed species also host insects; volunteer wheat is a host for the Russian wheat aphid and the wheat curl mite, which is the vector of wheat streak mosaic virus disease. Aphids, thrips, viruses are mainly transmitted from weeds to wheat plants and cause great yield loss. Climber weeds including *Convolvulus* spp. can reduce light interception by crop plants and under heavy infestation may substantially cover wheat crop and prevent photosyntheisi in addition to their direct competition effects. Hard, unpalatable and difficult to digest woody parts and stems of certain weed species harvested with wheat and contaminating hay may cause problems during storage period, and decay very fast spoiling stored hay. The off-flavor that many weed species cause to wheat flour and bread is well known including wild onions (*Allium* spp.), *Gallium* spp. and *Asperula odorata* seeds. Spiny weeds cause injuries to labors and increase cost of harvesting (e.g..*Carthanus*, *Centaurea*, *Onopordon*, *Gundelia*, *Cirsium* and others). In general, weed seeds contamination of crop seeds deteriorate crop yield quality and separation of these from crop grains increase the cost of final yield production. Seeds of many weed species contaminating crop seeds decay fast during storage, cause problems and loss of stored wheat seeds while undecayed weed seeds are re-sown in the field with crop seeds and thus reduce next season wheat yield or crops in rotation.

8. Some Management Practices Favor Crop Canopy Development

The following are some management practices and precaution measures enable early development of crop canopy and enhance crop competitive ability with weeds. These include:

- Selection of a wheat cultivar that early germinated rapidly forms a good vegetative canopy that could cover weeds once emerged above the soil.
- Use of suitable and ideal fertilizers rates and mange their application to benefit crop over weeds
- Sowing wheat at narrow rows that reduce the space left between crop plants which may occupied by weeds.

- Careful selection of sowing dates in fields heavily infested by weeds based on weed history in the area. Wheat may be sown after rain when weeds may be effectively removed by tillage. This is more suiting spring sown cultivars at which germination may be very rapid due to relatively high soil temperature. In low weed- infested fields wheat may be sown early in the growing season and thus crop plant could substantially benefit from the long rainy season.
- Use of agronomic practices that favor crop development such as application of fertilizers, good seedbed preparation, planting at optimum seeding depth and implementing irrigation, and other practices at suitable time.
- Use of selective herbicides and herbicides mixtures that kills weeds without harming crop plants.
- Timing of irrigation, pest control and tillage practices.

General rules must be taken into consideration before starting weed control

- Weed competition is serious at early growth stages of crop plants.
- Weeds similar in growth habit to crop plants are more competitors and difficult to control than weeds of different growth habit.
- Weeds compete with crop plants for water, light, nutrients and space. In addition they release toxic chemicals to the soil or surrounding environment that inhibit seed germination or growth or both.
- Moderate densities of weeds may have the same effects on crop plants as high densities.
- Competition occurs when there is a limitation in the supply of one or more of growth factors and the supply are below the demand of competing individuals.

Factors determine the outcome of weed/crop competition

- Early germination and emergence
- Initial large photosynthetic area
- High net assimilation rate of energy and mineral nutrients
- High rate of leaf production and vertical growth
- High rate of stem elongation and light trapping in response to shade effect.
- Ideal leaves arrangements on the stem to intercept light, horizontally displayed
- High phenotypic plasticity
- Rapid production of roots, dense root per unit volume of occupied soil
- High accumulation of dry matter to support stem
- Production of extensive and deep root system
- High ability to absorb water and nutrients
- High efficiency in absorbing and translocating nutrients to aerial parts
- High rate of water uptake
- Presence of high percentage of active roots
- High root/shoot ratio
- High root length per root weight

- High proportion of actively growing roots
- Long and abundant root hairs
- High tolerance to drought and severe environmental conditions
- High rate of O_2 and nutrients uptake
- Strategically placed root system
- Presence of dense and long root hairs
- Allelopathic influence

9. Methods of Weed Control

Different methods of weed control in wheat are available to farmers' worldwide including traditional and aimed to improve crop productivity. These depend upon cropping pattern, economic status of the farmer, technology adopted, crop growing area and its climatic conditions and consisting cultural, mechanical, biological, chemical and integrated weed management practices. However, cultural and chemical methods seem more preferred and are the most common in many countries. In weed control, fields should be checked for weeds regularly after planting and until before jointing stage for broadleaf weeds. Broadleaf weed problems encountered in the stubble after winter wheat harvest can be largely eliminated by a timely spring herbicide application in the growing winter wheat.

Methods of weed control in wheat include the following:

I. Prevention Methods

These include those implemented through governmental laws and regulations and part of these are implemented as farmer's precautions in cooperation with the agricultural extension services and public media. These include the following:

Governmental Measures

1) Strict implementation of quarantine regulations, inspection of samples and all agricultural materials including seeds, fertilizers, soil samples and samples for research purposes.
2) Prevention of seed exchange between local farmers especially contaminated crop seeds with weed seeds within and between regions of the same country. Seeds for cultivation must be certified, licensed and proved as weed-free seeds.
3) Animal movement from place to another must be controlled and prevented from grazing in weed- infested fields with free movements.
4) Organic fertilizers must be checked against weed seed contamination before applied in the fields. These need to be treated for weed seed control before application.
5) There should be a law/regulations force farmers to control weeds in their fields during fallow periods, or control weeds of uncultivated fields.

Farmer's Precautions

1) Selection of a cultivar that is well adapted to field conditions, highly efficient in utilizing water, nutrients and of a high yield and strong competitive ability.
2) Use crop seeds free from weed seeds.
3) Keep uncropped areas (fencelines and field borders) weed free.
4) Selection of optimum seeding rate that allow highest level of weed smothering effects, high yield and high interspecific competition effect.
5) Selection of proper sowing method (broadcast or drilling) and reducing as much as possible space between plants and rows without reaching the intraspecific competition.
6) Selection of the best sowing time and seeding rate that speed crop germination and growth and enhance crop competitiveness.
7) Selection of the best methods of irrigation, fertilizer application and other agricultural practices.
8) Use of other agronomic practices that enhance crop growth (e.g. fertilizers, plant hormones, soil preparations, optimum seeding depth, suitable irrigation, and pest control).
9) Application of selective herbicides or herbicide mixtures that control weeds without harming crop plants.
10) Treatment of animal and organic fertilizers in general before applied in the field.
11) Be sure that agricultural machines are cleaned before moving these from field to another and especially from weed-infested to clean fields.
12) Weed control before reach seeding stage and prevention of weed seed production in all areas.
13) Avoidance of weed seeds burying into deep soil layers since giving these a protection value for a long period.
14) Prevention of vegetative parts dissemination of perennial weeds.
15) Weed control in waste places and field borders.
16) Uses of spot weed control method for weed species grown in patches and not widely scattered in the field.

The Media Role

1) Familiarize farmers and public in general with weed problems, losses caused in agriculture and crop yield and their ecological negative impacts (water loss, spread of diseases and insects, fire hazard, allelopaty.... etc.)
2) Familiarize farmers that different crops have different competitive abilities with weeds and these differences are also applied to weeds.
3) Reminding farmers by the proper dates of each agricultural practice implementation in their fields (tillage, fertilization, irrigation and others) and by control time of each weed group.(annuals, biennials or perennials).
4) Asquint farmers that weed competition is severe at early growth stages of crop plants and weeds similar to crop plants in requirements and morphology are more effective and problematic to crop plants than other weed species.

5) Familiarize farmers with critical densities of weeds and critical period of weed control in their crop at which weed control must be carried out for no significant yield loss. In addition, farmers should know that moderate weed densities may have similar effect on their crop as high densities and thus immediate weed control.
6) Familiarize farmers that competition is severe when there is a limitation in the supply of one or more of growth factors (nutrients, water, light) and it may not be necessary to overcome weed competition effects on crop plants by higher supply from these factors since weeds and crop plants are different in their responses to the increased levels of different growth factors.
7) Reminding farmers that certain methods of weed control may be of no value if implemented at wrong time, against none target weed species, or misused. Examples including
   a) Control of perennial weeds vegetatively reproduce by tillage or hoeing
   b) Control of perennial weeds by contact herbicides unless repeated for many times and at short intervals.
   c) Control of weeds at or after start seeding or when crop is sensitive.
8) Reminding farmers by dates of herbicides application and best application methods.
9) Familiarize farmers by the harmful effect of incorporated weed residues on crop plants since yield allelochemicals inhibit crop germination and/or growth
10) Instruct farmers that weed control should be complete and implemented over the whole area in all fields at the same time.
11) Technical bulletins must be delivered to all farmers which enable farmers to know on weed species in their fields, their critical densities and competition periods and suitable control methods.
12) Familiarize farmers by the periods in crop life cycles during which crops have no physiological or growth activities and crop plants became highly sensitive to weed invasion. Weed control at these periods are strongly recommended.
13) Farmers should know that weed populations are unstable and affected by many interrelated factors. Crop species and soil type in addition to weed management practices employed and play important roles in this regard.

To conclude, prevention, or stopping heavy weed infestations, is an important part of a total weed management program. It requires diligence from agriculture extension services and the producers and offers a low cost and effective weed control.

II. Mechanical Methods

One type of these methods is hand weeding although is coasty, time consuming and ineffective against perennial weeds. Shallow tillage is another implementation of these methods and usually practiced as pre-cultivation and may be considered for soil preparation prior to crop sowing. If weeds are possibly controlled using other methods thus it becomes unnecessary.

1. Hand Weeding

Hand weeding in crops and in wheat in particular is as old as agriculture itself. It is a traditional operation practiced in many populated countries world over especially where labor cost is reasonable or low. It is practiced early after weed emergence to avoid any yield loss due to weed competition. However, this operation like any others of weed control has advantages and disadvantages. Among the advantages that allows crop grain and straw yield production free from any herbicides residues and allow adoption of normal crop rotation system with no fear from residual herbicides. Wheat production system is much closer to the organic farming, more safe to crop plants than chemicals and can be carried out at any time after crop emergence and up to jointing stage with no harm to crop plants. In addition it works against all types of weeds either broad- or narrow-leaved, annuals or perennials (before establishment). In contrast, hand weeding is tedious, time consuming where labors number is limited and costly in industrial countries with low population and high labor cost. It may be impractical in large cultivated areas and labors must be well familiar with weeds and able to differentiate clearly between weeds (especially grasses) and crop plants at early growth stages. Weeds of wheat such as, wild oat and *phalaris minor* have close resemblance to the wheat crop, which could not be identified during early growth stages, affect crop growth and reduce wheat yields.

2. Tillage

Tillage is a widely practiced method of weed control and for seedbed preparation. This operation may be done at different dates prior to crop sowing. In most countries growing wheat under rainfed conditions, tillage is carried out early in autumn (October) and crop seeds are thereafter sown. This kind of tillage aimed mainly at seedbed preparation and may control some perennial weed species or late growing summer weeds. However, high risk on crop seed germination is to be expected due to delay in rainfall and thus absence or limited soil moisture available for seed imbibitions and germination. However, if rain was not delayed mostly crop plants are heavily infested with weeds and thus require use of other methods of weed control (mainly herbicides). An alternative tillage operation can be done after rain at which a high number of weed seeds take an advantage and start emergence. In this case, all of emerged weed seedlings are easily and effectively removed by tillage operation primarily carried out to prepare the seedbed. However, it has been reported that annual tillage in the fall prior to barley planting caused a progressive increase from 1 to 400 plants m^{-2} in wild oat density, while spring cultivation and planting with barley decrease the wild oat's density from 500 plants m^{-2} to 5 plants m^{-2} in 4 years experiment. This indicates that timing of cultivation has a significant advantage and is a better cultural practice to reduce weed infestation and competition effects. Results also reflects the value of a timely tillage as a means of reducing weed seedling density and the value of changing cultural practices to disfavor existing adapted ecotypes. Monthly cultivation resulted in higher germination than quarterly, annual or no cultivation (Thomas *et al.*, 1976). Summer fallow increases the number of seeds that break dormancy. A new stand of wild oats will emerge after each tillage operation. Summer fallow can be grazed to control wild oats. It is important to grow fall crops in the same year as the fallow to provide competition against wild oats that germinate in the spring. Growing smother fallow crop during this period will benefit the crop next season. Pre-seeding tillage as early as possible after spring thaw to warm, aerate the soil and stimulate early germination of wild oats. Early tillage of heavy, wet soils is particularly important.

Tillage prior to seeding should be done before weeds reach the three-leaf stage. This minimizes the moisture and nutrients loss by weeds. Tillage should be done to a depth of 2.5 to 5 cm during warm, dry weather so the wild oats cannot re-establish. Post-seeding tillage is valuable to delay date of emergence of wild oat because wild oats that emerge before the crop are more competitive than those emerge later. This type of tillage should be carried out before crop emergence to prevent any crop damage. A shallow cultivation with a rod-weeder or harrows will eliminate wild oat seedlings.

Fall tillage is useful if weed seeds have been exposed to two or three weeks of warm, dry weather. A shallow tillage will lightly cover weed seeds and promote early germination in the spring. A cultivator is more suitable than a discer or harrow. If fall weather is cool and moist, tillage can be avoided so that dormant wild oat seeds remain on the surface exposed to these weather conditions. In winter cereals, *Avena fatua* populations increased most rapidly under cultivation without straw burning. Straw burning and cultivation is important for decreasing field contamination over time. Cultivation without straw burning increases the cost and amount of herbicides needed. Control of weed seed dissemination in cultivation without burning also needs longer time of cultivation (Wilson *et al.,* 1984).

However, types of tillage may be adopted and plowers used for each in wheat field are listed below:

- No-till - 75% of the weed seed bank is in the upper 5 cm (2 in.) of soil. The use of burndown herbicides has been effective for controlling many perennial weeds such as quackgrass or thistle.
- Moldboard plow - the seed bank is more uniformly distributed over the depth of the plow layer.
- Blind harrowing - kills small weed seedlings just before crop emergence.
- Rotary hoe - at 10-20 km/hr, it has "fingers" that lift and mix soil, uprooting small weeds just before or shortly after crop emergence.
- Inter-row cultivation - or scuffling, of row crops uproots small weeds and cuts off larger ones and smothers weeds in the crop row. Relative size of crops to weeds and timing of cultivation will determine success.

3. Mowing and Rouging

When an infestation of wild oats is moderate to heavy, and the crop is of low density, mowing is an effective preventive measure. Wild oats should be mowed at the shoot-blade stage. Early mowing can result in wild oat regrowth and seeds can still be produced. Mowed wild oats can be used for green-feed or, if some of the seeds have set, for silage. The fermentation process in silage preparation will destroy wild oat seed viability. Rouging is most important to reduce *A. fatua* infestation (Couteny and Maccullits, 1976). The weed should be rouged within 7 days of head emergence in the field, because if it is delayed longer than that time it will shed and infest the field (Thomas *et al.*, 1976, Courteny and Maccallits, 1976)

III. Cultural Methods

These methods include soil sweeping lead to soil compaction and thus prevent seed germination of many sensitive weed species. Implementing these methods should lead to a good crop population that effectively competes with weeds (Gail *et al.* 1999). Cultural practices that improve weed control include crop rotation, fallow weed control and timely weed control during summer fallow and crop growth, seedbed preparation, variety selection, seed quality, seeding date, seeding rate, seeding depth, row spacing, row direction, soil fertility, cover crops, use of quality seeds free of weed seeds, and seeding techniques.

Cultural weed control method aimed at manipulating of crop/weed environment and thus balance in favor of crop plants but not for weeds. Crop competitiveness and crop rotation are two important control practices in winter wheat production.

Crop competition involves establishing a vigorous crop which can compete more effectively with weeds for water, light, space and nutrients. Several factors contribute to the competitiveness of crop plants, including proper seedbed preparation, adequate fertilizer, high quality crop seed, careful variety selection, method of sowing (hand-broadcasting or drilling) and proper rate, date and depth of sowing. In a properly established, healthy winter wheat crop, it is difficult for summer annual weeds to survive, and the result is high winter wheat yields.

Crop rotations that include late spring seeded crops break the life cycle of problematic winter annual broadleaf weeds, and allow the use of tillage or herbicides that may not be feasible in a winter wheat monoculture. Adapted rotational grain crops include corn, grain sorghum, proso millet, soybean and sunflower, depending on location.

A good stand of vigorously growing winter wheat has fewer weeds than a thin stand. Poor wheat stands and lack of plant vigor often result from poor soil moisture management during the fallow period before winter wheat seeding.

Precipitation and temperature greatly influence crop and weed growth in the semi-arid areas. The purpose of fallow is to store water in the soil to stabilize winter wheat yields and to manage weeds. With good fallow practices, sufficient soil water is usually available to establish winter wheat. However, below are cultural practices that enable weed control in wheat fields.

1. Seedbed Preparation

This operation include pre-tillage weed removal in cultivated land and the surrounding area, tillage, soil sweeping and lay of straw sheets/residues of allelopathic plant species that prevent weed seed germination, or inhibit their seedlings emergence and growth and prevent crop seeds erosion by wind or rain and their loss from the soil by birds and animals. These practices allow removal of any source of plant diseases, reduce seed introduction to the cultivated site, destroy insects eggs and larvae hidden in the soil, prepare a good and suitable seed bed that allow rapid crop seed germination and seedlings emergence and the extension of crop roots into deep soil layers. Fast growing and early maturing crop has a competitive advantage to suppress growth of weeds and their seed production. Plant straw mulches have many advantages which will be mentioned later in this section.

If soil was tilled in the previous fallow years, then rodweeder may be used to control weeds 2-4 weeks before sowing wheat and to prepare the seedbed. Rainfall prior to crop sowing allows weed seeds to germinate and then their control. After rain, rodweeding and wheat sowing should be delayed for a week or so to allow inter the field. This delay has effectively controlled downy brome and prevented grain yield losses up to 69% (http://www.ianrpubs.unl.edu/epublic/pages/publicationD.jsp?publicationId=596), in addition it is important to allow the time for wild oat seedlings to be destroyed by tillage before or at seeding.

2. Variety/Cultivar Selection

Varieties are different in their germination time and emergence above the soil level, late germinated and emerged varieties are usually suffered from weed competition. In addition, varieties are different in there stature (Zemer *et al.*, 2008), growth and development (Table 3), and also in their phenotypic plasticity. Tall growing varieties compete with weeds better than short varieties while those having high relative growth rates of shoots or roots have clearly better chance to utilize light, shade weeds, and to capture soil particles and thus below ground resources better than varieties with low relative growth rates. Varieties are also different in their leaf display, tillering ability (Figure 6), tolerance to unfavorable environmental conditions, responses to shortage/ excessive supply of growth factors such as nutrients and water and in their responses to herbicides (Qasem and Tell, 1993).

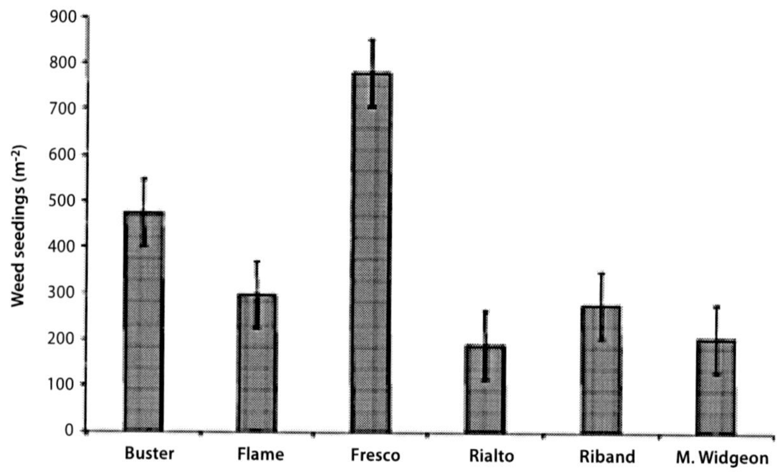

Source: Weed Research 42: 417-428.

Figure 6. Effects of cultivars on total number of weed seedlings at final harvest (total number of seedlings across three crop Densities and weedy subplots. Vertical bars represent SED (d.f. ¼ 72).

3. Seed Quality

Contamination of crop seeds with weed seeds is usually expected especially when weeds resemble crop plants in germination, growth and maturation time and also when both are morphologically similar. Therefore it is highly expected that wheat grains are contaminated with grassy weed (wild barley, wild oats, downy brooms, darnel and many others) seeds that easily pass through sieves and thus re-sown with crop seeds. This problem of weed infestation may be widening in certain areas due to seed exchange between local farmers which lead later to the build up of extensive populations of certain weeds species. Using weed-contaminated wheat seeds also reduces the seeding rate, resulting in a lower wheat stand and a less competitive crop

Farmers however, should have their seed cleaned at certified seed conditioners. Seeds also must be storage treated against contamination of plant diseases or insect attack. More recently produced and viable wheat seeds with good size should be always selected.

4. Crop Rotation

Crop rotation is an important component of integrated weed management since prevent the build up of weed populations. In addition, rotation is usually accompanied with chemical (herbicide) rotation and thus herbicide resistance, weed races, dominance of certain weed species similar in life cycles to crop and tend to be the most common weeds in winter wheat. (Downy brome, hairy chess, jointed goatgrass, feral rye) may not be possible. Several crop rotation sequences are usually followed including legumes, vegetables, and cereals and fallow years. In winter wheat, winter wheat-fallow, winter wheat-corn or grain sorghum-fallow, winter wheat-corn-soybean, proso millet and sunflower rotation with wheat and continuous winter wheat are the most common rotations. In general crop rotation duration in humid regions are longer than those followed in dry areas since weed density and weed species richness in the former are far exceeding those in arid regions.

Winter annual broadleaf weeds also increase but can be readily controlled in winter wheat with herbicides. Inclusion of a warm-season crop such as corn, grain sorghum, proso millet, soybean, or sunflower into a winter wheat-fallow rotation can break the life cycle of these important weed species.

A rotation of winter wheat-corn-fallow is an excellent choice for managing winter annual weeds in winter wheat and for providing good yields. Extending a wheat- fallow rotation to include grain sorghum or grain sorghum and sunflower reduced jointed goatgrass populations more than any other cultural practices tested (White *et al.*, 2004). An important reason for rotating winter annual and summer annual crops is to deplete the soil weed seed bank. With two or more years between winter wheat crops, soil weed seed banks decline to levels where the weeds that do grow may be more easily managed. Due to the increased longevity of jointed goatgrass seed, a four-year rotation is encouraged if jointed goatgrass is a problem (http://www.ianrpubs.unl.edu/epublic/pages/publicationD.jsp?publicationId=596)

Continuous cereal cropping and new production technology favored the growth and distribution of *A. fatua*. In Italy it was found that management systems with a long rotation, considerable soil disturbance and high levels of fertilizer applications were associated with low *A. fatua* infestation of low diversity. Seed dormancy is the main contributing factor in the continued increase of the weed in the continuous cropping system. Rotations of crops and herbicides resulted in a gradual decrease in the number of weeds (Zemanek, 1977). Land seeded to perennial forage for three or more years can provide good control for heavy infestations of wild oats. However, some wild oat seeds can survive under sod and germinate when the sod is broken. Use annual crops in the rotation, especially the more competitive crops minimized yield loss and the wild oats are suppressed. In descending order, barley, canola and wheat are the most competitive small grains. Seed reserves of *A. fatua* were reduced by 95% after 3 consequentive fallow years and were increased by 654 % with continuous *A.fatua* and no crop. The seeds of *A. fatua* in the seasons of rye fallow reduced but barley, spring wheat and flax grown with no herbicide treatment increased the number of wildoat seeds in the soil (Miller and Nalewaja, 1983).

5. Cover Crops

The use of cover crops to smother weeds under field conditions may act through allelochemicals, competition or other mechanism including stimulation of microbes' allelochemicals, physical barriers, shading associated with debris, and changes in soil physical factors (Lehman and Blum, 1997). Cover crops are characterized by strong abilities to cover the soil surface and to effectively smother weeds (Qasem, 2003). If these are legumes they add beneficially to the soil and crop plants by elevating the nitrogen level and make this mineral element more available to crop plants. Legumes are also less competitive than other crops such as cereals and crucifers, and their root systems with the associated bacterial nodules could modify the soil to be fluffier, enable better penetration of crop roots to deep soil layers, and exploiting more water and nutrients. Soil mulch with living *Vicia villosa* has been reported to improve various soil physical properties including increase moisture, stabilization of soil temperature, increase water permeability and drainage, and decrease soil hardness (Fujihara and Yoshida, 1999). In contrast, other workers reported largest weed dry weight in mulched plots although mulch reduced number of weeds compared with the untreated control (Araki and Hatano, 2002). Changes in physical environment were found responsible for the reduction in *Circium vulgare* seedlings emergence in the presence of leaf litter, although chemical effects were also possible (Dowson, 1998). In another study carried out by Kojima and Ohkubo (1999), good summer weed suppressions was achieved using *Cortalaria juncea*, *Cortalaria spectabilis*, *Glyine max*, *Mucuna prurience*, and *Panicum maximum* as green manure while the effect was explained due to the quick growth and good ground cover of these species. Effect of allelopathy on both crops and weeds is also implicated in the effect of cover crops or their residues. Bradow (1996) speculated that decomposed legume cover crop residue emit inhibitory chemicals to crop plants, while residues of *Sinapis alba* controlled weeds adequately but decreased emergence of *Pisum sativum* by 90% at 2% residue rate, and was strongly phytotoxic to emerged seedlings (Jaakkola, 2002). The green manure of *Sinapis alba* was more toxic to *Spinacia oleracea* and *Pisum sativum* than to weeds (Jaakkola, 2005). In another study Fujiyoshi (1998) found that low density of *Cucurbita pepo* intercropped with *Zea mays* reduced weed biomass while high density was detrimental to both weeds and *Zea mays*, while some or all benefits of soil mulch were eroded by phytotoxic leachates from residues of *Secale cereale* and *Trifolium subterranean* as cover crops (Stirzaker and Bunn, 1996).

Dealing with role of allelopathy using cover crops for weed management, all of the above mentioned possibilities should be considered. Most used cover crops of beneficial effect on crop plants are legumes (Fujii and Heradata, 2005), while comparative effect of these on crop plants should be weighed against the weed-free crop treatment with no cover. However, many of the allelopathic crops (Table 4) are used under field condition in straw mulch form at which space between crop lines is mulch-covered or in some cases crop plants may be grown in straw mulched soil. Among the widely used straw mulch are those of *Triticum aestivum*, *Hordeum vulgare*, *Secale cereale*, which yield allelochemicals. *Secale cereale* and *Triticum aestivum* as cover crops have been reported to reduce emergence, height and yield of *Zea mays* (Burgos and Ronald, 1996), and water soluble toxic substances of wheat straw mulching leached into soil under natural conditions exerted allelopathic effects on *Zea mays* growth, and the effect was more pronounced under wet conditions (Yong Qing, 1994). In both field and laboratory studies, Narwal (1996) found that germination and growth of summer weeds were stimulated using wheat residues and those of forage crops smothered weeds and their effect persisted up to 45 days in the next crop.

It is worth indicating that value of soil cover may be better observed in arid and semi-arid regions. Mulching deny light, and prevent photosynthesis. The ability of emerged seedlings to establish depends on the thickness of straw mulch layer; therefore, it is difficult to separate the mechanical or physical effect of mulch from that of allelopathy. Residue from desiccated cover crops is influenced by many factors and generally cover crop residue at natural levels alone is not sufficient to provide full-season weed control, but requires integration with other management practices for optimum control (Teasdale, 2002). Cover crops can also antagonize the efficacy of some weed management practices such as pre-emergence herbicides and may increase the incidence of certain agricultural pest, and for this reason untreated weed-clean crop should be always included for comparison. Cover crops or residues may also increase the incidence or population of certain agricultural pests, or the persistence of one or more of these. Therefore, species of different families or their residues may be recommended for use in rotation. Conklin *et al.* (2002) reported an increase in the incidence of *Pythium* spp. infection of wild mustard seedlings as a result of amended soil with *Trifolium pratense* manure and compost. Intercropping of *Cajanus cajan, Glycine max*, *Gossypium hirsutum*, *Macrotyloma uniflorum*, or *Vigna radiata* with sorghum increased the infestation of *Striga asiatica*, enabled stimulated parasite seeds establish on the true host sorghum (Prabhakarasetty, 1980). However, certain living mulch crops may be undersown in wheat fields, these are preferred to be legumes of short stature, not interfering with wheat plants, smother weeds and provide nitrogen to crop plants. However, when choosing a cover crop, it is important to consider how this crop and its management will affect the succeeding crops.

Table 4. Most commonly reported allelopathic crops and other plant species, targeted weeds controlled and most commonly cultivated crops in which allelopathiy was used

Allelopathic plants	Target weed species	Cultivated crops
Brassica (accessions)	*Amaranthus* spp.	*Brassica* spp.
Cajanus cajan	*Avena ludoviciana*	*Glycine max*
Echinochloa colonum	*Chenopodium album*	*Helianthus annuus*
Festuca arundinaceae	*Cyperus* spp.	*Hordeum vulgare*
Helianthus annuus	*Echinochloa colona*	*Lactuca sativa*
Hordeum vulgare	*Phalaris minor*	*Lycopersicon esculentum*
Lolium spp.	*Portulaca oleracea*	*Medicago sativa*
Oryza sativa and accessions	*Trianthema*	Orchards
Secale cereale	*portulacastrum*	*Oryza sativa* (most frequent)
Sorghum bicolor	Unidentified weeds	Stored potato tubers
Sorghum halepense		*Triticum aestivum* (most
Sorghum hybrids		frequent)
Sorghum sudanense		*Vitis vinifera*
Trifolium incarnatum		*Zea mays* (most frequent)
Triticum aesivum and		
accessions		
Vicia villosa		
Zea mays		

6. Fallow Weed Control

This is an important operation for weed control since allow farmers controlling weeds in absence of crop plants by different herbicides. In this treatment many of the weed species could be eliminated, seed bank and density of weed population may be significantly and seriously reduced to the level below economic injury. Herbicides usually recommended are general (nonselective) translocated herbicides which could easily remove all type of weeds. However, where weed population is mainly composing annual weeds contact herbicides (e.g. paraquat) are normally recommended. It is always advised that weeds during the fallow year must be controlled at pre-flowering stage to prevent their seed production and enrichment of the seed bank which has been already depleted because of germinated withdrawn seeds. In all cases the herbicides used during fallow year should be nonresidual, persist not longer than the fallow year or a season, otherwise may harm sensitive crops in rotation. Weeds in fallow year may be also controlled by tillage. This mechanical operation must be timed when weeds are small in size (>10 cm in height) and always at pre-flowering stage. If perennial weeds are dominating then translocated herbicides (e.g. glyphosate) are more appropriate while tillage must be implemented before perennials establishment and may be repeated. In this case soil must be dry enough to avoid any regeneration from segmented vegetative parts of these species. Herbicides maintain greater residue cover than tillage, which helps to reduce soil erosion and increase soil water storage. If tillage is used, it should preserve as much residue on the soil surface as possible.

Effective weed management during the ecofallow and pre-wheat fallow periods is critical to preserve soil water, eliminate weed seed production and disrupt insect and disease pests. Volunteer wheat should be controlled throughout the fallow periods, but it is critical that it be completely eliminated for at least a week between wheat harvest and wheat seeding.

When herbicides option is excluded during the fallow period, tillage should be carried out while the soil moisture is reasonable and allow tillage operation. If tillage is the only option, the first tillage is usually done before July in certain regions in the pre-wheat fallow period. Sweep plow, tandem-disk harrow, offset disk, chisel plow, and/or herbicides may be used. The number of tillage operations needed depends on precipitation, the weed species present, slope, susceptibility to erosion, and how much crop residue the drill can handle. In moist soil; therefore, disks are often used although they destroy more crop residue. In wet soil, herbicides are the proper control option.

7. Date of Seeding

Sowing winter wheat during the optimum sowing date is usually recommended. Sowing wheat earlier than the optimum seeding date may result in lower winter wheat yield because wheat may be under risk if delay in rainfall occurred (Table 5). This may lead to loss of seeds before germination or death of germinated seedlings in the soil or after emergence. If rain came on time this date of sowing could result in a high wheat yield because of the high amount of rain received providing that weeds are controlled. Corp plants however, will be more vulnerable to crown and root rot infection. Additionally, weeds are more prevalent in wheat that is seeded before the optimum date. Even the following summer crop was found to have more weeds. Fall-seeded crop emerges early the next spring and can smother the emerging wild oats. .

Winter wheat sown after rain may grow relatively free of weeds but seed filling may not be possible because of short rainy season and high temperature and drought conditions (seasonal hot wind) prevailing in spring time in different regions in the world. Yields of crop may be reduced if seeding is delayed by more than 10 to 14 days as compared to normal planting dates. Wheat sown too late may not tiller enough to suppress weeds in the spring and yield may be reduced. Therefore it is important to consider both yield loss due to delay plating time and weed control so that economical and beneficial weed control date will be selected depending on the growing condition in the area.

Table 5. Effect of winter wheat planting date on density of summer annual grasses in wheat and the following grain sorghum crop

	Summer annual grasses		
Planting date	*Winter wheat*	*In wheat*	*In sorghum*
	bu/ac	Number/meter^2	
1-Sep	18	14	270
15-Sep	37	7	140
25-Sep	39	8	100
LSD (5%)	5	5	120

Source: the University of Nebraska–Lincoln Extension Publications Index: Weeds Field and Pasture 1999, Revised August 2006, (Adapted from Weed Science 43: 434-444). Available at: http://www.ianrpubs.unl.edu/epublic/pages/publicationD.jsp?publicationId=596

8. Seeding Rate

Optimum seeding rate is essential in determining the size of weed infestation in the field (Figure 7). This rate means ideal number of wheat plants per unit area that allow highest yield and smothering of weed. It is well established that yield increased with seeding rate up to a certain rate level after which no more increase obtained per plant basis while increasing rate per unit area is greatly decreased or kept steady or unchanged after certain plant density. The concept was termed as "the constant final yield". Optimum seeding rate minimize the space available between plants and rows which may be occupied by weeds. However, farmers should not exceed this rate in order to avoid any intraspecific competition effects of crop plants.

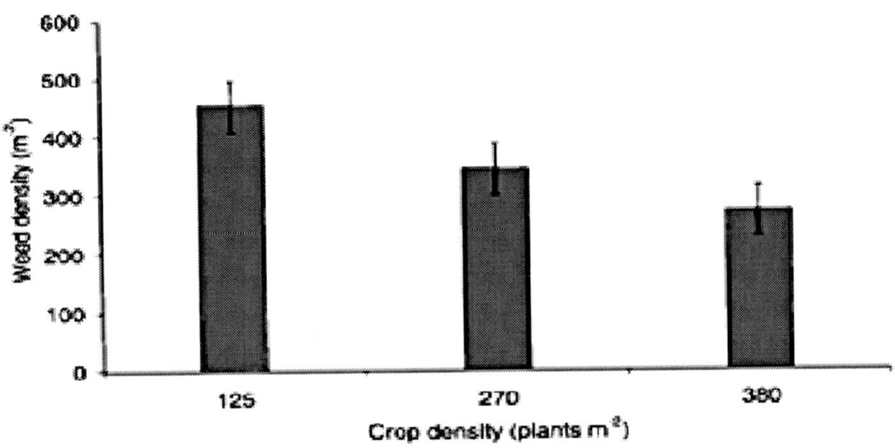

Source: Weed Research 42: 417-428.

Figure 7. Effects of crop density on weed infestation (total number of plants across six cvs. and four sampling occasions). Vertical bars represent SED (d.f. ¼ 30).

When winter wheat is planted at the optimum time, the appropriate seeding rate is about 60 lb/ac with average seed size. Planting fewer seeds may increase weed growth. Generally, seeding rates should be increased when seeding is delayed beyond the optimum dates to compensate for reduced tillering.

9. Row Spacing

Row spacing is in direct link with seeding rate and availability of water and nutrients. Under low soil moisture and fertility level planting distance between plants and rows may be increased to allow enough space to be exploited by individual crop plant. This however, opens the way for weed invasion and occupation of space available. Weed control by hoe or herbicides should be implemented between rows. Normal distance between wheat rows range between 10-15 cm. Wide rows associated with low seeding rate allows more crop tillering and facilitate mechanical weed control. Removed weeds vegetation may be laid into between rows (soil cover) to prevent any subsequent weed seed germination or growth. However, when moisture and nutrients are unlimited, narrow rows and increased crop density offer advantages for weed control. (Figure 6). Emerged weed seedling strongly suffered form both shading and competition effects of tall growing wheat plants. Narrow row spacing can improve weed control during the fallow periods because weeds are smaller and more easily controlled with herbicides than they are in wide row spacing (http://www.ianrpubs. unl.edu/epublic/pages/publicationD.jsp?publicationId=596).

10. Row direction

Tillage direction either north-south or west-east is affected by the land-slope direction. In all cases tillage and rows should be crossed with land slope to avoid seed loss by water erosion and for better water harvesting. Plants facing the sunrise are better grown and can do photosynthesis more efficiently compared with other direction grown plants. Shade of plants facing sunrise direction may effectively cover the soil and thus shade germinated weed seedlings and reduce weed growth and competitiveness. This may clearly demonstrated in plants grown on wide and deep rows at which differences in growth of plants grown in the two sides of the filth are quite clear and also on the growth of weeds in both sides. The influence of row direction on weed growth is well demonstrated in Table (6). In wheat rows of north-south direction, weeds were greatly reduced in growth compared with crop rows east-west directed. It is hypothesized that the north-south rows shade the ground better than east-west rows and reduce weed emergence. In fields where soil erosion is not a concern, north-south rows are preferred.

Table 6. Effect of row direction on weed density when an 11- to 14-month fallow period precedes winter wheat

Direction of rows	*Stinkgrass*	*Tumble pigweed*
North-south	0.8b	0.3b
East-west	4.5a	1.7a

Source: Weed Technology 17: 467-474.

11. Seeding Depth

Planting depth is affected by soil type, wheat cultivar and their seed size. Soil mechanical, structural and texture influence the speed of seedlings emergence. Fine texture of soils allow rapid seedlings emergence and the opposite is true for dry, hard ban and stony soils. Wheat cultivars are different in their time of germination, stem extension and emergence above the soil surface. Large sized grains have more stored food and better ability to emerge from deep soil layers, compared with small sized grains that contain less stored food which may be completely consumed before seedlings emergence. However, depth of soil over the seed should be 2 to 4 cm in medium to fine textured soils and 5 cm in coarse textured soils. Cultivars with large seeds and long coleoptiles may withstand more seeding depth but never deeper than 7 cm in all cases. If the top 2 to 4 inches of soil are dry at sowing, a hoe drill is preferred over a disk drill to place the winter wheat seeds into firm moist soil while hand broadcasting should be avoided.

12. Fertilization and Hormone Application

Fertilizers represent one of the main growth factors necessary for wheat growth and development at early stages after emergence. Phosphorus is usually applied during seedbed preparation prior to crop sowing or at sowing time with wheat seeds. Nitrogen and to certain extent potassium are applied at early growth stages to offer crop plants an advantage over weeds and encourage early canopy development and fast tillering. However, before fertilizer application soil analysis for its contents from different mineral elements must be conducted since fertilizer factor is an expensive one which farmers pay for. Any higher level beyond crop needs will go directly for weed uses and weeds can accumulate substantially higher amounts than wheat crop. Appropriate application timing will increase the vigor and competitiveness of winter wheat. In general, weed control is better when nitrogen is applied in the fall rather than in the spring. Fall fertilization improves the competitiveness of winter wheat and reduces summer annual weed growth; however, fall-applied nitrogen is more susceptible to leaching than spring-applied. The effective chemicals such as the nitrate of potassium, ammonium and sodium, cytokinins (Benyyladenie, kinetin) and thioupea can be used to simulate germination and seedlings emergence. In addition potassium and sodium azide are tended in general to stimulate *A.fatua* germination depending on time and rate of application (Fay, 1976). Nitrogen fertilizers broadcast in the early spring can stimulate wild oat emergence. Application of GA to stimulate the germination of dormant seeds can be used as control measure if it is applied before crop sowing and then cultivate the emerged seedling, which can relief the crop from early competition of the weed. Using different stimulating substances to stimulate germination and seedling emergence prior to crop seeding can facilitate easy control of weed seedlings either by herbicides or tillage.

Nitrogen applied in the spring requires adequate and timely rain to move it into the root zone. If rainfall is not adequate, late germinating weeds can benefit from the nitrogen. Weeds grown after crop harvest may be also more difficult to control when nitrogen was applied late in the spring. The excess weed growth is due to wheat's incomplete utilization of nitrogen as a result of the late application. Spring nitrogen applications should be carried out early in the season. It may be better splitting the application of nitrogen in wild oat-infested fields, and apply the wild-oat control measures before the second applications of fertilizer.

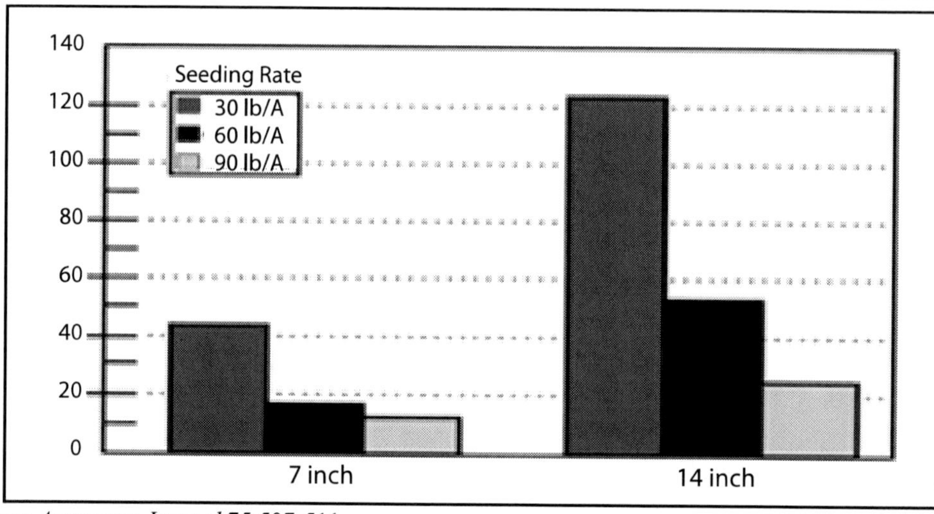

Source: *Agronomy Journal* 75:507-511.

Figure 8. Summer annual weed density five days after winter wheat harvest is affected by row spacing and wheat seeding rate.

Phosphorus may be provided through band application at planting to stimulate crop growth. This fertilizer placement can increase wheat yield and reduce weed density after crop harvest. Row-applied phosphorus is very beneficial to wheat seeded after the optimum planting date, even for soils containing high levels of phosphorus. However, ammonium thiosulfate (12-0-0-26) should not be applied with crop seeds.

Phosphorous fertilizer drilled in with the seed promotes more vigorous crop growth and it helps to smother wild oat plants and reduce its seed production. With application of nitrogen, crop that has been subjected to previous competition is able to compensate, and thereby maintain its numbers of fertile stems (Peters, 1983) in barley- wild oat competition.

It should be noted that excessive application of fertilizers may create new weed problems at which new weed species may became better adapted to high fertilizer application and better tolerate shading effect offered by tall growing crop cultivars.

Allelopathy

The allelopathic plant products are known to offer a vast array of secondary compounds, which have the potential to be used directly as herbicide substitutes or as structural leads for new synthetic herbicides. Certain plant species may promote wheat growth (Table 7) while many other species expressed herbicidal activities on different weed species in wheat fields (Table 8).

The residues from mature harvested crops of sorghum, sunflower, oilseed rape, wheat and field pea exhibited selective effects on *A fatua* seed germination and growth under field condition. The residues of all crops, especially of wheat and pea promoted the germination and growth of wild oat (Purvis *et al*, 1985). Water extractable phytotoxins from senescing and dead paquilinum fronds inhibited radicle extension of *A. fatua*. It is suggested that these phytotoxins could be leached in sufficiently high concentrations to inhibit seed germination after the first spring rains and have contributed to the slow rate of colonization of the weed (Everson and Breer, 1983). Leachates from the litter of the understorey spp. *Pholistome aulitum* inhibited the growth of *A. fatua* (Parker and Miller, 1979). Continuous evaluation and extraction chemical compounds of the residue of the above mentioned crops may synthesized to produce organic herbicide to control wild oat and other weed species. Allelopathic plant water extracts have shown encouraging result for use alone or in mixture with herbicides. Cheema et al (2003) reported that isopruturon dose could be reduced by 50-60% when combined with sorghum aqueous extract in wheat. Canary grass (*Phalaris minor*) and wild oat (*Avena fatua*) are effectively reduced using this combination and Shahid *et al.*, (2007) came to similar conclusion.

Table 7. Some of the recently reported promoting plant species to wheat crop

Donor species	Form used	Promotion type	Reference
Agrostemma githago	Mixed culture	Growth	Mallik & Williams, 2005
Malva sylvestris	Shoot Residues	Growth	Qasem, 2010
Populus deltoids	Extract	Growth	Nandal & Singh, 2002
Pseudomonas capacia	Seed inoculation	Growth & yield increase	Mallik & Williams, 2005
Sisymbrium irio	Shoot residues	Growth	Qasem, 2010

VI. Chemical Control

Chemical weed control is carried out using selective and nonselective, contact and translocated herbicides applied in preemergence or postemergence treatments. It proved effective in controlling some weed species difficult to control by other methods of weed control. Weeds may be effectively controlled within a short time after herbicide application and considered as the most important control method in areas where tillage is abandoned or not possible. However, among weakness encountered using this method is the absence of herbicide rotation resulted imbalanced weeds population. In addition some other problems detected in wheat fields are the following:

- Increase development of weed population that withstand herbicides effects
- Difficulties in elimination of weedy races of crop plants
- Tolerance to a single herbicide automatically leads to weed tolerance of other herbicides
- Wheat has strong relationships with other weedy relatives with which may exchange genes and hybrids became strongly associate with crop plants
- Hybridization with crop plants may lead to hybrids characterized with a long seeds dormancy period
- Winter injury, hail or disease occurrence and previously applied residual herbicides may limit the choices for recropping.

Table 8. Plant species of herbicidal activity and their significant role in weed management in wheat fields

Donor species	Receiver species	Form used	Rate of application	Reference
Artemisia annua (source of Artemisinin)	*Amaranthus retroflexus* *Chenopodium album*	Extract	66% weed emergence & 80% biomass	Delabays & Mermillod, 1999
Helianthus annuus	Weeds	Cover crop, Mulch	Significant reduction in population	Gawronski *et al.*, 2002; Gawtonski, 2004
Helianthus annuus	*Trianthema portulacastrum*	Cover crop	94% population 96% biomass	Dharmaraj *et al.*, 1994
Helianthus annuus/ Legumes (*Pisum arvense, Pisum sativum, Vicia sativa*) mixture	Weeds (mustard)	Mulch	Management	Bernat *et al.*, 2004; Gawtonski, 2004
Oryza sativa & rice germplasm (PI 312777, XL8, 4593, Damagung)	*Alopecurus aequalis* *Ammannia coccinea* *Bidens tripartite* *Cyperus difformis* *Cyperus iria* *Cyperus serotinus* *Dinebra retroflexa* *Echinochloa crusgalli* *Eleocharis kuroguwai* *Heteranthera limosa* *Leersia japonica* *Leptochloa fascicularis* *Monochoria vaginalis* *Persicaria hydropiper* *Phalais minor* *Portulacastrum* sp. *Trianthema portulacastrum* Weeds	Living & Dead mulch Cultivars Extract exposure Accessions (TONO BERA 439, CICA4, TANG GAN, PI 312777)	Weed suppression Suppressing cultivars 37% - 97%	Ma & Han, 1995; Park, 1996; Dilday *et al.*, 1996; Dilday *et al.*, 1999; Kim *et al.*, 1999b; Olfosdotter *et al.*, 1999; Okuno *et al.*, 1999; Fujihara & Yoshida, 1999; Lovelace *et al.*, 2001; Hassan *et al.*, 2002; Ahn *et al.*, 2005; Gealy *et al.*, 2003; Inderjit & Kaushik, 2005; Kong , 2005

Donor species	Receiver species	Form used	Rate of application	Reference
Pennisetum glaucum	Weeds *Parthenium hysterophorus* *Trianthema portulacastrum*	Rotation	Smothering 31-63%	Narwal *et al.*, 2004a
Pseudomonas isolates	*Bromus tectorum*	Metabolites	50%	Mallik & Williams, 2005
Pseudomonas isolates	*Phleum pratense* *Trifolium pretense*	Metabolites	Suppression	Mallik & Williams, 2005
Sorghum bicolor genotypes (Giza 115, Giza 15, Enkath, JS.263)	*Alternanthera tenella* *Avena fatua* *Chenopodium album* *Cyperus rotundus* *Echinochloa crus-galli* *Ipomoea grandifolia* *Lolium temulentum* *Parthenium hysterophorus* *Phalaris minor* *Rumex dentatus* *Sinapis arvensis* *Trianthema portulacastrum* Weeds	Cover crop Extract Rotation, Residue 0.5-8.5 t/ha Sorgaab Varieties	Density 67% Smothering 13-80%	Weston, 1996; Sene *et al.*, 1999; Cheema & Khaliq, 2000; Gavazzi & Paris, 2000; Narwal, 2000; Cheema & Khaliq, 2002; Correia *et al.*, 2002; Narwal *et al.*, 2004b; Alsaadawi *et al.*, 2005; Cheema *et al.*, 2005; Irshad & Cheema, 2004; Urbano *et al.*, 2006
Sorghum bicolor Plus *Helianthus annuus*	*Avena fatua* *Phalaris minor*	Extract	Density and growth	Mushtaq *et al.*, 2010
Sorghum sudanense	*Sinapis arvensis*	Cover crop	-	Urbano *et al.*, 2006

Table 8. (Continued)

Donor species	Receiver species	Form used	Rate of application	Reference
Triticum aesivum & accessions	*Amaranthus retroflexus* *Ammania coccinea* *Avena ludoviciana* *Chenopodium album* *Cirsium arvense* *Cyperus esculentus* *Digitaria ciliaris* *Echinochloa crus-galli* *Heteranthera limosa* *Imperata cylindrica* *Ipomoea hederace.* *Melilotus album* *Phalaris minor Rumex retroflexus* *Sida spinosa* *Stellaria media* Weeds	Accessions Cover crop Extract Rotation Straw mulch 6 t/ha	Excellent control Reduced infestation 87-96% density 78-100% 100% control	Ma & Han, 1995; Pereira *et al.*, 1996; Weston, 1996; Jordan *et al.*, 1999; Li-Xiang *et al.*, 2000; Blum *et al.*, 2002; Narwal *et al.*, 2002; Ni & Zhang, 2005; Kong , 2005; Li *et al.*, 2005
Zea mays	Weeds *Digitaria* sp. *Trifolium* sp. *Taraxacum* sp.	Rotation *Zea mays* gluten meal	Smothering 58-91%	Christians, 1995; Narwal, 2000

Success of chemical weed control and crop tolerance to herbicides depends on the following:

- Crop cultivar and growth stage.
- Safe treated crop and its vegetative growth since diseased plants suffer from herbicides effects.
- Rate of herbicide application and implementation of herbicide use information instructions by producers.
- Correct use of herbicide with regard to volume of spray solution, solution pressure and sprayer calibration.

On the other hand sensitivity of weeds to applied herbicides depends on:

- Weed species.
- Stage of weed growth since seedling stage is most sensitive.
- Weed growth conditions, rapid growing weeds are more susceptible to herbicides.
- Climatic conditions, rain, light intensity, temperature, wind and relative humidity.
- Crop growth and status.
- Interaction of the herbicide with other pesticides.

Chemical weed control is one of the improved method and recent technology that encounter weed problems. In wheat, the most easy and cheap method is use of herbicides, which take less time and is effective input to control weeds on a large scale. There are number of herbicides proved highly selective and effective against different types of weeds in wheat. The use of herbicides in wheat fields is expanding by time and this operation is well known in technical term as chemical weed control technology. Therefore, specific or different types of herbicides are formulated and under use, which proved very effective in controlling annual and perennial weeds. (Table 9). Weed control technology (herbicides) in wheat must be reflected in terms of its benefits and losses on crop productivity. The longer the weed-free period the higher is the crop yield. Several herbicides provide excellent broadleaf weed control with minimal or no wheat injury. However, some varieties are more sensitive to chemical weed control than others (Duwayri and Saghir, 1983). Injury varies with herbicide, variety, location conditions and growth stage. However, not all currently used cultivars were tested for herbicide tolerance while herbicides are considered effective and efficient in terms of their selectivity, volume, application on specific weed and/ or crop.

Herbicide application would be essential when the losses of weeds increase at economic threshold level. In wheat, weed density and species, which cause growth and yield losses must be monitored before herbicides application. Herbicides are categorized based on different factors among which is herbicide selectivity. Based on that, herbicides kill or inhibit weeds and don't harm crop plants beyond point of economic recovery.

The nonselective herbicides referred to those chemicals kill all plants when applied at enough rates and mainly used in wheat borders or to eliminate weed colonies where no vegetation is required. Selectivity however, depends on rate and formulation of the herbicide applied, time of and method of application, environmental conditions and stage of weeds and crop plants.

Table 9. Registered herbicides for use in wheat fields, their trade and common names, formulation and manufacturer

Trade name	Common name/ application method	Formulation	Manufacturer	Remarks
A. Narrow-leaved weeds				
1. Preplant incorporated				
AAtrex 4L	Atrazine	4 lb/gal PSE	Syngenta	
Amber	Triasulfuron	75WDG	Syngenta	ppi, pre, post
Buckle	Triallate+ Trifluralin			
For-go	Triallate			
2. Preemergence incorporated				
Amber	Triasulfuron	75WDG	Syngenta	ppi, pre, post
Finesse	Chlorsulfuron + Metsulfuron	62.5 + 12.5 WG	DuPont	ALS
Glean	Chlorsulfuron	75DF	DuPont	pre, post
Maverick	Sulfosulfuron	75WG	Monsanto	pre, post
3. Postemergence				
Stam	Propanil	4E	Dow AgroSciences	
Assert	Imazamethabenz			
Cleanwave	A premix of Aminopyralid (Milestone™) and fluroxypyr (Starane™).			Best suited for fields where kochia and/or wild buckwheat are the main weeds. Although fluroxypyr is very safe to wheat, aminopyralid can cause injury similar to dicamba or 2,4-D if applied after wheat stem elongation (jointing) or under adverse climate conditions.
Avenge	Difenzoquat		Amvac Chemical Corporation.	postemergence herbicide for wild oat control. Avenge herbicide should be applied when a majority of the wild oats are in the 3 - 4 leaf stage to minimize early wild oat competition and provide maximum yield increases. AVENGE herbicide will not control wild oat plants that are in the 1 - 2 leaf stage, AVENGE herbicide, applied at the 4 - 5 leaf stage will give very good control but yield increases may be reduced

Table 9. (Continued).

Trade name	Common name/ application method	Formulation	Manufacturer	Remarks
Puma	Fenoxaprop-p-ethyl 92 g/L	1EC	Bayer CropScience LP	Post-emergent provides excellent grassy weed control and when mix with Affinity TankMix it controls many broadleaf weeds Application beyond the 6 leaf stage in Spring and Durum Wheats, or the 5-leaf stage in Spring Barley, may result in injury. - Application of PUMA SUPER Herbicide must take place within the leaf stages Outlined for all weeds on this label. Late application will result in reduced weed control.
Beyond (Clarified wheat only)	Imazamox		BASF	Amino Acid Inhibitors: (ALS inhibitors), a broad-spectrum herbicide (grass and broadleaf weeds) that provides post-emergence and some in-season residual weed control. Grasses controlled by Beyond include jointed goatgrass, feral rye, downy brome, Italian ryegrass, wild oat, and volunteer cereals. also controls several winter annual broadleaf weeds including flixweed, henbit, chickweed, shepherd's purse, field pennycress, and other mustard species. Spring applications of Beyond will control or suppress summer annual broadleaf weeds such as common lambsquarters, pigweed, and wild buckwheat. a jointed goatgrass population resistant to Beyond.

Table 9. (Continued).

Trade name	Common name/ application method	Formulation	Manufacturer	Remarks
Clearmax (Clarified wheat only)	a co-pack of imazamox (Beyond herbicide) and MCPA ester r.			Clearmax herbicide should only be used with Clearfield wheat cultivars such as Above, Infinity CL, or Bond CL. These cultivars contain the gene or genes for tolerance to imazamox. Other wheat cultivars treated with Clearmax herbicide will be seriously injured or killed., primarily used to control Troublesome winter annual grass weeds such as jointed goatgrass, downy brome, and feral rye. Tillering should be initiated in the wheat before applying the herbicide. Spring applications should be made prior to stem elongation (jointing) in the wheat.
Finess G + B	Chlorsulfuron + Flucarbazone	25% + 46.7%	DuPont	ALS + ALS
Hoelon	Diclofop –methyl	3E	Bayer CropScience	
Maverick	Sulfosulfuron	75WG	Monsanto	pre, post
Olympus 70WDG	Propoxycarbazone	70%	Bayer	ALS
Olympus Flex	6.75% Propoxycarbazone-sodium and 4.5% Mesosulfurone methyl	WG	DuPont	Control grasses and broadleaf
Osprey 4.5WDG	Mesosulfuron	4.5%	Bayer	ALS
Sencor	Metribuzin	75DF	Bayer CropScience	pre, post
Topic				Control wild oat & *Phalaris minor*
B. Broad-leaved control				
1. Pre-emergence				

Table 9. (Continued).

Trade name	Common name/ application method	Formulation	Manufacturer	Remarks
	Tribunal combi			Jhil, Senjh, Naro and Wild oat, *Phalaris minor*
Methabenzthiazuron	Tribunal		AgNova Technologies	
Ally	Metsulfuron	60DF	DuPont	Post
Amber	Triasulfuron	75WDG	Syngenta	ppi, pre, post
Finesse	Chlorsulfuron + Metsulfuron		DuPont	preplant/preemergence weed control in winter and spring wheat control broadleaf and grasses
Finesse	Chlorsulfuron		DuPont	
Glean	Chlorsulfuron	75DF	DuPont	pre, post
Gramoxone	Paraquat	2L	Syngenta	Post
Karmex	Diuron	80DF	Griffin	pre, post
Maverick	Sulfosulfuron	75WG	Monsanto	pre, post
Olympus Flex	6.75% Propoxycarbazone-sodium and 4.5% Mesosulfurone methyl	WG	DuPont	Post , Control grasses and broadleaf
Orion	Florasulam	SE (suspo-emulsion) formulation	Syngenta	Florasulam, a member of the triazolopyrimidine sulfonanilide class of herbicides, inhibits the plant enzyme acetolactate synthase (ALS), which is essential for the synthesis of the amino acids valine, leucine and isoleucine necessary for plant growth. Florasulam is a Group 2 herbicide. MCPA, a member of the phenoxy class of herbicides, mimics the growth regulator indole acetic acid causing uncontrolled growth and eventual death in susceptible plants. MCPA is a Group 4 herbicide.
Peak 57DG	Prosulfuron	57%	Syngenta	ALS

Table 9. (Continued).

Trade name	Common name/ application method	Formulation	Manufacturer	Remarks
PowerFlex	Pyroxsulam (sulfonamide)		Dow AgroSciences	ALS PowerFlex controls a broad spectrum of difficult grasses and broadleaf weeds in winter wheat, including downy brome, cheat, Italian ryegrass, wild oats, Russian thistle, pigweed, chickweed and cleavers. PowerFlex also controls many mustard species, such as field pennycress, flixweed (tansy mustard), shepherdspurse, Jim Hill (tumble) mustard and blue mustard. Can be applied post-emergence to winter wheat in the fall or spring from the three-leaf to the jointing stage with excellent crop safety. Application Rate: 3.5 oz. per acre.
Rage D-Tech	Carfentrazon & 2,4-D ester	4EC	FMC Corporation	Early preplant
Rave	Triasulfuron		Syngenta	Post, Post Broadleaf control
Glyphosate Original	Glyphosate (dimethylamine salt)	3 lb/gal EPSP	Griffin	
Sencor 4L	Metribuzin	4 lb/gal PSE	Bayer	
Treflan	Trifluralin	4E	Several	ppi
2. Post-emergence				
Turbutryn	Igran	80% WP		Post
Peak	Prosulfuron	57WG	Syngenta	Post
Affinity BroadSpec	Thifensulfuron (Harmony) and Tribenuron (Express).			ALS inhibitor, controls many of the common broadleaf weeds found in winter wheat can be achieved with. Apply Affinity BroadSpec to wheat from the two-leaf stage to before flag emergence
Affinity TankMix	Thifensulfuron			

Table 9. (Continued).

Trade name	Common name/ application method	Formulation	Manufacturer	Remarks
Agility SG	a premix of three sulfonylurea herbicides [Thifensulfuron (Harmony), Tribenuron (Express), and Metsulfuron (Ally)] plus Dicamba (Banvel)			inhibitors of the ALS enzyme (Group 2) Apply Agility SG herbicide at a rate of 2.4 to 3.2 oz/A from the two-leaf stage of wheat to before wheat stem elongation (jointing) controls kochia, prickly lettuce, and Russian thistle.
Aim	Carfentrazone			Carfentrazone is a protox inhibitor (Group 14). It is the only wheat herbicide with this mode of action. Aim herbicide may be applied to wheat from prior to planting up to leaf stem elongation (jointing). Aim herbicide will cause leaf spotting on emerged leaves, but the plants quickly outgrow the injury. Adding Aim herbicide to 2,4-D improves control of weeds like lambsquarters, mustards, kochia, Russian thistle, and wild buckwheat. Aim is a contact herbicide so good plant coverage is essential to good weed control. Apply in at least 10 gallons of water/A. There are no crop rotation restrictions following the use of Aim herbicide.
Ally	Metsulfuron	60DF	DuPont	Post
Ally + 2,4-D	Metsulfuron + 2,4-D			

Table 9. (Continued).

Trade name	Common name/ application method	Formulation	Manufacturer	Remarks
Ally Extra SG	A premix of Thifensulfuron (Harmony), Tribenuron (Express), and Metsulfuron (Ally XP)			inhibitors of the ALS enzyme Ally Extra can be applied from the 2-leaf stage of wheat until the flag leaf appears.
Amber	Triasulfuron	75WDG	Syngenta	ppi, pre, post
Arelon	Isoproturon.	Suspension concentrate containing 500 g/L	Nufarm	A contact and soil acting herbicide for control of black-grass, for the control of annual grasses and annual broad-leaved weeds in winter and spring wheat. applied at 2-3 leaves fully unfolded up to the end of tillering varieties of winter and spring wheat (except durum)
Assert	Imazamethabenz-methyl	67% SG	BASF	DO NOT spray ASSERT if rain is predicted within 6 hours. a selective postemergence herbicide for use in cereals, is absorbed through the foliage and roots of wild oat plants and other susceptible weeds. ASSERT herbicide used alone and in combination should only be applied when the majority of wild oat plants are at the 1-4 leaf stage of growth. broadleaf weed control assert herbicide used alone and applied at the cotyledon to six leaf stage of weed growth
Banvel 4E	Dicamba	4 lb/gal GR	MicroFlo	
Banvel + 2,4-D				
Clarity 4L	Dicamba	4 lb/gal GR	BASF	
Beyond (Clarified wheat only)				

Table 9. (Continued).

Trade name	Common name/ application method	Formulation	Manufacturer	Remarks
Bison	Ethylhexyl ester of MCPA	SC		a systemic, postemergence herbicide for the selective control of wild oats, volunteer tame oats, green foxtail, yellow foxtail, barnyard grass and Persian Darnel in spring wheat, durum wheat, winter wheat, control of annual grasses
Bison Advanced	Contains 2.5 pounds Bromoxynil and 2.5 pounds MCPA acid per gallon			Postemergence broadleaf weed control can be applied to small grains from the 3-leaf stage until before the boot stage.
Branate	Bromoxynil + MCPA			
Buctril, Moxy, Bromox	Bromoxynil			Post-emergence photosynthesis inhibitor, controls many annual broadleaf weeds, including black nightshade, cocklebur, ragweeds, lambsquarters, and smartweed
Buctril 4EC	Bromoxynil	4 lb/gal PSE	Bayer	Post emergence contact herbicide
Clearmax (Clarified wheat only)				
Curtail	(Clopyralid plus 2,4-D amine).			Curtail should be applied from the time of wheat tillering up to wheat stem elongation (jointing), control a wider range of broadleaf weeds
Curtail M	Clopyrlid + MCPA			
2,4-D amine or ester	2,4-D	Several, GR	Many	
MCP amine 4L	MCPA	4 lb/gal GR	Dow, others	
Express TotalSol	Tribenuron-methyl	50%	DuPont	ALS
Finesse	Chlorsulfuron + Metsulfuron	62.5 + 12.5 WG	DuPont	ALS
Finesse G & B				

Table 9. (Continued).

Trade name	Common name/ application method	Formulation	Manufacturer	Remarks
Glean	Chlorsulfuron			
Harmony Extra	Thifensulfuron + Tribenuron	50 + 25%75DF	DuPont	ALS + ALS For selective postemergence broadleaf weed control in Wheat (including durum), Barley, Oat, Triticale and Fallow. Controls tough weeds such as henbit and chickweed
Harmony TotalSol	Thifensulfuron	50% SG	DuPont	ALS
Huskie	a premix containing pyrasulfotole and bromoxynil (Buctril) plus a crop safene			an HPPD inhibitor, effective control of many of the common broadleaf weeds, provided good to excellent control of kochia and Russian thistle Good coverage is essential to good control
Logran Extra		WG	Syngenta Crop Protection Pty	Post-emergence , It is a selective and translocated herbicide that is absorbed through the leaves and roots control of a large variety of grass and broad-leaved weeds as pre-emergence application on wheat or as a post-emergence application on wheat
Maverick	Sulfosulfuron	75WG	Monsanto	pre, post
Olympus 70WDG	Propoxycarbazone	70% WDG	Bayer	ALS
Sencor	Metribuzin	75DF	Bayer CropScience	pre, post
Stinger	Clopyralid	3E	Dow AgroSciences	Post
Strane Ultra				
Though	Pyridate			
Tordon	Picloram	2L	Dow AgroSciences	Post
Widematch	A Premix of Clopyralid (Stinger) and Fluroxypyr (Starane).			To control Canada thistle. WideMatch can be applied to winter wheat from three-leaf wheat to flag leaf emergence,

Herbicides are also grouped based on their method of action to contact which only kill plant parts that became in contact with and most are effective against annual weeds, but if used on perennials another application may be necessary. On the other hand, systemic (translocated) herbicides are those absorbed by leaves, shoots or roots, transport from the site of application to where toxic action take place. These are useful against established perennials; kill slowly while contact herbicides are rapidly acting. Herbicides are also classified based on method of application to soil-applied absorbed by roots of large plants mainly affect seed germination, germinated seeds or small seedlings and thus used to eliminate weeds before emerge and can selectively eliminate germinating weeds in established crops. However, these may or not be foliage active. Foliage applied herbicides are contact or translocated Considering time of application, herbicides are either preplanting applied at which crop should resist herbicide treatment; pre-emergence used after crop is sown (seeds) but before emergence and crop tolerance is necessary; post-emergence applied after crop emerged, may be soil or foliage applied but crop resistance is essential. If soil applied, they kill newly germinated seedlings while foliage applied herbicides kill emerged weeds. Persistence of soil-applied herbicide may be required since offer a long period of weed control and may be through out the whole growing season or until harvest. However, herbicide carry-over problem should be taken into account using such chemicals (Schnoor, 1992).

Of the most widely used herbicides for broad-leaf weed control are the 2,4-D and MCPA (phenoxys) and their combinations with Dicamba (Besnzoic) (Table 10). Both are selective foliage applied and transloated herbicides that kills all broad-leaved weeds including annuals, biennials and perennials. They control weeds in wheat fields at a reasonable price. However, different forms of 2,4-D are available in the world markets including amine and low volatile ester formulations. Ester 2,4-D is the most used formulation that provides better broadleaf weed control than amine formulations because of oil compatibility with leaf surface waxy layer. Amine formulations are polar not easily penetrate through leaf surface and therefore need surfactants otherwise water resulting in reduced control of weeds. However, amine formulations are more safe to crop plants and more stable than ester formulation especially at high temperature and light intensity. Wheat is considered as the most tolerant to 2,4-D treatment among all grass crops. However, cultivars are widely different in their responses to this herbicide treatment (Appleby, 1967; Duwayri and Saghir, 1983). Generally, spring wheat is sensitive for this herbicide from emergence until reach 4 leaf stage then tolerate the herbicide from 5-leaf stage until stem elongation stages, then becomes sensitive at jointing until flowering and spiklets formation then tolerant after dough stage until maturation. However, 2,4-D and MCPA are best applied at full tillering stage at which crop plants are at resting stage and exhibit low physiological activities. Weeds at the same time are physiologically active; their cell division and cell elongation rates are high and thus accumulate lethal doses from both herbicides in their embryonic tissues.

Winter wheat cultivars must be between four tillers and the joint stage when 2,4-D is applied. Winter wheat generally is in the proper growth stage for 2,4-D application in March to early May, depending on planting date, the season, and location.

Wheat injury and yield loss can be significant if 2,4-D or other herbicides are misapplied. Winter wheat yields were reduced more than 20% when 2,4-D was applied in the fall to winter wheat with two to four leaves (Robinson and Fenster, 1973). Winter wheat yields were also reduced with spring applications.

Dicamba controls weeds tolerating 2,4-D and MCPA treatments. Mixture of Dicamba and 2,4-D is used to control a wider spectrum of broadleaf weeds, including some mustard family members, and wild buckwheat. Dicamba plus 2,4-D must be applied to well tillered wheat and before jointing to avoid crop injury... application of the herbicide at 3 leaf stage or at late crop growth stage at high rates resulted in severe crop injuries and lodging, panicle twisting and spikes not emerged or seperated from coleoptiles (Quimby and Nalewaja, 1966). The herbicide was found effective against *Cirsium arvensis* (Keys, 1975) and its tank mixture with MCPA was very effective in controlling *Salvia syriaca* (Table 12) in spring wheat (Qasem and Abu-Irmaileh, 1983).

Table 10. Effect of different herbicides treatments on Syrian sage control, as estimated visually, and on the " cv. Stork" wheat grain and straw yield, during 1980/1981 growing season

Treatment	Rate of application (kg a.i/ha)	Grain yield (Kg/ha)	Straw Yield (T/ha)	Rate of Syrian sage control
No-weeding	-	624 c	0.9 c	10
Hand weeding	-	1475 a	2.0 a	0.0
2,4-D + Dicamba	0.5 + 0.15	599 c	1.1 bc	3.0
2,4-D + Dicamba	1 + 0.15	801 bc	1.5 abc	2.0
2,4-D	1	645 c	1.2 bc	5.0
MCPA	1	1327 a	2.0 a	4.0
MCPA + Dicamba	0.5 + 0.15	776 bc	1.2 bc	4.0
MCPA + Dicamba	1 + 0.15	1202 ab	1.7 ab	3.0

0-10 scale where the lower score (0) denotes weed-free plots while the higher score (10) denotes the maximum weed infestation.

Means followed by the same letter were not significantly different at the 5% level by DMRT test. (Adopted from Dirasat 5, 1983:123-128.

The sulfonylurea herbicides and Curtail are effective against germinating broadleaf weeds up to 4 weeks after application. A surfactant should be added to the spray solution whenever they, unless liquid fertilizer is combined with the herbicide. Different wed species became resistant to sulfonylurea herbicides. Sulfonylurea herbicides are persistent in the soil for three years after application and thus restrict crop rotation options at which sensitive crops such as cucurbits should be introduced after their application and their effects is especially important when the crop is lost to hail or other crop failures. Some of the sulfonylurea herbicides should not be applied to soils with a pH higher than 7.9 to avoid the risk of rotational crop injury.

Liquid Nitrogen Fertilizers and Herbicides

Manufacturers have combined liquid nitrogen fertilizers (UAN - 28 and 32%) and herbicides to control weeds and fertilize the crop with one application. In some situations the winter wheat showed crop injury using this combination. When wheat is under stress, spraying herbicides with UAN may cause yield reduction regardless of the crop stage. Adding sulfur increases crop injury.

An alternative to this program may be to strip band (20 inches) nitrogen fertilizer, if needed, as soon as field conditions permit in the spring and apply the herbicide later. The advantages of strip banding over broadcast for nitrogen fertilizer application are probably great enough to pay for the second application.

Harvest Aids

Ally + 2,4-D amine, 2,4-D ester and glyphosate are registered for spraying winter wheat prior to harvest .Winter wheat must be treated after the dough stage and treated at least seven days before harvest. Wheat sprayed with 2,4-D while the wheat nodes are green may result in stem breakage. A 10-day preharvest interval is needed for wheat sprayed with Ally + 2,4-D amine.

Herbicide Resistance

Herbicide resistant weeds are weeds that survive and grow normally at dosages that usually control the weed effectively and that resistance results from a change or genetic shift within a population that impairs control in the field.

Herbicide resistance was first identified in populations of green foxtail in wheat fields frequently treated by trifluralin for many years. Herbicide resistance has been also found in wild oats, kochia, chickweed, Russian thistle and wild mustard. A herbicide resistant weed may be present initially in a field in very small numbers within the weed population. As selection pressure, from the continued use of the same herbicide or herbicides with similar modes of action, acts on the weed population, a shift in the overall weed population from susceptible to resistant weeds occurs. This resulted when the resistant weed population is allowed to set seed and flourish under conditions of reduced competition due to the continued control of the susceptible weed population. Kochia, Russian thistle and chickweed populations have been also identified to resist the sulfonylurea family. Wild mustard has shown resistance to auxin type herbicides such as 2,4-D.

The best way to avoid or prolong the build up of herbicide resistance is to use a rotation of herbicides with different modes of action. The use of a good crop rotation will help make alternative herbicide selection easier. Cultural control methods such as the use of weed-free seed and tillage may be utilized to combat herbicide resistance. One should only use herbicides when needed.

If one suspect's weeds to be resistant to a herbicide one should first be sure that the problem is not a result of environmental conditions, spray timing, rates, application procedures or a late flush of weeds. Herbicide resistance is suspected when:

- Weed escapes occur in irregular-shaped patches.
- Weeds previously controlled by a herbicide escape treatment, while other labeled weeds are controlled.
- Your records show repeated use of the same herbicide or group of herbicides.
- You can't attribute escapes to environmental conditions or (with post-emergent products) weed emergence after application.
- The same herbicide or herbicide group failed in this area of the field in the previous year.

V. Integrated Weed Control

Any successful and effective weed control program in wheat should consider integration of weed control methods including prevention, cultural, mechanical and chemical. Prevention of weed seeds introduction to wheat fields may solve half of the problem while combination of other methods could lead to effective weed control and highest yield. Intensive grazing, mowing and burning to prevent any shedding of weed seeds. However, any successful backage of two or more methods of weed control must be considered.

Wild Oats Control: A Case Study

1. Avena Fatua

The use of suitable herbicide at appropriate time with appropriate surfactant to susceptible parts of the weed prevented the spread of resistant wild oat even when cereals were grown for 3 years in succession (Zemanek, 1977). Site of application was found to be important in determining herbicide performance and where wild oats (*Avena fatua*) was used (Moser *et al.*, 1976). Differential sensitivity of wild oat (*A. fatua*) and wheat to barban was highest when barban was applied to the coleoptile. The coleoptile acts as a physical and physiological barrier to reduce the concentration of free barban in the stem section where the sensitive meristematic are located and thus enhance the differential sensitivity to barban between wild oat and wheat (Bukuro, *et al.,* 1975). The greater decrease in dry weight and shoot length and the most damage to the growing point occurred when the herbicide, Hoe 23 408, was applied to the base of the leaf blade. The structure and amount of wax at the base area of the leaf blade may help to increase penetration and, with this in mind attention must be made to achieve a more precise application of herbicide to the base of *A. fatua* (Waller and Bischof, 1976). In the presence of surfactant 93% of applied C^{14} difenzoquat was absorbed by the leaves of *A. fatua*. Surfactants at 0.01 to 0.3% at high relative humidity and in temperature from 10-30° C greatly enhanced the penetration of herbicides in to leaves. Asulam carbamate with octoxynol reduced wild oat fresh weight when plants were in high than low humidity regardless of temperature by increasing toxicity to wild oat. Asulam toxicity to wild oats with the low humidity increased as the temperature increased from 20 to 30 ^{o} C. The increased phytotoxicity of assulam with high temperature and humidity relate to increased herbicide absorption (John *et al.*, 1988).

Application of difenzoguat increases in cell membrane permeability, swollen guard cells. There was accumulation and then disappearance of starch granules, swelling oftrets, fusion of granal thylakoids, detachment and then rupture of the outer membrane of the envelope and clumping of ribosome's (Thai *et al.*, 1989). The herbicides inhibit carbon assimilation and translocation to the roots and incorporation into the starch (Mannova, and Taim, 1985). The activity of barban and Dizenzoquat appeared to be dependent on growth stages. Both of the chemicals are effective at active growth stages of weeds. Combination of triallate with diclofop or difenzoquat reduced seeds of *A. fatua* in wheat by 98% and 95%, respectively. Herbicides are effective in mixtures than when used separately/. Continuous application of similar herbicide can increase the development of resistant biotype.

2. Avena Sterilis

The most commonly used pre-plant or pre-emergence herbicide is triallate (Aibar *et al.*, 1988; Balyan *et al.*, 1988; Antonelli and Formigoni, 1976). This increased wheat grain yield and protein content when used alone (Gonzaliz-Ponce *et al.*, 1989) or supplemented with isoproturon (Balyan and Malik, 1992). However, the granular form of this herbicide was better than its EC formulation (Terekhova and Nesterenko, 1977).

Among the effective post- sowing or emergence herbicides reported are methabenzthiazuron at 2-3 weed leaf-stage (Thakur and Singh, 1990); tralkoxydim (Balyan *et al.*,1992; Yaduraji *et al.*, 1992) that reduced wild-oat population (Balyan *et al.*, 1992) and increased wheat yield (Yaduraji *et al.*, 1992).

The combinations of tralkoxydim with isoproturon were also effective against many narrow-leaved weeds including *A. sterilis* (Michalopoulos *et al.*, 1990; Balyan and Malik, 1992; Samunder and Malik, 1992; Villiers and Du Tott, 1992).

Isoproturon has been reported as an excellent herbicide for wild-oats control in wheat (Sharma *et al.*, 1989, 1991; Walia *et al.*, 1990; Boparia *et al.*, 1991; Johri and Govindra, 1991) and resulted in grain yield equivalent to those of the weed-free control either alone (Balyan *et al.*, 1988) or in mixture with ioxynil (Cortes *et al.*, 1976). It increased protein content in wheat (Gupta *et al.*, 1990) and reduced *A. sterilis* competition for NPK fertilizer (Johri *et al.*, 1992). Application of metoxuron increased wheat grain yield and was most effective on wild-oats (Panwar *et al.*, 1990). However, the highest wheat yield and the best *A. sterilis* control were obtained using isoproturon applied 15 days after sowing and metoxuron after 25 days (Gill and Walia, 1988). Pendimethalin and isoproturon applied 30 days after sowing but before emergence resulted in the greatest control of *A. sterilis* subsp. *ludoviciana*, while isoproturon and metoxuron plus difenzoquat were more economic than the hand weeding (Singh and Gosh, 1992). Isoproturon, metoxuron and diclofop-methyl sprayed alone or with non-ionic surfactant greatly increased herbicides toxicity to *A. sterilis* subsp. *ludoviciana* (Malik *et al.*, 1989).

Imazamethabenz has been reported to reduce *A. sterilis* emergence (Skorda *et al.*, 1991). The continuous application of this herbicide in winter barley gave 57-100% control of the emerged *A. sterilis* subsp. *ludoviciana* while difenzoquat was much less effective (Navarrete and Fernandez-Quintanilla, 1990). Chlorsulfuron, and tribenuron-methyl mixtures with imazamethabenz-methyl or dichlofop-methyl (Efthimiadis, *et al.*, 1987) were found effective.

Excellent control was achieved with diclofop-methyl (Breidert, 1978; Mirkamali, 1993). It was the best among several used (Aibar, *et al.*, 1988), provided better *A. sterilis* subsp. *ludoviciana* control than isoproturon and was more effective with a surfactant (Malik *et al.*, 1988). Fenoxaprop-p-ethyl alone (Manning *et al.*, 1993) or with diclofop-methyl gave excellent control at 1-2 leaf weed stage (Everaere and Thiebault, 1990). Effective control of *A. sterilis* was also obtained with barban (Eshel, 1978) and clodinafop-propargyl resulted in 90-100% control and increased wheat yield by 130% (Mirkamali, 1993).

Benzoylprop-ethyl and flamprop-isopropyl both gave 80-92% control of *A. sterilis* and increased yield by more than 100%, and flamprop-isopropyl reduced the number of weeds from 102 to 15 plants m^{-2} (Gicheru, 1978).

Difenzoquat, flamprop-ethyl, diclofop-methyl and barban all reduced *A. sterilis* spikelet production (Wilson, 1979; Madeira and Dordio, 1981) and best control was obtained using difenzoquat, benzoylprop-ethyl, flamprop-m-isopropyl and chlorfenprop-methyl (Macek, 1990). Other herbicides recommended included mesosulfuron-methyl for the control of a wide spectrum of grassy (including *A. sterilis*) and some broadleaved weeds (Hacker *et al.*, 2001); Mesosulfuron-methyl and mefenpyr-diethyl mixture has a wide spectrum including some of the ACCase resistant species (Hacker *et al.*, 2001); and. sulfosulfuron for bread wheat and was efficient in controlling *A. sterilis* by 87% and *Sinapis arvensis* L. by 95% (Hamal *et al.*, 2000); and clodinafop-propargul (Lopez and Sugranes 1997). *A. sterilis* control was possible using imazamethabenz-methyl or its mixture with isoproturon, and tralkoxydim increased grain yield by 21.4%. (Qasem, 2007) and similar results were reported by other woekers (Johri and Govindra, 1991; Skorda *et al.*, 1991; Balyan *et al.*, 1992).

Performance of difenzoquat was not satisfactory. It failed to increase wheat yield beyond the weed-infested control (Qasem, 2007). Efthimiadis *et al.*, (1987) reported that diclofop-methyl in a tank mixtures containing imazamethabenz-methyl showed phytotoxic symptoms on wheat, in contrast to the other imazamethabenz-methyl mixtures.

Other workers reported phytotoxicity of certain herbicides to wheat plants (Chhokar *et al.*, 2006) or failure of others to control weeds or improve crop yield (Wilson, 1979; Duwayri and Saghir, 1983), while the effects of environmental factors on herbicide performance are well documented (Gonzalez-Ponce and Rodriguez, 1983; Gonzalez-Ponce and Salas, 1992)

Dimethanamid encouraged wild-oat population and growth over the weed-infested control and lowered crop yield compared with weedy plots (Qasem, 2007).

Other workers (Boparia *et al.*, 1991; Johri and Govindra, 1991; Sharma *et al.*, 1991) reported that isoproturon alone failed to increase crop yield, to the level of the weed-infested control. However, a synergistic effect of its combination with imazamethabenz–methyl proved effective and increased crop yield especially at lower doses (Qasem, 2007). However, as indicated above continuous use of single spectrum herbicides will lead to shift in weed population toward more problematic weed individuals. Efthimiadis *et al.*, (1987) reported that applications of wild-oat herbicides alone increased the population of broadleaved weeds, while treatments with the broadleaved-weed herbicides increased the population of wild-oats.

Certain chemicals affected wild-oat plants population and panicle number, others affected its population and *Vice versa*. Imazamethabenz-methyl/ isoproturon mixture was the best in reducing wild-oat as well as broadleaved weeds. (Thakur and Singh, 1990, Qasem, 2007), while its combination with isoproturon is highly recommended for a wide- spectrum weed control.

VI. Perennial Weed Control

Perennial weeds may be easily and better controlled during the fallow years using translocated herbicides. Timing of herbicide treatement at which food reserve in the below ground organs at the lowest level. This occurs at pre-f lowering at which most reserved food is directed toward flower production and thus moved upward. Probably in absence of herbicide treatment deep plowing during dry conditions may be carried out while perennating organs may be collected and burned. However, when there is a need to control perennial weeds growing with wheat plants translocated herbicides are also the best option. Many perennial weed species tend to form colonies in the field and may be treated with translocated general or selective herbicides to completely eliminating these patches. Many of these herbicide are available in the world markets including different form of 2,4-D, MCPA and glyphosate. As a general rule, most effective control will occur when perennials are at least 20 to 30 cm tall at the time of herbicide application. The herbicide is to be applied prior to the onset of plant senescence, or by about mid-bloom for any species that are flowering in the early fall.

CONCLUSION

Wheat is the first food crop grown in the world; it is a strategic human food and source of energy for millions of wealthy and poor people worldwide. This crop suffers from different agricultural pests which add negatively to the severe environmental conditions (drought, salinity) under which this crop is grown in different parts of the world. Weeds stand first among all factors affect wheat growth and yield. They compete with wheat for different growth factors and their competition effect is becoming more severe under drought (arid and semi-arid regions) and low fertility levels. In addition weed may exert an allelopathic influence and harbor insects and diseases. They affect crop yield quantity and quality and their effects are well demonstrated in the field and on stored products. Control of weeds in wheat is as important as grain yield production and human food itself. It is possible increasing area cultivated by wheat but vertical yield increase and high yield horizontal production in large areas is not possible in absence of weed control. Many methods of weed control are available and in hands of farmers including prevention, mechanical, cultural and chemical. None of these however, is better than the other while all or more than one method is always recommended to collectively solve the effect of all annual and perennials, broad and narrow-leaved weeds at the same time and in the same field.

REFERENCES

Afentouli, Catherine G.; and Eleftherohorinos, I. G. (1996). Littleseed canarygrass (*Phalaris minor*) and short-spiked canarygrass (*Phalaris brachystachys*) interference in wheat and barley. *Weed Science*, 44(3), 560-565.

Agostinetto, D., Rigoli, R.P., Schaedler, C.E. *et al.* (2008). Critical period of weed competition with wheat. *Planta Daninha* [online]. 26 (2). 271-278.

Ahn, J.K., Hahn, S.J., Kim, J.T., Khanh, T.D. and Chung, I.M. (2005). Evaluation of allelopathic potential among rice (*Oryza sativa* L.) germplasm for control of *Echinochloa crus-galli* P. Beauv in the field. *Crop Protection*, 24: 413-419.

Aibar, J., Ochoa, M.J., Sopena, J.M., Lorente, M. and Zaragoza, C. (1988). Wild-oat [*Avena sterilis ssp. Ludoviciana* (Dur.) Nym., *Avena fatua* L.] herbicide evaluation in Aragon (Spain). *ITEA, Information Tecnica-Economica Agraria*, 19, 57-64.

Alam,S.M. and A.R.Azmi. (1989). Influence of *Prosopis glandulosa* water extract on seedling growth of wheat cultivars. *Pakistan Journal of Scientific and Industrial Research*, 32: 708.

Allard, R. (1965). Genetic systems associated with colonizing ability in predominatly self-pollinated species. P. 49. in H.Baker and G.Stebins eds. The genetic of colonizing species. Academic press, New York.

Alsaadawi, I.S., Al-Ekelle, M.H.S. and Al-Hamzawi, M.K. (2005). Allelopathic potential of *Sorghum bicolor* L. (Moench) genotypes against weeds. In *Proceedings of Fourth World Congress on Allelopathy "Establishing the Scientific Base"* (eds. Harper, J.D.I., An, M., Wu, H., Kent, J.H), 21-26 August, 2005, Charles Strut University, Wagga Wagga, NSW, Australia, pp. 254-257.

Al-Sahooki, M.M., Obaid, A. and Al-Muhammadi, A.F. (2009). Crop management and breeding for drought tolerance. *Araqi Agricultural Journal*, 40: 1-28. (in Arabic).

Altieri, M.A. and Doll, J.D. (1978). The potential of allelopathy as a tool for weed management in crops. *Pest Articles and News Summaries*, 24: 495-502.

Amritphale, D. and Mall, L.P. (1978). Allelopathic influence of *Saccharum spontaneum* L. on growth of three varieties of weed. *Science and Culture*, 44: 28-30.

Anaya,A.L. and Comez-Pompa, A.. (1971). Inhibition del crecimiento producida por el "piru" (*Schinus molle* L.). Revista. *Sociedad Mexicana de Historia Natural*, 32: 99-109.

Anderson,R.C., A.J.Katz and M.R.Anderson (1978). Allelopathy as a factor in the success of *Helianthus mollis* Lam. *Journal of Chemical Ecology*, 4: 9-16.

Andrew, J.C., Kells, J.J., Siler, L.F. and Word, R.W..(2005). Response to winter wheat to post-emergence herbicides applied in the fall. *North Central Weed Science Society Proceeding,* 60:6-12.

Antonelli, C., and Formigoni, A. (1976). Experiments with granular tri-allate for control of wild-oats in winter wheat. International Congress on Control of Weeds and Algae in Mediterranean Countries. Sardinia, 1973. *Notiziario sulle Malattie delle Piante*, no. 94-95, 241-246.

Appleby A. P. (1967). Weed control in wheat and barley. Asain-Pacific Weed Control Interchange. Honolulu, Hawaii.

Araki, H. and Hatano, Y. (2002). Weed control effect of cover crop mulch in the soil with different weed density. In *Abstracts, Third World Congress on Allelopathy Challenge for the New Millennium* (eds. Fujii, Y., Hiradata, S., Araya, H.). 26-30 August, 2002. Tsukuba, Japan. Abstract No. S11-03, pp.116.

Ashiq, M., Muhammad, N. and Ahmed, N..(2007).Comparative efficacy of different herbicides against broad leaf weeds in wheat. *Pak J. Weed Sci. Res.*13 (3-4):149-156.

Badhani, R., Sarika, Rao, S and Rao, P.B. (2008). Alleopathic effect of five weed species on germination, seedling growth and protein in different varieties of wheat (*Triticum aestivum* L.). *Indian Journal of Environmental Sciences*, 12: 67-74.

Balyan, R.S, and Malik, R.K (1992). Chemical control of wild-oat (*Avena ludoviciana*) and other weeds in wheat (*Triticum aestivum*). *Beitrrage-zur-Tropischen-Landwirtschaft-und-Veterinarmedizin*, 30, 169-175.

Balyan, R.S., Malik, R.K. and Bhan, V.K. (1988). Effect of time of application of isoproturon on the control of weeds in wheat (*Triticum aestivum*). *Indian J. Weed Sci.* 20, 10-14.

Balyan, R.S., Malik, R.K. and Panwar, R.S. (1992). Effects of rates, times and methods of application of tralkoxydim on wild-oat (*Avena ludoviciana*) in wheat. *Trop. Pest Manag.* 38, 411-415.

Balyan, R.S., Malik, R.K., Panwar, R.S. and Singh, S. (1991). Competitive ability of winter wheat cultivars with wild-oat (*Avena ludoviciana*). *Weed Sci.* 39, 154-158.

Barroso, J., Fernandez-Quintanilla, C., Ruiz, D., Hemaize, P. and Rew, L.J. (2004). Spatial stability of *Avena sterilis ssp. Ludoviciana* populations under annual applications of low rates of imazamethabenz. *Weed Res.* 44, 178-186.

Beatty, K.D., Eldridge, I..L.and Simpson, Jr. A.M. (1982). Soybean response to different planting pattens and dates. *Agro. J.* 74: 859-862.

Becker,Y. and L.Guyot. (1951). Sur les toxines racinaires des sols incultes. *C.R. Academia des Sciences*, 232: 105-107.

Becker,Y., L.Guyot and J.Montegut. (1951). Sur quelques incidences phytosociolgiques du probleme des excretions racinaires. *C. R. Academia des Sciences*, 232: 2472-2474.

Bell, A.R. and Nalewaja. J.D. (1968). Competition of wild oat in wheat and barley. *Weed Sci.* 16: 505- 508.

Bernat, W., Gawtonska, H. and Gawtonski, S.W. (2004). Effectiveness of different mulches in weed management in organic winter wheat production. In *Abstracts, II European Allelopathy Symposium* 2004, Pulaway, Poland. pp.118.

Bhowmik,P.C. and J.D.Doll. (1979). Evaluation of allelopathic effects of selected weed species on corn and soybeans. *Proceedings of North Central Weed Control Conference*, 34: 43-45.

Bhowmik,P.C. and J.D.Doll. (1983). Growth analysis of corn and soybean in response to allelopathic effects of weed residues at various temperature and photosynthetic flux densities. *Journal of Chemical Ecology*, 9: 1263-1280.

Blackman, G.E. and Templeman, W.G. (1938). The nature of competilion between cereal crops and annual weeds. *J. Agric. Scl.* 28 : 247-271.

Blez, R. G. (2007). Allelopathy in crop/weed interactions-an update. *Pest Management Science*, 63, 308-326.

Blum, U., King, L.D. and Brownie, C. (2002). Effects of wheat residues on dicotyledonous weed emergence in a simulated no-till system. *Allelopathy Journal*, 9: 159-176.

Boparia, B.S., Dhaliwal, H.S. and Devinder, S. (1991). Chemical control of small canary grass (*Phalaris minor*) and broad-leaf weeds in wheat (*Triticum aestivum*). *Indian J. Agronomy*, 36, 198-201.

Bowden B.A. and Friesen, G. (1967). Competition of wild oats (*Avena fatua* L.) in wheat and flax. *Weed Reasech*, 7: 349-359.

Bradow, J.M.(1996). Plant residue and ephemiral allelochemicals in agriculture. In *Abstracts, First World Congress on Allelopathy, a Science for the Future* (eds. Macias, F.A., Galindo, J.C.G., Molinillo, J.M.G., Gutler, H.C.) 16-20 September, 1996. Cadiz, Spain. Abstract, pp.71.

Breidert, D. (1978). Hoe 23408 36 e.c. (diclofop-methyl) a new post-em. selective grasskiller in cereals and broadleaved crops. In: Proceedings of the Mediterranean Herbicide Symposium, Madrid, 1978. Madrid, Spain; Ministerio of Agricultura. (1978) Vol. 2, 147-154.

Bukolova,T.P. (1971). A study of the mechanism of action of water-soluble substances of weeds on cultivated plants. In: *Physiological-Biochemical Basis of Plant Interactions in Phytocenoses*, ed. A.M. Grodzinsky. Naukova Dumka, Keiv. Vol.2, pp. 66-69. (In Russian, English summary).

Bukuro.S., Walson, W.S. and. Hourauf., R. (1975). The role of coleoptile on barban sensitivity between wiild oat and wheat. *Pesticide Chemistry and Physiology*, 6 (2): 115-125.

Burgos, N. R. and Ronald, E.T. (1996). Weed control and sweet corn (*Zea mays* var. *rugosa*) response in a no-till system with cover crops. *Weed Science*, 44: 355-361.

Carlson, H. L., and Hill, J.E. (1985). Wild oat (*Avena fatua*) competition with spring wheat: plant density effects. *Weed Sci.* 33: 176-181.

Caussanel, J.P. (1979). Effect of non competitifs entre le chenopode blanc (*Chenopodium album*) de la mais (INRA 258). Weed Research 19: 123-125Chancellor R.J., and Peters, N.C.B. (1974). The time of onset of competition between wild oat and spring cereals. *Weed Research*. 14: 197-202.

Chaudary, S.U; Hussain, M.; Ali, M.A., and Iqbal, J. (2008). Effect of weed competition period on yield and yi~ld components of wheat. *J.Agric. Res*. 46(1):47-52.

Chaudhary, S.U., Hussain,M, Ali, M.A. and Iqbal.J. (2008). Effect of weed competition period on yield and yield components of wheat. *J. Agric. Res*., 2008, 46(1): 47

Cheema, Z.A. and Abdul-Khaliq, Igbal, N. (2005). Use of allelopathy in field crops in Pakistan. In *Proceedings of Fourth World Congress on Allelopathy "Establishing the Scientific Base*" (eds. Harper, J.D.I., An, M., Wu, H. Kent, J.H.), 21-26 August, 2005, Charles Strut University, Wagga Wagga, NSW, Australia, pp. 550-553.

Cheema, Z.A. and Khaliq, A. (2000). Use of sorghum allelopathic properties to control weeds in irrigated wheat in a semi arid region of Punjab. *Agriculture, Ecosystems and Environment*, 79: 105-112.

Cheema, Z.A. and Khaliq, A. (2002). Use of sorghum water extract (sorgaab) for weed suppression in irrigated wheat and maize. In *Abstracts, Third World Congress on Allelopathy Challenge for the New Millennium* (eds. Y. Fujii, S. Hiradata and H. Araya). 26-30 August, 2002. Tsukuba, Japan. Abstract No. S13-09, pp.132.

Cheema, Z.A. and Khaliq, A. (2003). Reducing isopruturon with sorghaab for weed control in wheat. *Pakistan Journal of Weed Science Research*, 9: 153-160.

Chhokar, R. S., Sharma, R. K., Chauhan, D. S. and Mongia, A. D. (2006). Evaluation of herbicides against *Phalaris minor* in wheat in north-western Indian plains. *Weed Res*. 46, 40-49.

Chopra, N., Singh, H. and Tripathi , H. P. (1999). Critical Period of Weed Crop Competition in Wheat (*Triticum aestivum L.*). *Indian Journal of Weed Science* 31, Issue: 3&4, Print ISSN : 0253-8040.

Chou, C. H. (1990). The role of allelopathy in agroecosystems: studies from tropical Taiwan. In: Gliessman, S. R. (ed) 1990. Agroecology: Researching the ecological basis for sustainable agriculture. Ecological studies #78. Springer- Verlag. Berlin. p.105-121.

Chou, C.H. (1989). Allelopathic research of subtropical vegetation in Taiwan. IV. Competitive phytotoxic nature of the leachate from four subtropical grasses. *Journal of Chemical Ecology*, 15: 2149-2159.

Christians, N. (1995). A natural herbicide from corn meal for weed free lawns. *IPM Practitioner*, 17: 5-8.

CimmyT. (1996).World wheat facts and trends understanding global trends the use of wheat diversity and international flow of wheat. Genetics. Resource. Cimmy T. Mexico. D. F.

Cochran,V.L., L.F.Elliott and R.I.Papandick. (1977). The production of phytotoxins from surface crop residues. *Soil Science Society of America Journal*, 41: 903-908.

Conklin, A.E., Erich, M.S., Liebman, M., Lambert, D., Gallandt, E.R.and Halteman, W.A. (2002). Effects of red clover (*Trifolium pratense*) green manure and compost soil amendments on wild mustard (*Brassica kaber*) growth and incidence of disease. *Plant and Soil*, 238: 245-256.

Conn, J.S. (1990). Seed viability and dormancy of 17 weed species after burial for 4-7 years in Alaska. *Weed sci.* 38: 134-138.

Correia, N.M., Souza, I.F. and Klink, U.P. (2002). Sorghum straw associated with imazamox for weed control in soybeans planted in succession under no-till system. In *Abstracts, Third World Congress on Allelopathy Challenge for the New Millennium* (eds. Fujii, Y., Hiradata, S., Araya, H.). 26-30 August, 2002. Tsukuba, Japan. Abstract No. P-040, pp.183.

Cortes, A.J., Alecada, I., Lopes, J., Gervasio, J. and Cavaco, A. (1976). Isoproturon in mixtures for weed control in wheat. In: Proceedings II Symposium National of Herbology, Oeirras, 1976, Vol II (1976) 223-239, Lisbon, Portugal.

Courteny, A.D. and Maccallits, J. (1976). Incidence awareness and action against the wild oat in northen Ireland. *Weed abstract. Agriclture in Northern Ireland*. 51: 70-74.

Cudney, D.W., Jordan, L.S. and A.E. Hall. (1991). Effects of wild oat (*Avena fatua*) infstations on light interceptions and growth rate of wheat (*Triticum aestivum*). *Weed Sci.* 39: 175-179.

Cussans, G.W. and Wilson, B.J. (1975). Some effects of crop row width and seed rate on competition between spring barley and wild oat (*Avena fatua* L.) or common couch[(*Agropyron repens* (l) Beauv.]. Proc.Eur Weed res. Soc. Symp. Status and control of grass weeds in Europe. Pages 77- 86.

Das, T.M. and Pal, S.K. (1970). Effects of volatile substances of aromatic weeds on germination and subsequent growth of rice embryos. *Bulletin. Botanical Society of Bengal*. 24: 101-103.

Datta, S.C. and Dasmahapatara, S.. (1984). Allelopathy in two legumes: Inhibition of mustard (*Brassica*) germination and seedling growth. *Comparative Physiology and Ecology*, 9: 285-289.

Delabays, N. and Mermillod, G. (1999). Assessment of the allelopathic properties of a selected strain of *Artemesia annua*. In *Abstracts, Second World Congress on Allelopathy, Critical Analysis & Future Prospects* (ed. Mallik, A.U.). Lakehead University, Canada. Abstract, pp. 82.

Dharmaraj, G.., Sheriff, M.M., Nagarajan, M. and Kannaiyan, S. (1994). Allelopathic effects of sunflower (*Helianthus annuus* L.) on carpet weed (*Trianthema Portulacastrum* L.). In *Abstracts, International Symposium on Allelopathy in Sustainable Agriculture, Forestry and Environment*, (eds. Narwal, S.S., Tauro, P., Dhaliwal, G.S., Prakash, J.). 6-8 September, 1994. New Delhi, India. Abstract No. 2.24, pp. 89.

Dhima, K.V. and Eleftherohorinos, I.G. (2001). Influence of nitrogen on competition between winter cereals and sterile oat. *Weed Sci.* 49, 77-82.

Dilday, R.H., Mattice, J.D., Moldenhauer, K.A. and Gealy, D.R.(1999). An overview of rice allelopathy. In *Abstracts, Second World Congress on Allelopathy, Critical Analysis & Future Prospects* (ed. Mallik, A.U.). Lakehead University, Canada. Abstract, pp. 84.

Dilday, R.H., Yan, W.G. and Liu, J. (1996). Allelopathic activity in rice (*Oryza sativa* L.) to major weed species. In *Abstracts, First World Congress on Allelopathy, a Science for the Future*, (eds. Macias, F.A., Galindo, J.C.G., Molinillo, J.M.G., Gutler, H.C.), 16-20 September, 1996. Cadiz, Spain. Abstract, pp. 253.

Dordio, J.J., Dordio, M.F. and Madeira, J.M. (1981). Wild oats- a weed of wheat fields-levels of infestation. In: Congresso Portugues de Fitiatria e de Fitofarmacologia e III Simposio Nacional de Herbologie, Volume 1 : 223-236.

Dowson, M.P. (1998). *Effect of Leaf Litter on Seedling Emergence of Bull Thistle* [Circium vulgare (Savi) Ten].M.Sc. Dissertation. The University of Western Ontario, Canada. Pp. 173.

Duwayri, M. and Saghir, A.R. (1983). Effect of herbicides on weeds and winter cereals in Jordan. *Dirasat.* 10, 115-122.

Efthimiadis, P.; Skorda, E. and Adamidis, T. (1987). The combined control of wild- oat and broadleaved weeds in Greece in wheat with new herbicides. Agric. Coll., Athens, Greece. Mededelingen van de Faculteit Landbouwwetenschappen, Universiteit Gent (1987), 52(3B), 1167-76

Einhelling, F. A. (1987). Interactions among allelochemicals and other stress factors of the plant environment. In: Waller, G. R. (ed) 1987. Allelochemicals: Role in agriculture and forestry. ACS symp ser 220. Wash. DC. Amer. Chem. Soc.

Eshel, Y. (1978). Selective control of graminaceous weeds in spring wheat. *Phytoparasitica*, 6, 71-77.

Everaere, L. and Thiebault, P. (1990). What is dopler?. *Defense des Vegetaux*, 44, 12-14.

Everson, C.S. and Breer, C.M. (1983). Influence of seasonal on the suppression by pteridim leaf extracts of root growth in *Avena fatua*; a possible influence on the distribution of Philippia evansii. *S.Africa Journal of botany*. 2 (4): 297-300.

Fay, D.K. (1976). The effect of germination stimulators on wildoat (*Avena fatua*) emergence in the field. Proceedings N.C. *weed contro;*. Conf. 30: 110-111.

Felton, W.L. (1976). The influence of row spacing and plant population on the effect of weed competition in soyabeans, *Aus. J.Exp. Agri. Anim. Husb.* 16: 926-932.

Fernandez-Quintanilla, C., Navarrete, L., Torner, C. and Sanchez Del Arco, M.J. (1997). *Avena sterilis* en cultivos de cerea. In: Biologia de las Malas Hierbas de Espana (eds. X Sanz; C. Fernandez-Quintanilla), 4-17. Phytoma, Valencia, Spain.

Fletcher, A.R. and A.J. Renney. (1963). A growth inhibitor found in *Centaurea* spp. *Canadian Journal of Plant Science*, 43: 475-481.

Fujihara, S. and Yoshida, M. (1999). Allelopathy of hairy vetch, *Vicia villosa* Roth. and its application for crop production as mulching material. *Bulletin of Shikoku National Agricultural Experiment Station*, 65: 17-32.

Fujii, Y. and Heradata, S. (2005). A critical survey of allelochemicals in action-the importance of total activity and the weed suppression equation. In *Proceedings of Fourth World Congress on Allelopathy "Establishing the Scientific Base"* (eds. Harper, J.D.I., An, M., Wu, H., Kent, J.H), 21-26 August, 2005, Charles Strut University, Wagga Wagga, NSW, Australia, pp. 73-76.

Fujiyoshi, P.T. (1998). Mechanisms of weed suppression by squash (*Cucurbita* spp.) intercropped in corn (*Zea mays* L.). Ph.D. Dissertation. University of California Santa Cruz. Pp. 89.

Gail A. Wicks_ Alex Martiny and Drew J. Lyonz (1999). G99-1389 Cultural Practices to Improve Weed Control in Winter Wheat. Extension Historical Materials from University of Nebraska-Lincoln Extension University of Nebraska - Lincoln Year 1999. University of Nebraska – Lincoln Available at: http://digitalcommons.unl.edu/extensionhist/1507.

Gavazzi, C. and Paris, P. (2000). Allelopathy: do plants hate each other too? *Informatore Agraria* 56: 53-55.

Gawronski, S.W., Bernat, W. and Gawronska, H. (2002). Allelopathic potential of sunflower mulch in weed control. In *Abstracts, Third World Congress on Allelopathy Challenge for the New Millennium* (eds. Fujii, Y., Hiradata, S., Araya, H.). 26-30 August, 2002. Tsukuba, Japan. Abstract No. P-017, pp. 160.

Gawtonski, S.W. (2004). Allelopathic pre-crop mulch as a tool of weeds managements in winter organic farming.In *Abstracts, II European Allelopathy Symposium* 2004, Pulaway, Poland. pp. 26.

Gealy, D. R., Wailes, E.J., Estorninos, L.E. and Chavez, R.S.C. (2003). Rice cultivars different in suppression of barnyardgrass (*Echinochloa crus-galli*) and economics of reduced propanil rates. *Weed Science*, 51: 601-609.

Gicheru, F.F.(1978). Benzoylprop-ethyl and flamprop-isopropyl for the control of wild-oats in wheat and barley in Kenya. In Proceedings of the Sixth East African Weed Science Conference (1976) publ. 1978), 17-24. Shell Chem. Co. East Africa Ltd., Nairobi, Kenya.

Gill, H.S. and Walia, U.S.(1988). Studies on the control of wild-oats (*Avena ludoviciana* Dur.) in irrigated wheat (*T. aestivum* L.) with substituted urea herbicides. *Indian J. Weed Sci.* 20, 1-4.

Gonzalez Ponce, R. (1979). Response of a wheat-wild oat association to potassium fertilization. *Anales de Edafologia y Agrobiologia*, 38 (5-6): 939-944.

Gonzalez Ponce, R. (1982). Competition of Avena sterilis L. against several wheat varieties. *Anales de Edafologia y Agrobiologia*, 41 (9/10): 1971-1982.

Gonzalez Ponce, R. (1987). Competition for nitrogen and phosphorus between wheat and wild oat (Avena sterilis L.) according to the proximity of their time of emergence. *Plant and Soil*, 102: 133-136.

Gonzalez Ponce, R. and Fernandez, H. (1978). Effect of time of sowing Triticum vulgare L. and level of Avena sterilis L. infestation in competition between the two species. *Anales de Edafologia y Agrobiologia*, 37 (9-10): 839-852.

Gonzalez Ponce, R. and Hernando, V. (1977). Wild oat competition in wheat at different levels of infestation and different stage of development of both species. Paper given at: Escarda Quimica en Cultivos Herbaceos Extensivos: V!!! Jornada de Estudio de la Association Interprofessional para el Desarrollo Agrario, Saragossa 1976. Inst. de Edafologia y Biologia Veg.

Gonzalez Ponce, R. and Rodriguez Senas, J. (1983). Experiments with isoproturon for the control of weeds in winter cereal crops. *Weed Res.* 23, 39-43.

Gonzalez, Ponce R. (1980). National responses of *Triticum vulgare* L. and *Avena sterilis* L. in competition. In: Proceedings of the 6th International Colloquium on Weed Ecology, Biology and Systematic, COLUMA-EWRS Montpellier, 1980, volume II, 433-441.

Gonzalez, Ponce R. and Correa, M.(1979). Effect of *Avena sterilis* in *Triticum vulgare*. In Compte Rendu de la 10 Conference du COLUMA, 1979, (1979) 401-406.

Gonzalez-Ponce, R. and Salas, M. L. (1992). Effects of tri-allate and nitrogen fertilization on wheat grain yield and its components. Cent. Cienc. Medioamb., CSIC Serrano, Madrid, Spain. Plant Protection Quarterly (1992), 7(3), 140-141.

Gonzaliz-Ponce, R. and Santin, I. (2001). Competitive ability of wheat cultivars with wild-oats depending on nitrogen fertilization. *Agronomie*, 21, 119-125.

Gonzaliz-Ponce, R.., Lamela, A. and Salas, M.L. (1989). Effects of nitrogen and herbicide on grain yield and protein content of tall and semi-dwarf wheat (*Triticum aestivum*) cultivar in semi arid conditions. Plant Nutrition: Physiology and Applications. Proceedings of the Eleventh International Plant Nutrition Colloquim, Wageningen, Netherlands, 30 July- 4

August 1989 (Edited by Beusichem, M.L. van). 1990, 556-568. Development in Plant and Soil Sciences Vol. 41. Dordrecht, Netherlands; Kluwer Academic Publisher.

Gounay, X.de. (1964). Data pertaining to wild oat control in spring barley. *Ann. Epiphyt.* 15: 285-320.

Gupta, V.K., Malik, R.K., Balyan, R.S. and Bhan, V.M. (1990). Effect of growth of wild-oat on the efficacy of urea herbicides applied post-emergence. *Haryana Agricultural University Journal of Research*, 20, 57-65.

Hacker, E.; Bieringer, H.; Willms, L.; Lorenz, K.; Koecher, H.; Huff, H. P.; Borrod, G. and Brusche, R. (2001). Mesosulfuron-methyl - a new active ingredient for grass weed control in cereals. Research Frankfurt, Aventis Crop Science GmbH, Frankfurt, Germany. BCPC Conference--Weeds (2001), Vol. 1, 43-48. British Crop Protection Council, UK.

Hamal, A.; Benbella, M.; Rzozi, S. B.; Bouhache, M.; Msatef, Y. and Bulcke, R. (2000). Sulfosulfuron used for chemical control of *Bromus rigidus* in wheat in the Sais area of Morocco. Laboratoire de Malherbologie, CRRA. Sais et Moyen Atlas, Meknes, Morocco. Mededelingen-Faculteit Landbouwkundige en Toegepaste Biologische Wetenschappen. Universiteit Gent, 65, 59-67. Universiteit Gent, Faculteit Landbouwkundige en Toegepaste Biologische Wetenschappen,

Harper, J. L. (1977). The populatin biology of plants. Academic Press, London. 892 pp.

Harper, J.L.(1964). The nature and consequences of interference among plants. Proceedings of the 11th International Conference of Genetics, The Hague, 1963, pp. 465-481.

Hassan, S.M., Draz, A.E., Sheble, S.M., Abou-Yousef, M.I., Bastawisi, A.O. and Aidy, I.R.(2002). Allelopathic activity of Egyptian rice varieties around most troublesome weeds in rice. In *Abstracts, Third World Congress on Allelopathy Challenge for the New Millennium* (eds. Fujii, Y., Hiradata, S., Araya, H.). 26-30 August, 2002. Tsukuba, Japan. Abstract No. S05-05, pp. 74.

Heather,E.M.,A.Navabi., B.LErick.,T.Onavan, and D.M.Spancer. (2007). The weed competitive of western spring cultivars grow organic management. *Crop.Sci.* 3:505-508.

Henson, J. and L.S. Jordan. (1982). Wild oat (*Avena fatua*) competition with wheat (*Triticum aestivum* and *T.turgidum*) for nitrate. *Weed Sci.* 30: 297-300.

Hoesle, U.M. (1981). *Studies on The Weed Competition in Moroccan Wheat Crops With Paticular Regard to the Water Factor*. Dissertation, Universitat Hohenheim. 123 pp. German Fedral Republic.

Hussain, F., T.W.Khan and A.Hussain. (1987). Allelopathic effects of *Cirsium arvense* (L.) Scop. In: *Modern Trends of Plant Science Research in Pakistan*, eds. I.Ilahi. and F.Hussain. Proceedings Third National Conference of Plant Scientists, Dept. of Botany, University of Peshawar, Peshawar, pp. 24-28.

Inderjit and Kaushik, S. (2005). Allelopathic properties of rice straw. In *Proceedings of Fourth World Congress on Allelopathy "Establishing the Scientific Base"* (eds. Harper, J.D.I., An, M., Wu, H.,. Kent, J.H)., 21-26 August, 2005, Charles Strut University, Wagga Wagga, NSW, Australia, pp. 549.

Infanger, C.A. (1956). Economic consequences of 2,4-D in controlling Canada thistle in irrigated spring wheat. M.S. thesis, Montana State College.

International Allelopathy Society (IAS). (1994). Consitution and Bylaw. NewDelhi, India.

Irshad, A. and Cheema, Z.A. (2004). Effect of sorghum extract on management of barnyardgrass in rice crop. *Allelopathy Journal*, 14: 205-213.

Jaakkola, S. (2002). Weed control with cruciferous plant material. In *Abstracts, Third World Congress on Allelopathy Challenge for the New Millennium* (eds. Fujii, Y., Hiradata, S., Araya, H.). 26-30 August, 2002. Tsukuba, Japan. Abstract No. S06-02, pp. 81.

Jaakkola, S. (2005). White mustard mulch is ineffective in weed control. In *Proceedings of Fourth World Congress on Allelopathy "Establishing the Scientific Base"* (eds. Harper, J.D.I., An, M., Wu, H., Kent, J.H), 21-26 August, 2005, Charles Strut University, Wagga Wagga, NSW, Australia, pp. 227-232.

Jannie,O; Krisiensen, L.and Weiner, J.(2005). Effect density spatial pattern of winter wheat on suppression of different species.*Weed Sci.* 53(5):690-694.

Jimenez-Osornio, J .J. and Gliessman. S.R. (1987). Allelopathic interference in a wild mustard (*Brassica campestris* L.) and broccoli (*Brassica oleracea* L. var. *italica*) intercrop agroecosystem. In: *Allelochemicals Role in Agriculture and Forestry*, ed.G.R.Waller, ACS Symposium Series 330. American Chemical Society, Washington D.C. pp. 262-274.

John, D.N, and Noznica, Z.. (1988). Effect of environment and adjuvant on asulam phytotoxicity. *Weed sci.* 36: 367-372.

Johri, A.K. and Govindra, S. (1991). Herbicide cum-cultural weed control in wheat (*Triticum aestivum*) *Indian Journal of Agronomy*, 36, 54-59.

Johri, A.K., Govindra, S., Sharma, D. and Singh, G. (1992). Nutrient uptake by wheat and associated weeds as influenced by management practices. *Tropical Agriculture*, 4, 391-393.

Jordan, D.L., Bollich, P.K., Braverman, M.P. and Sanders, D.E. (1999). Influence of tillage and *Triticum aestivum* cover crop on herbicide efficacy in *Oryza sativa*. *Weed Science*, 47: 332-337.

Joseph, O O. ., Hobbs, L. S.A. and Jana, S. (1990). Diclofop resistance in wild oat (*Avena fatua*) *weed sci.* 38: 475-477.

Keys, C. H. (1975). Evaluation of Dowco 290 for the control of annual and perennial weeds. *Down to Earth*, 32 (1). 4-7.

Kiely, T., Donaldson, D. and Grube, A. (2004). *Pesticides Industry Sales and Usage: 2000 and 2001 Market Estimates*. EPA's Biological and Economic Analysis Division, Office of Pesticide Programs, and Office of Prevention, Pesticides, and Toxic Substances. http://www.epa.gov/pesticides.

Kim, K.U., Shin, D.H. and Kim, H.Y.(1999). Study on rice allelopathy. I. Evaluation of allelopathic potential in rice germplasm. *Korean Journal of Weed Science*, 19: 105-113.

Koch, W. (1967). Competition between crop plants and weeds. II. Effect of annual weeds on cereals). *Weed research*, 7: 22-28

Kohli, R.K. Singh, H.P. and Batish, D.R. (eds.) (2001). *Allelopathy in Agroecosystem*, page 43-119. (). Haworth Press, USA.

Kohlmuenzer, S. (1965). Botanical and chemical studies of collective species Gallum mollugo L; with reference to Karyotypes growing In Poland. VI. Effect of extracts and some other chemical componenfs of Galium mollugo L; on seeds and growth of selected plants. Diss. pharm. 17, 369, see Chem. Abstr. 64, 10085 g (1966).

Kojima, K. and Ohkubo, Y. (1999). Weed suppression of several green manure crops planted in upland field converted from rice paddy. In *Abstracts, Second World Congress on Allelopathy, Critical Analysis & Future Prospects* (ed. Mallik, A.U.). Lakehead University, Canada. Abstract, pp.113.

Kommedahl,T., Kotheimer, J.B. and Bernardini, J.V. (1959).The effects of quackgrass on germination and seedling development of certain crop plants. Weeds 7: 1-12.

Kong, C. (2005). Allelopathy in China. In *Proceedings of Fourth World Congress on Allelopathy "Establishing the Scientific Base"* (eds. Harper, J.D.I., An, M., Wu, H.,. Kent, J.H), 21-26 August, 2005, Charles Strut University, Wagga Wagga, NSW, Australia, pp. 314-317.

Korres, N.E. and Willams-Froud, R.J. (2002). Effect of winter wheat cultivars and seed rate on the biological characteristics of naturally occurring weed flora. *Weed Research*, 42: 417-428.

Kropf, M.J. (1988). Simulation models for crop weed competition. *Weed Research*, 28: 453-459.

Le Tourneau, D. and Heggeness, H.G.. (1957). Germination and growth inhibitors in leafy spurge and quackgrass rhizomes. *Weeds*, 5: 12-19.

Le Tourneau, D., Failes, G.D. and Heggeness, H.G. (1956). The effects of aqueous extracts of plant tissue on germination of seeds and growth of seedlings. *Weeds*, 4: 363-368.

Leather, G.R. (1987). Weed control using allelopathic sunflowers and herbicides. *Plant and Soil*, 98: 17-23.

Lehman, M. E. and Blum, U. (1997). Cover crop debris effects on weed emergence as modified by environmental factors. *Allelopathy Journal*, 4: 69-88.

Li, H.H., Nishimura, H.; Hasegawa, K. and Mizutani, J. (1992). Allelopathy of *Sasa cernua*. *Journal of Chemical Ecology*, 18: 1785-1796.

Li, X., Wang, G., Li, B. and Blackshaw, R.E. (2005). Allelopathic effects of winter wheat residues on germination and growth of crabgrass (*Digitaria ciliaris*) and corn yield. *Allelopathy Journal*, 15: 41-48.

Li-Xiang, J., Lu, D.Z. and Li-Yang, H. (2000). Study on allelopathic effect of wheat straw on emergence of crabgrass (*Digitaria ciliaris*). *Journal of Agricultural University*, 23: 74-77&81.

Locken, L.J. and Kalsey, R.G. (1987). Cnicin concentrations in *Centaurea maculosa* spotted knapweed. *Biochemical Systematics and Ecology*, 15: 313-320.

Lopez, F. and Sugranes Alsina, J. M. (1997). Topik Plus, a postemergence herbicide for use on wheat and triticale. Martinez. Novartis Agro S.A., Barcelona, Spain. Actas - Congreso, Sociedad Espanola de Malherbologia, Valencia, Nov. 24-26, 1997 (1997), 377-382. Publisher: Sociedad Espanola de Malherbologia, Lleida, Spain

Lovelace, M.L., Talbert, R.E., Dilday, R.H., Scherder, E.F. and Buchring, N.W. (2001). Use of allelopathic rice with reduced herbicide rates for control of barnyardgrass (*Echinochloa crus-galli*). Research Series, Arkansas agricultural Experimental Station 485: 75-79.

Lovett, J.V. and Lynch, J. A.. (1979). Studies of *Salvia reflexa* Hornem. I. Possible competitive mechanisms. *Weed Research*, 19 : 351-357.

Ma, Y.Q. and Han, Q. H. (1995). Effect of wheat straw mulching on the growth, development and yield of maize. Acta agriculturae Boreali-Sinica 10: 106-110.

Macek, J. (1990). Wild-oat (*Avena fatua* L. and *A. ludoviciana* Dur.) in rye fields for artificial production of sclerotis of *Claviceps purpurea* Tul. in Istria, Yugoslavia.Symposium on Integrated Weed Management in Cereals. Proceedings of an EWRS Symposium, Helsinki, Finland, 4-6 June 1990, 289-296. Wageningon, Netherlands; European Weed Research Society.

Madeira, J.M. and Dordio, M.F. (1981). A study of the weed flora in the rotation: fallow-chickpea-wheat-barley. In I Congresso Portugues de Fitiatria e de Fitofarmacologia e III Simposio Nacional de Herbologia, 1980. (1981) Vol. 3, 47-68. Direccao Regional de Agricultura, Alentejo, Portugal.

Malik, R.K., Balyan, R.S. and Bhan, V.M. (1988). Effect of surfactants in the efficacy of some post-emergence grass herbicides. *Haryana Agricultural University Journal of Research*, 18, 276-283.

Malik, R.K., Malik, R.S., Bhan, V.M. and Panwar, R.S. (1989). Influence of time of application of urea herbicides and diclofop-methyl in wheat. *Indian Journal of Agronomy*, 34: 312-315.

Mallik, M.A.B. and Williams, R.D. (2005). Allelopathic growth stimulation of plants and microorganisms. Allelopathy Journal 16: 175-198.

Manners, G.D. and D.S.Galitz. (1986). Allelopathy of small everlasting (*Antennaria microphylla*): Identification of constituents phytotoxic to leafy spurge (*Euphorbia esula*). Weed Science 34: 8-12.

Manners,G.D. (1987). The role of phytochemistry in attacking the leafy spurge (*Euphorbia esula*) problem. In: *Allelochemicals Role in Agriculture and Forestry*, ed. G.R.Waller. ACS Symposium Series 330. American Chemical Society, Washington D.C. pp. 228-237.

Manning, T.H., Palmer, J.J. and Hewson, R.T. (1993). Weed control in wheat with HOE 1028- a combination of fenoxaprop-p-ethyl and isoproturon. In: Brighton Crop Protection Conference, Weeds. Proceedings of an International Conference, Brighton, UK, 22-25 November. Farnham, UK; British Crop Protection Council (BCRC) Vol. 2: 597-600.

Mannova, B. and Taim, R.C. (1985). Effect of Chlorfenprop-methyl, flamprop-isopropyl and benzoyl prop- Ethylon CO_2 fixation in *Avena fatua, Tritcum aestium* L., and *Hordeum Vulgarel. Weed research*. 25 (5): 347- 353.

Martin,P. and Rademacher, B.. (1960). Studies on the mutual influences of weeds and crops. Symposium. *British Ecological Society*, 1: 143-152.

Mason, H.E., Navabi, A., Fric, B.L., O'Donova, T. and Spaner, D.M. (2007). The weed-competitive ability of Canada western red spring wheat cultivars grown under organic management. *Crop Sci*. 47:1167-1176.

Mason-Sedun,W., Jessop, R.S. and Lovett, J.V. (1986). Differential phytotoxicity among species and cultivars of the genus *Brassica* to wheat. 1. Laboratory and field screening of species. *Plant and Soil*, 93: 3-16.

McCuen, G.W. and Silver, E.A. (1943). Combine harvest investigation. *Ohio Agri. Exp. Sta. Bull*. 643.

Menges, R.M. (1987). Allelopathic effects of palmer amaranth (*Amaranthus palmeri*) on other plant residues in soil. *Weed Science*. 35: 339-347.

Mersie,W. and M.Singh. (1988). Effects of phenolic acids and ragweed parthenium (*Parthenium hysterophorus*) extract on tomato (*Lycopersicon esculentum*) growth and nutrient and chlorophyll content. Weed Science 36: 278-281.

Michalopoulos, G., Papasotriou, K., Ghioulbasanis, A. and Milio, P. (1990). Evaluation of tralkoxydim in Greece. *Zizaniology*, 2, 155-160.

Miller, S.D. and Nawaja, J.D. J.D. (1983). Wildoat seed longivity and production. Abstract. In proceedings of NorthenCenteral Western Conference.80.

Mirkamali, H. (1993). Chemical control of grasses in wheat. In: Brighton Crop Protection Conference, Weeds. Proceedings of an International Conference, Brighton, UK, 22-25 November. Farnham, UK; British Crop Protection Council (BCRC) Vol. 2, 579-584.

Mondol, MRH., Khan, MM , Islam, R., Islam, MI. and Hossain. MH . (2007). Studies on competition between wheat and *Chenopodium album* L. *Bangladesh Journal of Agricultural Research*, Vol 32, No 2. 191-203

Moser, E. W and Koch, H. Ganzelmeir. (1976). Determination of optimal conditions for applications in weed control for cereal production. Grundlagender landtechnik. 26: 17-121. *In weed abstract.* 26 (5): 1449.

Moss, S.R. (1985). The influence of crop variety and seed rate on *Alpecurus ,mysuroides* competition in cereals. In: Proceedings 1985 British Crop Protection Conference-Weeds, Brighton, UK, 701-708.

Muller, W. H., Lorber, P., Haley, B. and Johnson, K. (1969). Volatile growth inhibitors produced by SaLvia leucophylla Greene: Effect on oxygen uptake by mitochondrial suspensions. Bull-Torrey Bot. Club 96 : 89-95.

Mushtaq, M.N., Cheema, Z.A. and Khaliq, A. (2010). Effect of mixture of allelopathic plant aqueous extracts on *Trianthema portulacastrum* L. weed. *Allelopathy Journal*, 25: 205-212.

Musick, T. J. (1969). Weed Biology and Control. McGraw-Hill, New York. 273 P.

Nandal, D.S. and Singh, R.R. (2002). Allelopathic effect of (*Populus deltoids* Bartr.) on wheat varieties. In *Abstracts, Third World Congress on Allelopathy Challenge for the New Millennium* (eds. Fujii, Y., Hiradata, S., Araya, H). 26-30 August, 2002. Tsukuba, Japan. Abstract No. P-025, pp. 168.

Narwal, S.S. (2000). Weed management in rice-wheat rotation by allelopathy. *Critical Reviews in Plant Sciences*, 19: 249.

Narwal, S.S.(1996). Allelopathic strategies for weed control in sustainable agriculture. In *Abstracts*, *First World Congress on Allelopathy, a Science for the Future*, (eds. Macias, F.A., Galindo, J.C.G., Molinillo, J.M.G., Gutler, H.C.) 16-20 September, 1996. Cadiz, Spain. Abstract, pp.73.

Narwal, S.S., Palaniraj, R., Sati, S.C. and Dahiya, D.S. (2004b). Field screening of sorghum varieties for allelopathic potential against *Parthenum hysterophorus* and *Trianthema portulacastrum*. In *Abstracts. IV International Conference Allelopathy in Sustainable Terrestrial and Aquatic Ecosystems*, (eds. S.S.

Narwal, S.S., Palaniraj, R., Singh, H., Sati, S.C. and Rawat, L.S. (2002). Field screening for allelopathic potential of wheat accessions against major winter weed in North India. In *Abstracts, Third World Congress on Allelopathy Challenge for the New Millennium* (eds. Fujii, Y., Hiradata, S., Araya, H). 26-30 August, 2002. Tsukuba, Japan. Abstract No. S06-05, pp. 84.

Narwal, S.S., Sati, and Palaniraj, R. (2004a). Allelopathic weed suppression of pearlmillet accessions against on *Parthenium hysterophorus* and *Trianthus portulacastrum*.In *Abstracts. IV International Conference Allelopathy in Sustainable Terrestrial and Aquatic Ecosystems*, (eds. Narwal, S.S., Barbara, P.). 23-25 August, 2004. International Allelopathy Foundation, 8/15 Haryana Agricultural University, Hisar 125 004, India, pp. 14

Navarrete, L. and Fernandez-Quintanilla, C. (1990). The population dynamics of *Avena sterilis subsp. ludoviciana* under various cropping systems in a Mediterranean environment. Symposium on integrated Weed Management in Cereals. Proceedings of an EWRS symposium, Helsinki, Finland, 4-6 June 1990. 1990, 173-178. Wageningen, Netherland; European Weed Research Society.

NI, H. and Zhang, C. (2005). Use of allelopathy for weed management in China- a review. *Allelopathy Journal*, 15: 3-12.

Nieto, J.H; M.A. Brondo and J.T. Gonzalez. (1968). Critical periods of the crop growth cycle for competition from weeds. PANS 14 : 159-166.

Okuno, K., Ebana, K., Yan, W.G. and Dilday, R.H. (1999). Rice allelopathy: Screening strategy and genetic variation. In *Abstracts. Second World Congress on Allelopathy, Critical Analysis & Future Prospects* (ed. Mallik, A.U.). Lakehead University, Canada. Abstract, pp. 140.

Olfosdotter, M., Navarez, D., Rebulanan, M. and Streibig, J.C.(1999). Weed suppressing rice cultivars-does allelopathy play a role? *Weed Research*, 39: 441-454.

Oudhia,P. and Tripathi, R.S.(1999). Allelopathic effects of *parthenium hysterophorus*, *Lantana camara* and *Ageratum conyzoides* on germination and seedling vigour of wheat and selected rabi weeds. Second World Congress on Allelopathy: Critical Analysis and Future Prospects. Lakehead University, Thunder Bay, Ontario, Canada. August 8-13. Abstract no. 142.

Panwar, R.S., Malik, R.K. and Malik, R.S. (1990). Influence of tank mixture of herbicides on weed control in wheat. *Haryana Agricultural University Journal of Research*, 20, 191-194.

Parasher, V. and Singa, O.S.. (1984). Pysiology of anaerobic in *Phalaris minor* and *Avena fata* L. *Seed Research*. 12 (2): 1- 7.

Park, K.H.(1996). A potential weed management by allelopathic rice germplasm in Korea.In *Abstracts, First World Congress on Allelopathy, a Science for the Future*, (eds. Macias, F.A., Galindo, J.C.G., Molinillo, J.M.G., Gutler, H.C.) 16-20 September, 1996. Cadiz, Spain. Abstract pp. 269.

Parker, V.T. and Muller, C.K. (1979). Allelopathic dominance by a tree associated herb in a California annual grassland. *Oecologia*, 37: 315-320.

Pavlychenko T. K. and Harrington, J.T. (1934). Competitive efficiency of weeds of cereals crops. *Canadian Journal of Research*, 10, 77-94.

Pavlychenko, T.K. (1949). Plant competition and weed control. Agrlc Inst. Rev. 4 : 42-145. (C.F. Arnon, I. 1972. Crop production in dry regions Vol. 1, Leonard Hill).

Pereira, M. M., das Neves, C.H. and Gaspar, E.M. (1996). Potential allelopathic sterols and ketosteroids from wheat straw (*Triticum aestivum*).In *Abstracts, First World Congress on Allelopathy, a Science for the Future,* (eds. Macias, F.A., Galindo, J.C.G., Molinillo, J.M.G., Gutler, H.C.), 16-20 September, 1996. Cadiz, Spain. Abstract, pp. 155.

Peters, N.C.B and B. J. Wilson. (1983). Some studies on the competition between *Avena fatua* L. and spring barley. II. Variation of *Avena fatua* emergence and development and its influence on crop yield. *Weed Research*, 23: 305-311

Peters, N.C.B, (1983). Time of onset of competition and effects of various fractions of an *Avena fatua* L. population on spring barley. *Weed Research*. 24: 305-315.

Ponce, R.G.Lamela, A. and Salas, M.L. (1990). Effects of nitrogen and herbicides on grain yield and protein content of a tall and semi-dwarf wheat (Triticum aestivum) cultivar in semi-arid conditions. In: Physiology and applications. Proceedingsof the eleventh

International Plant Nutrition colloquim. Wageningen, Netherlands30 July-4 August1989 (Beusichem, M.L. van edt.), 1990, 565-568

Porwal, M.K. and Gupta, O.P. (1986). Allelopathic influence of winter weeds on germination and growth of wheat. *International Journal of Tropical Agriculture*, 4: 276-279.

Prabhakarasetty, T.K.(1980). Studies on the Biology and Control of *Striga asiatica* (L.) Kuntze. Ph.D Thesis. Bangalore, India, Department of Agronomy, University of Agricultural Science, pp.277.

Purvis, C.E., Jessop, R.S. and Lovett, J.V. (1985). Selective regulatin of germination and growth of annual weeds by crop residue. *Weed research*. 25 (6): 415-421.

Putnam, A.R.(1988). Allelochemicals from plants as herbicides. *Weed Technology*, 2: 510-518.

Qasem J. R. (1982). The competitive and allelopathic effects and chemical control of Syrian sage (*Salvia syriaca* L.) in Wheat. M. Sc. Thesis, University of Jordan.

Qasem J. R. and Abu-Irmaileh B. E. (1983). The competitive effect and chemical control of Syrian sage (*Salvia syriaca* L.) in wheat. *Dirasat*, 10 (2)7 123-128.

Qasem J. R. and Abu-Irmaileh B. E. (1985). Allelopathic effect of Salvia syriaca L. (Syrian sage) in wheat. *Weed Research*, 25747-52.

Qasem J.R. (2007). Chemical control of wild-oat (*Avena sterilis* L.) and other weeds in wheat (*Triticum durum* Desf.) in Jordan. *Crop Protection*, 26, 1315–1324

Qasem, J. R. (1990). Aqueous extract effect of nettle-leaved goosefoot (*Chenopodium murale*) on wheat and barley. Research Journal of Aleppo University, 14, 37-53.

Qasem, J. R. (1993). Aqueous extract effect of hoary pepperwort (*Cardaria draba* (L) Desv.) on wheat (*Triticum durum* L.) and barley (*Hordeum vulgare* L.). *Mu`tah Journal for Research and Studies*, 8(3) , 63-79.

Qasem, J. R. (1994 a). Allelopathic effects of white top (*Lepidium draba*) on wheat and barley. *Allelopathy Journal*, 1(1), 29-40.

Qasem, J. R. (1994b). Allelopathic effects of white top (*Lepidium draba*) on wheat and barley. *Allelopathy Journal*, 1(1), 29-40.

Qasem, J. R. (1995 a). The allelopathic effects of three *Amaranthus* spp. (pigweeds) on wheat (*Triticum durum*). Weed Research, 35 (1), 41-49.

Qasem, J. R. (1995b). Allelopathic effect of some arable land weeds on wheat (*Triticum durum* L.), A survey. Dirasat 22B(4), 81-97.

Qasem, J. R. (1999 a). Biological activity of corn buttercup (*Ranunculus arvensis* L.). In, *Recent Advances in Allelopathy, A science for the Future*. Vol. 1. (F.A. Macias, J.C.G. Galindo, J.M.G. Molinillo and H.G. Gutler, Edts.).Servicio De Publicaciones Universidad De Cadiz, Cadiz, Spain. 287-300.

Qasem, J. R. (1999 b). Research on allelopathy in Jordan. (Chapter) In , *Allelopathy Update*. Vol. 1 (S. S. Narwal, Edt.). Science Publishers, Inc. New Hampshire, USA. 221-246.

Qasem, J. R. (2003). *Weeds and Their Control*. University of Jordan Publications. Amman, Jordan. 628 pp.

Qasem, J. R. and Abu- Irmaileh, B. E. (1983). The competitive effect and chemical control of syrian sage (*Salvia syriaca* L.) in wheat. *Dirasat*, 10 (2), 123-128.

Qasem, J. R. and Tell, A. M. (1993). Response of some wheat cultivars to weeding and fertilizer treatments. *Dirasat*, 20B, 63-74.

Qasem, J.R. (2001). Allelopathic potential of White top and Syrian sage on vegetable crops. *Agronomy Journal*, 93, 64-71.

Qasem, J.R. (2009). Weed competition in cauliflower (*Brassica oleracea* L. var. *botrytis*) in the Jordan Valley. *Scientia Horticulturae*, 121, 255-259.

Qasem, J.R. (2010). Differences in the allelopathy results from field observation to laboratory and glasshouse experiments. Allelopathy Journal 26 (1), 45-58.

Qasem, J.R. and Foy, C.L. (2001). Weed allelopathy, its ecological impact and future prospects. In, *Allelopathy in Agroecosystem*, page 43-119. (R.K. Kohli, H.P., Singh, and D.R. Batish eds.). Haworth Press, USA.

Qasem, J.R. and Hill, T.A. (1989). On difficulties with allelopathy methodology. *Weed Research* 29: 345-347.

Quimby P. C. and Nalewaja J. D. (1966). Effect of dicamba and wild buckwheat at various stages of development. *Weeds*, 14, 229-232.

Rajaram.S. (2002).Prospect and promise of wheat the 21st century. 6th. Intern.Wheat.Conf.Budapest. Hungary.P:24.

Rassmussen, J.A. and Einhellig, F.A. (1975). Non competitive effects of common milkweed, *Asclepias syriaca* L. on germination and growth of sorghum. *American Midland Naturalist*, 94: 478-483.

Rassmussen, J.A. and Einhellig, F.A. (1979). Allelochemic effects of leaf extracts of *Ambrosia trifida* (Compositae). Southwest. *Nature*, 24: 637-644.

Regnier, E.E. (1986). Aspects of velvetleaf (abutilon theopharasti medik.), Jimsonweed (*Datura stramonium*) and commoncocklebur (*Xanthium strumarium L.*) interfrence with soyabean. Ph.D thesis . univ. Illionis, Urbana, IL. 116p.

Regnier, E.E., E.W. Stoller, and E. Nafziger. (1989). Common cocklebur (*Xanthium Strumarium*) Root and shoot interfrence in soybeans (*Glycine max)*. *Weed sci.* 37: 308-313.

Rice, E.L. (1974). *Allelopathy*. Academic Press. New York.

Robison, L. R. and Fenster, C. R. (1973). Winter wheat response to herbicides applied postemergence. *Agron. J.* 65:749-751

Sahid,I.B. and J.B.Sugau. (1993). Allelopathic effect of lantana (*Lantana camara*) and Siam weed (*Chromolaena odorata*) on selected crops. *Weed Science*, 41: 303-308.

Samunder, S. and Malik, R.K.(1992). Evaluation of tralkoxydim against weeds in wheat. Tests of Agrochemicals and Cultivars 13, 60-61.

Sankhla,N., D.Baxi, and U.N.Chatterji. (1965). Eco-physiological studies on arid zone plants. I. Phytotoxic effects of aqueous extracts of mosquite (*Prosopis juliflora* DC.). Current Science 34: 612-614.

Sarma,K.K.V. (1974). Allelopathic potential of *Digera arvensis* Forsk. On *Pennisetum typhoides* Stapf. Et Hubb. Geobios (Jodhpur) 1: 137.

Schnoor, J.L. (1992). *Fate of Pesticides and Chemicals in the Environment*. John Wiley-Interscience, New York.

Schumacher,W.J., Thill, D.C. and Lee, G.A. (1983). Allelopathic potential of wild oat (*Avena fatua*) on spring wheat (*Triticum aestivum*) growth. *Journal of Chemical Ecology*, 9: 1235-1246.

Selleck,G.W. (1972). The antibiotic effects of plants in laboratory and field. *Weed Science*, 20: 189-194.

Sene, M., Dore, T. and Gallet, C.1999. Allelopathic potential increase with biomass accumulation in grain sorghum [*Sorghum bicolor* (L.) Moench]. In *Abstracts. Second*

World Congress on Allelopathy, Critical Analysis & Future Prospects (ed. Mallik, A.U.). Lakehead University, Canada. Abstract, pp. 163.

Shahid, M., Ahmad, B., Khattak, R.A. and Arif, M. (2007). Integration of herbicides with aquous alelopathic crop extracts for weed control in wheat, African Crop Science Conference Proceedings, 8: 209-212/

Sharma, M.L., Bhardwaji, S.K. and Bhardwaji, G.S. (1989). Comparative study of chemical and mechanical weed control in wheat. *Haryana Journal of Agronomy*, 5, 177-178.

Sharma, M.L., Bhardwaji, S.K. and Bhardwaji, G.S. 1991. Comparative efficiency of chemical and mechanical weed control methods in wheat. *Indian Journal of Agronomy*, 36, 260-261.

Sharma, M.P, McBeath, D.K. and VandenBorn. W.H. (1976). Studiies on the biology of wild oats I. Dormancy, germination and emergency. *Canadian Hournal of plant Science*, 56: 611-618.

Sharma,V. and Nathawat, G.S. (1987). Allelopathic effecst of *Argemone mexicana* L. on species of *Triticum*, *Raphanus* and *Pennisetum*. *Current Science*, 50: 427-428.

Shashkov, V. P.; Kolmakov, P. P.; Volkov, E. D.; Trifonova, L. F. (1977), Effect of offset weeds in spring wheat plantings on the removal of nitrogen

Shebeski, L.H. and Frieson, G (1955). Weed competition as affected by time of spraying. Proc. West. Can. Weed Conf. pp. 1-5.

Shorma, M.P., Born, W.H., Vanden, W.H, Friesen, H.A and McBeath, D. K. (1976). Penetration, translocation and metabolism of 14 C ditenzoqunt on wild oats and barley. *Weed science*. 24: 379-384..

Singh, R.D. and Gosh, A.K.(1992). Evaluation of herbicides for control of wild-oat (*Avena ludoviciana*) in wheat (*Triticum* species). Indian Journal of Agronomy 37, 327-331.

Singh, V.K. and Bajpai, R.P. (1992). Crop-weed competition in dwarf wheat under hill zone of Chhattisgarh-MP. Current Research-University of Agricultural Sciences (Bangalore) 21: 43-44.

Skorda, E.A., Efthimiadis, P. and Adamidis, T. (1991). Wild-oat seed production and seed fate affected by herbicides. Proceedings of the Brighton Crop Protection Conference, Weeds. 1991, Vol. 3, 929-936.

Song,Q., Fu, Y.; Tang, J.; Feny, Z. and Yang, C.. (1999). Allelopathic potential of *Eupatorium adenophorum*. Second World Congress on Allelopathy: Critical Analysis and Future Prospects. Lakehead University, Canada. August 8-13. Abstract no. 170.

Soufizadeh, S., Zand, E., Baghestani, M.A., Bena Kashani, F., Nzama-badi, N., and Sheibany, K. (2007). Integrated weed management in saffron (*Cocus sativus* L.) *Acta Hort.* 739, 133-138.

Spitters, C.T. and Bergh, J.P. van den. (1982). Competition between crop and weeds. A Systems Approach. Pages 137-148 IN Holznerand M. Numata, eds. Biology and Ecology of weeds. Dr. W. Junk Publishers.

Spring, O. and Hager, A. (1982). Inhibition of elongation growth by two sesquiterpene lactones isolated from *Halianthus annuus* L. *Planta*, 156: 433-440.

Stachon,W.J. and Zimdahl, R.L.. (1980). Allelopathic activity of Canada thistle (*Cirsium arvense*) in Colorado. *Weed Science*, 28: 83-86.

Stirzaker, R.J.N. and Bunn, D.G. (1996). Phytotoxicity of ryegrass and clover cover crops and a lucerne alley crop for no-till vegetable production. *Biological Agriculture and horticulture*, 13: 83-101.

Stivers, R.K. and Sweaingin, M.L. (1980). Soybean yield compensation with different plant populations and missing plant patterns. *Agron. J.* 72: 98-102.

Sugha, S.K. (1979). Effect of weed extracts on wheat germination. *Science and Culture*, 45: 65-66.

Teasdale, J.R. (2002). Weed management with cover crop mulches. In *Abstracts, Third World Congress on Allelopathy Challenge for the New Millennium* (eds. Fujii, Y., Hiradata, S., Araya, H). 26-30 August, 2002. Tsukuba, Japan. Abstract No. S11-02, pp. 115.

Terekhova, M.A. and Nesterenko, A.M. (1977). The use of triallate in spring wheat crops. Khimiya v Sel'skom Khozyaistrve 15, 43-45.

Tessema, T., Tanner, D.G. and M. Hassena.(2001). Grass weeds competition with bread wheat in Ethiobia.ll. Prediction of grain yield loss and implications for economic weed control. *African Crop Science Joumal.*4(2):411-421.

Thai, K., Jana, S. and Fowke, L.C. (1989). Cell membrane permeability and ultra structural effect by difenzoquat on wild oats (Avena fatua). *Weed sci.* 37: 98-106.

Thakur, D.R and Singh, K.K. (1990). A note on wed control efficiency of isoproturon and methabenzthiazuron in wheat under mid-hills of Himachal Pradesh. *Himachal Journal of Agricultural Research*, 16, 163-165

The University of Nebraska–Lincoln Extension Publications Web site. Index: Weeds Field and Pasture (1999), Revised August 2006 Availableat: http://www.ianrpubs.unl.edu/ epublic/pages/publicationD.jsp?publicationId=596.

Thomas, P.E., Schwerzei, L., Oosterman, P.J., Buder, H.M., Lindert, G.D. and Van. H.J.A. (1976). *Weed Biology*. Henderson, Res.Sta. Salisbury, Rhodesia.

Thurston J.M. and Philipson, A. (1976). Distribution. Pages 19-64 in D.P. Jones, ed. *Wild oats in World Agriculture*. Agric. Res. Counc., London.

Timmons, F.L. (1970). A history of weed control in the United States and Canada. *Weed Sci,*. 18 : 294-307.

Tollennar, M., Dibo, A.A., Aquilera, A. Weise, S.F. and Swanton, C.J.(1994). Effect of crop density on weed inteference in maize. *Agronomy Journal*, 86: 591-595.

Tripathi,R.S., R.S.Singh and J.P.N.Rai. (1981). Allelopathic potential of *Eupatorium adenophorum*, a dominant ruderal weed of Meghalaya. Proceedings Indian National Science Academy, Part B, *Biological Science*, 47: 458-465.

Urbano, B., Gonzalez-Andres, F. and Ballesteros, A. (2006). Allelopathic potential of cover crops to control weeds in barley. *Allelopathy Journal*, 17: 53-64.

Van Acker, R.C., Lutman, P.J.W. and Froud-Williams, R.J. (1997). The influence of intraspecific interference on the seed production of *Stellaria media* and Hordeum vulgare (volunteer barley). *Weed Research*, 37: 277-286.

Vander Vorst, P. B., G. A. Wicks, and O. C. Burnside. 1983. Weed control in a winter wheatcorn-ecofarming rotation. *Agron. J.* 75:507-511.

Vezina,L. and Doyon, D.. (1983). A note on the inhibitor effect of black knapweed (*Centaurea nigra*) residues. *Phyto Protection*, 64: 77-81.

Villiers, B.L. De. and Du Tott, D. (1992). Antagonistic effect of MCPA on *Avena sterilis* with tralkoxydim. *S. Afr. J Plant Soil*, 9, 1-3.

Vizantinopoulos, S.; Katranis, N. (1998). *Management of blackgrass (Alopecurus myosuroides*) in winter wheat in Greece. *Weed Technology*, 12(3), 484-490.

Walia, V.S., Brar, L.S. and Mehra, S.P. Interaction effects of nitrogen levels and weed control measures in wheat. *Indian J. Weed Sci.* 22, 28-33.

Waller, H., and F. Bischof. (1976). The effectiveness of new post emerrgency herbicides against wildoat (*Avena fatua* L.) in relation to site of application. *Weeds absract.* 26 (6): (1665). GDR.

Weaver, S. E., Kropff, M. and Groeneveld, R. M. W. 1992. Use of ecophysiological models for crop–weed interference: the critical period of weed interference. *Weed Sci.* 40:302–307.

Welsh J P., Bulson, J. H. A., Stopes, C. E. , Froud-Williams, R. J.and Murdoch, A. J. (1999). The critical weed-free period in organically-grown winter wheat. *Ann. Appl. Biol.* 134:315-320

Weston, L.A.(1996). Utilization of allelopathy for weed management in agroecosystem. *Agronomy Journal*, 88: 860-866.

Weston,L.A. and Putnam, A.R.. (1986). Inhibition of legume seedling growth residues and extracts of quackgrass (*Agropyron repens*). *Weed Science*, 34: 366-372.

White, A.D., Stahlman, P.W., Phillip, W., Northam, F.E. (2004). Impact of integrated management systems on jointed goatgrass (Aegilops cylidrica) populations. *Weed Science*, 52:1010-1017.

Willis, R.J. 1996. The history of allelopathy 1. The first phase 1785-1845: The era of A.P. de Candolle. *Allelopathy Journal*, 3: 165-184.

Wilson, B.J. (1979). Post-emergence control of wild-oats in Queensland with difenzoquat, flamprop-methyl, diclofop-methyl and barban. *Aust. J. Exp. Agric. & Animal Husb.* 19, 108-117.

Wilson, B.J., Cousens, R. and Cussans, G.W. (1984). Exercises in modeling populations of *Avena fatua* L. to aid strategic planning for the long term control of weed in cercus. *EWRS*. 287-294.

Wilson, B.J., Peters, N.C.B., Wright, K.J. and Atkins, H.A. (1988). The influence of crop competition on seed production of Lamium purpureum, Viola arvensis and papaver rhoeas in winter wheat. In: Weed control in cereals and the impact of legistlation on pesticide application. *Aspects of Applied Biology*, 18: 71-80.

Wit C.T. De. and Bergh, J.P. van den. (1965). Competition among herbage plants. *Neth. J. Agric. Sci.* 13: Sci.13: 212-221.

Wu, H. (2005). Molecular approaches in improving wheat allelopathy. In *Proceedings of Fourth World Congress on Allelopathy "Establishing the Scientific Base"* (eds. Harper, J.D.I., An, M., Wu, H.,. Kent, J.H), 21-26 August, 2005, Charles Strut University, Wagga Wagga, NSW, Australia, pp. 201-208.

Yaduraji, N.T., Ahuja, K.N. and Gautam, K.C. (1992). Bioefficacy and selectivity of tralkoxydim with and without 2,4-D and fluroxypyr. *Pesticide Research Journal*, 4, 159-163.

Yang, C. and Mu, X. (2006). Allelopathy of *Descurainia Sophia* as weed on wheat. Yingyong Shengtai Xuebao 17: 2389-2393.

YongQing, M. (1994). Allelopathic effects of wheat straw mulching on corn seedlings growth and development. In: Narwal, S.S., Tauro, P., Dhaliwal, G.S., Prakash, J. (Eds.), Abstracts, International Symposium on Allelopathy in Sustainable Agriculture, Forestry and Environment, 6-8 September, 1994, New Delhi, India, p.13.

Zanin G, Berti A & Toniolo L (1993) Estimation of the economic thresholds for weed control in winter wheat. *Weed Research*, 33, 459–467.

Zemanek. J. (1977). The influence of annual application of herbicides on the changes of weed communities on plough ground. *Weed abstract.* 26: 10(13).

Zemer, M.C., G.S.Gill, and R.K. Vandeleur. (2008). Effect of height on the competitive ability of wheat with oat.*Agron.J.*100 (6):1594-1599.

WEBSITES

htpp://www.weedscience.org/summary/MOASummary.asp
http://www.ianrpubs.unl.edu/epublic/pages/publicationD.jsp?publicationId=596
http://www.ianrpubs.unl.edu/live/g1389/build/g1389.pdf
http://www1.agric.gov.ab/ca/$department/deptdocs.nsf/all/crop1280)

In: Wheat: Genetics, Crops and Food Production
Editor: Maria T. Almeida, pp. 81-119
ISBN 978-1-61209-307-9

Chapter 2

WHEAT: GENETIC IMPROVEMENT AND QUALITY IN BREAD MAKING PRODUCTS IN MEXICO AND BRAZIL

Maria Teresa Pedrosa Silva Clerici[1*], Elizabeth Harumi Nabeshima[2†], Jorge Minoru Hashimoto[3‡], Fernando Martinez Bustos[4], Arikani Soberanes-Felix[5] and Yoon Kil Chang[6]

[1]Federal University of Alfenas (UNIFAL-MG)
Rua Gabriel Monteiro da Silva 700, Centro, zip code 37130-000, Alfenas, MG, Brazil
[2]Food Technology Institute (ITAL)
Av. Brasil, 2880, Bairro Chapadão, zip code 13070-178, Campinas, SP, Brazil
[3]Food Technology Institute (ITAL)
Av. Brasil, 2880, Bairro Chapadão, zip code 13070-178, Campinas, SP, Brazil
[4]CINVESTAV Unidad Querétaro, Apdo. Postal 1-798, Querétaro, Qro., 76230, Mexico
[5]Facultad de Ciencias Químico Biológicas. Universidad Autónoma de Sinaloa
Ciudad Univeritaria, Culiacán, Sinaloa, México, CP 80010
[6]State University of Campinas (UNICAMP) Department of Food Technology – Food Engineering Faculty, Campinas, SP, Brazil

ABSTRACT

Wheat, a noble cereal of high global consumption is, produced in quantity and quality by a few countries and imported by developing countries, which in most cases do not have enough production to supply the internal market. In this chapter, we will present the efforts of countries as Brazil and Mexico, which require large volumes of wheat import for bread making products, in production and evaluation of cultivars adapted to

* E-mail: maria.pedrosa@unifal-mg.edu.br, mtcleric@fea.unicamp.br; Tel. +55-35-3299-11399
† E-mail: nabeshima@ital.sp.gov.br;Tel. +55-19-3743-1952
‡ E-mail: jorge@ital.sp.gov.br; Tel. +55-19-3743-1830

environmental conditions, as in development and evaluation of products with the use of additives and ingredients to improve the quality of bread obtained. Initially, we will focus on research conducted to improve technological and commercial quality of Mexican wheat cultivars aimed at improving the production processes for the producer and milling industries, which increasingly require good quality wheat for bread. In this chapter, we will detail relevant work done in Mexico. These studies have been done considering that technological quality of wheat depends on genetic fitness of the variety, climatic conditions during the crop cycle, soil resources to be cultivated, technological resources applied in cultivation, post-harvest management of field production, and industrial flour processing. Finally, the chapter will detail the work of Research Institutes and Universities in the area of Cereal Technology being directed at improving nutritional and technological quality of wheat products, with major studies in recent years involving the use of gamma radiation and infrared radiation on whole wheat flour as a method to extend the shelf-life, the effect of enzymes in whole grain breads using the enzyme transglutaminase to strengthen the gluten network in bread making products, the use of mixed flours, and application of thermoplastic extrusion, among other works that will be described in this chapter. With this chapter, it is expected that the results and trends pointed out can contribute to other countries that also need large volumes of wheat import for bread.

1. Introduction

The wheat consumption in Brazil and Mexico has increased in recent times, and flour produced in these countries comes from grains of wheat in the country of origin, imported wheat grain and that processed in the country and imported flour.

Many efforts are being made to improve the quality of wheat products, and research is being performed in every area: genetic improvement together with the knowledge of science, technology, and food engineering that has shown promising results. Research institutes and universities are focused on research that can meet the requirements of market, as economy and researcher training programs on study and development of quality products to fulfill the commitment of beneficial effects on health. This chapter intends to present the main scientific impact after the year 1970, which represented a landmark year in changing the habit of eating wheat products in Brazil and Mexico.

2. World Wheat Production

In the 2008/09 harvest, the world wheat production was 682.399 million tons, 10.47% higher than the previous crop (610,951,000 tons). Among the major producing countries, three showed a decrease in production: Pakistan, Kazakhstan, and Iran (Table 1). The European Union, composed of 27 countries (Austria, Belgium, Bulgaria, Cyprus, Czech Republic, Denmark, Slovakia, Slovenia, Spain, Estonia, Finland, France, Greece, Hungary, Ireland, Italy, Latvia, Lithuania, Luxembourg, Malta, Netherlands, Poland, Portugal, UK, Romania, and Sweden) had the highest production followed by China, India, United States, Russia, Canada, and Ukraine (118).

In Table 2, the main producing countries (Table 1) have been shown to be the main exporters, except for China and India whose production is almost entirely destined to supply the domestic market (118).

Table 1. Major wheat-producing countries in the world (thousand metric tons)

Countries	2002/03	2003/04	2004/05	2005/06	2006/07	2007/08	2008/09
European Union	132,579	110,578	146,886	132,356	124,870	120,235	151,638
China	90,290	86,490	91,952	97,445	108,466	109,298	112,500
India	71,810	65,100	72,150	68,640	69,350	75,810	78,600
United States	43,705	63,805	58,698	57,243	49,217	55,821	68,026
Russia	50,550	34,100	45,400	47,700	44,900	49,400	63,700
Canada	16,198	23,049	24,796	25,748	25,265	20,054	28,610
Ukraine	20,556	3,600	17,500	18,700	14,000	13,900	25,900
Australia	10,132	26,132	21,905	25,173	10,822	13,838	21,500
Pakistan	18,227	19,183	19,500	21,173	21,277	23,300	21,500
Turkey	16,800	16,800	18,500	18,500	17,500	15,500	16,800
Kazakhstan	12,600	11,000	9,950	11,000	135,00	16,600	12,500
Iran	12,450	13,440	14,568	14,308	145,00	15,000	10,000
Others	71,983	80,236	83,772	81,138	819,53	82,195	71,125
Total	567,880	553,513	625577	619,563	595,620	610,951	682,399

Source: USDA (118)

Table 2. Main exporters of wheat (thousand metric tons)

Countries	2002/03	2003/04	2004/05	2005/06	2006/07	2007/08	2008/09
United States	23,139	31,524	29,009	27,291	24,725	34,363	27,637
European Union	18,132	9,834	14,745	15,701	13,816	12,271	24,500
Russia	12,621	3,114	8,456	10,514	10,584	12,552	18,393
Canada	9,432	15,843	14,872	16,020	19,434	16,375	18,000
Australia	9,146	18,031	14,722	16,012	8,728	7,487	14,000
Ukraine	6,569	66	4,403	6,461	3,366	1,236	12,950
Others	2,6703	30,241	25,461	24,603	30,983	33,188	21,918
Total	105,742	108,653	111,668	116,602	111,636	117,472	137,398

Source: USDA (118)

By analyzing Table 3, which presents the main consuming countries, a 3.2% increase in global consumption of wheat crops between 2007/08 to 2008/09 can be observed (118).

Table 4, which presents the ending stock of wheat by the main producer countries, shows that the increase in production was greater than the increase in consumption, bringing the ending stock at 28.14% for the period. Imports and exports of wheat increased overall by 14.5%, respectively, for the period considered (118).

A comparison of data from Tables 1 and 3 shows that Australia, a large producer, is not one of the largest consumers, while Brazil, which is not a large producer, can be considered as one of the largest consumers.

Table 3. Major wheat-consuming countries in the world (thousand metric tons)

Countries	2002/03	2003/04	2004/05	2005/06	2006/07	2007/08	2008/09
European Union	124,991	115,095	123,220	127,525	125,500	116,536	127,500
China	105,200	10,4500	102,000	101,550	102,000	106,000	102,500
India	74,294	68,258	72,838	69,980	73,365	76,345	70,770
Russia	39,320	35,500	37,400	38,400	36,400	37,700	38,900
United States	30,448	32,498	31,783	31,320	30,940	28,614	34,014
Pakistan	18,380	19,100	19,600	20,900	21,900	22,400	22,800
Turkey	16,800	16,800	16,800	16,100	16,650	16,800	16,900
Egypt	12,800	13,300	14,200	14,800	15,450	15,950	16,550
Iran	14,800	14,550	14,550	14,800	15,300	15,500	15,600
Ukraine	14,500	9,025	11,700	12,500	11,700	12,300	11,900
Brazil	9,690	9,700	9,900	10,450	10,300	10,300	10,700
Canada	7,972	7,179	8,240	8,242	8,986	6,373	8,450
Others	132,963	135,030	143,184	150,409	149,703	147,474	156,039
Total	602,158	580,535	605,415	616,926	618,194	612,292	632,623

Source: USDA (118)

Table 4. Ending stock of wheat in major producing countries (metric tons)

Ending stock	2002/03	2003/04	2004/05	2005/06	2006/07	2007/08	2008/09
China	60,378	43,293	38,821	34,387	38,450	38,963	48,721
European Union	18,491	11,514	27,496	23,384	14,075	12,445	19,083
United States	13,374	14,872	14,699	15,545	12,414	8,323	18,154
India	15,700	6,900	4,100	2,000	4,500	5,800	13,440
Russia	6,133	2,645	3,386	3,454	2,231	1,819	8,426
Canada	5,725	5,985	7,922	9,698	6,865	4,561	7,071
Others	46,835	46,230	53,237	58,196	48,446	49,899	54,609
Total	166,636	131,439	149,661	146,664	126,981	121,810	169,504

Source: USDA (118)

Brazil stands out as one of the three largest wheat importers (Table 5): in the last three years, it has imported about six million tons, more than 50% of its domestic needs, which accounted for 10.7 million tons in 2008/09 (118).

Table 5. Major wheat-importing countries (metric tons)

Countries	2002/03	2003/04	2004/05	2005/06	2006/07	2007/08	2008/09
Egypt	6,327	7,295	8,150	7,771	7,300	7,700	9,800
European Union	19,686	7,374	7,061	6,758	5,137	6,942	7,000
Iran	2,075	766	200	380	1,100	200	7,000
Brazil	6,563	4,898	4,904	6,348	8,048	6,711	6,000
Algeria	5,792	3,746	5,358	5,476	4,874	5,904	5,800
Indonesia	3,984	4,535	4,661	4,981	5,711	5,458	5,300
Others	68,053	71,864	79,394	79,254	82,357	80,727	94,416
Total	103,480	100,478	109,728	110,968	114,527	113,642	135,316

Source: USDA (118).

The importing countries are prevalent in *Mercado Comum do Sul* (MERCOSUL), composed of countries of South America (Brazil, Argentina, Uruguay, Paraguay, and Venezuela), with the aim of facilitating trade agreements agreed upon by them. Imports from countries outside MERCOSUL are taxed by the Common External Tariff (CET), currently at 10%, which is applied only when countries belonging have a lack of availability.

3. Production and Consumption of Wheat in Brazil

In the 2008/09 harvest, Brazil produced 6,015,600 tons of wheat (Table 6), a value 31.89% higher than the previous period. The consumption increased from 10.4 million tons to 10.7 million tons for the period 2007/08 and 2008/09, respectively. Imports for the same period declined from 8,048 million tons to 6,711 million tons (28,29). From the 2007/08 crop year, when the harvested area increased by 27% resulting in a record volume of production of six million tons, the area has remained between 2.4 and 2.5 million ha, although the yield of two subsequent harvests was decreased by 20% and 17% (Table 6) (28,29).

Table 6. Production, imports, consumption, per capita consumption, and final stock of wheat in Brazil

Harvest	Area (ha)	Production	Imports	Consumption	Consumption Per capita*	Final Stock	Stock (consumption)%
2001/02	1,710,200	3,194,200	7,055,400	10,059,200	58.57	1,002,100	10
2002/03	2,051,600	2,913,900	6,853,200	9,851,500	56.70	912,700	9
2003/04	2,464,200	6,073,500	5,707,500	9,947,100	56.60	1,370,700	14
2004/05	2,756,300	5,845,900	5,311,000	10,196,000	57.36	2,324,900	23
2005/06	2,361,800	4,873,100	6,266,100	10,683,700	59.41	1,992,900	19
2006/07	1,758,000	2,233,700	7,809,900	10,260,700	56.41	1,753,700	17
2007/08	1,851,800	4,097,100	6,895,700	10,381,000	56.42	1,614,900	16
2008/09	2,422,512	6,015,600	6,160,300	10,813,000	56.90	2,577,900	24

Source: CONAB (28,29). * kg/inhabitant/year.

The Brazilian production is concentrated in the Southern area: 53.3% for Paraná, 33.5% in Rio Grande do Sul, and 5.6% in Santa Catarina (28,29).

The main suppliers of wheat to Brazil in 2008 were Argentina (4,234,385 tons), United States (906,752 tons), Paraguay (527,752 tons), Canada (272,889 tons), Uruguay (90,869 tons), and Lebanon (2 tons), as shown in Table 7. Recently, Brazil has imported considerable quantities of wheat flour from the countries of MERCOSUL (Table 8), mainly Argentina, Paraguay, and Uruguay (108).

Table 7. Brazilian wheat imports by Country of Origin, 2004-2009 (tons)

Countries	2004	2005	2006	2007	2008	2009
Argentina	4,653,223	4,519,655	5,974,222	5,629,890	4,234,385	5,630,214
United States	73,948	29,799	16,499	354,065	906,793	354,065
Paraguay	120,613	408,926	337,763	163,355	527,752	153,405
Canada	0	0	71,525	340,991	272,889	340,991
Uruguay	27	3,725	131,169	149,377	90,869	146,219
Lebanon	2	6.1	0	7.9	2	0
Others	0.6	2.4	0	7.9	0	0
Total	4,847,852	4,988,125	6,531,178	6,637,686	6,032,691	6,624,894

Source: SECEX (108).

Table 8. Brazilian imports of wheat flour by country of origin, 2004-2009 (tons)

Countries	2004	2005	2006	2007	2008	2009
Argentina	9,329.0	4,271.0	109,881.0	604,683.0	638,120.0	604,683.0
Uruguay	8,784.0	17,837.0	21,111.0	19,504.0	40,785.0	18,099.0
Paraguay	8,971.0	4,580.0	2,112.0	1,012.0	2,360.0	962.0
United Kingdom	0.0	12.1	237.0	566.0	923.0	0.0
Italy	660.0	198.0	243.0	52.1	60.4	0.0
Mexico	0.0	0.0	0.0	17.4	23.1	0.0
France	5.9	20.2	159.0	6.4	8.0	0.0
Others	6,325.0	1,278.0	1,927.0	0.2	0.0	47.0
Total	34,075.0	28,196.0	135,671.0	291,845.0	682,283.0	623,791.0

Source: SECEX (108).

3.1. History of Wheat Consumption

The fact that Brazil currently imports over 50% of their consumption needs reflects some remarkable historical facts related to wheat:

Before 1960—Brazilians consumed many products of maize and tubers, since wheat products were mostly offered in South (higher production) (5,76).

The milling industry passed through a major economic crisis and, in 1962, the monopoly of national wheat operations was created aiming to strengthen the milling industry (Ordinance n. 820 in 11/22/62) (5,76). In 1967, the Government officially interfered with the wheat politics, resulting in the Decree n.210 of 02/27/67, which drained miller national park, then consisting of 489 mills with registered grinding capacity around 10 million tons, but that used only 2.5 million tons showing an idleness of 75%. Since manufacturing capacity was higher than the market need, the measurement of grinding capacity for each mill was established by Law. The analysis in each of the nearly 500 mills has resulted in a sum so high that the number of mills needed to be reduced to 178. In addition, the kind of flour to be produced was fixed, and the selling price of wheat and products was established by the government. Over 23 years, the Government has bought domestic and foreign wheat and sold in weekly share to the mills (5,76).

In late 1960s and early 1970s, the countries producing good quality wheat for bread production, which had a large production surplus, used as strategy the sale of wheat and export of appropriate technology (for mounting bakeries, research centers, mills' installation and others) at low cost to other countries, resulting in a habit of consuming products made with wheat, like breads, cakes, pasta and biscuits. Consequently, there was the need for massive imports of wheat by non-producers of grain, leading to a decrease in the consumption of flour from natives and a consequent reduction in planting (5,76).

At the end of 1972, due to fall of the Russian wheat crop in 1972/73, which led to an increased international demand, the price of wheat rose and the Brazilian government began subsidizing wheat to maintain inflation rates at 12% per annum. This subsidy lasted about 15 years and, in 1990, a regulation ended the quota system of grinding and the Federal monopoly on buying and selling wheat. Such regulation, although approved in November 1990, only came into force at the end 1991 (5,76).

In 1991, the market began to rule the sector decisions with value added, making milling industries manager of their own business. This freedom opened a new market, providing a modernization in industry and requiring a more aggressive trade policy, since there was the risk of being off this new competitive market. Today, the number of mills is around 200 (5,76). In brief, this allowance provided from 1973 to 1991 by the Brazilian government has resulted in:

- a large increase in consumption of grain by all social classes;
- an expansion of national production;
- a modification of dietary habits of the population, displacing rice, beans, cassava and maize for its traditional positions and becoming one of the wheat grains of greater consumption per capita in the country;
- a worsening trade imbalance due to the large wheat import (12).

3.2. Current Consumption

The market for bakery products in Brazil is growing year after year, with sectors' revenues in 2009 having reached U.S. $43.98 billion, surpassing the 63,000 bakeries in operation (4,5).

Much of the consumption in Brazil is represented by a line consisting of bread with thin crust and much crumb, such as loaves, hamburger, hot dog rolls, and tubes. The success of these types of bread is mainly characterized by the softness and easiness of chewing, combined with the use of fillings such as creamy mayonnaise, cream cheese, margarine, pasta, and others (45).

In 2007, the average consumption of bread in Brazil was 29 kg per person per year, with its maximum amount in the Southeast and South, with 35 kg/inhabitant/year, and the minimum amount in the Northeast with 10 kg/capita/year (111). In the Northeast, the consumption of cooked or processed roots and tubers contributes to reducing the consumption of bread, since the change in eating habits in the Southeast and South have reduced the consumption of typical Brazilian foods, contributing to increased bread consumption.

Some studies published in the 1990s (35,47,48,49,100) showed concerns about the consumption of wheat and its products by countries that do not manufacture good quality wheat for bread and pasta. These studies indicated several solutions, such as quality breeding of wheat flour mixed use, back to the native products. In addition, the use of additives, ingredients and enzymes that can improve the quality of wheat flour and its products is currently under research.

Next in this chapter, the efforts of Mexico in genetics, and Brazil and Mexico in technology and wheat products will be presented.

4. Tests Used to Evaluate Wheat Genotypes by Technological Quality

Currently, one of the aspects that has had a significant influence in commercialization of wheat grain is industrial quality. The Mexico industry demands wheat with quality of strong gluten for the mechanized industry of bread and hard wheat with high quality for the elaboration of semolina flour and pastas. The breeding national program of wheat in Mexico uses the same methodology used by International Maize and Wheat Improvement Center (CIMMYT) (95).

The Quality parameters used by bread wheat breeders for quality screening in the United States, Europe, at CIMMYT (Mexico) (95 to 99), and Brazil (16,17,18,89) are:

4.1. United States

Early generation (F_2-F_5): Protein, Hardness, Mixograph: Advanced generation (F_6 and higher): Milling (extraction, ash), Flour protein, Water absorption, Farinograph (time, stability, etc.), Dough characteristics, Loaf volume, Bread characteristics.

4.2. Europe

Early generation (F_2-F_5): protein Zeleny sedimentation, sodium dodecyl sulphate (SDS) sedimentation and high molecular weight glutenin subunits. Advanced generation (F_6 and higher): Hagberg falling number, alveograph (W, P/L), gluten elastic recovery, extensometer, oaf volume, baking score, and machinability test.

4.3. Mexico (CIMMYT)

Early generation (F_2-F_5): protein, hardness, SDS sedimentation. Advanced generation (F_6 and higher): Hagberg falling number, grain and flour protein, high molecular weight glutenin subunits, flour SDS sedimentation, mixograph (time, tolerance), alveograph (W, P/L), loaf volume, and crumb structure.

4.4. Brazil

At the beginning of the selection process, that is, in the first segregated generation, individual plants are selected and seed quantity is reduced, precluding decisive or definite technological quality tests. From the homozygosis in F_7-F_8, new wheat lines are selected this time with a bigger seed quantity (500-900g), when more complete wheat quality tests can be performed (89).

Embrapa's wheat-breeding program (89) does complete quality analyses are carried out only after varietal lines evaluation essays, when more than 2 kg of seeds are produced (F_8-F_9). For evaluation of wheat genotypes' technological quality, the tests performed are: test weight (16,17); thousand kernel weight (16,17); hardness index in single-kernel characterization system (SKCS); falling number of grain; experimental milling or flour yield in Brabender Quadrumat Senior mill; flour moisture content (1); grain moisture content in Dickey-John equipment; wheat flour color (Minolta colorimeter) with parameters: L* (lightness), a* and b* (chromaticity coordinates: +a= red and –a= green, +b= yellow and -b= blue); gluten analysis: wet gluten, dry gluten, and gluten index (1). In the advanced stages of breeding with larger amount of kernel sample, two other analyses can also be performed: alveograph (1) with parameters: T, tenacity; E, dough extensibility; ratio T/E; W, gluten strength; SI, swelling index and EI, elasticity index (pressure ratio P200/Pmax of alveogram, that predict rheological behavior of dough used in baking); and a 50g sample farinograph method for flour (1), with parameters: water absorption, dough development time, tolerance index, and stability. This last analyses, as it takes longer and demands more sample amount, is performed only in some samples of Embrapa wheat-breeding program assays.To investigate the performance of final products, baking test (39) and cookie test (1) can be used.

5. Quality of Wheat Produced in Brazil

Wheat is cultivated in different states, being adaptable to regions in Brazil; however, approximately 90% of Brazilian production is concentrated in the states of Paraná and Rio Grande do Sul. This wheat production is originated from sowing of different cultivars and harvested and stored in different silos, which gives origin to the commercial sample lots that in turn are sent to the mills for evaluation and possible commercialization. These lots present differences in terms of test weight, falling number, gluten strength, color, among other parameters that define the wheat quality (89).

The current wheat Brazilian legislation in 2010, Normative Instruction (NI) no. 7 of August 15th, 2001 named “Technical Rules of Identity and Wheat Quality,” classifies the wheat according to its alveography gluten strength (W) and falling number (FN) values, in Extra-strong Wheat (minimum 300 x 10^{-4}J of W, and 250 s of FN), Bread Wheat (minimum 180 x 10^{-4}J of W, and 200 s of FN), “Brando” Wheat (minimum 50 x 10^{-4}J of W, and 200 s of FN), Other Uses Wheat (any W and FN below 200 s), and Durum Wheat (minimum 250 s of FN) (18).

However, to wheat commercialization in general, additional quality parameters or differentiated from NI no. 7 are used, which are established to fulfill the wheat flour specifications demanded from different segments of chain supply, as for milling, bakery, pasta, cake, and cookie industries (89).

In traditional production regions of Brazil, the main problem facing wheat culture is still grain yield and wheat technological quality for baking instability (baking is the biggest market share in Brazil). When the meteorological conditions encourage cultivatation of wheat without irrigation, the grain mean yield reaches 2,300 $kgha^{-1}$, while in the 1970s, this was not superior to 1,000 $kgha^{-1}$ (29).

The quality of Brazilian wheat has significantly increased in the last few years. Nevertheless, cultivar mixtures and the varying features of Brazilian wheat are still a challenge that should be overcome in order to increase the value of this cereal.

6. Quality of Wheat Produced in Mexico

The quality of the wheat grain is affected by environmental factors; the harvests from the cycle Autumn-Winter have a high performance in the districts of irrigation of the Northwest of the country, whereas the harvests of dry lands of Spring-Summer are affected by the lack of rains. The industrial quality of the wheat is characterized by means of physical, chemical and rheological analysis of related parameters with milling and bread making quality. The rheological analysis of the dough includes gluten strength (alveogram-w), gluten extensibility (alveograms-P/L and P/G), kneading properties (kneading time and tolerance to the kneading measured in mixograph), among others, since they are related to the viscoelastic properties of the dough that determine their properties for baking. The bread making analysis measures aspects such as bread volume, structure, and appearance of the crumb, which are important variables for the characterization of bread making quality of wheat (98).

These rheological properties of gluten can diversify their use in the baking, biscuit, and pastry industry (81,82). Canimolt (23) classified Mexico wheat (*Triticum aestivum* L.) according to industrial quality in:

Group 1: Texture of Wheat Grain Semi-Hard to Hard and Strong and Extensible Gluten

These types of wheat are used mainly by mechanized bread making industry, producing flour for sliced bread, as well as to improve weak wheat.

Group 2: Gluten with Hard to Semi-Hard Texture and Medium Strong and Extensible

These types of wheat are used for handmade bread or semi-mechanized bread industry, being used to improve wheat with weak gluten.

Group 3: Soft Texture and Weak and Extensible Gluten

These types of wheat are used by the cookie industry and for flour tortillas, doughnuts and others; although it can be used in the handmade bread-making industry as well as to improve wheat with very hard gluten.

Group 4: Gluten with Tenacity and Little Extensible

This wheat is not used in bakery products due to its high toughness. If mixed with strong wheat, it can be used in confectionery industry (pastry and biscuit).

6.1. Mexico Wheat Sector

6.1.1. Production

In Mexico, wheat is planted in two agricultural cycles during the year—the cycle Autumn-Winter and Spring-Summer. The most important producing regions of wheat are the Northwest of Mexico, with approximately 85% of the national production in the States of Sonora (35%), Sinaloa (9.2%), Baja California (11.5%), and Central zone in the States of Guanajuato (17.5%), Michoacán (6.4%) and Jalisco (4.4%). Approximately 18.4% is planted in weather areas, mainly in the Mexican highland (105).

The wheat yield (ton/ha) has been substantially increased in the last years. During four decades, the national yield of wheat in Mexico increased 772 kg/ha in 1940, and up to 4,357 kg/ha in 1984. In 2000, 5,300 kg/ha was due to in irrigation conditions and 2,000 kg/ha in rain-fed environments, whereas in 2006, it was 5,750 kg/ha in irrigation and 2,300 kg/ha for rain-fed. This increase is attributed to genetic improvement and to the suitable use of better practices of culture. In the Sonora State, crystalline wheat is cultivated only in the cycle Autumn-Winter, which is classified in the group V (production of pastas). In 2004, the production of crystalline wheat in Sonora was 301,350.74 tons. It corresponded to 42.4% of the national production of wheat, and 75% corresponded to crystalline wheat (105).

The preference of crystalline wheat over hard wheat must be due to a greater yield of the first and to a greater resistance to the partial coal and rust. The partial coal is a latent disease in the South of Sonora that prevents free wheat commercialization in the region. In spite of the great volume of production that represents the crystalline wheat, this one is not commercialized in its totality; of 1,532,850 tons harvested in agricultural cycle 2000/2001, only 22.6% were exported. Of the remaining amount, a proportion was destined to the milling for semolina production and another one destined to animal feed. In these years, Mexico exported crystalline wheat by nearly two million tons, one million tons to Switzerland, followed by Holland (151 thousand tons), the United States (149 thousand tons), Bermudas (118 thousand tons), and Algeria (109 thousand tons) (23).

The planted surface of wheat in Mexico in 2007, was 706,000 ha, 818,000 ha in 2008, and 848,000 ha in 2009. For the rest of the period, there was a slight diminution in the surface planted due to the relative low prices of the wheat related with other grains. The area of culture for wheat fell considerably from 4,660,000 tons in 1985 to 1989, to 3,509,241 tons in 2007 (105).

In addition, in Mexico, the irrigation surface of wheat (*Triticum aestivum L.*) has been reduced drastically due mainly to the deficiency of water and the reconversion towards more profitable cultures and exportation. This has contributed so the industry could import wheat to replace the national demand.

Planting wheat on rain-fed areas in Mexico has a potential to reduce the deficit experienced by the country at the moment, since there is around a million hectares of rain-fed areas where the wheat is a viable option. A factor that has prevented the increase of the planting of wheat in irrigation areas is that the producers frequently face problems to commercialize the grain, since the milling industry prefers to import wheat from the international market with the argument that Mexican wheat does not have the quality required of wheat for bread making (121).

6.1.2. Importations of Wheat

Growth in domestic demand has induced Mexico to increase its dependence on wheat imports. The United Nations Food and Agriculture Organization reports that Mexican wheat imports increased over 800% from 442,800,000 tons in 1990, to 4,066,500 tons in 2005 (48). A factor contributing to the increase in wheat imports is the disparity between the type of wheat primarily produced in Mexico and the type of wheat demanded by consumers. Mexico produces primarily durum-type wheat, which is used to make noodles and pasta. By contrast, most of the wheat imported and used by domestic consumers is hard wheat, suitable to make bakery products. For example, from 2000 to 2006, durum wheat accounted for about 91% of all wheat produced in Mexico (105). Mexico is the third largest wheat importer of U.S. wheat, ranked after Egypt and Japan, and is the largest single-country buyer of U.S. hard red winter wheat produced in the U.S. Southern Plains. From 1997 to 2007, Mexico imported 24,525,756 MT of wheat from the U.S. and 10,206,818 MT from Canada (115,118). Despite the transportation advantages, the U.S. maintains over Canada in serving Mexican market, although Canadian wheat exhibits consistent quality, primarily as a result of a strong export regulatory board (77,88).

In 2005, the wheat national production represented 32% and the importations 68%, whereas in 2007, the national production increased to 37% and the importations, 63%. The total consumption in 2005, was 5,475,000 tons and in 2007, this value fell to 5,200,000 (23). In Mexico, the annual per capita consumption of wheat was estimated in 48 Kg/year, considering that 78% of the national production is destined for food and the remaining is used mainly for animal fed; thus, this represents half of world-wide average (55).

6.1.3. Wheat Milling

The installed milling capacity in 2007 was 8,027,000 ton. The three industries that processed wheat flour into end final products are cookies and pastes, bakeries (traditional), and the industrial bread manufacture. The destination in 2007 was 2,399,000 tons for bread; 1,028,000 tons for cookies, flour tortillas, coatings and pastas; and total production of flour and semolina of hard wheat 3,869,000 tons. Sweet cookies and pastas manufacture represented 55.6% and 28.0% of the total, respectively. Unlike the milling of wheat and the manufacture of pastas and cookies, the industrial bakery is much more diversified, with sweet bread representing 24.3% of total production, followed by white bread (16.5%) and small cakes (11.3%); the remaining is distributed between more than eight generic products (58,23).

6.1.4. Wheat Consumption

National statistics reported that in 2007, the milling of hard wheat for bread making was 398,000 tons: 1,371,000 for soft wheat and 614,000 for crystalline wheat, and the imported wheat for bread making in this year was 3,253,000. The manufacture of bread and cakes registered a participation of 62%, followed by cookies and tortillas (26%), and snacks, pastas gluten and starch (12%), bran, germ, shorts and middlings (feed). The world-wide production of wheat will increase approximately in a 1.3 annual percent during the period of projections to 679 million tons for 2010. In wheat milling, two types of flour are generated: a) type A or good for making bread with three degrees of quality (common or standard, fine, and superfine), and b) type B or semolina (non-good for making bread), between which the special flours are included to make pastas. The produced flour is destined mainly to the bread making and, to a lesser extent, to the manufacture of tortillas, cakes, cookies, and pastas (23).

6.2. National Breeding Wheat Program México

The national breeding wheat program belongs to the National Research Institute for Forestry, Agriculture and Livestock (INIFAP). Its objective is to develop new cultivars of higher yield and rust resistance than the actual commercial cultivars in use. INIFAP had 30 Experimental stations in all the country. Before initiating genetic improvement of wheat (*Triticum aestivum* L.) in Mexico, the varieties that breeders planted were mixtures of different types of wheat very susceptible to stem rust and leaf.

6.2.1. History

The genetic improvement formally begun in 1945, and in 1948, the breeders already had an availability of improved varieties, with some of them being introduced of other countries. In 1951, varieties more resistant and yielded to the rust started to be used in commercial production. From then, the genetic improvement program has developed and released a great number of varieties. The progress obtained in genetic improvement of wheat during the decade 1950 to 1960 was a very important factor in the production, because the improved varieties have shown to be more efficient to produce grain. That limitation to increase yield imposed by El Bedridden had the highest influence in the decision to obtain semi-dwarf varieties. Until 1960, the increase in national yield average was slow; nevertheless, with the obtaining and use of varieties semi-dwarves, the application of higher amounts of fertilizer and the use of better practices of culture, the total production increased to the degree in which it enabled the country to be self-sufficient in this cereal production. The rain-fed wheat genetic improvement program in Mexico began in 1969 by INIA (National Agricultural Research Institute), an institution that later became INIFAP (National Forestry, Agriculture and Livestock Research Institute). By 1963, 95% of Mexico's wheat fields were growing the new seeds. Yields were much higher. The 1964 harvest was six times larger than in 1944. Whereas Mexico had imported half of its wheat in 1943, by 1964, it was exporting 500,000 tons a year. (Since that time, the combined effects of growing population and farmers changing to other crops have returned Mexico to its previous status of being a net importer of wheat) (80). Nowadays, it is progressively difficult to obtain increases in wheat (*Triticum aestivum* L.) yields like those achieved from 1960 to 1980 through genetic improvement. Increases of only 5 to 10% are common with experimental lines or new varieties of wheat obtained using traditional improvement methods (113).

6.2.2. International Wheat-breeding Program

CIMMYT is working with the national program (INIFAP) to develop wheat varieties that perform well in hot areas during the cooler winter periods of the year when, up to now, the land has usually been left uncultivated. These new heat-resistant varieties let farmers produce a second crop each year and earn extra income. Wheat stands virtually alone as a robust, water-efficient, winter-season crop that farmers can grow efficiently in rotation with such summer crops as rice, maize, sorghum, cotton, and soybeans.

The CIMMYT conserves and handles 150,000 unique collections of seed of Tritici (wheat and its relatives and ancestors) originated from more than 100 countries: the greatest unified collection of the world for a single culture. Nevertheless, the production is affected by factors like irregular and misdistribution of rains present at early and delayed frosts, droughts, attack of plagues, diseases, and weeds. In the last 35 years, more than 95% of wheat varieties seeded at a national level derived directly from forward lines of CIMMYT genetic improvement program. The wheat varieties generated by the CIMMYT contributed with increases in average of 53.8 kg/ha. At this moment, the emphasis is concentrated in the combination of genes associated with high potential of yield, resistance to diverse diseases, reduction of stress caused by environmental factors (such as drought and heat), and industrial quality of the grain. CIMMYT scientists reported a rapid progress in developing newly disease-resistant wheat varieties. Sixty disease-resistant varieties have been developed via conventional breeding, some of which are thought to be not only resistant to Ug99, but higher yielding than today's most popular varieties. The introduction of high-yielding varieties was combined with modern agricultural production techniques to Mexico, Pakistan, and India. As a result, Mexico became a net exporter of wheat by 1963. These collective increases in yield have been labeled the Green Revolution. By 1963, ninety-five percent of Mexico's wheat crops used semi-dwarf varieties developed by Borlaug. In that year, the harvest was six times larger than in 1944 (9).

6.2.3. Research to Improve Quality of Mexican Wheat

The viscoelastic properties of gluten are mainly determined by their composition of subunits α, β, γ, and ω-gliadins, high molecular weight glutenin (HMW-GS), and low molecular weight glutenins (LMW-GS) (127). In some Mexican varieties released for temporary wheat, HMW-G subunits have been identified as well as their effect on dough rheology and bread volume (32,33,34,81,82).

The knowledge of individual effect of gliadins and LMW-G is necessary to understand their influence on quality, as well as to manipulate them under a breeding program. But it is difficult to identify subunits of LMW-G and gliadins (complex *Glu-3* and *Gli-1 loci*) in polyacrylamide gels in the presence of sodium dodecyl sulfate, due to the number of genes, 35 to 40, that form the *Glu-3 loci* (25), by gliadins of high molecular weight, and LMW-G showing among them patterns of similar electrophoretic mobility, making difficult their identification (64). There is also a genetic linkage between LMW-G (*Glu-3 locus*) and γ and ω-gliadins (*Gli-1 locus*), located on the short arm of the chromosomes 1A, 1B and 1D (93). Martinez-Cruz et al. (85) characterized the composition of glutenin subunits with high (HMW-G) and low (LMW-G) molecular weight, ω-gliadins in 72 parents used in the breeding program for rain-fed bread wheat of the Instituto Nacional de Investigaciones Forestales, Agrícolas y Pecuarias (CEVAMEX-INIFAP-Mexico), and in 600 F6 lines derived from crosses between varieties of different quality groups and their parents, which were analyzed by electrophoresis in polyacrylamide gels with sodium dodecyl sulfate. These authors reported that HMW-G and LMW-G characterization will allow making crosses specifically tailored to obtain combinations of desired glutenins, as well as to carry out a more efficient selection in breeding program. Moreover, the lines which are product of crosses analyzed will allow better understanding of the genetic effects of LMW-G, ω-gliadins and 1BL/1RS translocation (secaline proteins) in the bread-making quality.

De la O et al. (33) determined the effect of different high molecular weight glutenin alleles on viscoelastic properties of bread wheat dough. Tests were conducted on a group of 69 lines derived from the cross Rebeca F2000 by Salamanca S75 by single-seed descent from F2 to F6, along with the parental varieties. The lines were sown in Roque, State of Guanajuato, Mexico (Fall-Winter 2003 to 2004) with normal irrigation (five irrigations) and restricted irrigation (three irrigations). The combinations 1-17+18-5+10, 2*-7+8-5+10, and 1-7+8-5+10 presented the best viscoelastic properties of the flour. In genome A, subunit 1 was characterized by inducing greater kneading time and mixographic type. In genome B, subunits 17+18 and 7+8 had a positive effect on the viscoelastic properties of the dough. In genome D, subunit 5+10 was associated with a positive effect on the viscoelastic properties, while subunit 2+12 was associated with a negative effect. When the combinations were compared, it could be noted that the effect of substituting 5+10 with 2+12 was very significant, since where the allele 5+10 appeared high values were found for alveogram-W, alveogram-P/L and alveogram-P/G, which increased the strength and tenacity, whereas the extensibility of the dough has reduced. In contrast, when subunit 2+12 appeared, strength has reduced, but extensibility increased, perhaps due to a pleiotropic effect of the genes that control subunits 5+10 and 2+12.

Both genotype and environment significantly affected the quality parameters of the dough and glutenin- and gliadin-rich fractions and their ratio (34). In the first phase, ten varieties of Mexican bread wheat were evaluated in 22 rain-fed environments in the states of Mexico, Tlaxcala, Jalisco, Oaxaca, Puebla, Guanajuato, Morelos, and Queretaro, during the Spring-Summer/1999 and 2,000 cycles. At a second stage, 24 lines of the cross Rebecca F200 × Verano S91 were evaluated in five environmental conditions, as generated by agronomic management (irrigation, fertilization, application of sulfur, and planting date) during the Fall-Winter/2006-2007 cycle in Roque, Gto., Mex., using a randomized complete blocks experimental design with two replications. Rheological properties of the dough, kneading time, W-alveogram, P/L-alveogram, P/G-alveogram, glutenin-rich fraction, gliadin-rich fraction, and the gliadin-rich fraction/glutenin-rich fraction ratio were evaluated. Highly significant differences were found in all viscoelastic properties for G-APM and BPM combinations. The estimate of the stability parameters showed that the 1,17+18,5+10 combination was the most stable one on rheological properties of dough.

7. Subjects Related with Wheat Technological and Nutricional Quality Produced in Brazil and Mexico

7.1. Nutritional Quality of Commercial Flours

7.1.1. Brazil

Wheat flour currently commercialized in Brazil should comply with Resolution - RDC No. 344 of December 13, 2002, which makes mandatory the addition of iron and folic acid in wheat flour and corn flour pre-packaged in the absence of client and ready to offer for the consumer and for industrial use, including the bakery flour and the flour added to premixes. Each 100g of wheat flour and corn flour should provide at least 4.2 mg of iron and 150 mcg of folic acid. These two nutrients are recommended by the World Health Organization-WHO, Pan American Health Organization-PAHO, and by the Inter-institutional Commission of Conduction and Implementation of the Actions of Wheat Flours and Maize Flours Fortification, coordinated by the Ministry of Health. The following products are excluded from this Regulation due to limitations in their technological processes: *Bijú* flour or corn flour produced by maceration; large flake; whole wheat flour, and durum wheat flour (18).

7.1.2. Mexico

Almost all of Mexican wheat flours are fortified with vitamins as niacin (B3), riboflavin (B2), thiamin (B1), folic acid, iron, and zinc. The Federal Government and Mexican wheat millers make an effort to increase the fight against malnutrition and anemia, which are the most serious sufferings of the country, in addition to fighting against other specific diseases like anencephaly, bifid column that affects nursling during the process of gestation or pregnancy, or even trying to take care of problems during growth. With the addition or fortification of flours, essentially women and children will improve its feeding when consuming bread, cakes, pastas soups, and wheat flour tortillas (23).

7.2. Post-Harvest Processes in Wheat Grain

Due to Brazil's climatic conditions, many studies have been conducted to increase the stability of post-harvested wheat, which has a reduced shelf life due to its high enzymatic activity and the presence of molds and yeasts. The main studied processes are:

7.2.1. Use of Irradiation in Whole Grain Flour

International studies have demonstrated that whole wheat flour shelf-life can be improved through stabilization methods, such as gamma radiation or infra-red radiation under adequate conditions (62). Costa (31) tested the application of gamma radiation doses (in flour and in grain) and infra-red radiation (in bran) and verified that all gamma radiation doses have negatively affected the rheological characteristics and bread quality, due to modifications caused in the starch and protein fractions; however, the use of low doses improved stability, but this technique is not recommended for flour intended for baking.

This study suggested that the use of infrared radiation increased the stability of whole wheat flour and improved bread quality, depending on time and temperature conditions used.

The exposure of bran to 90 °C for 20 minutes was efficient to reduce lipase activity and yeast and mold count, improving stability of whole wheat flour and bread quality (31).

7.2.2. Microwave

Microwave has been studied as an effective method to control grain moisture during storage. Marsaioli Júnior (80) developed microwave equipment with a cylindrical rotatable applicator, in which the movement of the product inside the cavity permits a more uniform distribution of heating. Campos (22) studied the viability of use of microwave radiation for inactivation of α-amylase in wheat (BH 1146) and evaluated the effect of this treatment on technological quality of the flour. The treatment with microwaves for 180 seconds at all moisture levels studied (15, 18 and 21%) and for 270 seconds at 15% moisture resulted in the increase in dough resistance to extension, while the extensibility was reduced. This indicates an increase in binding forces of the gluten molecules, with a consequent improvement in technological properties. According to the author, these conditions were adequate to improve overall technological quality of flour, as evidenced by a 3% to 20% increase in bread quality.

Another study applied the microwave energy in drying of short pasta, which resulted in a significant reduction in processing time compared to conventional processes. Under the conditions utilized, microwave power of 3.2 kW and density power of 25.3 Wh/kg, the best results were obtained starting with moisture contents of approximately 15%, since the drying of pasta with higher moisture contents resulted in the emergence of bubbles in the dough. Higher moisture content increases the material's ability to absorb microwave energy (91).

8. BREAD PRODUCTION

For a long time, Brazilian legislation (15) classified the wheat flour available in the market as special flour and common flour, which differed by ash content, the first with the limitation of up to 0.385% and the second with up to 0.85%. Later, governed by Decree 354/96 (DETEN/MS), the limits went up to 0.65% ash for special flour, while common flour should present an ash level between 0.66 to 1.35% (14). This fact hindered the production of bread, since there was no guarantee of final product quality. Currently, the mills provide appropriate flour for each specific use (bread, pasta, cakes, and cookies), since the classification of wheat flour meets new legislation (18). Contributing to the improvement of product quality, the mills have provided premixes suitable for each kind of product: French bread, confectionery, and cakes. These pre-mixes already have the additives, ingredients, and specific instructions for mixture, adding water, fermentation time and baking time (Bunge® do Brasil, Emulzint®, Vilma® Alimentos, Puratos® do Brasil). This fact has greatly contributed to improving the quality of bread when produced in small bakeries with no infrastructure or technical personnel to perform the quality control of flour, as to provide the necessary ingredients and additives for the production of breads and cakes.

In addition to all investments in genetic improvement of wheat quality, cereal technology has made studies to reduce the need to import wheat for bread production, which are focused on two distinct lines: the first one is based on the use of mixed flours and full use of wheat and its by-products, and the second one is based on the strengthening of gluten network by means of studies focused on the use of some additives and enzymes.

8.1. Mixed Flours

The production of loaf breads in Brazil is continuously growing. As the majority of bakery products, its manufacture requires the usage of flour with specific characteristics to result in sensory acceptable products. Until today, wheat flour is the only flour capable of producing loaf breads with those characteristics. Research for wheat flour substitutes face technical limitations that interfere with the formation of a stable protein network, capable of playing a satisfactory role in the bread structure and gas retention produced during fermentation (61). The production of breads with composed flours can be an alternative to reduce the dependence of wheat import, and many studies have been accomplished (41 to 44,53,61). Table 9 shows that the flour can be replaced at 25 to 50%, depending on native flour. The use of raw materials for the development of alternative products should meet the following criteria: compatibility and availability. The flour that has good compatibility to replace the wheat must be white, free of strong odors, and with a protein level close to that of wheat. Flour with good availability presents prices and costs that are competitive and lower than the wheat.

Table 9. Levels suggested for composite mixture of wheat flour and others in bread formulation (% of total flour)

Flour	Protein content	Used level	Suggested maximum	Reference
Corn	8-10	18-30	30	El-Dash & Germani (43)
Sorghum	8-10	15-50	30	El-Dash et al. (42)
Millet	8-10	10-30	30	Dendy (35)
Barley	8-10	10-30	30	Bortolotti (13)
Rice	8-10	10-25	25	Dendy (35)
Cassava	1-2	13-30	30	El-Dash et al. (44)
Cassava starch	0	10-50	50	Dendy (13)
Amaranth	13,3	26	26	Capriles et al. (24)
Oat	14,0	5-20	10	Gutkoski et al. (64)
Yam	2-4	10-30	20	Dendy (35)
Potato	2-4	6-30	30	Dendy (35)
Sweet potato	2-4	15-30	30	Dendy (35)
Soy (defatted)	52,0	-	5	El-Dash et al. (41)
Triticale	10-15	20-30	30	Jardine (73); Gragnani (61)

Flours obtained from cereals have advantages such as high yield during production (Table 10) and processing, being able to be stored for long periods and processed away from the place of production. The same cannot be said about the flour from tubers and roots due to their high water content. They have low income and are highly perishable to such an extent that they should be processed at the production site (35).

Many composed flour programs had failed due to the lack of regular supply of raw materials. The cultivation of a culture with a new purpose demands appropriate planning: infrastructure for the production of tons of a new culture, storage, processing, and transport (35, 41 to 44). These facts can be exemplified by production information of some of these raw materials that cannot meet the market if added at the rate of replacement of wheat flour, from 25 to 50%.

Roots and Tubers

Cassava is the most produced and industrialized root in Brazil (Table 11), with its domestic consumption either cooked or processed as flour widely accepted by the population. A great part of cassava is destined to starch production, which has been used in several products as soups, creams, and the bread of cheese, a Brazilian typical product.

The remaining tubers have a higher domestic consumption, with few studies focused on industrialization.

With respect to sweet potato, Brazil is the fifteenth world-wide producer, with nearly 500,000 tons/year in a planted area of 48,000 hectares, with an average productivity of 11,219 kg/hectare (49).

According to Murilo (90), sweet potato is the fourth-most consumed vegetable by the Brazilian population, with an average of 3.6 kg/capita/year, exceeded only by potatoes, tomatoes, and squash.. In 2006, the yam had a production of 99,640 tons, sweet potato reached 273,126 tons, and cassava reached 43,093 tons (49).

Amongst the grains, the triticale, the cereal resultant from the crossing of wheat and rye, has very similar characteristics to its progenitors, without, however, the ability to substitute wheat in great quantities in loaf bread production.

The culture of triticale was officially approved in Brazil in 1985, with the launching of two cultivars. The country was considered with great potential for production of this cereal (119). The variety IAC 5 Canindé, launched by Campinas Agronomic Institute in the year of 2005, is indicated for the conditions of culture of São Paulo State. This variety presents excellent characteristics of culture and possible more promising industrial applications (71).

According to the official notice of National Supply Company (CONAB) CONAB/MOC N.º 015, of 07/31/2009 (28) that brings "Specific Norms for wheat and Triticale - Harvest of 2009," the minimum price for purchase of triticale grains is 0.28 Brazilian *reais* (BRL) per kg (net weight) of grain, while the wheat with quality for bakery has minimum price average of 0.54 Brazilian *reais* per kg (net weight) of grain being, therefore, 46.8% cheaper than the wheat. This difference of cost can bring great benefits to the industry, mainly to the bakery industry, where the wheat is the majority ingredient in formulations.

Gragnani (61) showed that it is possible to substitute up to 71.28% of wheat flour by triticale using microbial transglutaminase to make a loaf of bread, maintaining its quality without increasing the industrial costs.

Associated with this fact of using mixing flours, the development of products with integral grains (breads with the addition of cereals as rye, barley, oats, rice, sorghum, and other grains as flaxseed, soy, sesame, quinoa, amaranth, etc.), breads with reduction and without addition of fat and sugar, breads with soluble fibers, etc., can be produced in some bakeries due to sales of proper daily pre-mixtures for more beneficial-to-health bread production, following the worldwide functional food trends.

8.2. Whole Wheat

The most important research in this area emphasizes the use of flour of entire wheat and bran.

The demand for whole products is increasing day by day, both in Brazil and around the world, mainly because such foods are related to health. To improve the nutritional value of wheat and to have greater economic profit, one of the simplest and most effective ways is through the increase of the fee of extraction and/or flour use of wheat of entire grain. However, this product presents reduced stability to the storage, characterized by the development of flavors and undesirable odors and loss of quality for bakery. This mainly occurs due to the presence of lipases and glutathione in bran and germ, as presented in the already mentioned research in this chapter (31,99). The use of radiation gamma and infrared have already shown promising results for stability increase.

Silva & Chang's (110) results showed that the whole-wheat flour used with the addition of xylanase, glucose oxidase, and ascorbic acid presented considerable potential for the elaboration of loaf bread with high nutritional and functional value.

8.3. Wheat Bran

Bran has similar protein content to the one of wheat, even though its proteins are not of good quality for bakery. It possesses a bigger content of leached ashes, lipids, and particularly pentosans. The starch content of bran normally remains around 12 to 15%. The wheat bran is not as rich in β-glucanas as the oat's bran, but it has high dietary fiber content (approximately 45%) (30).

The wheat bran, compared with other fiber sources, has the advantage of not supplying a strange taste to bread (102). A great number of researchers have studied the use of this fiber source in bakery, with levels varying between 1 and 30% (6).

The use of wheat bran is being stimulated for human feeding, aiming to promote one better exploitation of this by-product, which is greatly used in the animal ration.

Fonseca Filho et al. (52) developed and assessed bread with different fiber concentrations (5, 10, 15 and 20% of wheat bran). The gradual addition of wheat bran increased the absorption of water and the rate of tolerance to the mixture, while the stability of the dough and specific volume of bread were reduced. The addition of 5% of wheat bran led to production of bread with characteristics close to those obtained with the standard dough. Bread with 10 and 15% of wheat bran showed a total count of more than 70 points in the baking test.

Table 10. Production of grain in Brazil (in thousand tons) in the period of 2002 to 2010

Grain	2002/03	2003/04	2004/05	2005/06	2006/07	2007/08	2009	2010 Estimate
Triticale	239.9	228.6	220.5	306.3	203.8	211.9	184.7	172.2
Oat	390.1	411.0	433.3	516.5	378.0	221.8	232.2	266.3
Barley	303.7	367.2	386.7	399.4	205.8	263.7	237.0	267.2
Rye	3.4	3.5	3.4	6.6	5.9	5.0	6.1	6.1
Rice	10,367.1	12,960.4	13,355.0	11,721.7	11,315.9	12,074.0	12,602.5	11,260.3
Corn	13,226.2	12,783.0	12,208.2	12,963.9	14,054.9	14,765.7	14,171.8	12,932.4

Source: CONAB (29).

Table 11. Area and production of tubers in Brazil between 2005 and 2010

	Harvest 2005		Harvest 2006		Harvest 2007		Harvest 2008		Harvest 2009		Harvest 2010	
Tubers	Area (ha)	Production (t)	Area (ha)	Production (t)	Area (ha)	Production (t)	Area (ha)	Production (t)	Area (ha)	Production (t)	Area (ha)	Production (t)
Potato	142,111	3,128,488	140,731	3,137,782	142,327	3,375,054	144,829	3,676,046	140,987	3,452,454	142,638	3,628,645
Cassava	1,886,424	25,725,207	1,901,561	26,713,038	1,912,925	26,920,521	1,860,800	26,336,652	1,872,812	26,030,909	1,872,812	26,190,371

Source: GCEA/IBGE (57)

Gandra et al. (56) studied the application of lipase and monoglyceride in fiber-enriched bread (20% of wheat bran) to verify the possibility of substituting the emulsifier by the enzyme. The instrumental firmness varied from 955.94 to 2073.91 g.f during the time of storage of breads. For control breads, a firmness range from 1355.6 to 2190.82 g.f was observed, which are higher values than the ones found in tests with addition of lipase and the monoglyceride. For appearance and texture attributes, all assays presented acceptance better than the control. According to the authors, under the conditions of the assays and for the studied range of monoglyceride (0 to 2%) and lipase (0 to 50 ppm), the possibility of replacing the monoglyceride for lipase in bread formulation enriched with fibers was verified.

Although importing more than 50% of wheat, all ingredients addition to the bread that promote the substitution of wheat and/or use of whole wheat flours or addition of wheat bran result in products of higher economic value (Table 12). This fact opposes the perspectives of some researchers, who had always proposed the mixed flour use as a suggestion for reduction of the import needs.

Table 12. Prices searched in the market for diverse types of Brazilian breads

Product	Weight (g)	Price in U.S. dollars	Price in Real (R$)-Brazil
Loaf bread	500g	1.60	2.52
Corn flour bread	500g	4.00	6.41
Multigrain bread	500g	4.65	7.44
Oat bread	500g	3.68	5.89
Rye bread	500g	3.91	6.27
Flaxseed bread	370	4.78	7.65

8.4. Use of Additives and Enzymes

Flour of good quality and with baking potential is uniformly capable of producing an attractive final product at a competitive cost. If the flour does not show good results in the final product, it can be supplemented with an additive or additive package (enhancers) that will make its treatment correct functional characteristics. Additives are substances that inhibit, praise, complement, optimize, or change components or characteristics of wheat flour (30,100).

Mechanization, large-scale production and growing consumer demand for high quality, convenience and a longer shelf life created the need to stabilize and bring uniformity to baked goods. The use of additives such as emulsifiers, antioxidants, and enzymes are applied to meet these goals. For the industry, optimization of dough properties and quality improvement of finished product are procedures of high interest. For consumers, sensory appeal of finished product is the most important attribute. After baking, the freshness of bread is rapidly reduced due to starch retrogradation, generating significant losses to baking industry: 80 to 10% of bread production has become unmarketable due to starch retrogradation (114,116). In 2005, Brazil imported US$20.3 million in ascorbic acid and US$27.1 million in enzymes to be used as functional additives in bakery (19).

The main additives and enzymes interacting in gluten strength that are being researched and used are:

8.4.1. Oxidants

The flour recently milled suffers oxidation caused by oxygen from the air during storage, producing a more malleable mass and breads with larger volume (66,37). Ewart (46) proved the reduction of SH in wheat flour stocked for six months and that this process happens in flours stored for up to six years.

Currently, it is impracticable to store flour for enough time so that the natural maturation occurs (aging or maturation), due to the great milling capacity of modern mills and to avoid expenses with storage. These factors were decisive in the introduction of flour enhancements that provide the same effects obtained through natural maturation in a significantly shorter time (66), adopting the addition of oxidants to the majority of leavening bread formulations as a common practice (75).

Potassium bromate was one of the most common oxidizing agents; however, its use was forbidden due to questions regarding security (120). Therefore, toxicological studies have indicated that it could be carcinogenic (60). The biggest challenge is to find a substitute for $KBrO_3$ (120).

The use of antioxidant compounds in bakery products provides benefits such as acceleration of flour maturation process; a change in rheological and sensory properties of freshly ground flour, providing the same production characteristics achieved through natural maturation; effects of flour bleaching due to the reduction of yellow-pale color promoted by carotenoid pigments naturally present in flour; and finally the reinforcement of gluten matrix necessary to support stress from high-speed mechanical handling systems and to grant better final volume to the bread (74).

8.4.2. Benzoyl Peroxide

The Benzoyl peroxide (PBZ) produces oxidation of carotenoid through a typical free radical mechanism, breaching the system of double-conjugated connections. According to Saiz et al. (106), its action of discoloration or bleaching is shown without influencing baking properties. It is added in the form of a diluted pre-mixture containing 32% of active ingredient (7). Its use is only allowed in some countries, such as Brazil.

PBZ is reactive and degrades at low temperatures (<40 °C) due to instability of O_2-binding with a binding energy of about 30 kcal/mol. PBZ degradation form benzoate derive radicals in initial phase. These radicals propagate the reaction or react with PBZ or with the solvent to form solvent radicals and more radical as benzoate (26). In baking, PBZ is widely used due to its efficiency in the degradation of carotenoids under reduced water activity.

Junqueira Júnior (74) studied lipoxygenase action in the presence of both oxidant agents (ascorbic acid and PBZ) in rheological and sensory properties of bread. A combination of 4.8 ppm of PBZ, 027% active soybean flour and 325 ppm of ascorbic acid in national wheat flour freshly ground resulted in maximum elasticity and extensibility of flour with a higher whiteness. This study showed that benzoyl peroxide acted negatively on rheology, extensibility increasing and decreasing the elasticity of dough, but had a positive effect on bleaching of breadcrumb.

According to Inchem (72), PBZ is generated as a product of its benzoic acid reaction, which can damage consumer's health; therefore, its addition should be reduced.

8.4.3. Ascorbic Acid

Ascorbic acid is one of the most used ingredients in the formulation of mixtures for additive improves at baking industry (3). Carbons 2 and 3 of ascorbic acid have pKa of 11.8 and 4.2, respectively, and are more stable in two tautomeric forms. The dehydroascorbic acid is formed by reversible oxidation of ascorbic acid and may exist in hydrated form before releasing water and then reaching its final form. In the initial stage of mixing, ascorbic acid is rapidly oxidized to dehydroascorbic acid helped by the enzyme ascorbate oxidase (AAO) through the intermediary monoanionic ascorbate ion (54).

Dehydroascorbic acid oxidizes GSH (reduced glutathione) endogenous to its oxidized form, GSSG (oxidized glutathione). GSH (reduced glutathione) oxidizes gluten proteins (PSSP) that have broken its disulfide bonds, forming protein-glutathione disulfide (PSSG). However, oxidation of GSH by dehydroascorbic acid occurs at a higher speed by inhibiting the reduction of GSH by PSSP. The phenomenon of gluten proteins reduction is known as "gluten depolymerization," and is responsible for dough weakening (63).

The ascorbic acid is an additive with fast speed of action that acts by reducing the extensibility and increasing gases retention, consequently resulting in breads with bigger volume and improvement in granulosity and texture. The recommended amount technologically ranges from 10 to 200 $mgkg^{-1}$ (with respect to flour weight) (50).

Lopes et al. (78) evaluated the simultaneous use of ascorbic acid and azodicarbonamide in French bread. The application of ascorbic acid and azodicarbonamide using wheat flour concentrations above 75 $mgkg^{-1}$ and $30mgkg^{1}$, considering the studied levels, provided French bread a production with higher specific volume. Economically, there is an advantage of replacing ascorbic acid per azodicarbonamide when used in concentrations higher than 30 $mgkg^{-1}$ for wheat flour, and not exceeding the maximum limit (40 $mgkg^{-1}$) established by Brazilian legislation.

8.4.4. Enzymes

The replacement of enzymes by chemical oxidants is a very interesting option despite ascorbic acid being commonly used (37).

Industrial enzyme preparations have been used for a long time in the baking industry, but in recent years, as a result of consumer demand for healthier products, its use has greatly increased, completely or partially replacing the most controversial additives such as oxidizing agents or emulsifiers (83).

Resolution RDC No. 205 of 11/14/2006 approves the "Technical Regulation on enzymes and enzyme preparations for use in food production for human consumption." The use of enzymes and enzyme preparations in the manufacture of some food depends on its technological need (19).

In addition to improving handling properties of the mass, the benefits of adding enzymes to collaborate with baking process include an increase in bread volume, improved crumb structure, and increased shelf life (67,83).

8.4.5. Glucose Oxidase (GOX)

The glucose oxidase (GOX) can be used to promote oxidation of protein matrix, attracting considerable interest as an improver of dough of wheat flour (37). It catalyzes the oxidation reaction of beta-D-glucose to beta-D-gluconolactone in the presence of oxygen, which is converted into hydrogen peroxide. Hydrogen peroxide promotes the oxidation of SH groups of two cysteine residues for SS crosslinks in gluten network (101). Cross-links in proteins of wheat flour resulted in significant improvements in functional properties of baked products technology.

The positive effect of glucose oxidase in flour has been attributed, by many authors, to the production of H_2O_2 during oxidation of glucose. Most studies report that improvements are related to the texture and strength of the crumb and that loaf volume was not changed with the addition of glucose oxidase.

Although H_2O_2 is hypothetically the responsible factor, the mechanism for the effect of strengthening the dough by glucose oxidase is not completely understood (70).

The hydrogen peroxide generated by GOX can also lead to cross-linking of arabinoxylans or even a gel of water-soluble arabinoxylan (120).

There is controversy about a positive (120) or negative (122,123) effect of arabinoxylan gels on the quality of bread. This effect can be corrected by co-adding xylanase, which cleaves arabinoxylan complex, generating small fragments (122,123).

Oxidation process that leads to gel formation is attributed to ferulic acid residues, esterified to arabinose in arabinoxylans. A possible mechanism is through the dimerization of ferulic acid residues on adjacent chains of arabinoxylans (37).

Silva & Chang (110) evaluated the effect of adding xylanase, glucose oxidase and ascorbic acid on the loaf quality using whole-wheat flour. The results showed that the addition of xylanase, glucose oxidase, and ascorbic acid changed the bread characteristics. With respect to loaf volume, the highest values were found in bread with lower concentrations of xylanase and larger amounts of glucose oxidase and ascorbic acid. Lower values for breadcrumb firmness were obtained for formulas with higher concentrations of ascorbic acid and glucose oxidase and intermediate levels of xylanase. The results showed that whole-wheat flour used with the addition of xylanase, glucose oxidase, and ascorbic acid presented considerable potential for the elaboration of loaf bread with high nutritional and functional value. Both formulations selected in the range of the design were designated as excellent (xylanase: 92.38 EDX/Kg, glucose oxidase: 148.95 SRU/Kg, ascorbic acid: 80ppm and xylanase: 92.38 EDX/Kg, glucose oxidase: 137.37 SRU/Kg, ascorbic acid: 70ppm).

8.4.6. Transglutaminase

The use of microbial transglutaminase (MTGase) in bakery products has been studied as an efficient empowering of gluten network, allowing the usage of weaker flours in products that requires strong flours as well as the incorporation of new protein sources in those foods (2, 10,21,61).

In bakery, MTGase causes the formation of covalent bonds between protein fractions leading to an improvement in viscoelastic properties of gluten with a positive effect on rheological behavior of dough in general (2, 10,27).

According to ANVISA resolution RDC nº 348, of December 02, 2003, MTGase is enclosed to the enzyme table, whose sole paragraph informs that the limit of use of this enzyme will be in "*quantum satis*," that is, amount enough to obtain the effect desired for the food (20).

Medina-Rodríguez et al. (86,87) studied the action of MTGase in the flour on rheological properties of three different cultivars of wheat: Altar (durum), Tacupeto (hard), and Cortazar (weak). The concentrations used were 0.5 and 10 Ug^{-1} of gluten. In practice, these concentrations corresponded to 0.00, 0.55, and 1.11% of enzyme (based on dry flour) for hard and soft wheat, and 0.00, 0.56, and 1.12% of enzyme (dry weight of flour) for durum wheat.

According to Medina-Rodriguez et al. (86,87), wheat flour can undergo a treatment with higher amounts of enzyme without showing the negative effects observed in flours stronger during similar treatments. Table 13 shows the flour farinographic parameter changes depending on MTGase use. With the exception of water absorption (reduced in three cultivars), the other farinographic parameters have different responses to the action of the enzyme.

Table 13. Farinographic responses of three wheat cultivars with the action of MTGase

Wheat	Water Absorption	Dough Development	Stability
Durum	Reduction	Reduction	No significant change
Soft	Reduction	No significant change	Increase
Hard	Reduction	Increase (5 Ug^{-1}); Reduction (10 Ug^{-1})	Increase (5 Ug^{-1}); Reduction (10 Ug^{-1}).

Source: Medina-Rodriguez et al. (86).

Gragnani (61) evaluated the technological viability of using microbial tranglutaminase to make loaf breads with the highest wheat flour substitution as possible, showing the same characteristics as wheat flour bread. This study showed that it is possible to substitute up to 71.28% of wheat flour by triticale, using 0.80% (w/w) microbial transglutaminase to make loaf bread maintaining its quality without increasing industrial costs. This formulation obtained breads with 105-minute fermentation, which were closer to standard loaf breads in instrumental analysis. A slight alteration on texture acceptance was evidenced in sensorial tests, getting "liked" score (70 minutes), while standards breads and 105-minute fermentation ones got "liked a lot."

9. Thermoplastic Extrusion with Wheat Flour

The use of wheat for processing new products that do not depend on gluten quality has been stimulated to increase the aggregate value of sales, once this wheat (weak flour) is often rejected for purchase by the mills.

Thermoplastic extrusion can be defined as a continuous process in which mechanical shear is combined with heat to gelatinize the starch and denature protein. As a consequence, materials are laminated and restructured to obtain products with new textures and shapes (40). The extruder promotes gelatinization, cooking, molecular break, mixing, sterilization, restructuring, inflating, and drying (103,112).

The use of extrusion process has many advantages: versatility, high productivity, low-cost products with various shapes, high quality by avoiding loss of nutrients, production of new products, and above all, lack of effluents (68). Simply passing raw materials can be converted into fully expanded products or semi-finished products, if necessary (51). The results have been dependent on the type of extruder used, the type of extruded product and extrusion parameters considered (8).

In Brazil, extruded products are obtained in larger quantities from corn flour. Some commercial products exist and some research of national wheat extrudates has been made, despite being a good alternative to new products, as demonstrated by Wang et al. (124) who obtained precooked wheat-soybean flour (90:10 ratio) with good technological characteristics for use in beef croquettes. Mixed flour containing 26% of moisture and extruded at 150 rpm and 130°C showed the best sensory characteristics (flavor and texture).

Another work was accomplished by Depiere (36), who used termoplastic extruder for production of instant pasta. In this work, a comparative study was made between pasta obtained from common wheat (*Triticum aestivum* L.) and durum wheat semolina (*Triticum durum* L). It could be verified that the quality and the amount of present proteins in wheat exert more influence on technological profile in the production of masses through conventional process. While using thermoplastic extrusion process, the protein content exerted less effect on product quality, but greater influence of gelatinized starch, as shown by SEM, DSC, and X-ray diffraction. According to the author, instant pasta durum wheat semolina considered as with better quality were obtained using extrusion temperature of 115°C, 26% humidity and 1% glyceryl monostearate emulsifier. Pasta with similar quality was obtained using common wheat flour by altering the process conditions and formulation and using extrusion temperature of 105 to 115°C, 24 to 25% humidity and from 1.15 to 1.50% of emulsifier.

10. Bread Products with Mexican Wheat

10.1. Bread

There is a wide variety of breads in Mexico, with unique names. Some are named after horns, bald ladies, policemen, keys, and pigs. Some bread is dedicated for special occasions such as Day of the Dead and Christmas. Other types of bread are especially made for breakfast. Some of these types of bread originate from many centuries ago. People in Mexico bake their own bread from scratch or purchase it at food markets. This tradition has its origin in Conquista, with the arrival of wheat to America. Indigenous and racially mixed contributions to the Spanish bread created unsuspected forms and flavors in the first times of the colony. Pulque (viscous alcoholic beverage made from fermented sap of maguey plant), anise, mead, sesame, coco, cinnamon, peanut, chocolate, and piloncillo (unrefined sugar commonly used in Mexican cooking) have adorned and given flavor and aroma to Mexican breads. Now the varicty is infinite. In any other side, there are 40, 50, 60 different pieces, in many places they have five at most, and in Mexico, there are at least 2,200 or 2,500 classes of bread (23).

Effect of Transglutaminase on Different Protein Fractions and on Dough Quality of Mexican Wheat

Medina-Rodríguez et al. (86) assess and compare the effect of microbial transglutaminase (MTGase), a protein-glutamine γ-glutamyl transferase (E.C. 2.3.2.13), on different fractions of dough proteins found in hard, soft, and durum Mexican wheat. Three different concentrations of the MTGase (0, 5, and 10U/g of gluten) were tested. Differences were observed in susceptibility to MTGase catalysis among dough proteins of cultivars studied: cultivar Cortazar (soft wheat) was the most susceptible one. Proteins of this cultivar had a characteristically higher amount of ω and α+β gliadins when compared with other cultivars. As reported earlier, solubility of high molecular weight glutenin subunits and ω-gliadins was reduced because of the MTGase treatment. However, all gliadin subclasses, including the γ and α+β gliadins, also participated in cross-linking. Proteins of cultivar Altar (durum wheat) were the least susceptible to the effects of MTGase. Albumins and globulins did not show any reduction in solubility, implying that they did not participate in cross-linking.

Microbial transglutaminase (MTG) can be used to increase wheat dough strength. Three enzyme dosages were used (0, 0.5 and 1.0 Ug^{-1} of gluten) in flours of six wheat cultivars. A principal component analysis showed a strong relationship of the 1.0 U dosage of MTG with higher values of insoluble polymeric protein, alveograph elasticity index, extensibility and work of deformation (dough strength), and specific volume of bread for both types of Mexican cultivars (strong and weak gluten), with higher values of farinograph dough stability for weak gluten cultivars. The improving effect of low dosage of MTG on dough strength and extensibility without increasing dough tenacity was reported, as well as the consequent bread volume increment of dough of weak and strong gluten cultivars (86).

10.2. Flour Tortillas

The majority of tortillas consumed in the U.S. are made from wheat flour, although traditional maize tortillas are also produced. Flour tortillas have become the most prevalent ethnic bread in the U.S., often replacing white pan bread in many products. As a result, tortillas are the fastest growing segment of the U.S. baking industry, with annual sales over US$6 billion and growth exceeding 10% per year. Wheat flour tortillas have been produced in the northern region of Mexico for centuries, and their recent popularity in other states and in the United States has led to a large-scale commercial production. Wheat tortillas can economically compete with corn tortillas as food carriers, being very popular in fast-food franchises and other restaurants. Some other wheat tortilla-related products are burritos, quesadillas, chimichangas, fajitas, buñuelos, and tacos. Commercial wheat flour tortilla products are prepared mainly by hot-pressing, although rolled and die-cut procedures are also used. Tortilla products are then baked in three-tier commercial ovens generally heated by gas, electricity, or hot air. The critical tortilla parameters are diameter, opacity, and rollability. Tortillas about 2 mm thick that are evenly opaque, with an ample diameter, and with at least a three-week shelf life are considered to be ideal (125,126). As in bread, wheat flour and gluten functionality significantly contributes to this shelf stability and the need for tortillas to resist breaking during consumption (92). The diameter of tortillas also requires extensible dough that resists shrink-back during processing (11).

The dough extensibility, in turn, depends again on gluten proteins. Thus, dough extensibility during hot pressing and retention of tortilla flexibility after baking requires a gluten functionality that is unique to the strong viscoelastic gluten functionality needed for bread. Good-quality tortillas require an extensible and mellow gluten network that differs from the strong gluten network needed for bread making. High or low gliadin content, depending on the allele, may positively or negatively influence tortilla quality. In previous studies, protein composition and dough mixing strength of the near-isogenic lines missing

Glu B3/*Gli B1* loci with its recurrent parent were compared. A reduction in polymeric proteins was observed in association with a reduction in dough strength (79). Deletions in the *Glu 1* loci reduced the glutenin to gliadin ratio and reduced dough strength (59), also significantly changing bread making quality. Whereas bread firms and stales in a week, tortilla flexibility and rollability are exhibited for more than three weeks, depending on flour properties, formulation, and processing. Waniska et al. (126) used near-isogenic wheat lines with deletions in *Gli 1* and *Gli 2* loci to relate qualitative and quantitative differences in gliadin composition to gliadin functionality in flour tortilla quality. In this study, gliadin functionality in tortilla quality was studied using near-isogenic wheat lines that have deletions in either *Gli A1*, *Gli D1*, *Gli A2*, or *Gli D2* gliadin loci. Data suggests that altering certain *Gli 2* loci through null alleles could be a viable strategy to develop cultivars improved for specific functionality requirements needed for rapidly growing tortilla market (117). Martinez-Bustos et al. (84) developed a faster method that reduced tortilla dehydration. In the infrared radiation (IR) method, tortillas were baked for 17 or 19 sec by IR using black-body radiation at a selected wavelength band and emission temperatures of 549 or 584°C. IR-baked tortillas showed good characteristics of rollability, puffing, layering, color, and texture. The loss of moisture during tortillas baking formed by hot-pressing and baked by IR was significantly lower than that of tortillas baked by traditional and commercial methods. The average energy used by the IR oven was inferior to that used in the commercial method which, in turn, is more efficient than traditional hot griddle method.

11. Future Prospects

Brazil and Mexico will have to intensify its research in this basic cereal feeding to continue the successful trajectory in research and development of wheat in all areas with aims including:

11.1. Genetic

Taking into account that cultivars developed by CIMMYT, Mexico had odd importance in the history of Brazilian wheat production from 1972 to 1990, resulting in more than fifty Brazilian cultivars being useful descendants of crossings using Mexican germplasm, the research should continue the great integration among researchers of Brazil and Mexico in search of new cultivars. As an example, the production of cultivar with characteristics of wheat durum will benefit the market of nutritious pasta in Brazil, while in Mexico and Brazil, cultivars will be developed to assist domestic consumption.

11.2. Agricultural

Increase the productivity of wheat with resistance to pests in the field, the productivity of wheat with higher gluten strength for bread and pasta market, and increase wheat stability in storage silos.

11.3. Technological

The biggest challenges are related to the final quality of the products, where integral mixing flour use with additives and enzymes have been researched for bread production with sensorial, nutritional and technological quality adjusted to the consuming market. Some perspectives can be highlighted:

- The increased use of whole grain flour in bread making products that, in addition to promoting nutritional benefit, promote a better exploitation of wheat and reduce the by-product formation. This consequently will diminish the import of wheat. The greatest challenge is to offer flour of entire grain, which is still limited, as well as stability of the flour and the respective products during storage.
- Increase of native flour using other cereals, roots and tubercles in flours of wheat with good strength of gluten for bakery emphasize the fact that these products have an incurred higher cost, probably with lesser availability or irregular supply of substitute raw materials of the wheat, in addition to the necessity of additive use to keep product quality, as the addition of other cereals or grains has deleterious action on the mass due to dilution of the gluten, incurring volume reduction. Another factor that interferes with the cost is the shorter shelf life of these products due to the occurred transformations in ingredients and mainly to the retrogradation of the starch, a reduction in the softness of the crumb of breads, and an increase in the compression force with an elapse of shelf life, limiting its acceptability.
- The use of additives for gluten enhancing, which will promote greater utilization of soft wheat with lower gluten strength already being used in bakery products, however, represents a higher cost to the product. Enzymes used, as transglutaminase and glucose oxidase, promote intercrossing and oxidation reactions, respectively, in order to obtain soft wheat flour with higher gluten strength.

CONCLUSION

In Brazil, the change in food habits that occurred from 1962 to 1991, led to a decreased consumption of native products such as corn and tubers and to an increase of bakery products based in wheat. This shift was facilitated by wheat-exporting countries and Brazilian government, which subsidized the wheat, making a policy of price control.

Nowadays, Brazil stands out as one of the three larger wheat importers, with Argentina, with an average of approximately six million tons/year, which corresponds to more than 50% of their domestic needs.

The oscillation in wheat production in Brazil is due to adverse effects caused by climatic instability in Southern Brazil in recent years, economic policies of prices discouraging positive genetic advances through new cultivars resistant to diseases, and more potential grain yield associated with the evolution of management practices (new fungicides, crop rotation, etc.), all of which have contributed to an almost linear increase in the average Brazilian yield per hectare.

However, Mexico's growth in domestic demand has influenced the country to increase its dependence on wheat imports. A factor contributing to the increase in wheat imports is the disparity between the type of wheat primarily produced in Mexico and the type of wheat demanded by consumers. Mexico produces primarily durum-type, and most of wheat imported and used by domestic consumers is hard wheat suitable for making bakery products. Currently, Mexico is the third largest wheat importer of U.S. wheat. Although wheat yield has been substantially increasing in the last years, this increase is attributed to genetic improvement and to suitable use of better practices of culture, with production of crystalline wheat for the pasta industry.

The diseases prevalent in Mexican cultivars have also been a key factor in increasing the production of strong gluten wheat. Moreover, regions with high technology for wheat are only located in the Northeast and Center of the country, limiting production to small regions.

In conclusion, the efforts of Brazil and Mexico to assist the reality of bakery market can serve as an example for countries that are non-producers of wheat.

REFERENCES

[1] AACC - American Association of Cereal Chemists. (2000). *Approved methods*. 10th ed. American Association of Cereal Chemists, Saint Paul, Minnesota.

[2] Aalmi, M.; Leelavathi, K. (2008). Effect of Microbial Transglutaminase on Spaghetti Quality. *Journal of Food Science, 73, 5,* 306-312.

[3] Aamodt, A.; Magnus, E. M.; Faergestad, E. M. (2003). Effect of flour quality, ascorbic acid, and DATEM on dough rheological parameters and hearth loaves characteristics. *Journal of Food Science, 68, 7,* 2201- 2210.

[4] ABIP. *Associação Brasileira da Indústria de Panificação e Confeitaria*. (2010) Available in: <http://www.abip.org.br/>.

[5] ABITRIGO- Associação Brasileira da Indústria do Trigo. (2010). *História do trigo. O papel do trigo na evolução da humanidade*. Available in: http://www.abitrigo.com.br/menu_historia_do_trigo.asp.

[6] Al-Saqer, J. M.; Sidhu, J. S.; Al-Hooti, S. N. (2000). Instrumental texture and baking quality of high-fiber toast bread as affected by added wheat mill fractions. *Journal of Food Processing and Preservation, 24, 1,* 1–16.

[7] American Ingredients Company. (1998). *Benzoyl Peroxide Technical Bulletin B060*, Kansas.

[8] Anderson, R. A.; Pfeifer, V. F.; Bookwalter, G.; Griffin Jr., E. L. (1971). Instant C.S.M. food blends for worldwide feeding. *Cereal Science Today, 16, 1*, 5-11.

[9] Barkley, A. P.; Nalley, L. L.; Crespi, J. (2008). *The Impact of the CIMMYT Wheat-Breeding Program on Mexican Wheat Producers and Consumers: An Economic Welfare Analysis*, 2008 Annual Meeting, February 2-6, 2008, Dallas, Texas 6931, Southern Agricultural Economics Association.

[10] Basman, A.; Koksel, H.; Atli, A. (2006). Effects of increasing levels of transglutaminase on cooking quality of bran supplemented spaghetti. *European Food Research and Technology, 223, 4*, 547-551.

[11] Bejosano, F. P.; Waniska, R. D. (2001). *Functionality of bicarbonate leaveners in wheat flour tortillas. Cereal Chemistry, 81, 1*, 77-79.

[12] Borges, J.M.; Ferreira, E.; Alvim, C.M.F. (1984). *Pão de mandioca e soja*. Viçosa. Boletim de Extensão. Universidade Federal de Viçosa.

[13] Bortolotti, C. R. (2009). *Caracterização de farinhas de cevadas e o efeito da sua incorporação sobre a qualidade do pão de forma*. Dissertação de mestrado em Ciência e Tecnologia de Alimentos. Santa Maria, Universidade Federal de Santa Maria, Centro de Ciências Rurais.

[14] Brasil (2010). Ministério da Saúde. Agência Nacional de Vigilância Sanitária. Portaria n. 354, de 18 de julho de 1996. Aporva a norma técnica referente a farinha de trigo. Available in: *<http://e-legis.anvisa.gov.br/leisref/public/showAct.php?id=69&word=>*. Access in February of 2010.

[15] Brasil. (1978). Ministério da Saúde. Secretária de Vigilância Sanitária. *Aprova normas técnicas especiais do estado de São Paulo, relativa a alimentos e bebidas*. Resolução da Comissão Nacional de Normas e Padrões para Alimentos- CNNPA n. 12, D.O.U. de 24 de julho de 1978. Seção 1, pt.1.

[16] Brasil. (1992). Ministério da Agricultura e Reforma Agrária. Secretaria Nacional de Defesa Agropecuária. *Regras de análises para sementes*. Brasília.

[17] Brasil. (1994). Ministério da Agricultura, do Abastecimento e da Reforma Agrária. *Norma de identidade, qualidade, embalagem e apresentação do trigo. (Triticum aestivum L.)*: texto legal. Porto Alegre. EMATER/RS.

[18] Brasil. (2005). Ministério da Agricultura, Pecuária e Abastecimento. Instrução Normativa nº 8 de 2 de junho de 2005. *Aprova o regulamento técnico de identidade e qualidade da farinha de trigo*. Diário Oficial da República Federativa do Brasil, Brasília, 3 de junho de 2005.

[19] Brasil. (2006). Ministério da Agricultura e do Abastecimento. RDC Nº 205, de 14 de novembro de 2006. Aprova o "Regulamento Técnico sobre Enzimas e Preparações Enzimáticas para Uso na Produção de Alimentos Destinados ao Consumo Humano". Available in: <http://www.anvisa.gov.br>.

[20] Brasil. (2003). RDC nº 348, of December 02. ANVISA. Available in: http://www.anvisa.gov.br.

[21] Caballero, P.A.; Bonet, A.; Rosell, C.M.; Gómez, M. (2005). Effect of microbial transglutaminase on the rheological and thermal properties of insect damaged wheat flour. *Journal of Cereal Science, 42*, *1*, 93-100.

[22] Campos, M. S. (1985). *Influência da energia de microondas na qualidade tecnológica de trigo*. Dissertação de Mestrado em Tecnologia de Alimentos. Universidade Estadual de Campinas, Faculdade de Engenharia de Alimentos, Campinas.

[23] Canimolt - Cámara Nacional de la Industria Molinera del Trigo. (2009). *Presentación estadística Información sectorial de la industria molinera*. Available in: http://www.slideshare.net/gromerol/presentacin-estadstica-canimolt.

[24] Capriles, V. D.; Coelho, K. D.; Matias, A. C. G.; Arêas, J. A. G. (2006). Efeito da adição de amaranto na composição e na aceitabilidade do biscoito tipo cookie e do pão de forma. *Alimentos & Nutrição, 17, 3*, 269-274.

[25] Cassidy, B. G.; Dvorak, J.; Anderson, O. D. (1998). The wheat low–molecular–weight glutenin genes: characterization of six genes and progress in understanding gene family structure. *Theoretical and Applied Genetics, 96, 6-7*, 743–750.

[26] Chellquist, E.; Gorman, W. (1992). Benzoyl peroxide solubility and stability in hydric solvent. *Pharmaceutical Research, 9, 10,* 1341-1346.

[27] Collar, C.; Bollaín, C.; Angioloni, A. (2005).Significance of microbial transglutaminase on the sensory, mechanical and crumb grain pattern of enzyme supplemented fresh pan breads. *Journal of Food Engineering, 70, 4*, 479-488.

[28] CONAB. (2009). *Companhia Nacional de Abastecimento.* Acompanhamento da Safra brasileira de Grãos, safra 2008/2009. Available in: http://www.conab.gov.br/conabweb/.

[29] CONAB. (2010). Companhia Nacional de Abastecimento. Acompanhamento da Safra brasileira de Grãos, safra 2009/2010. Available in: http://www.conab.gov.br/conabweb.

[30] Cornell, H. J.; Hoveling, A. W. (1998). *Wheat: Chemistry and Utilization.* Lancaster: Technomic Publishing Co., Inc.

[31] Costa, P. F. P. (2009). *Efeito da radiação gama e da radiação infravermelha na vida de prateleira e nas caracteristicas tecnológicas da farinha de trigo integral e do pão de forma integral.* Tese de Mestrado. Campinas, SP. Available in: http://www.fea.unicamp.br/alimentarium/ver_documento.php?did=709.

[32] De la O, O. M.; Espitia, R. E.; Molina, G. J. D.; Peña, B. R. J.; Santacruz, V. A.; Villaseñor, M. H. E. (2006). Efecto de diferentes subunidades de gluteninas–APM sobre la calidad panadera en trigos harineros mexicanos. *Revista Fitotecnia Mexicana, 29, 4,* 291–297.

[33] De la O, O. M.; Espitia, R. E.; Molina, G. J. D.; Peña, B. R. J.; Santacruz, V. A.; Villaseñor, M. H. E. (2006). Effect of different high molecular weight glutenin alleles on the viscoelastic properties of bread wheat dough. ***Agrociencia, 40*, 461-469.**

[34] De la O, O.M.; Peña, B.R.J.; Santacruz, V. A.; Villaseñor, M. H.E.; López, S. H.; Peña, B. R. J.; Santacruz, V. A. (2010). Estabilidad a través de ambientes de las propiedades reológicas de trigos harineros en función de sus gluteninas de alto peso molecular. *Revista Fitotecnia Mexicana, 33, 2,* 125–131.

[35] Dendy, D. (1992). Perspectives in composite and alternative flour products. In Cereal Chemistry and Technology: a long past and bright future. *9th International Cereal and Bread Congress*, Paris, 1-5, June.

[36] Depieri, M. (2004). *Desenvolvimento de pastas alimentícias instantâneas por processo de extrusão termoplástica: influência do tipo de trigo, teor de emulsificantes e parâmetros de processo na qualidade do produto.* Dissertação de Mestrado em Tecnologia de Alimentos. Universidade Estadual de Campinas, Faculdade de Engenharia de Alimentos, Campinas, SP.

[37] Dunnewind, B.; Van Vliet, T.; Orsel, R. (2002). Effect of oxidative enzymes on bulk rheological properties of wheat flour doughs. *Journal of Cereal Science, 36, 3*,357-366.

[38] Dutra, O. I. P.; Silva, A. E. L. (2010). *Produção de sementes de trigo no Rio Grande do Sul.* Porto Alegre: Apassul.

[39] El-Dash, A. A. (1978). Standardized mixing and fermentation procedure for experimental baking test. *Cereal Chemistry, 55, 4*, 436-446.

[40] El-Dash, A. A. (1981).Application and control of thermoplastic extrusion of cereals for food and industrial uses. In: Pomeranz, Y.; Munck, L. (Ed.) *Cereals – a renewable resource: theory and practice.* Saint Paul: American Association of Cereal Chemists, p. 165-216.

[41] El-Dash, A.; Cabral, L.C.; Germani, R. (1994). *Uso de farinha mista de trigo e soja na produção de pães*. Brasília: EMBRAPA – SPI.

[42] El-Dash, A.; Campos, J. E.; Germani, R. (1994). *Tecnologia de farinhas mistas: uso de farinha mista de trigo e sorgo na produção de pães*. Brasília: EMBRAPA - SPI.

[43] El-Dash, A.; Germani, R. (1994). *Tecnologia de farinhas mistas: uso de farinha mista de trigo e milho na produção de pães*. Brasília: EMBRAPA - SPI.

[44] El-Dash, A.; Mazzari, M. R.; Germani, R. (1994). *Tecnologia de farinhas mistas: uso de farinha mista de trigo e mandioca na produção de pães*. Brasília: EMBRAPA - SPI.

[45] Esteller, M. S. (2004). *Fabricação de pães com reduzido teor calórico e modificações reológicas ocorridas durante o armazenamento*. Dissertação (Mestre em Ciências Farmacêuticas) - Faculdade de Ciências Farmacêuticas, Universidade de São Paulo, São Paulo.

[46] *Ewart, J. A. D.* (1968). *A hypothesis* for the structure and rheology of glutenin. *Journal of the Science of Food and Agriculture, 19*, 617-623.

[47] FAO- Food Outlook – (2008). Wheat. Available in: http://www.fao.org/docrep/011/ai474e/ai474e03.htm.

[48] FAO. (2005). Food and Agriculture Organization of the United Nations. Available in: <http://www.fao.org/>.

[49] FAO. (2007). The Food and Agriculture Organization of the United Nations. *Available in:* www.fao.org.

[50] Fischer, F. (1985). Oxidation and reduction electron-transfer is key to dough improvements. *Bakers Digest, 59, 1*, 22.

[51] Flickinger, B. (1991). The market for extruded foods. *Food Engineering Intern*ational, *16, 6*, 42-44.

[52] Fonseca Filho, A. A.; Araújo, W. M. C.; Falcirolli, D. M. C.; Pilla, N.; Marques, O. O. (1997). Avaliação da qualidade tecnológica de pães enriquecidos com farelo de trigo. *Alimentos & Nutrição*, *8, 1*, 17-25.

[53] Freitas, R. E.; Stertz, S. C.; Waszczynskyj, N. (1997). Viabilidade da produção de pão, utilizando farinha mista de trigo e mandioca em diferentes proporções. *Boletim do CEPPA*, *15, 2,* 197-208.

[54] Frimer, A. A.; Gilinsky-Sharon, P. (1995). Reaction of superoxide with ascorbic acid derivatives: insight into the superoxide-mediated oxidation of dehydroascorbic acid. *Journal of Organic Chemistry*, *60*, 2796-2801.

[55] Gallardo, J. R. K.; Rodney, B L. L.; Holcomb. (2008). *Wheat Quality Information Availability and its Impacts on Mexican Millers' Welfare*. Seminars February 18.

[56] Gandra, K. M.; Del Bianchi, M.; Godoy, V. P.; Collares Queiroz, F. P.; Steel, C. J. (2008). Aplicação de lipase e monoglicerídeo em pão de forma enriquecido com fibras. *Ciência e Tecnologia de Alimentos*, *28, 1,* 182-192.

[57] GCEA/IBGE.- Grupo de Coordenação de Estatísticas Agropecuárias. (2007). *Levantamento Sistemático da Produção Agrícola*. Available in: http://www.ibge.gov.br/home/estatistica/indicadores/agropecuaria/lspa/lspa02200705.shtm.

[58] GEA- Grupo de Economistas y Asociados. (2003). Fundación Mexicana para la Innovación y Transferencia de Tecnología en la Pequeña y Mediana Empresa, A.C. (FUNTEC). *Estudio estratégico para elevar la competitividad y el desarrollo sustentable de la cadena productiva del sector. Trigo-harina-pan, pastas y galleta*s. Preparado para: Cámara Industrial Harinera de la Zona Centro.

[59] Gianibelli, M. C. (1998). *New proteins for improving wheat quality*. Ph.D. Thesis; University of Western Sydney: Sydney, Australia.

[60] Goesaert, H.; Brijs, K.; Veraverbeke, W.S.; Courtin, C.M.; Gebruers, K.; Delcour, J.A. (2005). Wheat flour constituents: how they impact bread quality, and how to impact their functionality. *Trends in Food Science & Technology, 16, 1*, 12-30.

[61] Gragnani, M. A. L. (2010). *Produção e avaliação de pão de forma com triticale e enzima transglutaminase microbiana*. Dissertação de Mestrado em Tecnologia de Alimentos. Universidade Estadual de Campinas, Faculdade de Engenharia de Alimentos, Campinas, SP.

[62] Gralik, J.; Warchalewski, J.R. (2006). The influence of gamma-irradiation on some biological activities and electrophoresis patterns of wheat grain albumin fraction. *Food Chemistry, 99, 2*, 289-298.

[63] Grosch, W.; Wieser, H. (1999). Redox reactions in wheat dough as affected by ascorbic acid. *Journal of Cereal Science, 29, 1*, 1-16.

[64] Gupta, R. B.; Shepherd, K. W. (1992). Identification of rye chromosome IR translocations and subunits in hexaploid wheats using storage proteins as genetic markers. *Plant Breeding, 109, 2,* 130–140.

[65] Gutkoski, L. C.; Velloso, C. B.; Dóro, C. T.; Silveira, A. A. E.; Bonafé, L. A. (1993). Uso de farinha mista de trigo e aveia em produtos de panificação: pães tipo forma, francês e pré-pizza. *Boletim do CEPPA, 11, 1*, 33-45.

[66] Gutkovski, L. C.; Pavanelli, A. P.; Miranda, M. Z.; Chang, Y. K. (1997). Efeito de melhoradores nas propriedades reológicas e de panificação da massa de farinha de trigo. *Ciência e Tecnologia de Alimentos, 7, 1*, 11-16.

[67] Haros, M.; Rosell, C.M.; Benedito, C. (2002). Improvement of flour quality through carbohydrases treatment during wheat tempering. *Journal of Agricultural and Food Chemistry, 50, 14,* 4126-4130.

[68] Harper, J. H. (1979). Food extrusion. *CRC Critical Reviews in Food Science and Nutrition, 11, 2,* 155-215.

[69] Hernández, S. A. (1988). Avance en el mejoramiento genético del trigo en Mexico. *Agronomie, 8, 7,* 633-638.

[70] Hilhorst, R.; Dunnewind, B.; Orsel, R.; Stegeman, P.; Van Vliet, T.; Gruppen, H.; Schols, H.A. (1999). Baking performance, rheology, and chemical composition of wheat dough and gluten affected by xylanase and oxidative enzymes. *Journal of Food Science, 64, 5*, 808-813.

[71] IAC-*Instituto Agronômico de Campinas*. (2010). Available in: http://www.iac.sp.gov. br/ UniPesquisa/GraosFibras/Cultivares/Triticale.asp.

[72] Inchem- International Programme on Chemical Safety. (2010). Concise international chemical assessment document no. 26. Available in: http://www.inchem.org/ documents/cicads/cicads/cicad26.htm. Accessed: October 09, 2010.

[73] Jardine, J. G. (1982). *Farinha mista de triticale e trigo para produção de pães*. Rio de Janeiro: EMBRAPA - CTAA.

[74] Junqueira Júnior, R. M. (2007). *Estudo da interação entre lipoxigenase da soja e ácido ascórbico nas propriedades reológicas e sensoriais de pães*. Dissertação de Mestrado em Tecnologia de Alimentos. Universidade de São Paulo, Faculdade de Ciências Farmacêuticas, São Paulo, SP.

[75] Kerr, C. L.; Faubion, J. M.; Hoseney, R. C. (1992). Effects of lipoxygenase and yeast on the rheological properties of dough. *Lebensmittel Wissenschaft und Technologie, 25, 3,* 244-247.

[76] Lagos, M. B. (1983). *História do melhoramento do trigo no Brasil.* Porto Alegre: Instituto de Pesquisas Agronômicas (IPAGRO. Boletim técnico, 10).

[77] Lavoie, N. (2005). Price Discrimination in the Context of Vertical Differentiation: An Application to Canadian Wheat Exports. *American Journal of Agricultural Economics, 87, 4,* 835-854.

[78] Lopes, A. L.; Ormenese, R. C. S. C.; Montenegro, F. M.; Ferreira Júnior, P. G. (2007). Influência do uso simultâneo de ácido ascórbico e azodicarbonamida na qualidade do pão francês The influence of simultaneous use of ascorbic acid and azodicarbonamide in the quality of French Bread. *Ciência e Tecnologia de Alimentos, 27, 2,* 307-312.

[79] MacRitchie, F.; Lafiandra, D. (2001). The use of near-isogenic wheat lines to determine protein composition–functionality relationship. *Cereal Chemistry, 78, 4,* 501-506.

[80] Marsaioli Junior, A. (1991). *Desenvolvimento de um protótipo de secador cilíndrico – rotativo a microondas e a ar quente para a secagem contínua de produtos sólidos granulados.* Tese (Doutor em Engenharia de Alimentos) – Faculdade de Engenharia de Alimentos, Universidade Estadual de Campinas.

[81] Martínez, C. E.; Espitia R. E.; Benítez R. I.; Peña B. R. J.; Santacruz, V. A.; Villaseñor M. H. E. (2007). Efecto de gluteninas de alto peso molecular de los genomas A y B sobre propiedades reológicas y volumen de pan en trigos harineros. *Agrociencia, 41, 2,* 153-160.

[82] Martínez, C. E.; Espitia, R. E.; Benítez, R. I.; Peña, B. R. J.; Santacruz, V. A.; Villaseñor, M. H. E. (2007). El complejo Gli-1/Glu-3 y las propiedades reológicas y volumen de pan de trigos harineros. *Revista Fitotecnia Mexicana, 30, 2,* 167-172.

[83] Martínez-Anaya, M.A.; Jiménez, T. (1998). Physical properties of enzyme-supplemented doughs and relationship with bread quality parameters. *Zeitschrift für Lebensmitteluntersuchung und-Forschung A, 206,* 2, 134-142.

[84] Martínez-Bustos, F.; Morales, S.E.; Chang, K.Y; Herrera-Gómez, A.; Martínez, M. J. L.; Baños, Rodríguez, L. M. E.; Flores, M. H. E. (1999). Effect of Infrared Baking on Wheat Flour Tortilla Characteristics. *Cereal Chemistry, 76, 4,* 491–495.

[85] Martínez-Cruz, E.; Espitia, R. E.; Villaseñor, M. H. E.; Molina, G. J. D.; Benítez-Riquelme, I.; Santacruz, V. A.; Peña, B. R. J. (2010). Diversidad genética de gluteninas y gliadinas en trigos harineros (Triticum aestivum L.) mexicanos. *Agrociencia, 44, 2,* 187-195.

[86] Medina-Rodríguez, C.; Torres, P.; Martínez-Bustos, F.; Ramirez-Wong, B.; Paredes-López, O.; Castaño-Tostado, E. (2009). Effect of Microbial Transglutaminase on Dough Proteins of Hard and Soft (*Triticum aestivium*) and Durum (*Triticum durum*) Wheat Cultivars. *Cereal Chemistry, 86, 2,* 127–132.

[87] Medina-Rodríguez, C.; Torres, P.; Ramirez-Wong, B.; Castaño-Tostado, E.; Martínez-Bustos, F.; Bello-Pérez, L. A.; Paredes-López, O. (2010). Low Dosage of Microbial Transglutaminase: Effect on Wheat (*T. aestivum*) Dough Rheological and Baking Qualities Studied Using Principal Component Analysis. *Cereal Chemistry, Accepted.*

[88] Mejia, M. L.; Rosales, S. (2004). *Analysis of Wheat Imports Behavior Considering Tendencies, Irregularities, Cycles, and Seasonalities.* Professional Degree Thesis, Universidad Autónoma del Estado de Mexico.

[89] Miranda, M. Z. ; Mori, C.; Lorini, I. (2009). *Qualidade comercial do trigo brasileiro: safra 2006*. Passo Fundo: Embrapa Trigo, 2009. Passo Fundo: Embrapa Trigo (Embrapa Trigo. Documentos Online, 112). Available in: <http://www.cnpt. embrapa.br/biblio/do/p_do112.pdf>.

[90] Murilo, D. V. (1999). *Aspectos econômicos da batata-doce*. In: 4º. Encontro de Professores, Pesquisadores e Extencionistas do Rio Grande Do Norte, Mossoró. Anais. Mossoró: ESAM, p. 21-28.

[91] Nitz, M. (1999). *Secagem de macarrão assistida por microondas*. Dissertação de mestrado em Engenharia de Alimentos. Universidade Estadual de Campinas, Faculdade de Engenharia de Alimentos, Campinas, SP.

[92] Pascut, S.; Kelekci, K.; Waniska, R. D. (2004). Effects of wheat protein fractions on flour tortilla quality. *Cereal Chemistry, 81, 1*, 38-43.

[93] Payne, P. I.; Jackson, E. A.; Holt, L. M. (1984). The association between y–gliadin 45 and gluten strength in durum wheat varieties: a direct causal effect or the result of genetic linkage? *Journal of Cereal Science, 2, 1*, 73–81.

[94] Peña, R. J. (2008). *Producción, consumo, usos y la calidad de la cosecha del trigo en Reunión SIAP*. OEIDRUS-Guanajuato.

[95] Peña, R. J.; Zarco-Hernández, J.; Amaya-Celis, A.; Mujeeb-Kazi, A. (1994). Relationship between chromosome 1B-encoded glutenin subunit compositions and bread-making quality characteristics of some durum wheat (*Triticum turgidum*) cultivars. *Journal of Cereal Science, 19, 3*, 243-249.

[96] Peña, R.J. (2000). Durum wheat for pasta and bread-making. Comparison of methods used in breeding to determine gluten quality-related parameters. International Maize and Wheat Improvement Center (CIMMYT). In: Royo, C.; Nachit, M.M.; Di Fonzo, N.; Araus, J.L. Options Méditerranéennes: Série A. Séminaires Méditerranéens; n. 40. Seminar on Durum Wheat Improvement in the Mediterranean Region: New Challenges, 2000/04/12-14, Zaragoza (Spain).

[97] Peña, R.J.; Amaya, A.; Rajaram, S.; Mujeeb-Kazi, A. (1990). Variation in quality characteristics associated with some spring 1B/1R translocation wheats. *Journal of Cereal Science, 12, 2*, 105-112.

[98] Peña, R.J; Trethowan, R.; Pfeiffer, W.H; Van Ginkel, M.. (2002). *Quality (end-use) improvement in wheat: compositional, genetic, and environmental factors/Quality Improvement in Field Crops*, ed. Basra, A. S. & Randhawa, L. S. 431 pp. New York: Haworth Press, Inc. ISBN 1 56022 101 1.

[99] Pereira, M.L.; Chang, Y.K. (1993). Contaminantes do trigo e farinha e medidas de sanitização na industria de moagem e panificação. *Higiene Alimentar, 7, 26*, 20-9.

[100] Pomeranz, Y. (1987). *Modern Cereal Science and Technology*, VCH Publishers, New York.

[101] Primo-Martín, C.; Wang, M.; Lichtendonk, W.J.; Plijter, J.J.; Hamer, R.J. (2005). An explanation for the combined effect of xylanase - glucose oxidase in dough systems. *Journal of the Science of Food and Agriculture, 85, 7*, 1186-1196.

[102] Rao, G.V.; Rao P.H. (1993). Methods for determining rheological characteristics of doughs: a critical evaluation. *Journal of Food Science Technology, 30, 2*, 77-87.

[103] Rossen, J. L.; Miller, R. C. (1973). Food extrusion. *Food Technology, 27, 8*, 46-53.

[104] Ryu, G. H.; Walker, C. E. (1995). The effects of extrusion conditions on the physical properties of wheat flour extrudates. *Starch, 47, 1*, 33-36.

[105] Sagarpa (2007). Secretaría de de Agricultura, Ganadería, Desarrollo Rural, Pesca y Alimentación. Available in: http://www.sagarpa.gob.mx.

[106] Saiz, A.; Manrique, G.; Fritz, R. (2001). Determination of benzoyl peroxide and benzoic acid levels by HPLC during wheat flour bleaching process. *Journal of Agricultural and Food Chemistry*, *49, 1,* 98-102.

[107] Scheeren, P. L. (1986). *Informações sobre o trigo (Triticum spp.).* Passo Fundo: EMBRAPA-CNPT, 34 p. (EMBRAPA-CNPT. Documentos, 2).

[108] SECEX. (2010). Secretaria do comércio Exterior. Available in: www.mda.gov.br/ portal/institucional/SECEX.

[109] Sekhon, K. S.; Singh, N.; Nagi, H. P. S. (1992). Effect of pearling and blending on the bread-making properties of sprout-damaged wheat. *Cereal Foods World, 37, 9,* 715-716, 721-724.

[110] Silva, C. B.; Chang, Y. K. (2007). *Efeito da adição de xilanase, glicose oxidase e ácido ascórbico na qualidade do pão de forma de farinha de trigo de grão inteiro.* Tese de Mestrado. Faculdade de Engenharia de Alimentos. UNICAMP.

[111] SINDPAN (2010). *Sindicato da Indústria de Panificação e Confeitaria.* Available in: http://www.sindipan.org.br/imprensa/20061003l.htm.

[112] Singh, N.; Sekhon, K. S.; Nagi, H. P. S. (1994). Effect of temperature on the extrusion behavior of flour from sound and sprouted wheat. *Journal of Food Science and Technology-India, 31, 3,* 233-235.

[113] Solís, M. E. (1996). Efecto de la roya de la hoja sobre cinco genotipos de trigo harinero Triticum aestivum L. en El Bajío, México. *Agricultura Técnica en México, 22*, 51-62.

[114] Stampfli, L.; Nersten, B. (1995). Emulsifiers in bread making. *Food Chemistry, 52,4* 353-360.

[115] Statics Canada (2010). Available in: www.statcan.gc.ca.

[116] Stauffer, C.E. (2000). Emulsifiers as anti-staling agents. *Cereal Foods World, 45, 3,* 106-110.

[117] Suchismita, M.; Hays, B. D.; Noviola, J. A.; Mason, R. E.; Tilley, M.; Waniska, R. D.; Scott, R. B.; Glover, K. D. (2009). Functionality of Gliadin Proteins in Wheat Flour Tortillas. *Journal of Agricultural and Food Chemistry, 57, 4*, 1600–1605.

[118] USDA.(2007). Foreign agricultural service. Available in: http://www.fas.usda.gov/.

[119] Varughese, G.; Pfeiffer, W.H.; Peña, J. (2006). Triticale: A successful alternative crop (part 1). *Cereal Foods World, 41, 6,* 474-482.

[120] Vemulapalli, V.; Miller, K.A.; Hoseney, R.C. (1998). Glucose oxidase in bread making systems. *Cereal Chemistry, 75, 4*, 439-442.

[121] Villaseñor, M. H. E.; Espitia, R. E. (2000). *El trigo de temporal en México.* INIFAP. Campo Experimental Valle de México, Chapingo, Estado de México.

[122] Wang, M.; Oudgenoeg, G.; Van Vliet, T.; Hamer, R.J. (2003). Interaction of water unextractable solids with gluten protein: effect on dough properties and gluten quality. *Journal of Cereal Science, 38,1*, 95-104.

[123] Wang, M.; Van Vliet, T.; Hamer, R.J.(2004). Evidence that pentosans and xylanase affect the re-agglomeration of the gluten network. *Journal of Cereal Science, 39, 3,* 341-349.

[124] Wang, S. H.; Nascimento, T. P.; Rocha, G. O.; Ascheri, J. L. R. (2007). Farinha de trigo e soja pré-cozida por extrusão para uso em croquete de carne. *Ciência e Tecnologia de Alimentos, 27, 3*, 572-578.

[125] Waniska, R. D. (1999). Perspectives on flour tortillas. *Cereal Foods World*, *44, 7,* 471-473.

[126] Waniska, R. D.; Cepeda, M.; King, B. S.; Adams, J. L.; Rooney, L. W.; Torres, P. I.; Lockhart, G. L.; Bean, S. R.; Wilson, J. D.; Betchel, D. B. (2004). Effects of flour properties on tortilla quality. *Cereal Foods World, 49, 4,* 237-272.

[127] Weegels, P.L.; Hamer, R.J.; Schofield, J.D. (1996). Functional properties of wheat glutenin. *Journal of Cereal Science*, 23, *1*, 1–18.

In: Wheat: Genetics, Crops and Food Production
Editor: Maria T. Almeida, pp. 121-151
ISBN 978-1-61209-307-9

Chapter 3

CEREALS AND ABIOTIC STRESSES: ROLES OF AQUAPORINS AND POTENTIAL IN WHEAT IMPROVEMENT

Runyararo Memory Hove and Mrinal Bhave*
Environment and Biotechnology Centre, Faculty of Life and Social Sciences, Swinburne University of Technology, Hawthorn, Victoria 3122, Australia

ABSTRACT

Wheat is the largest food crop that humans rely on, and as such, the world wheat production is expected to keep pace with demands of the growing population and changing climate. There are also new pressures on cereal grains from the bioethanol industries. However, wheat as well as other cereals are severely affected by climatic factors such as salinity, drought, and nutrient deficiency or toxicity in a number of areas around the world. These environmental stresses variously impact plant growth, health, nutritional value and crop yields in various geographic regions and climates. A number of genetic factors are known to participate in the responses of plants to abiotic stresses, and involve interacting networks of specific genes and proteins with the ultimate goal of stress tolerance. Aquaporins, members of the major intrinsic protein (MIP) family, are channel proteins located in cellular membranes, and their roles in water uptake in plants are now well established. In addition, recent years have seen an explosion in the number of substrates found to be transported by these channels, many of the substrates being of physiological importance. Further, aquaporins are also shown to contribute to processes such as germination, plant development and photosynthesis. In this work, the key biochemical properties of aquaporins will be summarised, followed by an overview of the main subfamilies of MIPs, the types of substrates that permeate through these channels and their effects on plant physiology, and perspectives on applications of these proteins in cereal biotechnology.

* Corresponding author: Mrinal Bhave, Environment and Biotechnology Centre, Faculty of Life and Social Sciences, Swinburne University of Technology, PO Box 218, Hawthorn, Victoria 3122, Australia. Tel: +61-3-92145759; Fax: +61-3-98190834. E-mail: mbhave@swin.edu.au

1. Introduction

1.1. Importance of Wheat as a Major Source of Nutrition

Bread wheat or common wheat (*Triticum aestivum* L.) is a member of the Triticeae family and one of the most valuable cereal foods in the world. It is an ancient crop, with the first cultivation thought to have occurred about 10,000 years ago, during the shift of humans from hunter-gathering to cultivation of plants (the 'Neolithic Revolution') in the Middle East (reviewed in Feuillet *et al.*, 2008). Wheat is a staple food for approximately 35% of the population and accounts for about 20% of caloric intake (reviewed in Vasil, 2007). It is a major food crop due to its many desirable characteristics such as ease of harvesting, transporting, storing seed from one season to the next, generally easy digestibility and its high nutritive value (>10% protein, 2.4% lipids and 79% carbohydrates) (Vasil, 2007). It is also arguably the most significant source of plant protein (Sahrawat *et al.*, 2003), containing more protein than rice or maize. It is also rich in complex carbohydrates, which are the most efficient source of energy for humans. Whole grain wheat also contains several B vitamins (thiamin, riboflavin and niacin), vitamin E and minerals such as iron magnesium, phosphorus, potassium and zinc (Henderson *et al.*, 2007). It is also an important source of fiber in the human diet. This shows that wheat has both economic and nutritive values; hence it is important to keep it both abundant and affordable.

1.2. World Wheat Production

Wheat is a major crop also because it is adjustable to a wide range of climatic conditions and has a short growing season. It is mostly grown in temperate climates (between latitudes 30° and 60° north and south), with an average growing temperature of 25°C (reviewed in Vasil, 2007). However, it can also be grown from within the arctic circle to close to the equator and from sea level to about 10,000 feet above sea level (Vasil, 2007). In 2007 world production of wheat was 607 million tonnes, making it the third most-produced cereal after maize (784 million tonnes) and rice (651 million tonnes) (FAO, 2007). It is one of the most important commodities in international trade, with remarkable increases in world production (from 200 million tonnes in 1950s to 626 million tonnes in 2000s) over the decades. The yields have improved from 1 tonne/ha in 1951 to 2.5 tonne/ha in 1995, owing largely to superior varieties and improved cultivation practices, without significant increases in the area under cultivation (Vasil, 2007).

1.3. Wheat Domestication and Genetics

The major cultivated species are the common or bread wheat (*Triticum aestivum*; hexaploid, AABBDD), durum (*Triticum turgidum* subsp. durum; tetraploid, AABB), einkorn (*Triticum monococcum*; diploid, AA), emmer (*Triticum turgidum* v. *dicocccum*; tetraploid, AABB) and spelt (*Triticum spelta*; hexaploid, AABBDD). Common wheat is most widely cultivated and used for human food, followed by durum, and small quantities of spelt (reviewed in Vasil, 2007). Domestication of einkorn and emmer is suggested to have occurred around the same time (9,600-9,000 BP), while that of bread wheat, more than six thousand years ago (reviewed in Feuillet *et al.*, 2008). Einkorn was likely domesticated from *T. boeoticum* and amplified fragment length polymorphism (AFLP) fingerprinting suggests the domestication occurred in south-east Turkey (reviewed in Feuillet *et al.*, 2008). It is cultivated in isolated regions in France, India, Italy, Turkey and Yugoslavia (Harlan, 1981) and used in dishes such as soups, salads, casseroles or sauces, and in animal feeds. Domesticated emmer likely resulted from wild emmer through mutations for a 'non-brittle rachis' (reviewed in Feldman and Kislev, 2007). The domestication of emmer is considered the starting point for evolution of durum and bread wheat, however, the site of its domestication remains elusive (Luo *et al.*, 2007). Emmer is cultivated in Ethiopia, India and Italy (Harlan, 1981) and used to make bread or pasta or eaten as whole grains. Durum is likely derived from selection of domesticated emmer, developing a naked free-threshing form, the place of domestication of durum being either Abyssinia or southern Mediterranean basin. It is mostly grown in the Middle East, North Africa, Russia, the North American Great Plains, India and Mediterranean Europe (Cantrell, 1987) and used to produce pasta and non-pasta foods.

Bread wheat, an allohexaploid (AABBDD; $2n = 6x = 42$), is most likely a product of two separate hybridization and chromosome duplication events, the first between *Triticum urartu* (AA; $2n = 14$) and a relative of *Aegilops speltoides* (SS; $2n = 14$), where chromosomal doubling of the sterile hybrid resulted in the formation of *Triticum turgidum* (durum wheat; $2n = 4x = 28$) (Kimber and Sears, 1987), and the second between durum and a D genome progenitor (DD, *Aegilops tauschii*; $2n = 14$), followed by chromosome doubling of the 3n hybrid to result in the fertile allohexaploid common wheat (Kimber and Sears, 1987). The complex genome of wheat makes breeding for specific traits difficult, and thus genetic modification may be a better option.

2. Abiotic Stresses and Effects on Wheat

2.1. Salinity stress

About 800 million ha (6% of world's total land area) is affected by salt (FAO, 2007) and increased salinity may result in 50% loss of arable land by 2050 (Wang *et al.*, 2003). Apart from natural salinity, some land has become saline due to clearing of native deep-rooted vegetation and irrigation (reviewed in Mahajan and Tuteja, 2005).

In Australia alone, about 2 million ha is affected by salinity and another 15 million ha is at risk in the next 50 years (National Land and Water Resources Audit 2001). It has resulted from mainly land clearing and little capacity to drain the continent of salt and water, and Western Australia, where the 'wheat belt' stretches, is at greatest risk. Wheat is a moderately salt-tolerant crop because of its low rate of Na^+ accumulation and improved K^+/Na^+ discrimination (reviewed in Munns and Tester, 2008). A survey by the US Salinity Laboratory (USDA-ARS, 2005) shows that wheat yields start declining at 60-80 mM NaCl, with decreases of up to 50% (Bray *et al.*, 2000). The section below describes the effect of high salinity in plants.

Studies on wheat and barley showed that the response of plant growth to salinity has two phases (reviewed in Munns, 2005). In the first stage, termed the 'osmotic or water-deficit effect', the decrease in availability of water to roots results in reduced leaf and root growth and a decline in stomatal conductance and thus photosynthesis. The response of plants to this stage has not been clearly elucidated but is suggested to be associated with water stress and hormonal signals that regulate growth reduction.

The second stage, the 'salt-specific or ion-excess effect' starts when the vacuoles become 'full', resulting in accumulation of salt to toxic levels in the cytoplasm of cells, leading to enzyme inhibition (Munns and Termaat, 1986) and leaves dying of dehydration (Munns *et al.*, 1995). Salt accumulation in leaves also results in reduced nutrient supply to the growing parts of the plant (Munns *et al.*, 1995). It is suggested that in salt-sensitive cultivars the leaves might die early because the cells are not able to efficiently sequester the salt into vacuoles (Greenway and Munns, 1980); e.g., after exposure to 200 mM NaCl, sugarbeet (salt-tolerant) would lose 20% in dry weight, cotton (moderately tolerant) would lose 60% and soybean (salt sensitive) would be dead (Greenway and Munns, 1980). Salinity stress in plants results in water deficit (Munns and Termaat, 1986), ion toxicity due to the accumulation of Na^+ and Cl^- and a reduced nutrient supply (Hasegawa *et al.*, 2000).

In response, plants attempt to develop adaptive mechanisms such as vacuolar compartmentalisation of Na^+, increase in supply of energy by plasma membrane ATPases (Hasegawa *et al.*, 2000), or accumulation of compatible solutes/ osmoprotectants such as glycinebetaine, proline, sorbitol or sugars (e.g, trehalose) in the cytoplasm (reviewed in Martinez-Ballesta *et al.*, 2006).

Increased synthesis and accumulation of proteins has also been reported; the proteins providing storage of nitrogen for utilization when the stress is alleviated and/or also be involved in osmotic adjustment (Singh *et al.*, 1987).

2.2. Drought Stress

Drought is another major abiotic stress limiting crop productivity around the world (reviewed in Mahajan and Tuteja, 2005). Wheat grows well on well-drained soils, receiving 250 to 1750 m of annual rain. However, most wheat growing areas only receive 375 to 875 mm annual precipitation on average (Briggle and Curtis, 1987) and world wheat production is regularly affected by up to 50% by various droughts (Pfeiffer *et al.*, 2005).

Drought stress results in the loss of leaf water and closure of the stomatal opening, causing a decline in CO_2 availability and photosynthetic carbon metabolism (Langridge *et al.*, 2006). These physiological changes elicit responses such as accumulation of metabolites (e.g abscisic acid), osmolytes (Ramachandra-Reddy *et al.*, 2004), or expression of late embryogenesis abundant (LEA) proteins and dehydrins (Bray *et al.*, 2000), lipid transfer proteins (LTPs) (Trevino and O'Connell, 1998) or aquaporins (discussed below).

2.3. Nutrient Stress

It is reported that about 60% of cultivated land worldwide has growth-limiting factors that cause nutrient stress (deficiency or toxicity) (Cakmak, 2002) and a decline in food production. The three main nutrients required for successful production of wheat are nitrogen, phosphorus and potassium, sulphur and zinc also being important but to a lesser extent. Nitrogen is vital as it has a role in utilizing absorbed light and photosynthetic carbon metabolism (Huang *et al.*, 2004), and a role during grain filling and increasing the protein content (Knowles *et al.*, 1991) resulting in good grain quality and vigor. Therefore, nitrogen deficiency results in chlorosis as a result of a break-down of the photosynthetic process (Snowball and Robson, 1991). Further, nitrogen deficiency in wheat also results in inhibition of cell growth and division and protein synthesis, thus reduced yields. To help increase the levels, nitrogen fertilizers are usually added to the field before sowing or prior to flowering to boost grain yield and level of protein in the grain (Laffan, 1999). Sources of nitrogen in plants are ammonia and urea, which are toxic at higher concentrations (Britto *et al.*, 2001; Wang *et al.*, 2008); therefore, there is a need for tight regulation of these into plants. Phosphorus is required for proper growth and reproduction of wheat (Laffan, 1999).

An estimated 5.7 billion hectares of land worldwide contains low levels of phosphorus, resulting in a decline in crop production (Batjes, 1997). For such soils, fertilizers are also applied to the field prior to sowing (Laffan, 1999); however, this in itself can result in pollution of water and the atmosphere, with nitrous oxide and ammonia resulting in the 'greenhouse effect' (Byrnes, 1990). Thus, it would be ideal to develop new methods for improved yield without adverse effects on the environment, e.g., by identifying and/or developing plants with the genetic ability to acclimatize to changes in nutrient supply.

2.4. Abiotic Stress Responses in Plants Including Wheat: A Summary

The response of plants to salinity and osmotic stresses is multigenic and involves multiple, often overlapping components. The stress-responsive proteins fall mainly into three groups: those that control salt uptake and transport, those with osmotic and protective functions, and those involved in regulation of gene expression and signal transduction (reviewed in Munns, 2005).

Examples of stress responsive genes include those encoding mitogen activated protein kinases (MAPK), salt overly sensitive (SOS) pathway proteins (Hasegawa *et al.*, 2000), calcium signalling (Kawasaki *et al.*, 2001) and abscisic acid signalling pathways, transcription factors (CBF/DREB and ABF/ABARE families) (Choi *et al.*, 2000), phospholipases (Frank *et al.*, 2000), heat shock proteins and chaperones, peroxidases such as ascorbate peroxidases (Yan *et al.*, 2005), osmoprotectants such as proline and glycine betaine (Ramachandra-Reddy *et al.*, 2004), late embryogenesis abundant (LEA) proteins and dehydrins (Bray *et al.*, 2000), lipid transfer proteins (LTPs) (Trevino and O'Connell, 1998) and ion transporters (Blumwald, 2000).

In the last 10 years or so, membrane channel proteins called aquaporins have been noted to transport water as well as a number of other substrates of physiological importance, and respond to drought and salinity. This work focuses on aquaporins the key properties of aquaporins and their potential applications for development of salinity and drought tolerant crops and modification of uptake/transport of other nutrients.

3. Aquaporins: Channels for Water and Other Substrates

3.1. Discovery of Aquaporins

Water is the most important molecule in all living cells and an efficient regulation of its supply is essential for all biological processes. Until recently it was believed that the trans-membrane flow of water through the membrane occurred by osmosis alone. Philip (1958) determined that diffusion rates across plasma membranes in oat coleoptiles did not account for water flow rates, nor did they match the demand of water for physiological processes, raising doubts about the theory of trans-membrane flow of water by osmosis only. These findings were followed by discovery of the protein CHIP28 (later renamed AQP1), an aquaporin in human erythrocyte membranes; heterologous expression of CHIP28 in Xenopus oocytes revealed the water permeability of this protein (Preston and Agre, 1991), forever changing our view of simple diffusion as the only means of trans-membrane flow of water. Water permeability of cellular membranes is now confirmed to be highly dependent on presence of aquaporins, members of the Major Intrinsic Proteins (MIP) family. Aquaporins have been confirmed in all other organisms, e.g., archaea (Kozono *et al.*, 2003), bacteria (Heller *et al.,* 1980), plants (Maurel *et al.,* 1993; Kammerloher *et al.*, 1994). The various MIP family members mediate the flux of water as well as substrates such as boron, silicon, ammonia and carbon dioxide (reviewed in Maurel *et al.*, 2008), making them good candidates for genotype selection and genetic engineering of cereal crops. The next sections give an overview on contributions of MIPs in the biology of cereals.

3.2. Key Structural Features and Pore Selectivity of MIPs

X-ray crystallography determination of atomic structures of microbial, animal and plant homologs points to highly conserved features in the aquaporin family (Tornroth-Horsefield *et al.*, 2006). Most MIP family members are 250–300 amino acids long (Park and Saier, 1996) and have a molecular mass of 27-31 kDa. An aquaporin monomer typically consists of six transmembrane helices (TM1-TM6), connected by three extracellular loops (A, C, E) and two intracellular loops (B, D), the N- and C-termini both being exposed to cytosol. Loops B and E contain symmetrically opposed, conserved Asn-Pro-Ala (NPA) motifs that form two short helices (HB and HE; Figure 1) (Park and Saier, 1996) which fold inwards to form a seventh pseudo-transmembrane helix that interacts with the other helices to form a pore, resulting in an hour glass structure (Jung *et al.*, 1994).

The six transmembrane alpha helices (TM1–TM6) (grey cylinders), connected by five loops. HB and HE (short grey cylinders) represent the two short helices containing the NPA motifs, shown in white ovals that form the pore. The four residues making up the ar/R filter are shown in black ovals, the P1-P5 positions in grey ovals and the AEFXXT motif in the white rectangle. The white triangle represents the glycosylation site predicted in NIPs (Hove and Bhave, unpublished); the grey diamond represents the position corresponding to Leu197 (Hove and Bhave, unpublished), previously implicated in aquaporin gating in SoPIP2;1 (Tornroth-Horsefield et al., 2006); the white stars show the residues corresponding to Lys51 and Asp185, reported to contribute to tetramer stability of HsAQP1 (Buck et al., 2007); the grey ovals, inscribed with an 'S' represent the phosphorylation sites that correspond to Ser262 in GmNOD26, Ser 115 and Ser274 in SoPIP2;1.

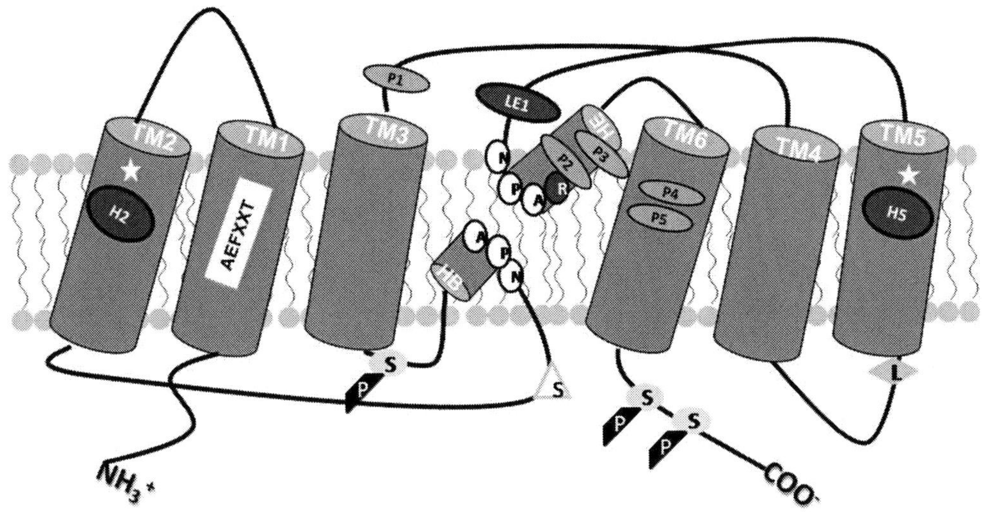

Figure 1. Secondary structure of aquaporins.

The substrate specificity of MIPs mainly occurs at two main constrictions, the NPA motifs and the Aromatic/Arg (Ar/R) selectivity filter. The NPA motif plays a role in proton exclusion (Fu *et al.*, 2000) and likely also as a size barrier, the pore being wider in aqua-glyceroporins (reviewed in Wallace and Roberts, 2004; Forrest and Bhave, 2007; Wu and Beitz, 2007) (Figure 2).

The NPA positions show a few variants, e.g., in some Arabidopsis and maize NIP and SIP isoforms (alanine replaced by leucine, valine, serine, threonine or cysteine) (Chaumont *et al.*, 2001; Johanson *et al.*, 2001). The ar/R selectivity filter is comprised of one residue each from TM2 and TM5 (residues TM2, TM5) and two from loop E (residues LE1, LE2) and likely serves as a size-exclusion barrier (Fu *et al.*, 2000; Beitz *et al.*, 2006; Wallace *et al.*, 2006).

In aquaporins or water-only transporters, the His and Arg provide a hydrophilic environment while for channels allowing transit of glycerol, other polyols as well as urea, the pore is more hydrophobic (Wu and Beitz, 2007). Many MIPs including human and plant also contain a conserved Ala–Glu–Phe (AEF or AEFXXT) motif (residues 16-18 in human AQP1; Preston and Agre, 1991) of unclear function, located within TM1 (Zardoya and Villalba 2001). Four other conserved positions (P2–P5; i.e., Thr116, Ser196, Ala200, Phe212, and Trp213 of AQP1) also seem to be involved in discriminating between aquaporins and glycerol transporters (Froger *et al.*, 1998), and the pairing of residues at positions equivalent of Phe24 and Leu149 of AQP1 is also characteristic of aquaporins (Heymann and Engel, 2000).

Aquaporins usually exist as tetramers in which each monomer forms an independent channel. The monomers cluster as homo- (Fu *et al.*, 2000) or hetero-tetramers (Fetter *et al.*, 2004), and the tetramer assembly may be important for protein folding, stability and/or transport of the MIPs to target cellular membranes (Fetter *et al.*, 2004).

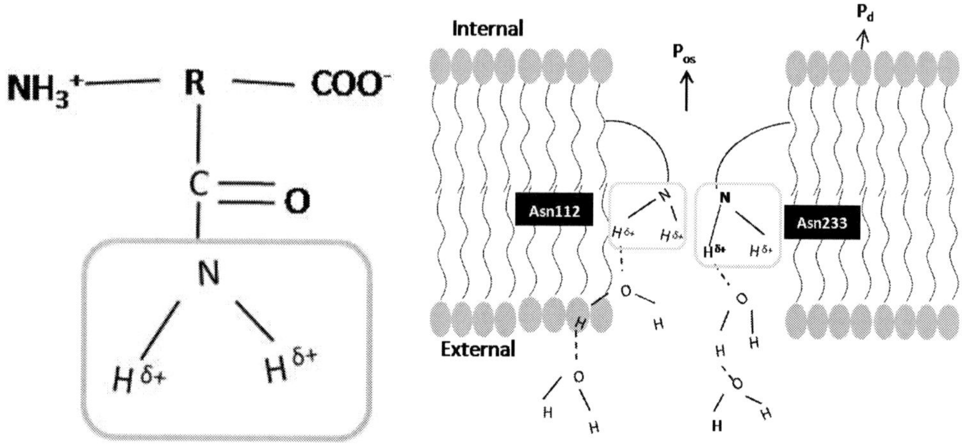

(A) (B)

Figure based on work of Sui *et al.* (2001) on human AQP1.

Figure 2. Interaction of water molecules and Asn residues of the NPA motifs during trans-membrane flow of water mediated by aquaporins. A) Structure of the Asn residue making up the NPA motif, with the molecules that interact with water shown in the grey rounded rectangle. **B)** The trans-membrane flow of water is usually via aquaporins (P_{os}, osmotic water permeability), or to a lesser extent via simple diffusion (P_d, diffusional water permeability). The two NPA (Asn-Pro-Ala) motifs and aromatic/Arg (ar/R) selectivity filter (not shown) serve as one of the means for pore selectivity of aquaporins, by Hydrogen bond formation between water molecules and the amide group of the Asn residues.

4. Diversity and Composition of MIP Families in Plants

In plants, the primary structure, transport substrate, functional regulation, gene expression profile, protein amount and intracellular localization are diversified (Katsuhara *et al.*, 2008; Maurel *et al.*, 2008). Arabidopsis γ-TIP (TIP1;1), a MIP member located in the tonoplast was the first plant MIP that showed water transport activity when assayed by expression in *Xenopus* oocytes (Johnson *et al.*, 1990; Maurel *et al.*, 1993). Since then, many MIP isoforms have been identified in various species, e.g., *Arabidopsis thaliana* (Johanson *et al.*, 2001), ice plant (Yamada *et al.*, 1995), sunflower (Sarda *et al.*, 1997), wheat (Forrest and Bhave, 2008), maize (Barrieu *et al.*, 1998; Chaumont *et al.*, 1998), barley (Katsuhara *et al.*, 2002), radish (Suga *et al.*, 2002) and rice (Lian *et al.*, 2004; Sakurai *et al.*, 2005). Plant MIPs show a high multiplicity, with 35 isoforms in Arabidopsis (Johanson *et al.*, 2001), 28 in *Vitis* (grapevine) (Fouquet *et al.*, 2008), 33-38 in rice (Sakurai *et al.*, 2005; Forrest and Bhave, 2007) and 36 in maize (Chaumont *et al.*, 2001) (Table 1). The greater diversity in plants has been attributed to more compartments in plant cells and a greater need for fine-tuning of water control to be able to adapt to changing environmental conditions, due to their sessile nature (Johanson *et al.*, 2001). Plant MIPs are divided into seven subfamilies, based on their common subcellular location (if known) and sequence similarities: plasma membrane intrinsic proteins (PIPs), tonoplast intrinsic proteins (TIPs), NOD26-like intrinsic proteins (NIPs), small, basic intrinsic proteins (Maurel *et al.*, 2008) (SIPs) and three novel subfamilies, i.e., GlpF-like intrinsic protein (GIP1;1), hybrid intrinsic protein (HIP) and the uncategorized X intrinsic protein

(XIP) (Danielson and Johanson, 2008). PIPs and TIPs have been studied extensively compared to NIPs and SIPs (discussed below), while the novel subfamilies are not yet reported in cereals.

5. MIP Sub-Families in Wheat and Other Cereals

Wheat, maize, rice and barley belong to the Poaceae family of grasses and are utilised worldwide as a significant part of human diet. However, worldwide production of cereals is threatened by stresses such as salinity, drought and nutrient deficiency (discussed earlier). Therefore, it is important to analyse the potential of genes that respond to stresses, and these include aquaporins. The MIPs in cereals comprise multigene families (Table 1). Thirty three MIP genes were identified from maize (Chaumont *et al.*, 2001) and 38 from common wheat (Forrest and Bhave, 2007). There is scant literature on barley MIPs; however, some PIP and TIP cDNAs are known (Hollenbach and Dietz, 1995; Schunmann and Ougham, 1996; Katsuhara *et al.*, 2002; Forrest and Bhave, 2007). Until recently, MIP genes from common wheat (*T. aestivum*) had not been characterised, possibly due to lack of genomic sequence data and its allohexaploid nature which makes gene analysis difficult. Jahn *et al.* (2004) demonstrated that wheat TIP2 could transport ammonia, Forrest and Bhave (2008) reported 24 PIP and 11 TIP genes, and Zhang *et al.* (2008) isolated six MIP genes from a drought-tolerant Chinese cultivar, comprised of PIP2s (TaAQP2-4), a PIP1 (TaAQP1), a TIP1 (TaAQP6) and a TIP2 (TaAQP5). TaAQP6 was shown to be involved in water transport and up-regulated by salt but not ABA, suggesting a role in an ABA-independent salt tolerance (Zhang *et al.*, 2008) and supporting a potential for MIPs in development of wheat lines resistant to abiotic stresses. The various subfamilies of MIPs and their roles are summarised briefly below.

Table 1. MIP family members in cereals

Cereal	PIPs	TIPs	NIPs	SIPs	Total	Reference
Wheat (*Triticum aestivum L.*)	24	11	-	-	35	Forrest and Bhave, 2008
	4	2	-	-	6	Zhang *et al.*, 2008
Barley (*Hordeum vulgare L.*)	5	3	-	-	7	Forrest and Bhave, 2007
	-	-	1	-	1	Chiba *et al.*, 2009
Maize (*Zea mays L.*)	13	12	5	3	33	Chaumont *et al.*, 2001
Rice (*Oryza sativa L.*)	11	10	10	2	33	Sakurai *et al.*, 2005
	13	10	13	2	38	Forrest and Bhave, 2007

5.1. The Plasma Membrane Intrinsic Proteins (PIPs)

The PIPs represent the subfamily with the largest number of members, 13 in maize (Chaumont *et al.*, 2001), 11 in rice (Sakurai *et al.*, 2005) and 24 in wheat (Forrest and Bhave, 2008). As the name indicates, a majority of PIPs are localized in the plasma membrane, and are mainly localized in organs and tissues distinguished by huge fluxes of water such as vascular tissues, flowers and guard cells. In leaf blades, PIPs have been shown to accumulate in mesophyll cells in rice (Sakurai *et al.*, 2008) and localization in guard cells was noted in spinach SoPIP1;1 (Fraysse *et al.*, 2005). In contrast, spinach SoPIP1;2 accumulated in phloem sieve elements of leaves, roots and petioles (Fraysse *et al.*, 2005), ice plant (*Mesembryanthemum crystallinum*) PIP mainly in vascular tissues (Yamada *et al.*, 1995), and AtPIP2;2 mainly in the roots, cortex, endodermis and stele (Javot *et al.*, 2003). The localization of PIPs in different organs or tissues could have important implications in water balance during different developmental processes.

The PIPs are further divided into PIP1 and PIP2 subgroups (Kammerloher *et al.*, 1994), based on the length of amino acid N- and C-terminus sequence and water-channel activity. The PIP1 members have a longer cytosol-protruding N-terminal region and a shorter C-terminal compared to PIP2 members (Chaumont *et al.*, 2001; Johanson *et al.*, 2001), and PIP2s have an additional 4-10 amino acids in the first extracytosolic loop. Although the amino acid sequences are similar, there are considerable differences in their water transport activity. PIP1s usually have little or no water channel activity *in vitro*, while PIP2s exhibit high water-channel activity, as confirmed for PIP1s and PIP2s from maize (Gaspar *et al.*, 2003; Fetter *et al.*, 2004), barley (Katsuhara *et al.*, 2002) and rice (Sakurai *et al.*, 2005; Sakurai *et al.*, 2008). Numerous reports show cereal PIPs play an important role in controlling water transport; e.g. over-expression of the barley PIP (HvPIP2;1) in transgenic rice demonstrated an increase in the radial hydraulic conductivity of roots (Katsuhara *et al.*, 2003), and several PIP2s, e.g., ZmPIP2;5 (Fetter *et al.*, 2004), HvPIP2;1 (Katsuhara *et al.*, 2002), ZmPIP2;1, ZmPIP2;4 (Fetter *et al.*, 2004), OsPIP2;4, OsPIP2;5 (Sakurai *et al.*, 2005, 2008), OsPIP2;1, OsPIP2;2 and OsPIP2;3 (Sakurai *et al.*, 2008) function as water channels. Although wheat PIP2s have not been functionally characterized, their comparison with other aquaporins known to transport water points to their potential (Forrest and Bhave, 2008). Although most PIP1s show little water transport, it can be enhanced through site-directed mutagenesis, and specific residues (Ile^{244} of RsPIP1s, Val^{235} of RsPIP2) may be responsible for the distinction between type 1 and type 2 of radish PIPs (Suga and Maeshima, 2004). Co-expression of PIP1 and PIP2 also increases water permeability of frog oocytes (the assay system), supporting the idea of a cooperative effect (Moshelion *et al.*, 2002; Fetter *et al.*, 2004).

5.2. Transport of Other Substrates by PIPs

A number of other substrates have also been suggested for PIP1s in plants including cereals (Table 2). These include glycerol (NtAQP1, Biela *et al.*, 1999), boric acid (ZmPIP1, Dordas *et al.*, 2000), urea (ZmPIP1;5, Gaspar *et al.*, 2003) and carbon dioxide (NtAQP1, Flexas *et al.*, 2006). PIP2-type HvPIP2;1 (Hanba *et al.*, 2004) and AtPIP1;2 (Aharon *et al.*, 2003) have also been shown to have CO_2 transport activity. PIP2;1 and PIP2;4 from Arabidopsis have also been reported to transport H_2O_2 (Dynowski *et al.*, 2008) (Table 2). Hanba *et al.* (2004) showed over-expression of barley aquaporin HvPIP2;1 in rice caused an increase in internal CO_2 conductance in leaf blades. Due to CO_2 transport in addition to water, Terashima *et al.*, (2006) suggested the term 'CO_2 aquaporin' or COOporin. Uehlein *et al.* (2008) suggested that NtAQP1 tends to change from a water-channel to transporting CO_2

depending on its cellular location or the intrinsic CO_2 permeability of the membrane, and showed that the plasma membrane is about 5 times more permeable to CO_2 than the chloroplast envelope. Thus, the chloroplast envelope could potentially meet the need for CO_2 permeation through aquaporins, as suggested by Hub and de Groot (2006), showing a need for more research on aquaporins in other plant organelles.

Table 2. Transport of other substrates by plant MIPs

Substrate	Aquaporin	Plant	Reference
PIPs			
Boron	ZmPIP1*	Maize	Dordas *et al.*, 2000
	HvPIP1;3*	Barley	Fitzpatrick and Reid, 2009
	HvPIP1;4*	Barley	Fitzpatrick and Reid, 2009
CO_2	NtAQP1	Tobacco	Flexas *et al.*, 2006
	AtPIP1;2	Arabidopsis	Aharon *et al.*, 2003
	HvPIP2;1*	Barley	Hanba *et al.*, 2004
Glycerol	NtAQP1	Tobacco	Biela *et al.*, 1999
	SsAQP1	Rain Tree	Moshelion *et al.*, 2002
H_2O_2	AtPIP2;1	Arabidopsis	Dynowski *et al.*, 2008
	AtPIP2;4	Arabidopsis	Dynowski *et al.*, 2008
Urea	NtAQP1	Tobacco	Eckert *et al.*, 1999
	ZmPIP1;5*	Maize	Gaspar *et al.*, 2003
TIPs			
Ammonia	TaTIP2;1*	Wheat	Jahn *et al.*, 2004
	AtTIP2;1 & AtTIP2;3	Arabidopsis	Loque *et al.*, 2005
	TaTIP2;2*	Wheat	Bertl and Kaldenhoff, 2007
Glycerol	NtTIPa	Tobacco	Gerbeau *et al.*, 1999
H_2O_2	AtTIP1;1 & AtTIP1;2	Arabidopsis	Bienert *et al.*, 2007
	AtTIP2;3	Arabidopsis	Dynowski *et al.*, 2008
Urea	NtTIPa	Tobacco	Gerbeau *et al.*, 1999
	AtTIP1;1 & AtTIP2;1	Arabidopsis	Klebl *et al.*, 2003
	AtTIP1;1, AtTIP1;2, AtTIP2;1, AtTIP4;1	Arabidopsis	Liu *et al.*, 2003
	AtTIP1;3 & AtTIP5;1	Arabidopsis	Soto *et al.*, 2008
NIPs			
NIPI			
Ammonia	GmNOD26	Soybean	Niemietz and Tyerman, 2000
Arsenite	AtNIP1;1	Arabidopsis	Kamiya *et al.*, 2009
Glycerol	GmNOD26	Soybean	Dean *et al.*, 1999
	LIMP2	*L. japonicus***	Guenther and Roberts, 2000
	AtNIP1;1, AtNIP1;2	Arabidopsis	Weig and Jakob, 2000
	PtNIP1;1	Loblolly pine	Ciavatta *et al.*, 2001
H_2O_2	PsNIP1;1	Pea	Schuurmans *et al.*, 2003
Lactic Acid	AtNIP1;2	Arabidopsis	Dynowski *et al.*, 2008
	AtNIP2;1	Arabidopsis	Choi and Roberts, 2007

Table 2. (Continued).

NIPII			
Antimonite	AtNIP5;1, AtNIP6;1, AtNIP7;1	Arabidopsis	Bienert *et al.*, 2008
Arsenite	LjNIP5;1, LjNIP6;1	*L. japonicus***	Bienert *et al.*, 2008
	AtNIP5;1, AtNIP6;1	Arabidopsis	Bienert *et al.*, 2008
	AtNIP7;1	Arabidopsis	Isayenkov and Maathuis, 2008
Boron	LjNIP5;1, LjNIP6;1	*L. japonicus***	Bienert *et al.*, 2008
	AtNIP5;1	Arabidopsis	Takano *et al.*, 2006
	AtNIP6;1	Arabidopsis	Tanaka *et al.*, 2008
Formamide	AtNIP6;1	Arabidopsis	Wallace and Roberts, 2005
Glycerol	AtNIP6;1	Arabidopsis	Wallace and Roberts, 2005
Urea	AtNIP6;1	Arabidopsis	Wallace and Roberts, 2005
NIPIII			
Antimonite	OsNIP2;1*	Rice	Bienert *et al.*, 2008
Arsenite	OsNIP2;1*	Rice	Bienert *et al.*, 2008
Boron	OsNIP2;1*	Rice	Mitani *et al.*, 2008
Silicon	OsNIP2;1*	Rice	Ma *et al.*, 2006; Mitani *et al.*, 2008
	OsNIP2;2*	Rice	
	HvNIP2;1*	Barley	Mitani *et al.*, 2008
Urea	ZmNIP2;1*, ZmNIP2;2*	Maize	Chiba *et al.*, 2009
	CpNIP1	Zucchini	Mitani *et al.*, 2009
	OsNIP2;1*	Rice	Klebl *et al.*, 2003 Mitani *et al.*, 2008

*Indicates cereal aquaporins. ***Lotus japonicus*.

5.3. Tonoplast Intrinsic Proteins (TIPs)

According to the classification by Heymann and Engel (2000), TIPs are structurally more closely related to the water-specific aquaporins than to aquaglyceroporins. TIPs are classified into TIP1 (previously γ-TIPs) (Maurel *et al.*, 1993; Chaumont *et al.*, 1998), TIP2 (e.g., δ-TIP; Daniels *et al.*, 1996), TIP3 (e.g., α-TIP; Johnson *et al.*, 1990), TIP4 (e.g., NtTIPa, Gerbeau *et al.*, 1999) and TIP5 (e.g. OsTIP5;1, Sakurai *et al.*, 2005). Members of TIP1, TIP2, and TIP4 groups have been localized to membranes of central vacuoles including pigment-containing vacuoles, and TIP3s to the protein body (vacuolar-derivative organelle) (Maeshima, 2001). Certain TIPs appear associated with developmental stages and organ specificity. For example, the expression of AtTIP2;1 is notable in developing vascular tissues of shoots (Daniels *et al.*, 1996) while AtTIP3;1 is mainly expressed in developing embryos and localized to the tonoplast of protein storage vacuoles of the embryo and endosperm (Hofte *et al.*, 1992). ZmTIP1 was highly expressed in the epidermal and conducting tissues (Barrieu *et al.*, 1998) and in zones of cell elongation and enlargement (Chaumont *et al.*, 1998). Some TIPs are differentially expressed during fruit maturation; e.g. pear TIP1;1 expression is highest in the young fruit (Shiratake *et al.*, 1997) while TIP protein levels in grape increase with ripening (Shiratake *et al.*, 2001). Recently, *Rosa hybrida* (rose) RhTIP1;1 was found mainly expressed in petals and maintained at high level during flower opening and decreased when the flowers were fully opened (Xue *et al.*, 2009).

Several isoforms of TIPs, in maize (ZmTIP1;1, Chaumont *et al.*, 1998; ZmTIP2;3, Lopez *et al.*, 2004), rice (OsTIP2;2, Sakurai *et al.*, 2008) and wheat (TaAQP6, Zhang *et al.*, 2008) have been shown to have water channel activity and although most of the wheat TIPs have not been functionally sequenced, they are potential water transporters (Forrest and Bhave, 2008). Sakurai *et al.*, (2008) reported that OsTIP2;2 had high water transport activity, in agreement with maize ZmTIP2;3 (Lopez *et al.*, 2004). However, OsTIP1;1 and OsTIP1;2 had low water channel activity (Sakurai *et al.*, 2008), and in contrast Arabidopsis AtTIP1;1 (Maurel *et al.*, 1993) and radish RsTIP1;1 (Suga and Maeshima, 2004) demonstrated high water activities. Thus, there is a need to investigate the water channel activities of TIP1s from plants, possibly using different expression systems. TIPs are suggested to be involved in the control of water exchange between cytosolic and vacuolar compartments and thus regulate cell turgor. This was supported by Sarda *et al.*, (1997), who demonstrated up-regulation of SunTIP7 in sunflower leaf guard cells during stomatal closure. The significance of TIPs to plant survival was demonstrated by the experiments of Ma *et al.*, (2004), where RNA(i)-targeting of TIP1;1 in Arabidopsis resulted in plant death, cell death, early senescence, lesion formation and difficulties in the distribution of metabolites. However, the relative water content of the plant was not notably affected. Thus it was suggested that AtTIP1;1 had a function in vesicle-based metabolite routing between the Golgi complex, pre-vacuolar compartments and tonoplast (Ma *et al.*, 2004). However, Schüssler *et al.* (2008), using an Arabidopsis transposon insertion line that completely lacked TIP1;1 protein showed that the absence did not have a noteworthy effect on metabolism. Therefore, it would be important to carry out further tests to confirm the functional role of TIP1;1 in plants.

5.4. Transport of other Substrates by TIPs

The substrates for TIPs apart from water are urea, glycerol and ammonia (Table 2). TIPs from tobacco (NtTIPa-TIP4) (Gerbeau *et al.*, 1999) and Arabidopsis (Klebl *et al.*, 2003; Liu *et al.*, 2003; Soto *et al.*, 2008) are reported to be permeable to urea, and NtTIPa is also permeable to glycerol (Gerbeau *et al.* 1999). In addition to water channel activity, AtTIP2;1 and AtTIP2;3 (Loque *et al.*, 2005) and TaTIP2 of wheat (Jahn *et al.*, 2004) have been confirmed to transport ammonia/ammonium and their analogs methylammonium or formamide. Liu *et al.* (2003) also showed AtTIP2;1 expression in Arabidopsis was up-regulated in response to nitrogen starvation. Expression of TIPs in Xenopus oocytes showed increased permeability for NH_4^+/NH_3 at rising pH, and TIPs are suggested to be involved in detoxification of the cytoplasm by an acid-trap mechanism (Loque *et al.*, 2005). Further, some Arabidopsis TIPs appear to transport H_2O_2 (Bienert *et al.*, 2007; Dynowski *et al.*, 2008). TIP1;1 is suggested to lower the level of H_2O_2 in cells surrounding developing vascular tissue wherein high levels of reactive oxygen species (ROS) including H_2O_2 are detected (Bienert *et al.*, 2007); this is of physiological significance.

5.5. Nodulin-26-like Intrinsic Proteins (NIPs)

The NIP subfamily represents a group of MIPs that are named after their sequence similarity to soybean (*Glycine max*) nodulin-26 (GmNOD26), an aquaporin expressed in the peribacteroid membrane of the N_2-fixing symbiotic root nodule bacteria. It has now become evident that NIPs are also present in non-leguminous plants, e.g., *Arabidopsis* (Johanson *et al.*, 2001), maize (Chaumont *et al.*, 2001) and rice (Sakurai *et al.*, 2005) (Table 2), indicating that their function is not limited to a role in nodule symbiosis. NIPs have also been found in non-vascular plants such as the moss *P. patens* (Danielson and Johanson, 2008) suggesting that plants probably developed NIPs early in evolution, and it has been proposed that the origin of NIPs in plants occurred by a "vertical gene transfer" of a microbial glyceroporin (Zardoya *et al.*, 2002).

NIPs constitute a unique group of MIPs that are permeable to small uncharged molecules such as glycerol (Dean *et al.*, 1999; Weig and Jakob, 2000; Ciavatta *et al.*, 2001), formamide (Wallace and Roberts, 2005), ammonia (Niemietz and Tyerman, 2000), lactic acid (Choi and Roberts, 2007) and urea (Klebl *et al.*, 2003), in addition to water (Dean *et al.*, 1999). Recently, NIPs have also demonstrated a fundamental role in the transport of metabolically important, as well as toxic metalloids such as boron (Takano *et al.*, 2006), silicon (Ma *et al.*, 2006) and arsenite (Isayenkov and Maathuis, 2008; Kamiya *et al.*, 2009) (Table 2). Compared to the generally more highly expressed TIP and PIP genes, NIP genes except NIP6;1 are generally expressed at lower levels (Alexandersson *et al.*, 2005); thus it may be difficult to detect their cDNA or EST. They display a great diversity in their Ar/R filter, suggesting specific roles in transport of several compounds (Wallace and Roberts, 2004). NIPs are divided into three groups according to the residues at their ar/R filter (Rougé and Barre, 2008): subgroup I (NIP1;1, NIP1;2, NIP2;1, NIP3;1, NIP4;1 and NIP4;2), subgroup II (NIP5;1, NIP6;1 and NIP7;1) (Wallace and Roberts, 2004; Wallace *et al.*, 2006) and subgroup III (OsNIP2;1, OsNIP2;2, ZmNIP2;1, ZmNIP2;2, CaNOD26 from *Cicer arietinum* - chickpea) and CpNOD26) (Mitani *et al.*, 2008; 2009; Rougé and Barre, 2008).

5.5.1. NIPI Subgroup

The predicted pore structure of subgroup I is similar to that of GmNOD26, whose ar/R region is composed of WVAR (Wallace and Roberts, 2004) and facilitates the transport of both water and glycerol (Dean *et al.*, 1999). Many NIPIs also exhibit a CDK phosphorylation site at the C terminus (Weaver and Roberts, 1992). NIP1s have been reported to transport mainly water, glycerol and lactic acid. Glycerol (Dean *et al.*, 1999) and ammonia (Niemietz and Tyerman, 2000) transport activities have been reported in GmNOD26. Arabidopsis NIP1;1 and NIP1;2 (Weig and Jakob, 2000), PtNIP1;1 (*Pinus taeda*, loblolly pine) (Ciavatta *et al.*, 2001) and PsNIP1;1 (*Pisum sativum*, pea) (Schuurmans *et al.*, 2003) also have glycerol transport activities. LIMP2, a NIP from *L. japonicus* with high similarity to AtNIP1;2, also expresses water and glycerol transport activities (Guenther and Roberts, 2000). AtNIP1;1 functions in As(III) transport (Kamiya *et al.*, 2009). AtNIP7;1 also plays a role in transport of As(III) (Isayenkov and Maathuis, 2008), however, NIP1;1 is dominant over NIP7;1 in determination of As(III) tolerance in Arabidopsis, and NIP1;1 is localized to the plasma membrane in roots (Kamiya *et al.*, 2009). Functional analyses of AtNIP2;1 have shown that it selectively transports protonated lactic acid. Moreover, it is extremely sensitive to oxygen deprivation, showing a rapid and large increase in transcript levels in response to water logging and anoxia, but shows extremely low permeability to both water and glycerol (Choi and Roberts, 2007). AtNIP2;1 is expressed in roots and localizes on the ER membrane (Mizutani *et al.*, 2006). In contrast, AtNIP2;1 appeared to localize to cell surface, presumably the plasma membrane (Choi and Roberts (2007). Further analysis is required to address this discrepancy.

5.5.2. NIPII Subgroup

NIPII subgroup proteins form glyceroporins with a low water permeability (Wallace and Roberts, 2005) and also transport other uncharged substrates including urea (Wallace and Roberts, 2005) and metalloid nutrients such as boron (Takano *et al.*, 2006) and silicon (Ma *et al.*, 2006). There is a substitution of Ala at H_2 for the highly conserved Trp residue (Wallace and Roberts, 2005), thus, NIPII members have a predicted pore of wider size (≤ 5Å) than NIPI subgroup (3.5Å), and are permeable to larger solutes such as urea (Wallace *et al.*, 2006). The proteins of the NIPII subgroup do not exhibit the conserved CDK phosphorylation site but the N and C terminal regions of NIP5;1, NIP6;1 and NIP7;1 share consensus phosphorylation sequences for MAP kinases. Arabidopsis NIP5;1, is localized to the plasma membrane, facilitates the diffusion of $B(OH)_3$ through root endodermis cells and provides the substrate for BOR1, an active $B(OH)_4^-$ extrusion transporter localized in the proximal plasma membrane domains of pericycle cells (Takano *et al.*, 2006). NIP5;1 is also crucial for plant growth under boron limitation and has low water permeability, and may be involved in other physiological roles such as water transport and transport of other small uncharged molecules (Takano *et al.*, 2006). AtNIP5;1 and LjNIP5;1 (from *Lotus japonicus:* a wild legume of the family Fabaceae, studied as a model plant) have recently been shown to be As(III) and antimonite (Sb(III)) channels (Bienert *et al.*, 2008). Arabidopsis NIP6;1 has an extremely low water permeability but transports glycerol, formamide, and larger solutes such as urea (Wallace and Roberts, 2005). The size and volume of boric acid are similar to those of urea (Dordas *et al.*, 2000). Takano *et al.*, (2006) suggested that similar to NIP5;1, NIP6;1 could also be a boric acid channel. This was proven by Tanaka *et al.*, (2008) who also showed NIP6;1 was completely impermeable to water and proposed that, during evolution, NIP5;1 and NIP6;1 may have diversified both in the specificity of their expression in plant tissues and water permeation properties, while maintaining their ability to be induced under low boron and to transport boric acid. It is thus important to investigate the involvement of NIP7;1 in boron transport as it also belongs to the same subgroup. NIP6;1 from Arabidopsis and *Lotus japonicus* are also As(III) and Sb(III) channels (Bienert *et al.*, 2008). AtNIP7;1 forms part of an As(III) uptake pathway in Arabidopsis (Isayenkov and Maathuis, 2008) and is involved in transport of Sb(III) (Bienert *et al.*, 2008).

5.5.3. NIPIII Subgroup

Recently, Rougé and Barre (2008) proposed a new subgroup, NIPIII, based on the ar/R filter of Nod26-like MIPs from *Cicer arietinim* (chickpea, CaNod26) and zucchini (CpNod26), both of which exhibited a highly hydrophilic and wide filter (≥ 6Å). The ar/R region of NIPIII is composed of GSGR, forming a larger constriction compared with other NIP groups (≤ 5Å and 3.5Å) (Wallace and Roberts, 2005), thus possibly allowing molecules such as silicic acid with a larger diameter (4.38Å) to permeate (Mitani *et al.*, 2008). Rice OsNIP2;1 (Ma *et al.*, 2006; Mitani *et al.*, 2008) and OsNIP2;2 (Mitani *et al.*, 2008), maize ZmNIP2;1 and ZmNIP2;2 (Mitani *et al.*, 2009) and barley HvNIP2;1 (Chiba *et al.*, 2009) have been demonstrated to have silicon transport activity. OsNIP2;1 is also a bidirectional As(III) channel (Bienert *et al.*, 2008) and can also transport [Sb(III)] (Bienert *et al.*, 2008), boric acid and urea (Mitani *et al.*, 2008), arsenite (Bienert *et al.*, 2008) and antimonite (Bienert *et al.*, 2008). Klebl *et al.* (2003) showed that CpNIP1 is permeable to urea. The similarity of AtNIP7;1 with OsNIP2;1 suggests AtNIP7;1 could be involved in silicon transport (Ma *et al.*, 2006). There is thus a need to investigate the involvement of NIP7;1 equivalents from other plants in silicon transport.

Recently, Li *et al.* (2009a) have shown that OsNIP2;1 notably increased the uptake of monomethylarsonic acid [MMA(V)] but dimethylarsinic acid DMA(V) to a lesser extent, and the uptake of both was accompanied with a decline in pH of the medium, consistent with the pH effect on dissociation of protons from these molecules (Li *et al.*, 2009a). Monomethylarsonic acid [MMA(V): $CH_3AsO(OH)_2$] and dimethylarsinic acid [DMA(V): $(CH_3)_2AsO(OH)$] are pentavalent methylated As species used as herbicides and pesticides (US Environmental Protection Agency, 2006) and have been detected from paddy soils.

In a nutshell, due to the low expression levels of NIPs in plants (Alexandersson *et al.*, 2005) intensive research on NIPs is a recent development and there is a need for further research in order to elucidate their function in plants. In particular, there is a need to investigate the solutes transported by NIPs (including NIP3;1, NIP4;2 and NIP4;1 which are little investigated as yet). It would also be interesting to investigate if NIPs are capable of transporting other metalloids not yet recorded in literature such as germanium (Ge), which is structurally similar to silicon.

5.6. Small Basic Intrinsic Proteins (SIPs)

The fourth subfamily of MIPs comprises of the small basic intrinsic proteins (SIPs) which has a small number of members in Arabidopsis (Johanson *et al.*, 2001), maize (Chaumont *et al.*, 2001) and rice (Sakurai *et al.*, 2005) (Table 2).

The SIP family was identified by database mining and phylogenetic analysis (Johanson and Gustavsson, 2002). Due to their low homology (less than 20%) with other MIPs and deviation in the NPA motif, SIPs have recently been classified as superaquaporins (reviewed in Ishibashi, 2006). SIPs have short cytosolic N-terminal tails and a short loop C (14-19 residues) between TM3 and TM4 compared to PIPs and TIPs (22-26 residues) (reviewed in Ishibashi, 2006). Most aquaporins have two NPA motifs; however, the first motif of SIP1;1, SIP1;2 and SIP2;1 is changed to NPT, NPC and NPL, respectively (reviewed in Ishibashi, 2006). SIPs also have highly basic C-termini (KKQKKA for AtSIP1;1 and AtSIP1;2, PKAKSE for AtSIP2;1) (reviewed in Ishibashi, 2006).

The di-lysine motif (KKXX) being an ER localizing signal (Cosson and Letourneur, 1994) and all SIPs have been shown to be localised on the ER membrane by expression of green-flourescent-protein (GFP)-SIP fusion proteins. However, the functional role of these is unknown (Ishikawa *et al.*, 2005). The water channel activity of SIPs was noted for SIP1;1 and SIP1;2 but not SIP2;1 (Ishikawa *et al.*, 2005). The transport substrate for SIP2;1 remains to be determined, and in addition, alternative substrates for SIP1;1 and SIP1;2 should be investigated in different plant species.

6. Physiological Roles and Regulation of Plant Aquaporins

6.1. Roles of Plant Aquaporins in Drought and Salinity Tolerance

The transcriptional regulation of aquaporin genes by phytohormones such as ABA, GA_3 and brassinolide, developmental stages, or environmental stimuli such as low temperature (Jang *et al.*, 2004), drought, salinity (Lopez *et al.*, 2004), anoxia (Klok *et al.*, 2002) and nutrient deprivation (Seki *et al.*, 2002) have been shown. The regulation of aquaporins during drought and salinity has been shown in plant species such as rice (Lian *et al.*, 2004), maize (Lopez *et al.*, 2004; Zhu *et al.*, 2005), barley (Katsuhara *et al.*, 2002) and wheat (Zhang *et al.*, 2008; Gao *et al.*, 2010). Literature on expression of aquaporins under drought and salinity stress conditions in different plant species show different results, with up-regulation in maize (Zhu *et al.*, 2005), rice (Lian *et al.*, 2004) and wheat (Zhang *et al.*, 2008), down-regulation in rice (Li *et al.*, 2000) or barley (Katsuhara *et al.*, 2002) or no notable changes, e.g., wheat (Morillon and Lassalles, 2002) (reviewed in Forrest and Bhave, 2007). The expression of aquaporins under water stress appears to depend on the length and intensity of the stress (Galmés *et al.*, 2007). In addition, differences in drought tolerant and susceptible cultivars affect aquaporin expression (Lian *et al.*, 2004). OsPIP1;3 in drought tolerant rice was up-regulated under drought (Lian *et al.*, 2004), and OsPIP1;1 and OsPIP1;2 were up-regulated while OsPIP2;1 was down-regulated during osmotic stress (leaf; Guo *et al.*, 2006). In another study, OsPIP1;1 was found down-regulated under osmotic stress induced by using different chemicals (Li *et al.*, 2000). Under salt stress, the expression of barley HvPIP2;1 was inhibited in roots but enhanced in shoots (Katsuhara *et al.*, 2002), but over- expression of HvPIP2;1 in transgenic rice decreased the salt tolerance (Katsuhara *et al.*, 2003). In wheat, experiments on durum and bread wheat in response to drought stress showed no changes in PIP1 and PIP2 protein levels (Morillon and Lassalles, 2002).

Studies on response of TIPs and NIPs to abiotic stresses are limited. However, OsTIP1;1 transcripts are found to be up-regulated in response to water stress and salinity (Liu *et al.*, 1994), and osmotic stresses affect the phosphorylation and transport activities of NIPs (Guenther *et al.*, 2003). Recently, a putative wheat NIP identified from gene chip expression analysis of a salt-tolerant wheat was up-regulated in response to salt stress treatment, and further, its over-expression in transgenic Arabidopsis resulted in higher salt tolerance (Gao *et al.*, 2010). Interestingly, the content of Ca^{2+} (involved in regulation of aquaporins and important in salt tolerance; Sheen, 1996) was higher in transgenic Arabidopsis plants compared to wild-type, as was the expression of other stress-associated genes such as RD29B and COR15a (Gao *et al.*, 2010). Gao et al. (2010) also suggested that, since aquaporins are capable of transporting water across membranes at a fast and regulated rate, transgenic plants can absorb water faster and dilute the Na^+ ions during salt stress, and showed this putative wheat NIP to be of paramount importance in tolerance to salt and also other abiotic stresses.

Given this background, it is evident that aquaporins are of significance in responding to abiotic stresses. However, while many studies have been done on PIPs, there is scant literature on TIPs, NIPs and particularly SIPs. Therefore, considering the potential summarised above for MIPs as candidates for drought and salinity tolerance, there is a need to intensify the research in this area. The studies also illustrate that different aquaporins have characteristic responses to stresses and possibly tissue-specific expression patterns; thus the nature of responses cannot be generalised.

6.2. Aquaporin Expression May be Regulated by Phytohormones

In plants abscisic acid (ABA) has a major role in mediating the adaptation of plants to stresses such as drought and salinity (Shinozaki *et al.*, 2003). The responses of aquaporins involve both ABA-dependent and ABA-independent signalling pathways (Shao *et al.*, 2008). In rice, ABA treatment increased the expression of OsTIP1;1 in the roots (Liu *et al.*, 1994). Similarly, in maize, ABA was also found to increase the expression of ZmPIP2;4 and ZmPIP1;2 in the roots (Zhu *et al.*, 2005). In wheat, ABA treatment did not change the expression levels of TaAQP6 in the shoots (Zhang *et al.*, 2008), while it increased the expression of a TaNIP in roots and leaves (Gao *et al.*, 2010). In addition, in Arabidopsis, ABA treatment was found to increase the expression of AtPIP2;2 and AtPIP2;4 in the aerial parts and expression of AtPIP2;3, AtPIP2;6 and AtPIP2;7 in roots, while it decreased the expression of AtPIP2;6 in the aerial parts (Jang *et al.*, 2004). Gibberellic acid (GA_3) is also involved in the regulation of plant aquaporin activity and has been shown to activate the promoter of PIP1b (Alleva *et al.*, 2006). GA_3 has been shown to suppress the levels of mRNAs of RsPIP2;1, RsPIP2;2 and RsPIP2;3 and the protein levels of RsPIP2;1 in the roots of radish. However, the protein levels of RsPIP1s and RsTIP were barely changed by GA_3 (Suga *et al.*, 2002). Despite having no effect on the mRNA and protein levels of aquaporins in the roots and shoots of radish, brassinolide has been demonstrated to change the water-transport activity of cell membranes (Fu *et al.*, 2000). Experimental evidence has also shown that ethylene significantly increases stomatal conductance, root hydraulic conductivity and root oxygen uptake in hypoxic seedlings (Kamaluddin and Zwiazek, 2002). Aquaporins were thus suggested to be involved in the ethylene-enhanced root water uptake in hypoxic seedlings and through the effect of ethylene on protein phosphorylation, and ethylene is likely involved in the regulation of aquaporins (Kamaluddin and Zwiazek, 2002). Further evidence was through the experiments of Li *et al.*, (2009b) which showed that ethylene reduced the water content of rose petals. Ma *et al.*, (2008) also showed an inhibition of RhPIP2;1 after ethylene application. It thus appears that ethylene and ABA, the major stress signalling pathways, have an influence on aquaporin activity or expression.

6.3. Roles of Plant Aquaporins in Transport of Other Substrates

Ammonia (NH_3) is a weak base and also exists as ammonium ion (NH_4^+). Both are substrates for amino acid biosynthesis and thus important for all living organisms; however, high levels of NH_4^+ are toxic to plants (Britto *et al.*, 2001), making it essential to tightly regulate the transport of ammonia and ammonium across cellular membranes. As mentioned earlier, nutrient stress is a major limiting factor for growth of wheat and other cereals. Therefore, permeability of TaTIP2;1 (Jahn *et al.*, 2004), TaTIP2;2 (Bertl and Kaldenhoff, 2007) and other aquaporins is significant in that it may enable use of aquaporins in boosting plant health and yield.

Boron is widely available in the earth's crust, generally as $[B(OH)_3]$, and is an essential element for eukaryotes and prokaryotes (reviewed in Tanaka and Fujiwara, 2008). In plants, boron is vital for the structure of cell walls where it cross-links rhamnogalacturonan molecules through borate-diol bonding (reviewed in Tanaka and Fujiwara, 2008). Other functions of boron in membrane structure have also been suggested (reviewed in Tanaka and Fujiwara, 2008). Boron deficiency results in stunted growth, inhibition of root elongation and reduction in leaf expansion (reviewed in Tanaka and Fujiwara, 2008). On the other hand, boron readily forms complexes with compounds containing cis-hydroxyl groups, leading to toxicity at higher concentrations (Bolaños *et al.*, 2004), which results in problems in food production, especially in arid areas (Tanaka and Fujiwara, 2008). Since boron is both essential and toxic, its uptake must be tightly regulated (reviewed in Bienert *et al.*, 2007). The permeability of some NIPs to boron (see above) provides a way to screen or develop crops that are tolerant to boron toxicity or deficiency. CO_2 is most significant as the major reactant in photosynthesis, on which almost all life depends directly or indirectly. Resistance to CO_2 diffusion into chloroplasts limits photosynthesis, the limiting factors being stomata (stomatal CO_2 conductance, g_s) and internal conductance (mesophyll CO_2 conductance, g_i) (reviewed in Katsuhara *et al.*, 2008). However, PIPs have now been reported to be involved in CO_2 uptake (see earlier), and it is suggested that improvement of aquaporin activity may help increase CO_2 assimilation in plants (Katsuhara and Hanba, 2008).

Silicon is the second most abundant element in the earth's crust and is associated with bone and connective tissue development in animals (Jugdaohsingh, 2007). In plants, silicon, in the form of poly-silicic acid, functions in defense to abiotic and biotic stresses (reviewed in Ma and Yamaji, 2006). In addition, it also helps improve canopy photosynthesis by improving light interception (reviewed in Ma and Yamaji, 2006). Recent reports have shown that silicon transport is unique to NIPIII members (see earlier). This is largely due to their large pore size ($\geq$ 6Å) (Rougé and Barre, 2008), thus allowing molecules such as silicic acid (diameter 4.38Å) to permeate (Mitani *et al.*, 2008). Further, the importance of NIPs as silicon transporters was confirmed by the fact that suppression of OsNIP2;1 expression resulted in a reduction in silicon uptake and grain yield and increased susceptibility of the plant to pests and disease (Ma *et al.*, 2006). Recent discovery of NIPIII members that are permeable to silicon makes it feasible to genetically screen or manipulate silicon uptake capacity of relevant NIPs in plants that generally do not accumulate it in large amounts, such as dicots (Ma and Yamaji, 2006). Urea is an excretory product in animals but an important N-source for plant metabolism. Its potential toxicity at high levels (Wang *et al.*, 2008) means its transport must be highly regulated. The discovery of permeability of some MIPs to urea has potential for improving its use as a source of nitrogen for plants (Wang *et al.*, 2008).

Arsenite and antimony are widely available in the biosphere (reviewed in Filella *et al.*, 2002). Arsenite is highly toxic (group 1 human carcinogen; International Agency for Research of Cancer), and there is evidence of carcinogenicity of certain forms of antimony by inhalation (Gebel, 1997). Some countries such as Argentina, Australia, Bangladesh, Chile, China, Hungary, India, Mexico, Peru and Thailand have reported cases of contaminated drinking water resulting in adverse effects on human health (reviewed in Filella *et al.*, 2002). Eating of contaminated food is another source of arsenic intake in humans and the accumulation of arsenic in rice may expose humans and animals to serious health risks (reviewed in Ali *et al.*, 2009). Thus the transport of arsenite and antimony by cereal NIPs could provide a vital step towards the development of less toxic plants and/or phytoremediation of these metals. H_2O_2 is a member of the ROS, which are generated as by-products of metabolic processes such as photosynthesis and respiration (reviewed in Quan *et al.*, 2008). H_2O_2 has been reported to be involved in signalling responses to biotic and abiotic stresses in plants and to be a regulator in processes such as photorespiration and photosynthesis, growth and development, cell cycle, senescence and stomatal movement (reviewed in Quan *et al.*, 2008). However, ROS are potentially toxic and can damage proteins, lipids and nucleic acids; hence the amount/transport of H_2O_2 is expected to be tightly regulated. Several studies now show some aquaporins are permeable to H_2O_2 (see earlier) and could provide ways of fine-tuning the regulation of H_2O_2 transport/ metabolism.

6.4. Aquaporins and Developmental Processes in Plants

PIPs are also thought to be involved in many developmental processes including cell expansion, organ movement and elongation (Tyerman *et al.*, 2002). In Arabidopsis, eight of the 13 PIP genes have been detected in flower buds (Quigley *et al.*, 2001) and in tobacco, transcriptional activity of NtAQP1 was associated with cells in rapidly expanding regions (e.g., young petioles and corollas; Siefritz *et al.*, 2004). Aquaporins are also involved in maintaining plant water balance during development processes (reviewed in Kaldenhoff *et al.*, 2008) and in physiological processes such as stomatal closure and opening (Sarda *et al.*, 1997). Although the function and regulation of aquaporins during developmental processes is not quite clear as yet, it is of importance to further investigate such function during developmental processes.

7. Transgenic Wheat for Improved Stress Tolerance

World population is growing at an annual rate of at least 1.5% and is expected to reach approximately 10 billion by 2050, placing a burden on commercial farmers to supply adequate amounts of food (Vasil, 2007). Owing to the ever-expanding use of land for non-agricultural purposes, new crop land will not be available; hence increases in food will need improving productivity on land that is already under cultivation (reviewed in Vasil, 2007). Furthermore, stresses such as drought, salinity and flooding adversely affect plant growth and can result in loss of yield of wheat by more than 80% (Bray *et al.*, 2000), increasing the urgency to identify or develop cultivars that are resistant to drought or salinity and able to maintain resistance in subsequent generations. Such significant increases are unlikely to occur through traditional methods or even marker-assisted breeding, as these processes requires a lengthy screening process (Vasil, 2007), and while some success has been achieved, aspects such as yield have not improved even with extensive breeding (Vasil, 2007). This points to a need to use new methods such as development of genetically modified (GM) plants.

There has been significant progress since the first successful wheat transformation was achieved in 1992 (reviewed in Bhalla *et al.*, 2006). The *Agrobacterium* method appears more suited for wheat because of the short time required for regeneration and production of a higher proportion of plants carrying a single copy of the transgene and free of additional vector sequences (reviewed in Bhalla *et al.*, 2006). Many useful genes to improve bread-making and nutritional qualities, yield or resistance to biotic and abiotic stresses have now been introduced (Sahrawat *et al.*, 2003). In Australia, a few applications involving GM wheat lines have been approved for limited and controlled release (Office of the Gene Technology Regulator (OGTR), Australian Government), for testing for salt tolerance (DIR 053/2004) and drought tolerance (DIR 071/2006, DIR 077/2007). However, there has been little progress as yet in development of wheat cultivars that can provide high yields under drought and high temperatures in saline soils, and there is an urgent need to work on development of drought and salt tolerant wheat and other cereal lines, in order to improve food production.

Conclusions and Future Perspective

Wheat is a major food source for both animals and humans and has significant economic value. However, wheat is under threat of abiotic stresses such as salinity, water stress and nutrient stress. Further, the rapidly increasing world population stresses the need for adequate amounts of food. It is therefore, important to increase wheat yields in order to keep wheat both abundant and affordable. Traditional methods such as fertilisers and breeding to develop crops that are more resilient to the abiotic stresses are often very successful; however, they have limitations such as water and air pollution by fertilizers (Byrnes, 1990) and a limited gene pool for breeding (reviewed in Bhalla *et al.*, 2006). Therefore, genetic modification provides a good option for developing tolerant crops. For this purpose, some stress-responsive genes have been identified such as those that encode signal transduction, or directly protect membranes or are involved in water and ion uptake. The ability of MIPs to transport other substrates (Maurel *et al.*, 2008) makes them very important in fundamental processes in plants such as photosynthesis, nitrogen fixation, nutrient acquisition and detoxification. Aquaporins are excellent candidates for development or selection of plants tolerant of stresses such as drought or salinity, and studies on manipulation of genes for stress-inducible transcription factors have shown their potential in transgenic technology.

References

Aharon R, Shahak Y, Wininger S, Bendov R, Kapulnik Y, Galili G (2003) Over-expression of a plasma membrane aquaporin in transgenic tobacco improves plant vigor under favorable growth conditions but not under drought or salt stress. *Plant Cell*, 15: 439-47.

Alexandersson E, Fraysse L, Sjovall-Larsen S, Gustavsson S, Fellert M, Karlsson M, Johanson U, Kjellbom P (2005) Whole gene family expression and drought stress regulation of aquaporins. *Plant Mol Biol*. 59: 469-484.

Ali W, Isayenkov SV, Zhao F-J, Maathuis FJM (2009) Arsenite transport in plants. *Cell Mol Life Sci*. 66: 2329-2339.

Alleva K, Niemietz CM, Maurel C, Parisi M, Tyerman SD, Amodeo G (2006) Plasma membrane of *Beta vulgaris* storage root shows high water channel activity regulated by cytoplasmic pH and a dual range of calcium concentrations. *J Exp Bot*. 57: 609-621.

Barrieu F, Chaumont F, Chrispeels MJ (1998) High expression of the tonoplast aquaporin ZmTIP1 in epidermal and conducting tissues of maize. *Plant Physiol*. 117: 1153-1163.

Batjes NH (1997) A world data set of derived soil properties by FAO/UNESCO soil unit for global modelling. *Soil Use Manage.* 1: 9-16.

Bertl A, Kaldenhoff R (2007) Function of a separate NH_3-pore in Aquaporin TIP2;2 from wheat. *FEBS Lett.* 581: 5413-5417.

Bhalla PL, Ottenhof HH, Singh MB (2006) Wheat transformation - an update of recent progress. *Euphytica*, 149: 353-366.

Biela A, Grote K, Otto B, Hoth S, Hedrich R, Kaldenhoff R (1999) The *Nicotiana tabacum* plasma membrane aquaporin NtAQP1 is mercury-insensitive and permeable for glycerol. *Plant J.* 18: 565-570.

Bienert GP, Møller ALB, Kristiansen KA, Schulz A, Møller IM, Schjoerring JK, Jahn TP (2007) Specific aquaporins facilitate the diffusion of hydrogen peroxide across membranes. *J. Biol. Chem.* 282: 1183-1192.

Bienert GP, Thorsen M, Schüssler MD, Nilsson HR, Wagner A, Tamás MJ, Jahn TP, (2008) A subgroup of plant aquaporins facilitate the bi-directional diffusion of $As(OH)_3$ and $Sb(OH)_3$ across membranes. *BMC Biol* 6: 26-41

Blumwald E (2000) Sodium transport and salt tolerance in plants. *Curr. Opin Cell Biol.* 12: 431-434.

Bolanos L, Lukaszewski K, Bonilla I, Blevins D (2004) Why boron? *Plant Physiol Biochem* 42: 907-912.

Bray EA, Bailey-Serres J, Weretilnyk E (2000) Responses to abiotic stresses. In: Buchanan B, Gruissem W, Jones R (eds) Biochemistry and molecular biology of plants. American Society of Plant Physiologists, Rockville, pp 1158-1203.

Briggle LW, Curtis BC (1987) Wheat worlwide. *In* E.G. Heyne, ed. Wheat and wheat improvement, 2nd ed. Agronomy 13: 1-32.

Britto DT, Siddiqi MY, Glass ADM, Kronzucker HJ (2001) Futile transmembrane NH_4^+ cycling: A cellular hypothesis to explain ammonium toxicity in plants. *Proc Natl Acad Sci USA*, 98: 4255-4258.

Buck TM, Wagner J, Grund S, Skach WR (2007) A novel tripartite motif involved in aquaporin topogenesis, monomer folding and tetramerization. *Nature Struct Mol Biol.* 14:762-769.

Byrnes BH (1990) Environmental Effects of N Fertilizer Use—An Overview. *Fertilizer Research*, 26: 209-215.

Cakmak I (2002) Plant nutrition research: Priorities to meet human needs for food in sustainable ways. *Plant Soil*, 247: 3-24.

Cantrell RG (1987) Breeding and genetics of durum wheat. ln 'Plant Breeding Reviews'. Janick, J. (ed.). Van Nostrand Reinhol Co., New York, Vol. 5, pp. 11-40.

Chaumont F, Barrieu F, Herman EM, Chrispeels MJ (1998) Characterization of a maize tonoplast aquaporin expressed in zones of cell division and elongation. *Plant Physiol.* 117: 1143-1152.

Chaumont F, Barrieu F, Wojcik E, Chrispeels MJ, Jung R (2001) Aquaporins constitute a large and highly divergent protein family in maize. *Plant Physiol.* 125: 1206-1215.

Chiba Y, Mitani N, Yamaji N, Ma JF (2009) HvLsi1 is a silicon influx transporter in barley. *Plant J* 57: 810-818.

Choi WG, Roberts DM (2007) *Arabidopsis* NIP2;1: a major intrinsic protein transporter of lactic acid induced by anoxic stress. *J. Biol. Chem.* 282: 24209-24218.

Choi HI, Hong JH, Ha J, Kang JY, Kim SY (2000) ABFs, a family of ABA-responsive element binding factors. *J Biol Chem.* 275: 1723-1730.

Ciavatta VT, Morillon R, Pullman GS, Chrispeels MJ, Cairney J (2001) An aquaglyceroporin is abundantly expressed early in the development of the suspensor and the embryo proper of loblolly pine. *Plant Physiol.* 127: 1556-1567.

Cosson P, Letourneur F (1994) Coatomer interaction with di-lysine endoplasmic reticulum retention motifs. *Science*, 263: 1629-1631.

Daniels MJ, Chaumont F, Mirkov TE, Chrispeels MJ (1996) Characterization of a new vacuolar membrane aquaporin sensitive to mercury at a unique site. *Plant Cell*, 8: 587-599.

Danielson JA, Johanson U (2008) Unexpected complexity of the Aquaporin gene family in the moss *Physcomitrella patens*. *BMC Plant Biol* 8: 45.

Dean RM, Rivers RL, Zeidel ML, Roberts DM (1999) Purification and functional reconstitution of soybean nodulin 26. An aquaporin with water and glycerol transport properties. *Biochemistry*, 38: 347-353.

Dordas C, Chrispeels MJ, Brown PH (2000) Permeability and channel-mediated transport of boric acid across membrane vesicles isolated from squash roots. *Plant Physiol.* 124: 1349-1362.

Dynowski M, Schaaf G, Loque D, Moran O, Ludewig U (2008) Plant plasma membrane water channels conduct the signalling molecule H2O2. *Biochem J.* 414:53-61.

Eckert M, Biela A, Siefritz F, Kaldenhoff, R., (1999) New aspects of plant aquaporin regulation and specificity. *J Exp Bot.* 50: 1541-1545.

FAO (2007) http://faostat.fao.org/ (retrieved 20 August 2009).

Feldman M, Kislev ME (2007) Domestication of emmer wheat and evolution of free-threshing tetraploid wheat. *Isr J Plant Sci.* 55: 207-221.

Fetter K, Van Wilder V, Moshelion M, Chaumont F (2004) Interactions between plasma membrane aquaporins modulate their water channel activity. *Plant Cell*, 16: 215-228.

Feuillet C, Langridge P, Waugh R (2008) Cereal breeding takes a walk on the wild side. *Trends Genet.* 24: 24-32.

Filella M, Belzile N, Chen YW (2002) Antimony in the environment: a review focused on natural waters. *I Occurrence Earth Sci Rev.* 57: 125-176.

Fitzpatrick KL, Reid R (2009) The involvement of aquaglyceroporins in transport of boron in barley root. *Plant Cell Envi.* 32: 1357-1365.

Flexas J, Ribas-Carbó M, Hanson DT, Bota J, Otto B, Cifre J, McDowell N, Medrano H, Kaldenhoff R (2006) Tobacco aquaporin NtAQP1 is involved in mesophyll conductance to CO2 *in vivo*. *Plant J.* 48: 427-439.

Forrest KL, Bhave M (2007) Major intrinsic proteins (MIPs) in plants: a complex gene family with major impacts on plant phenotype. *Funct Integr Genom.* 7: 263-289.

Forrest KL, Bhave M (2008) The PIP and TIP aquaporins in wheat form a large and diverse family with unique gene structures and functionally important features. *Funct Integr Genom* 8: 115-133.

Fouquet R, Lĕon C, Ollat N, Barrieu F (2008) Identification of grapevine aquaporins and expression analysis in developing berries. *Plant Cell Rep.* 27: 1541-1550.

Frank W, Munnik T, Kerkmann K, Salamini F, Bartels D (2000) Water deficit triggers phospholipase D activity in the resurrection plant Craterostigma plantagineum. *Plant Cell*, 12:111-124.

Fraysse LC, Wells B, McCann MC, Kjellbom P (2005) Specific plasma membrane aquaporins of the PIP1 subfamily are expressed in sieve elements and guard cells. *Biol Cell*, 97: 519-534.

Froger A, Tallur B, Thomas D, Delamarche C (1998) Prediction of functional residues in water channels and related proteins. *Protein Sci* 7: 1458-1468.

Fu D, Libson A, Miercke LJW, Weitzman C, Nollert P, Krucinski J, Stroud RM (2000) Structure of a glycerol-conducting channel and the basis for its selectivity. *Sci.* 290: 481-486.

Galmés J, Pou A, Alsina MM, Toma`s M, Medrano H, Flexas J (2007) Aquaporin expression in response to different water stress intensities and recovery in Richter-110 (Vitis sp.): relationship with ecophysiological status. *Planta*, 226: 671-681.

Gao Z, He X, Zhao B, Zhou C, Liang Y, Ge R, Shen Y, Huang Z (2010) Overexpressing a putative aquaporin gene from wheat TaNIP, enhances salt tolerance in transgenic Arabidopsis. *Plant Cell Physiol* 51: 767-775.

Gaspar M, Bousser A, Sissoëff I, Roche O, Hoarau J, Mahé A (2003) Cloning and characterization of *ZmPIP1-5b*, an aquaporin transporting water and urea. *Plant Sci.* 165: 21-31.

Gebel T, (1997) Arsenic and antimony: comparative approach on mechanistic toxicology. *Chemico-Biol Interact.* 107: 131-144.

Gerbeau P, Güçlü J, Pierre R, Maurel C (1999) Aquaporin Nt-TIPa can account for the high permeability of tobacco cell vacuolar membrane to small neutral solutes. *Plant J.* 18: 577-587.

Greenway H, Munns R (1980) Mechanisms of salt tolerance in non-halophytes. *Annu Rev Plant Physiol.* 31:149-190.

Guenther JF, Roberts DM (2000) Water-selective and multifunctional aquaporins from *Lotus japonicus* nodules. *Planta* 210: 741-748.

Guenther JF, Chanmanivone N, Galetovic MP, Wallace IS, Cobb JA, Roberts DM (2003) Phosphorylation of soybean nodulin 26 on serine 262 enhances water permeability and is regulated developmentally and by osmotic signals. *Plant Cell*, 15: 981-991.

Guo L, Wang ZY, Lin H, Cui WE, Chen J, Liu M, Chen ZL, Qu LJ, Gu H (2006) Expression and functional analysis of the rice plasma-membrane intrinsic protein gene family. *Cell Res.* 16: 277-286.

Hanba YT, Shibasaka M, Hayashi Y, Hayakawa T, Kasamo K, Terashima I, Katsuhara M (2004) Overexpression of the barley aquaporin HvPIP2;1 increases internal CO2 assimilation in the leaves of transgenic rice plants. *Plant Cell Physiol.* 45: 521-529.

Harlan JR (1981) The early history of wheat: Earliest traces to the sack of Rome. In: L.T. Evans and W.J. Peacock (eds.), Wheat science today and tomorrow. Cambridge Univ. Press, Cambridge UK.

Hasegawa PM, Bressan RA, Zhu J-K, Bohnert HJ (2000) Plant cellular and molecular responses to high salinity. *Annu Rev Plant Physiol Plant Mol Biol.* 51: 463-499.

Heller KB, Lin ECC, Wilson TH (1980) Substrate specificity and transport properties of the glycerol facilitator of *Escherichia coli. J Bacteriol.* 144: 274-278.

Henderson KN, Tye-Din JA, Reid HH (2007) A structural and immunological basis for the role of human leukocyte antigen DQ8 in celiac disease. *Immunity*, 27: 1-12.

Heymann JB, Engel A (2000) Structural clues in the sequences of the aquaporins. *J Mol Biol.* 295: 1039-1053.

Höfte H, Hubbard L, Reizer J, Ludevid D, Herman EM, Chrispeels MJ (1992) Vegetative and seed-specific forms of tonoplast intrinsic proteins in the vacuolar membrane of *Arabidopsis thaliana*. *Plant Physiol*. 99: 561-570.

Hollenbach B, Dietz KJ (1995) Molecular-cloning of *emip*, a member of the major intrinsic protein (MIP) gene family, preferentially expressed in epidermal-cells of barley leaves. *Botanica Acta*, 108: 425-431.

Hub JS, de Groot BL (2006) Does CO_2 permeate through Aquaporin-1? *Biophys J*. 93: 842–848.

Isayenkov SV, Maathuis FJ (2008) The *Arabidopsis thaliana* aquaglyceroporin AtNIP7;1 is a pathway for arsenite uptake. *FEBS Lett*. 582: 1625-1628.

Ishibashi K (2006) Aquaporin subfamily with unusual NPA boxes. *Biochimica et Biophysica Acta (BBA) – Biomembranes*, 1758: 989-993.

Ishikawa F, Suga S, Uemura T, Sato MH, Maeshima M (2005) Novel type aquaporin SIPs are mainly localized to the ER membrane and show cell-specific expression in *Arabidopsis thaliana*. *FEBS Lett* 579: 5814-5820.

Jahn TP, Møller ALB, Zeuthen T, Holm LM, Klærke DA, Mohsin B, Kühlbrandt W, Schjoerring JK (2004) Aquaporin homologues in plants and mammals transport ammonia. *FEBS Lett*. 574: 31-36.

Jang JY, Kim DG, Kim YO, Kim JS, Kang H (2004) An expression analysis of a gene family encoding plasma membrane aquaporins in response to abiotic stresses in *Arabidopsis thaliana*. *Plant Mol Biol*. 54: 713-725.

Javot H, Lauvergeat V, Santoni V, Martin-Laurent F, Guclu J, Vinh J, Heyes J, Franck KI, Schaffner AR, Bouchez D, Maurel C (2003) Role of a single aquaporin isoform in root water uptake. *Plant Cell*, 15: 509-522.

Johanson U, Gustavsson S (2002) A new subfamily of major intrinsic proteins in plants. *Mol Biol Evol* 19: 456-461.

Johanson U, Karlsson M, Johansson I, Gustavsson S, Sjövall S, Fraysse L, Weig AR, Kjellbom P (2001) The complete set of genes encoding major intrinsic proteins in Arabidopsis provides a framework for a new nomenclature for major intrinsic proteins in plants. *Plant Physiol*. 126: 1358-1369.

Johnson KD, Hoèfte H, Chrispeels MJ (1990) An intrinsic tonoplast protein of protein storage vacuoles in seeds is structurally related to a bacterial solute transporter (GlpF). *Plant Cell*, 2: 525-532.

Jugdaohsingh R, (2007) Silicon and bone health. *J Nutr Health Aging* 11: 99-110.

Jung JS, Preston GM, Smith BL, Guggino WB, Agre P (1994) Molecular structure of the water channel through aquaporin CHIP. The hourglass model. *J Biol Chem*. 269: 14648-14654.

Kaldenhoff R, Ribas-Carbo M, Sans JF, Lovisolo C, Hechwolf M, Uehlein N (2008) Aquaporins and plant water balance. *Plant Cell Environ* 31: 658-666.

Kamaluddin M, Zwiazek JJ (2002) Ethylene enhances water transport in hypoxic aspen. *Plant Physiol*. 128: 1-8.

Kamiya T, Tanaka M, Mitani N, Ma JF, Maeshima M, Fujiwara T (2009) NIP1;1, an aquaporin homolog, determines the arsenite sensitivity of *Arabidopsis thaliana*. *J Biol Chem*. 284: 2114-2120.

Kammerloher W, Fischer U, Piechottka GP, Schäffner AR (1994) Water channels in the plant plasma membrane cloned by immunoselection from a mammalian expression system. *Plant J.* 6: 187-199.

Katsuhara M, Hanba YT (2008) Barley plasma membrane intrinsic proteins (PIP aquaporins) as water and CO_2 transporters. *Pflugers Arch Eur J Physiol.* 456: 687-691.

Katsuhara M, Akiyama Y, Koshio K, Shibasaka M, Kasamo K (2002) Functional analysis of water channels in barley roots. *Plant Cell Physiol.* 43: 885-893.

Katsuhara M, Koshio K, Shibasaka M, Hayashi Y, Hayakawa T, Kasamo K (2003) Over-expression of a barley aquaporin increased the shoot/root ratio and raised salt sensitivity in transgenic rice plants. *Plant Cell Physiol.* 44: 1378-1383.

Katsuhara M, Hanba YT, Shiratake K, Maeshima M (2008) Expanding roles of plant aquaporins in plasma membranes and cell organelles. *Funct Plant Biol.* 35: 1-14.

Kawasaki S, Borchert C, Deyholos M, Wang H, Brazille S, Kawai K, Galbraith D, Bohnert H (2001) Gene expression profiles during the initial phase of salt stress in rice. *Plant Cell* 13: 889-905.

Kimber G, Sears E (1987) Evolution in the genus *Triticum* and the origin of cultivated wheat. In 'Wheat and Wheat Improvement'. (Ed. EG Heyne) pp. 154-164. (American Society of Agronomy: Madison, Winsonsin, USA).

Klebl F, Wolf M, Sauer N (2003) A defect in the yeast plasma membrane urea transporter Dur3p is complemented by CpNIP1, a Nod26-like protein from zucchini (*Cucurbita pepo L.*), and by *Arabidopsis thaliana* delta-TIP or gamma-TIP. *FEBS Lett.* 547: 69-74.

Klok EJ, Wilson IW, Wilson D, Chapman SC, Ewing RM, Somerville SC, Peacock WJ, Dolferus R, Dennis ES (2002) Expression profile analysis of the low-oxygen response in *Arabidopsis* root cultures. *Plant Cell* 14: 2481-2494.

Knowles TC, Dooerge TA, Ottman MJ (1991) Improved nitrogen management in irrigated durum wheat using stem nitrate analysis: Interception of nitrate-N contents. *Agron J.* 83: 353-356.

Kozono D, Ding X, Iwasaki I, Meng X, Kamagata Y, Agre P, Kitagawa Y (2003) Functional expression and characterization of an archaeal aquaporin. AqpM from *Methanothermobacter marburgensis*. *J Biol Chem.* 278: 10649-10656.

Laffan J, (1999) Cropping Systems for Sustainable Wheat Production. Continuing Education, CB Alexander Agricultural College - Tocal, Tocal, NSW.

Langridge P, Paltridge N, Fincher G (2006) Functional genomics of abiotic stress tolerance in cereals. *Brief Funct Genomic Proteomic* 4: 343-354.

Li L, Li S, Tao Y, Kitagawa Y (2000) Molecular cloning of a novel water channel from rice: its products expression in *Xenopus* oocytes and involvement in chilling tolerance. *Plant Sci.* 154: 43-51.

Li R-Y, Ago Y, Liu W-J, Mitani N, Feldmann J, McGrath SP, Ma JF, Zhao F-J (2009a) The Rice Aquaporin Lsi1 Mediates Uptake of Methylated Arsenic Species. *Plant Physiol.* 150: 2071-2080.

Li Y, Wu, Z, Ma N, Gao J (2009b) Regulation of the rose Rh-PIP2;1 promoter by hormones and abiotic stresses in Arabidopsis. *Plant Cell Rep.* 28: 185-196.

Lian H-L, Yu X, Ye Q, Ding X, Kitagawa Y, Kwak S-S, Su W-A, Tang Z-C (2004) The role of aquaporin RWC3 in drought avoidance in rice. *Plant Cell Physiol.* 45: 481-489.

Liu Q, Umeda M, Uchimiya H (1994) Isolation and expression analysis of two rice genes encoding the major intrinsic protein. *Plant Mol Biol.* 26: 2003-2007.

Liu L-H, Ludewig U, Gassert B, Frommer WB, von Wiren N (2003) Urea transport by nitrogen-regulated tonoplast intrinsic proteins in Arabidopsis. *Plant Physiol.* 133: 1220-1228.

Lopez F, Bousser A, Sissoeff I, Hoarau J, Mahe A (2004) Characterization in maize of ZmTIP2-3, a root-specific tonoplast intrinsic protein exhibiting aquaporin activity. *J Exp Bot.* 55: 539-541.

Loque D, Ludewig U, Yuan LX, von Wiren N (2005) Tonoplast intrinsic proteins AtTIP2;1 and AtTIP2;3 facilitate NH_3 transport into the vacuole. *Plant Physiol.* 137: 671-680.

Luo M-C, Yang Z-L, You FM, Kawahara T, Waines JG, Dvorak J (2007) The structure of wild and domesticated emmer wheat populations, gene flow between them, and the site of emmer domestication. *Theor Appl Genet* 114: 947-959.

Ma JF, Yamaji N, (2006) Silicon uptake and accumulation in higher plants. *Trends Plant Sci.* 11: 392-397.

Ma S, Quist TM, Ulanov A, Joly R, Bohnert HJ (2004) Loss of TIP1;1 aquaporin in Arabidopsis leads to cell and plant death. *Plant J* 40: 845-859.

Ma JF, Tamai K, Yamaji N, Mitani N, Konishi S, Katsuhara M, Ishiguro M, Murata Y, Yano M (2006) A silicon transporter in rice. *Nature* 440: 688-691.

Ma N, Xue JQ, Li YH, Liu XJ, Dai FW, Jia WS, Luo YB, Gao JP (2008) *Rh-PIP2;1*, a rose aquaporin gene, is involved in ethylene-regulated petal expansion. *Plant Physiol.* 148: 894-907.

Maeshima M (2001) Tonoplast transporters: organization and function. *Annu Rev Plant Physiol Plant Mol Biol.* 52: 469-497

Mahajan S, Tuteja N (2005) Cold, salinity and drought stresses: an overview. *Arch Biochem Biophys.* 444: 139-158.

Martinez-Ballesta MC, Silva C, Lopez-Berenguer C, Cabañero FJ, Carvajal M, (2006) Plantaquaporins: new perspectivesonwaterandnutrientuptakeinsaline environment. *Plant Biol.* 8: 535-546.

Maurel C, Reizer JU, Schroeder JI, Chrispeels MJ (1993) The vacuolar membrane protein -TIP creates water specific channels in *Xenopus* oocytes. *EMBO J.* 12: 2241-3347.

Maurel C, Verdoucq L, Luu DT, Santoni V (2008) Plant aquaporins: membrane channels with multiple integrated functions. *Annu Rev Plant Biol.* 59: 595-624.

Mitani N, Yamaji N, Ma JF (2008) Characterization of substrate specificity of a rice silicon transporter, Lsi1. *Pflugers Arch.* 456: 679-86.

Mitani N, Yamaji N, Ma JF (2009) Identification of Maize Silicon Influx Transporters. *Plant Cell Physiol.* 50: 5-12.

Morillon R, Lassalles J-P (2002) Water deficit during root development: effects on the growth of roots and osmotic water permeability of isolated root protoplasts. *Planta* 214: 392-399.

Moshelion M, Becker D, Biela A, Uehlein N, Hedrich R, Otto B, Levi H, Moran N, Kaldenhoff R (2002) Plasma membrane aquaporins in the motor cells of *Samanea saman*: diurnal and circadian regulation. *Plant Cell* 14: 727-739.

Munns R (2005) Genes and salt tolerance: bringing them together. New Phytol 167: 645-663

Munns R, Termaat A (1986) Whole-plant responses to salinity. *Aust J Plant Physiol.* 13: 143-160.

Munns R, Tester M (2008) Mechanisms of salinity tolerance. *Annu Rev of Plant Biol.* 59: 651-681.

Munns R, Schachthman DP, Condon AG (1995) The significance of two-phase growth response to salinity in wheat and barley. *Aust J Plant Physiol.* 22: 561–569.

National Land and Water Resources Audit; http://audit.ea.gov.au (Last accessed 08/2009).

Office of the Gene Technology Regulator http://www.ogtr.gov.au/internet/ogtr/publishing.nsf/Content/ir-1 (Last accessed 08/2009).

Niemietz CM, Tyerman SD (2000) Channel-mediated permeation of ammonia gas through the peribacteroid membrane of soybean nodules. *FEBS Lett.* 465: 110-114.

Park JH, Saier MH Jr (1996) Phylogenetic characterization of the MIP family of transmembrane channel proteins. *J Membr Biol.* 153: 171-180.

Pfeiffer WH, Trethowan RM, van Ginkel M, Ortiz-Monasterio I, Rajaram S (2005) Breeding for abiotic stress tolerance in wheat. p. 401–489. *In* M. Ashraf and P.J.C. Harris (ed.) *Abiotic Stresses: Plant Resistance through Breeding and Molecular Approaches*. The Haworth Press, New York.

Philip JR (1958) Osmosis and diffusion in tissue: half-times and internal gradients. *Plant Physiol.* 33: 275-278.

Preston GM, Agre P (1991) Isolation of the cDNA for erythrocyte integral membrane protein of 28 kilodaltons: member of an ancient channel family. *Proc Natl Acad Sci USA*, 88: 11110-11114.

Quan LJ, Zhang B, Shi WW, Li HY (2008) Hydrogen peroxide in plants: a versatile molecule of the reactive oxygen species network. *J Integr Plant Bio.* 50: 2-18.

Quigley F, Rosenberg JM, Shachar-Hill Y, Bohnert HJ (2001) From genome to function: the Arabidopsis aquaporins. *Genome Biology*, 3: 1-17.

Ramachandra-Reddy A, Viswanatha-Chaitanya K, Vivekanandan M, (2004) Drought-induced responses of photosynthesis and antioxidant metabolism in higher plants. *J. Plant Physiol.* 161: 1189-1202.

Rougé P, Barre A (2008) A molecular modeling approach defines a new group of Nodulin 26-like aquaporins in plants. *Biochem Biophys Res Commun.* 367: 60-66.

Sahrawat AK, Becker D, Lutticke S, Lorz H (2003) Genetic improvement of wheat via alien gene transfer, an assessment. *Plant Sci.* 165:1147-116.

Sakurai J, Ishikawa F, Yamaguchi T, Uemura M, Maeshima M (2005) Identification of 33 rice aquaporin genes and analysis of their expression and function. *Plant Cell Physiol.* 46: 1568-1577.

Sakurai J, Ahamed A, Murai M, Maeshima M, Uemura M (2008) Tissue- and cell-specific localization of rice aquaporins and their water transport activities. *Plant Cell Physiol.* 49: 30-39.

Sarda X, Tousch D, Ferrare K, Legrand E, Dupuis JM, Casse-Delbart F, Lamaze T (1997) Two TIP-like genes encoding aquaporins are expressed in sunflower guard cells. *Plant J.* 12: 1103-1111.

Schunmann PH, Ougham HJ (1996) Identification of three cDNA clones expressed in the leaf extension zone and with altered patterns of expression in the slender mutant of barley: a tonoplast intrinsic protein, a putative structural protein and protochlorophyllide oxidoreductase. *Plant Mol Biol.* 31: 529-537.

Schüssler MD, Alexandersson E, Bienert GP, Kichey T, Laursen KH, Johanson U, Kjellbom P, Schjoerring JK, Jahn TP (2008) The effects of the loss of TIP1;1 and TIP1;2 aquaporins in *Arabidopsis thaliana*. *Plant J.* 56:756-767.

Schuurmans JAMJ, van Dongen JT, Rutjens BPW, Boonman A, Pieterse CMJ, Borstlap AC (2003) Members of the aquaporin family in the developing pea seed coat include representatives of the PIP, TIP, and NIP subfamilies. *Plant Mol Biol.* 53: 655-667.

Seki M, Narusaka M, Ishida J, Nanjo T, Fujita M, Oono Y, Kamiya A, Nakajima M, Enju A, Sakurai T, Satou M, Akiyama K, Taji T, Yamaguchi-Shinozaki K, Carninci P, Kawai J, Hayashizaki Y, Shinozaki K (2002) Monitoring the expression profiles of 7000 *Arabidopsis* genes under drought, cold and high-salinity stresses using a full-length cDNA microarray. *Plant J.* 31: 279-292.

Shao H-B, Chu L-Y, Shao M-A, Zhao C-X (2008) Advances in functional regulation mechanisms of plant aquaporins: Their diversity, gene expression, localization, structure and roles in plant soil-water relations. *Mol Mem Biol.* 25: 179-191.

Sheen J (1996) Ca^{2+} dependent protein kinases and stress signal transduction in plants. *Sci.* 274: 1900-1902.

Shinozaki K, Yamaguchi-Shinozaki K, Seki M (2003) Regulatory network of gene expression in the drought and cold stress responses. *Curr Opin Plant Biol.* 6: 410-417.

Shiratake K, Kanayama Y, Maeshima M, Yamaki S (1997) Changes in H(+)-pumps and a tonoplast intrinsic protein of vacuolar membranes during the development of pear fruit. *Plant Cell Physiol.* 38: 1039-1045.

Shiratake K, Kobae Y, Suzuki Y, Nakaune M, Tanase K, Yamaki S (2001) Molecular cloning of a cDNA encoding tonoplast water channel of pear fruit and its expression during development. *J Jpn Soc Hortic Sci.* 70: 281-286.

Siefritz F, Otto B, Bienert GP, van der Krol A, Kaldenhoff R (2004) The plasma membrane aquaporin NtAQP1 is a key component of the leaf unfolding mechanism in tobacco. *Plant*, J 37: 147-155.

Singh NK, Bracker CA, Hasegawa PM, Handa AK, Buckel S, Hermodson MA, Pfankoch E, Regnier FE, Bressan RA (1987) Characterization of osmotin: a thaumatin-like protein associated with osmotic adaptation in plant cells. *Plant Physiol.* 85: 529-536.

Snowball K, Robson AD (1991) Nutrient Deficiencies and Toxicities in Wheat: A Guide for Field Identification. C1MMYT, Mexico, D.F.

Soto G, Alleva K, Mazella MA, Amodeo G, Muschietti JP (2008) AtTIP1;3 and AtTIP5;1, the only highly expressed Arabidopsis pollen-specific aquaporins, transport water and urea. *FEBS Letters*, 582(29): 4077-4082.

Suga S, Maeshima M (2004) Water channel activity of radish plasma membrane aquaporins heterologously expressed in yeast and their modification by site-directed mutagenesis. *Plant Cell Physiol.* 45: 823-830.

Sui H, Han B-G, Lee JK, Walian P, Jap BK (2001) Structural basis of water-specific transport through the AQP1 water channel. *Nature*, 414: 872-878.

Takano J, Wada M, Ludewig U, Schaaf G, von Wiren N, Fujiwara T (2006) The *Arabidopsis* major intrinsic protein NIP5;1 is essential for efficient boron uptake and plant development under boron limitation. *Plant Cell*, 18: 1498-1509.

Tanaka M, Fujiwara T (2008) Physiological roles and transport mechanisms of boron: perspectives from plants. *Pflugers Arch Eur J Physiol.* 456: 671-677.

Tanaka M, Wallace IS, Takano J, Roberts DM, Fujiwara T, (2008) NIP6;1 is a boric acid channel for preferential transport of boron to growing shoot tissues in Arabidopsis. *Plant Cell*, 20: 2860-2875.

Törnroth-Horsefield S, Wang Y, Hedfalk K, Johanson U, Karlsson M, Tajkhorshid E, Neutze R, Kjellbom P (2006) Structural mechanism of plant aquaporin gating. *Nature*, 439: 688-694.

Treviño MB, O'Connell MA (1998). Three drought-responsive members of the nonspecific lipid-transfer protein gene family in *Lycopersicon pennellii* show different developmental patterns of expression. *Plant Physiol.* 116: 1461-1468.

Tyerman SD, Niemietz CM, Bramley H (2002) Plant aquaporins: multifunctional water and solute channels with expanding roles. *Plant Cell Environ* 25: 173-194.

Uehlein N, Otto B, Hanson DT, Fischer M, McDowell N, Kaldenhoff R (2008) Function of Nicotiana tabacum aquaporins as chloroplast gas pores challenges the concept of membrane CO2 permeability. *Plant Cell*, 20: 648-657.

USDA-ARS (2005) GE Brown Jr Salinity Laboratory, Riverside, CA, USA (http://www.ars.usda.gov/Services/docs.htm?docid=8908).

U.S. Environmental Protection Agency (2006) Revised reregistration eligibility decision for MSMA, DSMA, CAMA, and cacodylic acid. http://www.epa.gov/pesticides/reregistration/REDs/organic_arsenicals_red.pdf (June 26, 2009).

Vasil IK (2007) Molecular genetic improvement of cereals: transgenic wheat (*Triticum aestivum* L.). *Plant Cell Reports*, 26: 1133-1154.

Wallace IS, Roberts DM (2004) Homology modeling of representative subfamilies of Arabidopsis major intrinsic proteins. Classification based on the aromatic/arginine selectivity filter. *Plant Physiol.* 135: 1059-1068.

Wallace IS, Roberts DM (2005) Distinct transport selectivity of two structural subclasses of the nodulin-like intrinsic protein family of plant aquaglyceroporin channels. *Biochem.* 44: 16826-16834.

Wallace IS, Choi WG, Roberts DM (2006) The structure, function and regulation of the nodulin 26-like intrinsic protein family of plant aquaglyceroporins. *Biochim Biophys Acta*, 1758: 1165-1175.

Wang W, Vinocur B, Altman A (2003) Plant responses to drought, salinity and extreme temperature: towards genetic engineering for stress tolerance. *Planta*, 218: 1-14.

Wang W-H, Kohler B, Cao F-Q, Liu L-H (2008) Molecular and physiological aspects of urea transport in higher plants. *Plant Sci.* 175: 467-477.

Weaver CD, Roberts DM (1992) Determination of the site of phosphorylation of nodulin 26 by the calcium-dependent protein kinase from soybean nodules. *Biochem.* 31: 8954-8959.

Weig AR, Jakob C (2000) Functional identification of the glycerol permease activity of *Arabidopsis thaliana* NLM1 and NLM2 proteins by heterologous expression in *Saccharomyces cerevisiae*. *FEBS Lett.* 481: 293-298.

Xue J, Yang F, Gao J (2009) Isolation of *Rh-TIP1;1*, an aquaporin gene and its expression in rose flowers in response to ethylene and water deficit. *Postharvest Bio Tech.* 51: 407-413.

Yamada S, Katsuhara M, Kelly WB, Michalowski CB, Bohnert HJ (1995) A family of transcripts encoding water channel proteins: tissue-specific expression in the common ice plant. *Plant Cell*, 7: 1129-1142.

Yan S, Tang Z, Su W, Sun W (2005) Proteomic analysis of salt stress-responsive proteins in rice root. *Proteomics*, 5: 235-244.

Zardoya R, Ding X, Kitagawa Y, Chrispels MJ (2002) Origin of plant glycerol transporters by horizontal gene transfer and functional recruitment. *Proc Natl Acad Sci USA*, 99: 14893-14896.

Zhang J, Deng Z, Cao S, Wang X, Zhang A, Zhang X, (2008) Isolation of six novel aquaporin genes from Triticum aestivum L. and functional analysis of TaAQP6 in water redistribution. *Plant Mol. Biol Rep*. 26:32-45.

Zhu C, Schraut D, Hartung W, Schaffner AR (2005) Differential responses of maize *MIP* genes to salt stress and ABA. *J Exp Bot*. 56: 2971-298.

In: Wheat: Genetics, Crops and Food Production
Editor: Maria T. Almeida, pp. 153-178
ISBN 978-1-61209-307-9

Chapter 4

ADVANCES IN WHEAT GENETIC ENHANCEMENT FOR GLOBAL FOOD PRODUCTION

Rodomiro Ortiz*
Martín Napanga 253, Miraflores Lima 18, Perú

ABSTRACT

Wheat provides about 19% of the food calories and 21% of the protein to the world population. In 2008, the total wheat area grownwas about 224 million ha and with a harvest of 690 million t of grain.The Green Revolution in wheat production, spawned by the development of short-stature, photoperiod-insensitive, high yielding wheat cultivars with host plant resistance to rusts and other pathogens, was able to increase wheat productivity at a 3.6% annual rate in the 1960s and 1970s. However, yield growth in wheat slip to 2.8% from the mid-1980s to the mid-1990s, and to 1.1% since the mid-1990s. In the period leading up to 2020, demand for wheat for human consumption in developing countries is expected to grow at 1.6 % per annum, and for feed at 2.6 % per annum. The global average of wheat yield will have therefore to increase in the coming years at least to 3.5 t ha^{-1}. This yield increase, essential to maintain global food security, requires a continuing supply of improved germplasm and appropriate agronomy in order to sustain enhanced productivity and preserve the natural resource base of the wheat farming. Global warming, as a result of climate change, may negatively affect wheat grain yields —potentially increasing food insecurity and poverty. Wheat breeders will depend on the crop´s genetic diversity and modern genetic enhancement methods (both genomics and transgenics) for improving wheat yield and its stability. They should conservewheat germplasm, study it, add value to it, and share it in enhanced form with clients worldwide. While changes in plant architecture (dwarfism and harvest index), photoperiod response and breeding for host plant resistance (especially for rusts) led to the productivity gains of the Green Revolution, adaptation to climate change (water and heat stresses, emerging pathogens and pests), efficient input use, food safety and quality, are additional global challenges to be tackled to fit newly bred-genotypes into sustainable and improved wheat farming in this 21st Century.

* Email: rodomiroortiz@gmail.com

OVERVIEW

In 2008, wheat was the most widely grown crop in the world (about 223.6 million ha) and only second after maize on total production (689.9 and 822.7 million, respectively) (FAO 2010). World wheat production more than trebled from the 1960s to date mostly due to grain yield increases (from1.1 t ha^{-1} in 1961 to 3.1 t ha^{-1} in 2008). International wheat breeding, led by the late 1970 Nobel Peace Prize Laureate Norman E. Borlaug (Ortiz et al. 2007b), contributed significantly to developing spring wheat germplasm that eventually spread throughout the world and its success became known as part of the Green Revolution (Borlaug 2007). The Green Revolution, which used genetically enhanced seed-embedded technology (or newly bred-cultivars), and inputs such as fertilizer and irrigation was regarded, in principle, scale-neutral because it could raise crop yields and incomes or both small- and large-scale farmers (Hazell and Ramasamy 1991). Furthermore, Evenson and Gollin (2002) clearly showed that crop yields in the developing world would have been at least 20% lower and the equilibrium prices for most important staples combined would have been about 19% higher than they were in 2000 without the Green Revolution. There would have been also a drop of 5% in calorie consumption and 2% more malnourished children in the developing world without bred germplasm on main food crops. Land spared for nature (instead of being used by agriculture) in the last half-century resulted from the significant crop yield increases(especially those of main cereals) worldwide (Milford and Runge 2007). For example, as already noted by Khush (2005), if Asia attempted to produce a 2000 harvest at the yield levels of the 1960s, most of the forests, woodlands, pastures and rangelands would have disappear, and mountain sides would have been eroded, with disastrous consequences for the upper watershed and productive lowlands, the extinction of wildlife habitats and the destruction of biodiversity.

As stated by Borlaug (2000, 2007), despite such successes of the Green Revolution, the fight to ensure food security for hundreds of millions of resource-poor people has yet to be won, and therefore there is no room for complacency. The world was witness of a significant food commodity price boom between 2006 and 2008. Several factors such as less past investment in extractive commodities (including those from agriculture), weak US dollar, fiscal expansion, lax monetary policy in many countries and investment fund activity coupled with adverse weather conditions, the diversion of some food commodities to the production of biofuels, and government policies (including export bans and prohibitive taxes) accounted for the lowering of global grain stocks to levels unseen sincethe early 1970s, which in turn accelerated the food price hikes that eventually led to the 2008 rally (Baffer and Haniotis 2010), e.g. wheat prices reached a record of nearly US$ 500 per t.

Total cereal demand is projected to grow by 1048 million t and 26% of this increase is expected for wheat (Hubert et al. 2010). Area expansion, however, cannot be expected to contribute significantly to future production growth of wheat and other main staples. Wheat prices are also projected to increase by 97% in 2050 vis-à-vis those of 2000 (Hubert et al 2010).A substantial food price increase will slow calorie consumption growth because of both direct price impacts and reductions in real incomes for poor consumers, who spend a significant share of their income on food, thereby leading to a little improvement in food security for the resource-poor in many regions of the developing world. Genetic enhancement of food crops through both conventional breeding as well as with biotechnology tools will be therefore needed to push the yield frontier higher and to increase yield stability due to better adaptation to both biotic and abiotic stresses (Borlaug and Dowswell 2005, Khush 2005), particularly with a declining availability of water and land that can be profitably cultivated. Conway and Toenniessen (1999) advocated a Doubly Green Revolution, which is both more productive and more "green" in terms of conserving natural resources and the environment than the first, to feed the world in this 21st Century. They also asked for a strong public sector agricultural research effort to ensure that the advances in crop biotechnology reach and benefit the resource poor.

Wheat farmers worldwide are confronting major challenges such as adaptation to climate change (particularly to heat), limiting water resources, emerging pests and pathogen epidemics, food safety (and fighting mycotoxins), grain quality for adding value in the commodity chain and more nutritious food, and fitting new cultivars into more sustainable agro-ecosystems (Ortiz et al. 2008a). For example, climate change could strongly affect the wheat crop because could reduce its grain yields in zones where optimal temperatures already exist. (Ortiz et al. 2008b). Advances in wheat genetic enhancement can flow to allow farmers to face such risks associated with climate change. This article gives an overview of recent development in wheat improvement, which will contribute to adapting the crop to the changing environment of this 21st century.

The Green Revolution and Wheat Breeding

Borlaug shaped international wheat improvement through innovations such as shuttle breeding, wide adaptation, durable host plant resistance to pathogens and pests, international multisite testing, and the wise use of genetic variability (Ortiz et al. 2007c). Under his leadership or mentorship, hundreds of modern wheat cultivars were released for use in excess of 50 million ha at both favorable and marginal environments (Reynolds and Borlaug 2006). As noted by Dixon et al. (2006), millions of wheat farmers decided to adopt or replace old wheat cultivars with these superior bred-materials worldwide. Such mass adoption of newly-bred wheat germplasm led to steady gains in average wheat yields during the past 50 years and helped country programs for achieving food security and reducing national poverty.

Although the Green Revolution displaced wheat landraces in favor of high yielding cultivars, these and other genetic resources were and are still utilized to genetically enhance new cultivars (Ortiz et al. 2008a). The genetic basis of the Green Revolution in wheat was initially due to radical changes in plant architecture and phenology owing to a few dwarfing genes with major effects on plant height and the elimination of photoperiod response (Trethowan et al. 2007). Further host plant resistance breeding greatly enhanced the yield potential and stability of wheat cultivars in low latitude wheat farming environments of the developing world.

Despite the successes of the Green Revolution, some recent assessments on wheat breeding gains suggest minimal genetic gains on yield potential or a likely yield plateau for some wheat farming areas in the last 1.5 decades(e.g. Graybosh and Peterson 2010). Hence, any further improvement of wheat grain yield through plant breeding should ensue from broadening its germplasm base with other gene pools such as crop wild relatives (Ortiz et al. 2008a), and advances in molecular breeding (Sorrell 2007). Wheat breeding for high yield may rely on biomass accumulation (a trait that could be available in crop wild relatives), yield potential*per se*resulting from a high harvest index, time to harvest to maturity adjusted to specific growing season, and adaptation to stress(es) at target location(s). Modern wheat breeding coupled with sustainable crop and resource management may buffer any yield instability due to unpredictable environments (Reynolds and Borlaug 2006b), which will be a major feature of this global changing climate.

Wild Crop Relatives for Evolutionary Wheat Breeding

Wheat genetic resources endowment includes modern cultivars in current use, obsolete cultivars —often the elite cultivars of the past and frequently found in the pedigrees of modern cultivars— landraces, crop wild relatives in the Triticeae tribe, genetic and cytogenetic stocks, and breeding lines (Ortiz et al. 2008a). Wild species are extremely valuable sources of untapped genetic variability for wheat improvement.

Wild emmer (*Triticum dicoccoides*) ensued from the hybridization between *Aegilops speltoides* and *T. urartu*about 10,000 year ago, and 2,000 years later its domesticated version (*T. dicoccum* or cultivated emmer) intercrossed naturally with *Ae. tauschii* (goat grass) to give rise to hexaploid bread wheat (*T. aestivum)*, which spread worldwide and became one of the major staples that feed the world. *T. dicoccum* itself evolved into *T. turgidum* ssp. *durum*; i.e., the modern tetraploid durum wheat. Polyploidization events leading to durum and bread wheats were extremely rare. Although, a genetic bottleneck could result from such domestication events and polyploidy speciation, today′s wheat adaptability to diverse environments and end-uses ensued from capturing part of the genetic diversity of its progenitors and by generating new diversity at a relatively fast pace (Dubcovsky and Dvorak 2007). Wheat polyploidy buffers gene deletions and disruptions, which are frequently generated by a fast replacement rate of repetitive sequences. Oksen et al. (2001) indicated that allopolyploidy also induced rapid genome evolution and provided the physical basis for the diploid-like meiotic behavior in wheat (resulting in disomic inheritance), which contributed to the successful establishment of this species. Furthermore, Pumphrey et al. (2009) showed that allopolyploidization *per se*could lead to a rapid initiation of differential expression of homoeologous loci andnon-additive gene expression in bread wheat.

Aegilops — the most closely related genus to *Triticum* — has been widely used in wheat improvement. Bread wheat has been artificially re-synthesized by intercrossing modern tetraploid durum wheat with present-day derivatives of goat grass. Derivatives of this synthetic hexaploid wheat are available in large numbers at breeding programs worldwide (Ortiz et al. 2008a, Trethowan and Mujeeb-Kazi 2008, van Ginkel and Ogbonnaya 2007). This evolutionary breeding strategy also broadens the genetic base of the wheat cultigen pool (Zhang et al. 2005). In this regard, Warburton et al. (2007) showed that exchange of materials between breeding programs, as well as the use of landraces and re-synthesized wheat derivatives for breeding new advanced materials led to a significant increase in the latent genetic diversity of recently bred lines and cultivars compared to the original Green Revolution wheats. Further analysis with DNA markers showed that some re-synthesized wheats with better adaptation to drought and salinity stresses are very distant from wheat cultivars and breeding lines (Das et al. 2007). Moreover, Dreisigacker et al. (2008) indicated that both re-synthesized wheats and their derivatives are genetically diverse at the DNA level when compared with traditional bread wheat cultivars. They also noted a preferential transmission of some alleles from the re-synthesized wheat parent in all genomes, indicating positive selection. This use of DNA marker-aided analysis reveals that wild ancestors are important as sources of variation for wheat breeders.

The re-synthesized wheats provide new genetic variation for adapting the crop to stress-prone environments due to drought, high temperature, salinity, waterlogging, and low soil micronutrients (Trethowan and Mujeeb-Kazi 2008), as well as host plant resistance to various major pathogens and pests affecting wheat farming elsewhere (Ogbonnaya et al. 2008a, Olivera et al. 2007, Thompson 2008). Re-synthesized wheat derivatives yielded very well in many stressed environments globally. For example, 'Vorobey' — a derived line from re-synthesized wheat — performed well across all environments compared with the best locally adapted check cultivar at each location (Lage and Trethowan 2008). Likewise, re-synthesized wheat-derived lines showed an increased partitioning of root mass to deeper soil profiles (between 60 cm and 120 cm) and increased ability to extract moisture from those depths (Reynolds et al. 2007). Wild emmer can be another source of several traits such as abiotic stress tolerances (salt, drought and heat), biotic stress tolerances (powdery mildew, rusts, and *Fusarium* head blight), grain protein quality and quantity, and micronutrient concentrations: namely Zn, Fe, and Mn (Xie and Nevo 2008).

Significant variation in other crop wild relatives gene pools remains unexploited to improve wheat adaptation to both abiotic or biotic stresses. For example, *Leymus racemosus*— a wild relative of wheat — shows biological nitrification inhibition (BNI) capacity; i.e., the plant roots ability to release chemical compounds that suppress soil nitrification. Although wheat does not have the ability to inhibit nitrification via exudates from its roots, when a *Leymus* chromosome containing the relevant gene(s) was introgressed into wheat, BNI compounds were produced (Subbarao et al., 2007). Likewise, other crop wild relatives could be sources of traits for perennial wheat (e.g. Cox et al. 2002) –an interesting alternative for sustainable agriculture in marginal lands with poor soil quality or that could be quickly depleted by a few years of intensive annual cropping (Glover et al. 2007). Some of these gene pools, with a few significant exceptions, are however more difficult to exploit due to complex inheritance, meiotic instability, and linked deleterious effects. Nonetheless, any hybridization barriers can be overcome after producing a hybrid and exploiting its potential with crop germplasm enhancement methods such as chromosome engineering or ploidy manipulations that lead to producing introgression lines for further use by wheat breeders.

Genomics-led Wheat Breeding

New tools ensuing from recent advances in genomics may enhance wheat breeding, and therefore are being added to the germplasm scientist's kit to genetically enhance wheat because they save time and resources. Table 1 provides a sample of some DNA marker systems, which can be useful for wheat genomics research, diversity assessment and further use in breeding. DNA marker-aided selection methods target better use of genetic resources, may improve the design of crossing blocks, and can facilitate early generation selection, thereby leading to enhanced breeding efficiency in wheat improvement.

Table 1. Some DNA marker systems available for wheat genomics research, diversity assessment and breeding

DNA marker system	Reference
Diversity arrays technology (DArT) for high-throughput profiling	Akbari et al. 2006
Expressed sequenced tags (EST)	Lazo et al. 2004, Zhang et al. 2004
EST-simple sequence repeat (SSR) markers	Balfourier et al. 2007
Single-nucleotide polymorphisms (SNP)	Ravel et al. 2006, Zhao et al. 2007
Targeting induced local lesions in genomes (TILLING) for reverse genetics and wheat breeding	Dong et al. 2009, Slade et al. 2005, Uauy et al. 2009

A genetic map with DNA markers is an important research tool for linkage analysis of target traits. Genetic linkage maps serve several purposes, e.g. in evolutionary genomics, for gaining genetics insights of complex traits or dissecting the genetic factors controlling trait expression. They can also assist in marker-aided breeding and map-based cloning. The first wheat genetic maps with DNA markers used restriction fragment length polymorphisms (RFLP), and thereafter were adopted PCR-based markers such as random amplified polymorphic DNA (RAPD) or amplified fragment length polymorphism (AFLP), and later on microsatellites (also known as simple sequence repeats or SSR), which became the main marker system for genetic analysis and mapping (Varshney et al. 2007). Table 2 gives some examples of recent genetic maps of wheat with DNA markers. Diversity arrays technology (DArT) has been used recently for genetic mapping and association genetics (Crossa et al. 2007).

Table 2. Some recent genetic maps of wheat with DNA markers

DNA marker system(s)	Number of DNA markers	Reference
Simple sequence repeats (SSR), Expressed sequenced tags (EST)-SSR, inter-SSR (ISSR), sequence-related amplified polymorphism (SRAP), target region amplification polymorphism (TRAP) and glutenin (*Glu*) loci	381 loci: 167 SSR, 94 EST-SSR, 76 ISSR, 26 SRAP, 15 TRAP, and 3 *Glu*	Li et al. 2007
Restriction fragment length polymorphism (RFLP), amplified fragment length polymorphism (AFLP), and diversity arrays technology (DArT)	1644 markers: 813 DART	Crossa et al. 2007
SSR and DArT	690 loci: 197 SSR and 493 DArT	Peleg et al. 2008
SSR, target region amplification polymorphisms (TRAP), RFLP and phenotypic markers	632 markers: 410 SSRs, 218 TRAP, 1 RFLP, and 3 phenotypic markers	Chu et al. 2008
EST	887 loci: 345 EST	Xue et al. 2008
Single feature polymorphisms (SFP)	1035 and 875 SFP	Banks et al. 2009

DArT has the advantages of simultaneously typing several thousands of loci in a single assay, and generating whole-genome fingerprints through scoring the presence or absence of fragments in samples of genomic DNA.

DNA markers tightly linked to target loci genes allow selection for host plant resistance in the absence of the pathogen. Microsatellites(or SSR) were used for mapping host plant resistance genes in wheat (Table 3). Genetic markers co-segregating with, or tightly linked to distinct loci could assist wheat breeding by pyramiding host plant resistance.

Table 3. Examples of DNA markers used for mapping host plant resistance genes in wheat

Pest	Alleles(s)	DNA marker type	Reference
Cyst nematode	*Cre3*	Simple sequence repeats (SSR)	Martin et al. 2004
Leaf rust	*Lr19*	Sequence characterized amplified region (SCAR)	Prabhu et al. 2004
	Lr22a	SSR	Hierbet et al. 2007
Powdery mildew	*Pm5d*	SSR	Nematollahi et al. 2007
	Pm16	SSR	Chen et al. 2005
Stem rust	*Sr2*	SSR	Hayden et al. 2004
	Sr9a	SSR	Tsilo et al. 2009
Stripe rust	*YrCH42*	SSR	Li et al. 2006
	YrAlp, Yr39	Resistance gene analog polymorphism (RGAP) and SSR	Lin and Chen 2007
	Yr26	SSR and sequenced tagges sites (STS)	Wang et al. 2008
	rC591	SSR and amplified length polymorphism (AFLP)	Li et al. 2009

Several DNA markers (e.g. RFLP, AFLP, SSR, STS or EST) and segregating population (mainly double-haploids and F_n recombinant inbred lines) were used for mapping quantitative trait loci (QTL) controlling various traits which are main targets for wheat breeding (Table 4), or to characterize the diversity of wheat germplasm with distinct host plant resistance to pathogens (e.g. Khlestkina et al. 2007 or Yu et al. 2006). QTL analysis allows breaking the expression of complex traits its genetic components. Of course, accurate QTL analysis depends on precise phenotyping for the target traits. Mapping individual target QTL with tightly linked DNA markers will further assist the manipulation of the wheat genome through marker-aided breeding approaches, which could lead to genetic gains for quantitative traits with low heritability, and significantly influenced by the environment and large genotype-by-environment interactions.

Table 4. Sample of DNA markers and segregating populations for mapping quantitative trait loci in wheat

Trait	DNA markers	Segregating population	Reference
Earliness components (photoperiod sensitivity, vernalization and intrinsic earliness)	Simple sequence repeats (SSR)	F_7 recombinant inbred lines (RIL)	Hanocq et al. 2004
Flag leaf senescence as yield determinant under drought stress	Amplified fragment length polymorphisms (AFLP) and SSR	Double-haploid (DH) population	Verma et al. 2004
Grain yield and components across environments	Restriction fragment length polymorphism (RFLP), AFLP and SSR	DH	Quarrie et al. 2005
	SSR	DH	Kuchel et al. 2007a,b
Grain yield and components (through physiological analysis)	SSR	DH	Snape et al. 2006
Grain yield under drought	SSR and expressed-sequenced tags (EST)-sequence-tagged site (STS)	RIL	Kirigwi et al. 2007
Host plant resistance to crown rust	SSR and STS	DH	Collard et al., 2005
Host plant resistance to cereal cyst nematode	SSR and STS	DH	Williams et al. 2006
Host plant resistance to *Fusarium* head blight	SSR	F_7 RIL	Jiang et al. 2007a.b
	AFLP and SSR	$F_{5:6}$ RIL	Zhou et al. 2004
Host plant resistance to stripe rust	SSR	DH	Imtiaz et al. 2004.
Low nitrogen input management	Mostly SSR	DH	Laperche et al. 2007
Seed dormancy and pre-harvest sprouting tolerance	SSR	F_8 RIL	Ogbonnaya et al. 2008b
Seedling allelopathy	RFLP, AFLP and SSR	DH	Wu et al. 2003
Slow leaf rusting	Expressed sequenced tags (EST)	F_5 lines and single chromosome recombinant lines	Schnurbusch et al. 2004
	AFLP and SSR	Single-seed descent lines	Xu et al. 2005
	RFLP, AFLP and SSR	RIL	Rosewarne et al. 2006

DNA markers can also assist to dissect QTL-by-environment interactions (QEI), understand adaptation patterns, and attribute such trait variation to edaphic and climatic factors affecting QEI. This kind of analysis will provide new insights on molecular and physiological paths and in-depth understanding of associations between genes and phenotypes, thereby giving more confidence and objectivity to wheat breeders when genetically enhancing complex traits.

The identification of DNA markers associated with phenotypic trait variation through linkage disequilibrium became an important new tool for crop breeding. This approach does not require the time-consuming and expensive generation of specific genetic populations as the conventional linkage mapping indicated in previous paragraphs. Linkage disequilibrium depends on the physical distance between loci across chromosomes and provides a useful means for dissecting complex traits (Arbelbide and Bernardo 2006, Crossa et al. 2007) or gene mapping in wheat (Maccaferri et al. 2009, Rhoné et al. 2007). The successful application of this association analysis requires comprehensive phenotypic data to model the genotype-by-environment interaction. Association analysis are often based on historical phenotypic data from breeding trials and may rely on linear mixed-model methods applied to plant phenotypic data, thereby allowing an accurate prediction of genotypic performance using covariance structures that consider genetic associations between relatives included in the experiment (Crossa et al. 2007). Christopher et al. (2006) indicate that pedigree-based whole genome marker application may also provide a vehicle for incorporating DNA marker technology into wheat (and barley) breeding programs by bridging the gap between marker–trait association and marker implementation.

MARKER-AIDED BREEDING

Marker-aided selection (MAS) needs robust genetic markers tightly linked to genes controlling significant variation that can be applied across a wide germplasm range and target populations of environments (Ortiz et al. 2007c). Such an approach seems particularly needed for traits that are not easy to select for, such as recessively inherited traits or traits expressed at the adult plant stage. Bonnet et al. (2005) point out that molecular markers allow more accurate selection in early generations than conventional screens, but their large numbers can make selection impracticable while screening in later generations may provide little or no advantage over conventional selection techniques. They further suggest that F_2 enrichment increased the frequency of selected alleles allowing large reductions in minimum population size for recovery of target genotypes (commonly around 90%) or selection at a greater number of loci, and that backcrossing was found to be a useful strategy to reduce population size and complementary to F_2 enrichment. Advanced backcross-QTL (AB-QTL) analysis will be a useful approach to identify QTL alleles from wild relatives of wheat (Huang et al. 2003) as well as in crosses between elite germplasm and re-synthesized wheat derivatives (Huang et al. 2004).

MAS should use marker allele(s) with a tight linkage to target gene(s) governing a qualitative or quantitative trait(s), thereby facilitating and accelerating wheat breeding, through a precise transfer of chromosome regions carrying the gene(s) of interest. The best method to avoid crossover between the marker(s) and target gene(s) is to identify a marker from the gene itself, which will significantly reduce false-positive/negative selection during screening. Landjeva et al. (2007) indicate that marker-aided breeding also depends on validation of marker/trait associationsacross diverse genetic backgrounds for their general application in MAS schemes, and on the availability of marker systems suitable for large-scale deployment MAS plus the ability to analyze vast numbers of plants in a time-adequate and cost-effective manner. Table 5 lists some successful examples of MAS in wheat.

Table 5. Some successful uses of marker-aided breeding in wheat (mainly aiming introgression of host plant resistance genes or a few quantitative trait loci)

Target trait or gene	Reference
Dwarfing genes *Rht1-Blb*, *Rht1-Dlb* and *Rht8*	Williams et al. 2007
Grain hardness and flour quality (swelling volume and dough strength)	Williams et al. 2007
High protein gene *Gpc-B1*	Williams et al. 2007
Heterodera avenae resistance genes *Cre1* and *Cre3*	Williams et al. 2007
Host plant resistance to Barley Yellow Dwarf Virus	Williams et al. 2007
Host plant resistance to crown rust	Williams et al. 2007
Host plant resistance to *Fusarium* head blight	Williams et al. 2007, Wilde et al. 2008, Buerstmayr et al. 2009
Host plant resistance to Hessian fly	Williams et al. 2007
Host plant resistance to Russian wheat aphid	Williams et al. 2007
Host plant resistance to *Septoria tritici*	Botez et al. 2009
Leaf rust resistance genes *Lr1*, *Lr9*, *Lr24*, *Lr47*	Nocente et al. 2007
Leaf and brown rust resistance genes *Lr34/Yr18* and *Lr46/Yr29*	Kuchel et al. 2007, Williams et al. 2007
Stem rust resistance gene *Sr2* Stem rust resistance genes *Sr24*, *Sr25*, *Sr26*, *Sr28* and *Sr35*	Hayden et al. 2004 Williams et al. 2007
Pairing homolog *ph1b* gene	Williams et al. 2007

Davies et al. (2006) indicated that phenotypic selection was as effective as MAS in selecting for high grain protein content (GPC) genotypes and more effective in some environments. They suggested that molecular markers could be advantageous for transferring the high GPC QTL from *T. dicoccoides* through backcrossing during parent development. Others showed, however, that MAS was an effective breeding approach for delivering a high frequency of desired outcomes and for combining host plant resistance, end use quality and grain yield alleles, as well as for increasing the genetic gains and reducing costs vis-à-vis its phenotypic alternative (Kuchel et al. 2005, 2007). In this regard, Brennan and Martin (2007) point out that replacing field screening with MAS would lead to an increase in costs, whereas using MAS in glasshouse screening could lead to a reduction in costs when the marker was applied in the most efficient combination of multi-pooling and multiplexing.

Marker-aided breeding has been useful for improving several traits in wheat. Marker-assisted backcrossing, forward breeding, MAS involving doubled haploid technology and F_2 enrichment have been successfully used for speeding and saving costs in wheat breeding.

Gupta et al. (2009) suggest that breeding approaches based on high throughput genotyping associated with marker systems such as DArT and SNP, and new selection strategies such as AB-QTL, mapping-as-you-go, marker-assisted recurrent selection and genome-wide selection will provide further genetic gains in wheat improvement. There are however, some important issues to be solved for marker-aided breeding to realize its full potential in wheat breeding programs elsewhere (Xu and Crouch 2008): high throughput precision phenotyping systems for QTL mapping, improved understanding of genotype by environment interaction and epistasis, and development of publicly available computational tools tailored to the needs of molecular breeding programs.

Transgenic Wheat Improvement

The first transgenic wheat plants (or GM-wheat) became available in 1992, and the crop was the last main cereal crop to be genetically engineered (Shewry and Jones 2005). Slight progress was made in the next 1.5 decades for the commercial molecular breeding of this crop (Vasil 2007), despite that several methods were developed for efficiently introducing transgenes into bread and durum wheat lines (Dunwell 2009, Wu et al. 2008). Nonetheless, genetic engineering of wheat still remains time consuming and genotype-dependent (Bhalla et al. 2007), and shows lower efficiency than that for maize and rice (Shewry and Jones 2005).

The early GM-wheat ensued from utilizing long-term regenerable embryogenic cultures derived from immature embryos and through the use of the biolistics method for the direct delivery of DNA into regenerable cells (Vasil 2007) whereas in the last decade most robust genetic transformation systems for wheat were developed (Bhalla et al. 2007). They rely on *Agrobacterium*-mediated wheat transformation that offers advantages such as single-copy gene insertion, minimal rearrangement of DNA, low cost and comparatively high efficiency. Likewise, new selection methods, removal of unnecessary DNA sequences, gene targeting and *in vivo* mutagenesis makes the process cleaner and more acceptable to regulatory processes and consumers than before (Shewry and Jones 2005). These recent developments in wheat genetic engineering may lead to an increased efficiency of molecular breeding of this crop.

The most promising recent progress has been on developing transgenic wheat plants showing significant tolerance to adverse water stress in comparison with checks under experimental greenhouse conditions (Ortiz et al. 2007a). There are also GM-wheat lines expressing transgenes that confer resistance to environmentally friendly non-selective herbicides, pests and pathogens in addition to other transgenic wheat lines with improved bread-making and nutritional quality (Vasil 2007). More arable land may become available by reducing wheat yield losses due to competitive transgenic plants against weeds, showing host plant resistance to pathogens and pests, performing well in other stressful environments due to drought, heat or salinity, and showing an enhanced use of inputs such as fertilizers and water. A major priority for further molecular breeding of wheat will be to improve its nutritional quality, including vitamin and mineral contents for the developing world, and starch digestibility and dietary fiber content and composition for industrialized countries (Shewry and Jones 2005). Other agronomic traits appear to be amenable to transgenic wheat breeding (Dunwell 2009), which also offers the opportunity for expanding as targets biofuel and biopharming traits.

Performance under Abiotic Stresses

Water and heat stresses will continues to be major limiting factors hindering wheat grain yield under adverse hot and dry environments, which will be exacerbated by global warming and drought due to climate change. Transcription factors have been shown to produce multiple phenotypic alterations that appear to be involved in stress responses, e.g. *DREB1A* is a transcription factor that recognizes dehydration response elements and plays a crucial role in promoting the expression of drought-tolerance genes in *Arabidopsis thaliana*. Pellegrineschi et al. (2004) showed that stress-induced expression in wheat of the *Arabidopsis thaliana DREB1A* gene delays water stress symptoms in greenhouse experiments. This *DREB1A* gene was placed under control of a stress-inducible promoter from the *rd29A* gene and transferred via biolistic transformation into bread wheat. GM-wheat plants of cultivar Bobwhite expressing the *DREB1A* transgene manifested by a 10-day delay in wilting when water was withheld. A cotton´s dehydration responsive element binding protein gene (*GhDREB*) also conferred enhanced tolerance to drought, high salt and freezing stresses in transgenic wheat, which was also bred through particle bombardment (Gao et al. 2009). The GM-wheat lines, which did not show a distinct phenotype than their non-transgenic source, accumulated higher levels of soluble sugar and chlorophyll in leaves after respective stress treatments.

There have been many efforts to isolate abiotic stress-inducible transcription factors in wheat because regulatory genes appear to be an important component of plants response to environmental stresses (Ortiz et al. 2007a). The *TaDREB1* of wheat was found to be induced by cold, salinity and drought, and was classified as a DREB2-type transcription factor based on phylogenetic analysis (Shen et al. 2003). Similarly, a gene named *TdDRF1* has been isolated and characterized in durum wheat (Latini et al. 2007). More recently, Wei et al. (2009) developed genome-specific and allele specific functional markers based on single nucleotide polymorphisms (SNPs) for gene mapping of *Dreb1* genes in bread wheat. The *Dreb1* genes were located on chromosomes 3A, 3B and 3D with the aid of these genome-specific primers and Chinese Spring nulli-tetrasomic lines. This finding may be helpful for wheat breeding programs aiming performance improvement of this crop under drought.

Accumulation of proline appears to be a promising approach to maintain the productivity of transgenic wheat plants under water stress (GruszkaVendruscoloa et al. 2007). The tolerance to water deficit observed in some transgenic wheat plants was mainly due to protection mechanisms against oxidative stress and not caused by osmotic adjustment. Likewise, transgenic GM-wheat plants expressing an oat arginine decarboxylase cDNA exhibited increases in polyamine content in vegetative tissue and seeds (Bassie et al. 2008). This significant polyamine accumulation was heritable in subsequent T_i generations. These results demonstrate that polyamine levels can be altered in transgenic wheat plants resulting in phenotypically normal and fertile plants.

Genetically-modified crops with enhanced environmental stress tolerance require substantial advances in biosafety assessment and regulatory approval that are very different to the first generation of commercial transgenic crops with traits such as delayed ripening, herbicide tolerance, host plant resistance to insects or viruses, or male-sterility (De Greef 2004). Genetically engineered crops for abiotic stress-prone environments pose new questions regarding safety and impact. This new biosafety framework needs to be based on a comprehensive molecular characterization of the expression of regulatory genes and their cascade effects, the very likely different composition of the transgenic plants than their control counterparts under the abiotic stress, the potential impact of such regulatory genes in human health, as well as an environmental impact assessment, which needs to address both the potential escape of the tolerant crop (i.e., increased direct weediness) and the escape of the gene into wild relatives (i.e., increased indirect weediness).

HOST PLANT RESISTANCE

Safeguarding the crop from important pathogens and pests remain a key target in wheat breeding (Ortiz et al. 2008a). Many pathogens and pests of wheat continue to pose major breeding challenges worldwide due to their ability to evolve continuously, to migrate long distances, and to overcome the deployed race-specific resistance genes. For example, *Fusarium* head blight (FHB) of wheat, caused by *Fusarium graminearum* and other *Fusarium* species, is a major problem for wheat production worldwide because it causes significant yield losses. Transgenic wheat breeding offers an alternative approach to enhance the level of host plant resistance to this pathogen, especially after progress made so far by conventional genetic improvement remains insufficient to withstand epidemics. Over-expression of defense response genes in transgenic wheat enhanced FHB resistance in both greenhouse and field experiments (Mackintosh et al. 2007). A β-1,3-glucanase transgenic line had enhanced resistance, showed lower FHB severity, deoxynivalenol concentration, and percent visually scabby kernels compared to the susceptible cultivar Bobwhite. Likewise, GM-wheat expressing *Arabidopsis thaliana's NPR1* gene (At*NPR1*), which regulates the activation of systemic acquired resistance, conferred a heritable, type II resistance to FHB (Makandar et al. 2006). It appears that the At*NPR1*-expressing transgenic wheat is more responsive to an endogenous activator of plant defense.

Zhao et al. (2006) were able to identify transgenic wheat plants exhibiting resistant to powdery mildew after introducing an *α-1, 3-glucanase* gene using an *Agrobacterium* inoculum in the basal portion of wheat seedlings, whereas Janni et al. (2008) used the biolistic particle delivery system for developing GM-wheat lines expressing a bean PGIP gene (*PvPGIP2*) having a wide spectrum of specificities against fungal endo-polygalacturonases (PG), which degrade plant cell wall polysaccharides to penetrate and colonize the host tissue. Some transformed plants expressing the *a-1, 3-glucanase* transgene had higher resistance to powdery mildew than the control lines.

Transgenic wheat tissue showed increased resistance to digestion by the PG of *Fusarium moniliforme*, and at the plant level by the fungal pathogen *Bipolaris sorokiniana*, illustrating thereby the feasibility of improving wheat's defenses against pathogens through the expression in transgenic plants of proteins with the ability to counteract those produced by the pathogens.

Very recently a new strain of stripe rust –caused by *Puccinia striiformis*– was found and it could cause billions of dollars in crop losses across North Africa, the Middle East, and Central Asia (http://www.sciencealert.com.au/news-global-food-crisis/20100206-21024.html). Key cereal production regions are affected by this new stripe rust epidemic, e.g. up to 80% of fields in Syria according to recent surveys. Fu et al. (2009) reported the map-based cloning of the gene *Yr36* (WKS1), which confers resistance to a broad spectrum of stripe rust races at relatively high temperatures (25°-35°C). This gene includes a kinase and a putative START lipid-binding domain. Five independent mutations and transgenic complementation confirmed that both domains are necessary to confer resistance. Such finding shows that genetic transformation of wheat provides a means also for research methods to identify genes and study their function *in planta*.

Plant Architecture and Yield Potential

Ectopic expression of a *gibberellin 2-oxidase* gene (*PcGA2ox1*) decreased the content of bioactive gibberellins (GAs) in transgenic wheat, producing a broad range of dwarf plants (Appleford et al. 2007).The dwarf transgenic phenotype, which included dark-green leaves, increased tillering and in severe cases a prostrate growth habit, was replicated by the application of a GA biosynthesis inhibitor to the wild type. Ear rachis length, grain set, and grain size were smaller in GM-wheat lines.

Meyer et al. (2007) assessed in field experiments transgenic wheat plants expressing a modified ADP-glucose pyrophosphorylase (AGP) large subunit. AGPcatalyzes a rate limiting step in seed starch biosynthesis. The hard spring wheat cultivar Hi-Line was genetically transformed with a modified maize AGP large subunit sequence (Sh2r6hs) to increase AGP activity. Their results indicate that significant yield increases were more likely to occur in space-planted, irrigated environments than densely planted, rainfed environments, suggesting that limited abiotic resources may subsequently limit Sh2r6hs-associated yield enhancement. Sh2r6hs appears to confer a yield advantage in elite lines only in non-limiting environments.

More Nutritious and Healthy Food

The quality of proteins, which have a large effect on end-use quality, is controlled by known high (HMW) and low molecular weight glutenins and gliadins in wheat. Alvarez et al. (2000) introduced wheat HMW glutenin subunit genes *1Ax1* and *1Dx5*into a commercial wheat cultivar that already expresses five subunits. Over-expression of *1Dx5* increased the contribution of HMW glutenin subunits to total protein up to 22%, whereas silencing of all the HMW glutenin subunits was observed in two different events of transgenic wheat expressing the *1Ax1* subunit transgene and over-expressing the *Dx5* gene. The over-expression of subunit *1Dx5* also resulted in about 2-fold increase in mixing time, associated with a significant improvement in dough strength, and a lower resistance breakdown, suggesting thereby an important increase in dough stability (Alvarez et al. 2001). The expression of *1Ax1* and *1Dx5* transgenes, associated with silencing of all the endogenous high-molecular-weight glutenin subunits, resulted in flours with lower mixing time, peak resistance and sedimentation volumes due to lower gluten strength.

Transgenic research also was undertaken to assess the potential for improving nutritional quality in wheat. Heat-stable phytases in transgenic wheat may improve phosphate and mineral bioavailability in this crop (Brinch-Pedersen et al. 2006). Tamás et al. (2009) used an amaranth albumin gene encoding the 35-kDa AmA1 protein of the seed, which shows a high content of essential amino acids, in the biolistic transformation of bread wheat. Their results indicated that essential amino acid content could be increased.

Biosafety and Food Safety. Genetic transformation of wheat may use selective agents such as herbicides and antibiotics, and marker genes for resistance to favor the multiplication of the initially transformed cells. Hence, Gadaleta et al. (2008) generated transgenic durum wheat lines free of herbicide and antibiotic resistance marker genes. The positive selectable marker was the *Escherichia coli phosphor-mannose-isomerase* (*pmi*) gene, whose product catalyzes the reversible inter-conversion of mannose-6-phosphate and fructose-6-phosphate, allowing plant cells to utilize mannose as a carbon source. Their research shows that it will be possible to combine biolistic transformation by minimal gene cassettes with genetic segregation to make marker-free transgenic wheat.

Gene flow from transgenic wheat was noted at frequencies below 0.01% over distances of less than 12 m. (Gatford et al. 2006). The results show that gene flow may occur at extremely low frequency and over very short distances in the field. Hence, separation of transgenic and non-transgenic wheat plots greater than 12 m should ensure that contamination of adjacent non-transgenic fields remains less than 0.02%. Schoenenberger et al. (2006) showed however that natural hybridization and backcrossing between the weedy *Aegilops cylindrical* and bread wheat can lead to introgression of wheat DNA into the wild species.

Comparative gene expression analysis is potentially a very powerful tool in the safety assessment of transgenic plants since it allows for the detection of differences in gene expression patterns between a transgenic line and its non-GM original source. For example, Gregersen et al. (2005) compared the gene expression profile in developing seeds of wild type wheat and wheat transformed for endosperm-specific expression of an *Aspergillus fumigates* phytase. Their results showed that the expression of the codon modified *A. fumigates* phytase gene in the wheat seed had no significant effects on the overall gene expression patterns in the developing seed. The safety of GM wheat lines was also investigated through detailed comparisons of their composition with those of conventionally bred crops in field and greenhouse experiments (Shewry et al. 2007). Results suggest the feasibility for developing transgenic wheat lines that are substantially equivalent to conventional cultivars.

Pleiotropic effects are regarded as other main concerns for GM-wheat, e.g. the unintended side effects of a transgene or its genome insertion site on the regulation of other endogenous genes, which could potentially cause the accumulation of different secondary metabolites that may impact on the diet or on the environment. Flavonoids are one the most prominent group of secondary metabolites in wheat. They function as signaling or defense molecules. Ioset et al. (2007) used a robust and reproducible analytical method to compare the flavonoid content of GM-wheat expressing genes that confer increased fungal resistance with their non-GM siblings. They found significant differences in flavonoid composition among wheat cultivars whereas transgenic wheat lines with increased antifungal resistance showed only minor differences in their flavonoid composition relative to their non-GM siblings. The transgenes used to increase wheat defense to fungal pathogens do not therefore interfere with the flavonoid biosynthesis pathway, whereas the genetic background resulting from conventional breeding has a direct impact on the biological composition of flavonoids.

The potential impact on non-target organisms remains also as a main concern for introducing transgenic crops into agro-ecosystems. Very recently, von Burga et al. (2010) showed that transgenic resistant wheat does not affect the clonal performance of the aphid *Metopolophium dirhodum*. There was no evidence for a genotype × environment interaction between aphid clones and wheat lines. The results imply that the GM-wheat lines used in their experiment were of similar host plant quality as the non-transformed control lines and that the introduced transgene had no major effect on the performance of individual aphid clones.

OUTLOOK

The Australian Office of Gene Technology Regulator issued in June 2010 a license approving the limited and controlled release of up to 150 lines of transgenic wheat showing enhanced carbon assimilation in drought- and heat-prone environments http://www.ogtr.gov.au/internet/ogtr/publishing.nsf/Content/dir102. These transgenic wheat lines will be included in field trials (using a maximum of 0.1 ha) in Redland (Queensland) from June 2010 to December 2013, though cannot be used yet as human food or animal feed or for the production of therapeutic goods. This decision came after an extensive consultative process of its risk assessment and risk management plan (RARMP) with national and state governments, other public agencies, relevant local councils and the public. Any issues regarding health and safety of people and the protection of the environment were taken into account by the RARMP and for issuing this license solely for field testing.

Bernwald et al. (2005) estimated that substantial economic rents were foregone in North America due to the lack of a conducive environment for transgenic wheat, which led to the shelving of this technology by a leading private sector multi-national seed company in the middle of the 2000s. There were then strong concerns from farmers, trade organizations and even state governments about market acceptance (Fox 2009). Very recently the National Association of Wheat Growers, the U.S. Wheat Associates, the North American Millers' Association, the Independent Bakers Association, and the Wheat Foods Council (2009) made, however, a compelling case for introducing GM-wheat for the crop to remain competitive, at a time when its production and area shows a long-term downtrend in the USA. They acknowledge that it will be essential a commitment to choice in the marketplace by customers who wish to procure non-GM wheat and for farmers who will be willing to provide grains from non-transgenic crop harvests. This appeal by the US wheat industry seems to be bringing back the interest on the private multinational seed sector in breeding transgenic wheat. Such efforts should also stimulate support for public molecular wheat breeding, and to establish private-public partnerships for modern genetic enhancement of this crop.

CONCLUSION

There has been significant progress in wheat breeding by incorporating crop wild relatives (through re-synthesizing bread wheat) and DNA markers for aided-selection. The first successes are already noted in above sections. Emerging wheat transgenic technology starts showing its potential for the genetic betterment of this crop. It was announced very recently the decoding of the wheat genome sequence, which will assist in the medium term to identify genetic variation for traits such ashost plant resistance, crop adaptation to abotic stresses, biomass and grain yield. The longer term impact of the wheat genome sequencing will be measured through its use for developing modern breeding approaches and new cultivars grown by farmers worldwide.

REFERENCES

Akbari, M., Wenzl, P., Caig, V., Carling, J., Xia, L., Yang, S., Uszynski, G., Mohler. V., Lehmensiek, A., Kuchel, H., Hayden, M.J., Howes, N., Sharp, P., Vaughan, P., Rathmell, B., Huttner, E. & Kilian, A. (2006). Diversity arrays technology (DArT) for high-throughput profiling of the hexaploid wheat genome. *Theoretical and Applied Genetics*, 113, 1409–1420.

Alvarez, M.L., Gómez, M., Carrillo, J.M. & Vallejos, R.H. (2001).Analysis of dough functionality of flours from transgenic wheat.*Molecular Breeding*, 8, 103–108.

Alvarez, M. L., Guelman, S., Halford, N.G., Lustig, S., Reggiardo, M.I., Ryabushkina, N., Schewry, P., Stein, J. & Vallejos R.H. (2000).Silencing of HMW glutenins in transgenic wheat expressing extra HMW subunits.*Theoretical and Applied Genetics*, 100, 319–327.

Appleford, N.E.J., Wilkinson, M.D., Ma, Q., Evans, D.J., Stone, M.C., Pearce, S.P., Powers, S.J., Thomas, S.G., Jones, H.D. Phillips, A.L., Hedden, P. & Lenton, J.R. (2007). Decreased shoot stature and grain α-amylase activity following ectopic expression of a *gibberellin 2-oxidase* gene in transgenic wheat. *Journal of Experimental Botany*, 58, 3213–3226.

Arbelbide, M. & Bernardo, R. (2006).Mixed-model QTL mapping for kernel hardness and dough strength in bread wheat.*Theoretical and Applied Genetics*, 112, 885–890.

Baffes, J. & Haniotis, T. (2010).Placing the 2006/08 commodity price boom into perspective.Policy Research Working Paper 5371. Washington DC: The World Bank Development Prospects Group. 40 pp.

Balfourier, F., Roussel, V., Strelchenko, P., Exbrayat-Vinson, F., Sourdille, P., Boutet, G., Koenig, J., Ravel, R., Mitrofanova, O, Beckert, M. & Charmet, G. (2007). A worldwide bread wheat core collection arrayed in a 384-well plate. *Theoretical and Applied Genetics*, 114, 1265–1275.

Banks, T.W., Jordan, M.C. & Somers, D.J. (2009).Single-feature polymorphism mapping in bread wheat.*The Plant Genome*, 2, 167–178.

Bassie, L., C. Zhu, I. Romagosa, P. Christou and T. Capell. (2008). Transgenic GM-wheat plants expressing an oat arginine decarboxylase cDNA exhibit increases in polyamine content in vegetative tissue and seeds. *Molecular Breeding*, 22, 39–50.

Berwald, D., Carter, C.A. & Gruère, G.P. (2005).Passing up new technology: an illustration from the global wheat market. Davis, CA: Univ. California. 25 pp.

Bhalla, P.L., Ottenhof, H.H. & Singh, M.B. (2007). Wheat transformation – an update of recent progress.*Euphytica*, 149, 353–366.

Bonnett, D.G, Rebetzke, G.J. & Spielmeyer, W. (2005).Strategies for efficient implementation of molecular markers in wheat breeding.*Molecular Breeding*, 15, 75–85.

Borlaug, N.E. (2000). The Green Revolution revisited and the road ahead. Special 30th Anniversary Lecture, 8 September 2000, Oslo, Norway. Oslo, Norway: The Norwegian Nobel Institute.

Borlaug, N.E. (2007). Sixty-two years of fighting hunger: personal recollections. *Euphytica*, 157, 287–297.

Borlaug, N.E. & Dowswell, C. (2005). Feeding a world of ten billion people: a 21st century challenge. In R.L. Tuberosa., R.L. Phillips & M. Gale (Eds.), Proc. Intl. Congress In the Wake of the Double Helix: From the Green Revolution to the Gene Revolution, 27-31 May 2003, Bologna, Italy. Bologna, Italy: Avenue Media. Pp. 3–23.

Brinch-Pedersen, H., Hatzack, F., Stöger, E., Arcalis, E..Pontopidan, K. & Holm, P.B. (2006). Heat-stable phytases in transgenic wheat (*Triticum aestivum* L.): deposition pattern, thermostability, and phytate hydrolysis. *Journal of Agricultural and Food Chemistry*, 54, 4624–463.

Botez, C., Pamfil, D:, Curticiu, D, Cota, L. & Saulescu, N.N. (2009). Marker assisted selection for *Septoria tritici* resistance in wheat dihaploid lines. Notulae Botanicae Horti Agrobotanici Cluj-Napoca, 37, 253-255.

Brennan, J.P. & Martin, P.J. (2007).Returns to investment in new breeding technologies.*Euphytica* 157, 337-349.

Buerstmayr, H., Ban, T. & Anderson, J.A. (2009). QTL mapping and marker-assisted selection for *Fusarium* head blight resistance in wheat: a review. *Plant Breeding*, 128, 1–26.

Chen, X.M., Luo, Y. H., Xia, X.C., Xia, L.Q., Chen, X. et al. (2005.) Chromosomal location of powdery mildew resistance gene *Pm16* in wheat using SSR marker analysis.*Plant Breeding*, 124, 225–228.

Christopher, M., Mace, E., Jordan, D., Rodgers, D., McGowan, P., Delacy, I., Banks, P., Sheppard, J., Butler, D. & Poulsen, D. (2006).Applications of pedigree-based genome mapping in wheat and barley breeding programs.*Euphytica*, 154, 307–316.

Chu, C.-G., Xu, S.-S., Friesen, T.L. & Faris, J.D. (2008).Whole genome mapping in a wheat doubled haploid population using SSRs and TRAPs and the identification of QTL for agronomic traits. Molecular Breeding, 22, 251–266.

Collard, B.C.Y., Grams, R. A. , Bovill, W. D., Percyl , R. Jolley, C. D., Lehmensiek, A., Wildermuth, G. & Sutherland, M.W. (2005). Development of molecular markers for crown rot resistance in wheat: mappingof QTLs for seedling resistance in a '2-49' · _Janz_ population. Plant Breeding 124, 532—537.

Conway, G. & Toenniessen, G. (1999).Feeding the world in the twenty-first century.*Nature*, 402, C55–C58.

Cox, C.M., Murray, T.D. & Jones, S.S. (2002). Perennial wheat germplasm lines resistant to eyespot, *Cephalosporium* stripe, and wheat streak mosaic. *Plant Disease*, 86, 1043–1048.

Crossa, J., Burgueño, J., Dreisigacker, S., Vargas, M., Herrera, S., Lillemo, M., Singh, R.P., Trethowan, R., Franco, J., Warburton, M., Reynolds, M., Crouch, J.H. & Ortiz, R. (2007). Association analysis of historical bread wheat germplasm using additive genetic covariance of relatives and population structure.*Genetics*, 177, 1889-1913.

Das, M.K., Bai, G.-H.& Mujeeb-Kazi, A. (2007).Genetic diversity in conventional and synthetic wheats with drought and salinity tolerance based on AFLP.Canadian *Journal of Plant Science*, 87, 691–702.

Davies, J., Berzonsky, W.A. &Leach, G.D. (2006).A comparison of marker-assisted and phenotypic selection for high grain protein content in spring wheat.*Euphytica*, 152, 117–134.

De Greef W. (2004).The Cartagena Protocol and the future of agbiotech.*Nature Biotechnology*, 22, 811–812.

Dixon, J., Nalley, L., Kosina, P., La Rovere, R., Hellin, H. & Aquino, P. (2006).Adoption and economic impact of improved wheat varieties in the developing world.*Journal of Agricultural Science* (Cambridge), 144, 489–502.

Dong, C., Dalton-Morgan, J., Vincent, K. & Sharp, P. (2009).A modified TILLING method for wheat breeding.*The Plant Genome*, 2, 39–47.

Dreisigacker, S., Kishii, M., Lage, J. & Warburton, M. (2008).Use of synthetic hexaploid wheat to increase diversity for CIMMYT bread wheat improvement.*Australian Journal of Agricultural Research*, 59, 413–420.

Dubcovsky, J. & Dvorak, J. (2007).Genome plasticity a key factorin the success of polyploidywheat under domestication.Science,316, 1862–1866.

Dunwell, J.M. (2009). Transgenic wheat, barley and oats: future prospects. Methods in Molecular Biology – *Methods Protocols*, 478, 333–345.

Evenson, R.E. & Golin, D. (2002).Assessing the impact of the Green Revolution, 1960 to 2000.*Science*, 300, 758-762.

FAO. 2010. FAOSTAT. Food and Agriculture Organization of the United Nations, Rome. Italy. http://faostat.fao.org/site/567/DesktopDefault.aspx?PageID=567#ancor(Accessed on 3 August 2010).

Fox, J.L. (2009). Whatever happened to GM wheat.*Nature Biotechnology*, 11, 974–976.

Fu, D., Uauy, C., Distelfeld, A., Blechl, A., Epstein, L., Chen, X., Sela, H., Fahima, T & Dubcovsky, J. 2009. A *kinase-START* gene confers temperature-dependent resistance to wheat stripe rust. *Science*, 323, 1357–1360.

Gao, S.-Q., Chen, M., Xia, L.-Q., Xiu, H.-J., Xu, Z.-S., Li, L.-C., Zhao, C.-P., Cheng, X.-G. & Ma, Y.-Z. (2009). A cotton (*Gossypium hirsutum*) DRE-binding transcription factor gene, *GhDREB,* confers enhanced tolerance to drought, high salt and freezing stresses in transgenic wheat. *Plant Cell Reports*, 28, 301–311.

Gadaleta, A., Giancaspro, A., Blechl, A.E. & Blanco, A. (2008).A transgenic durum wheat line that is free of marker genes and expresses *1Dy10.J. Cereal Science*, 48, 439–445.

Gatford, K.T., Basri, Z. Edlington, J., Lloyd, J., Qureshi, J.A., Brettell, R. & Fincher, G.B. (2006).Gene flow from transgenic wheat and barley under field conditions.*Euphytica*, 151, 383–391.

Glover, J.D., Cox, C.M. & Reganold, J.P. (2007). Future farming: a return to roots. Scientific American, August 2007, 82–89.

Graybosh, R.A. & Peterson, C:J. (2010). Genetic improvement in winter wheat yields in the Great Plains of North America, 1959–2008. *Crop Science*, 50, 1882–1890.

Gregersen, P.L., Brinch-Pedersen, H. & Holm, P.B. (2005). A microarray-based comparative analysis of gene expression profiles during grain development in transgenic and wild type wheat. *Transgenic Research*, 14, 887–905.

GruszkaVendruscoloa, E.C., Schuster, I., Pileggi, M., Scapim, C.A., Correa Molinari, H.B., Marure, C.J. & Esteves Vieira, L.G. (2007). Stress-induced synthesis of proline confers tolerance to water deficit in transgenic wheat. *Journal of Plant Physiology*, 164, 1367–1376.

Gupta, P. K., Langridge, P. & Mir, R.R. (2009). Marker-assisted wheat breeding: present status and future possibilities. *Molecular Breeding*, 26, 145–161.

Hanocq, E., Niarquin, M., Heumez, E., Rousset, M. & Le Gouis, J. (2004). Detection and mapping of QTL for earliness components in a bread wheat recombinant inbred lines population. *Theoretical and Applied Genetics*, 110, 106–115.

Hayden, M.J., Kuchel, H. & Chalmers, K.J. (2004). Sequence tagged microsatellites for the *gwm533* locus provide new diagnostic markers to select for the presence of stem rust resistance gene S*r2* in bread wheat (*Triticum aestivum* L.). *Theoretical and Applied Genetics*, 109, 1641–1647.

Hazell, P.B.R. & Ramasamy, C. (1991).Introduction. In P.B.R. Hazell, C. Ramasamy &P.K. Aiyasamy (Eds.), The Green Revolution reconsidered. Baltimore, MD: John Hopkins Univ. Press. Pp. 1–7.

Hiebery, C.W., Thomas, J.B., Somers, D.J., McCallum, B.D. & Fox, S.L. (2007). Microsatellite mapping of adult-plant leaf rust resistance gene *Lr22a* in wheat. *Theoretical and Applied Genetics*, 115, 877–884.

Huang, X.Q., Kempf, H., Ganal, M.W., Röder, M.S. (2003). Advanced backcross QTL analysis for the identification of quantitative trait loci alleles from wild relatives of wheat (*Triticum aestivum* L.). Theoretical and Applied Genetics, 106, 1379–1389.

Huang, X.Q., Cöster, H., Ganal, M.W., Röder, M.S. (2004). Advanced backcross QTL analysis in progenies derived from a cross between a German elite winter wheat variety and a synthetic wheat (*Triticum aestivum* L.). *Theoretical and Applied Genetics*, 109, 933–943.

Hubert, B., Rosegrant, M., van Boekel, M.A.J.S. & Ortiz, R. (2010). The future of food: scenarios for 2050. *Crop Science*, 50, S33-S50.

Imtiaz, M., Ahmad, M., Cromey, M.G., Griffins, W.B. and Hampton, J.G. (2004). Detection of molecular markers linked to the durable adult plant stripe rust resistance gene *Yr18* in bread wheat (*Triticum aestivum* L.).*Plant Breeding*, 123, 401—404.

Janni, M., Sella, L., Favaron, F., Blechl, A.E., De Lorenzo, G. & D'Ovidio, R. (2008).The expression of a bean PGIP in transgenic wheat confers increased resistance to the fungal pathogen *Bipolaris sorokiniana*. *Molecular Plant Microbe Interactions*, 21, 171–177.

Jiang, G-L., Dong, Y., Shi, J. & Ward, R.W. (2007a).QTL analysis of resistance to *Fusarium* head blight in the novel wheat germplasm CJ 9306. II. Resistance to deoxynivalenol accumulation and grain yield loss.*Theoretical and Applied Genetics*, 115, 1043–1052.

Jiang, G-L., Shi, J. & Ward, R.W. (2007b).QTL analysis of resistance to *Fusarium* head blight in the novel wheat germplasm CJ 9306. I. Resistance to fungal spread. *Theoretical and Applied Genetics*, 116, 3–13.

Khush, G.S. (2005). Green Revolution: challenges ahead. In R. Tuberosa, R.L. Phillips & M. Gale (Eds.), Proc. Intl. Congress In the Wake of the Double Helix: From the Green Revolution to the Gene Revolution, 27-31 May 2003, Bologna, Italy. Bologna, Italy: Avenue Media. Pp. 37–51.

Kirigwi, F. M.,Van Ginkel, M., Brown-Guedira, G., Gill, B.S., Paulsen, G.M. & Fritz, A.K. (2007). Markers associated with a QTL for grain yield in wheat under drought. *Molecular Breeding*, 20, 401–413.

Khlestkina, E.K., Röder, M.S., Unger, O., Meinel, A. & Börner, A. (2007).More precise map position and origin of a durable non-specific adult plant disease resistance against stripe rust (*Puccinia striiformis*) in wheat.*Euphytica*, 153, 1–10.

Kuchel, H., Fox, R., Reinheimer, J., Mosionek, L., Willey, N., Bariana, H. & Jefferies, S. (2007). The successful application of a marker-assisted wheat breeding strategy.*Molecular Breeding*, 20, 295–308.

Kuchel ,H., Williams, K.J., Langridge, P., Eagles, H.A. & Jefferies, S.P. (2007a). Genetic dissection of grain yield in bread wheat. II. QTL-by-environment interaction. *Theoretical and Applied Genetics*, 115, 1015-1027.

Kuchel ,H., Williams, K.J., Langridge, P., Eagles, H.A. & Jefferies, S.P. (2007b). Genetic dissection of grain yield in bread wheat. I. QTL analysis. *Theoretical and Applied Genetics*, 115, 1029-1041.

Kuchel, H., Ye, G., Fox, R. & Jefferies, S. (2005). Genetic and economic analysis of a targeted marker-assisted wheat breeding strategy.*Molecular Breeding*, 16, 67–78.

Lage, J. & Trethowan, R.M. (2008).CIMMYT 's use of synthetic hexaploid wheat in breeding for adaptation to rainfed environments globally.*Australian Journal of Agricultural Research*, 59, 461–469.

Landjeva, S., Korzun, V. & Börner, A. (2007). Molecular markers: actual and potential contributions to wheat genome characterization and breeding. *Euphytica*, 156, 271–296.

Laperche, A., Brancourt-Hulmel, M., Heumez, E., Gardet, O., Hanocq, E., Devienne-Barret, F. & Le Gouis, J. (). Using genotype · nitrogen interaction variables to evaluate the QTL involved in wheat tolerance to nitrogen constraints. *Theoretical and Applied Genetics*, 115, 399-415.

Latini, A., Rasi, C., Sperandei, M., Cantale, C., Iannetta, M., Dettori, M., Ammar, K. & Galeffi, P. (2007). *DREB* in durum wheat Identification of a *DREB*-related gene in *Triticum durum* and its expression under water stress conditions. *Annals of Applied Biology*, 150, 187–195.

Lazo, G.R. & 40 co-authors.(2004). Development of an expressed sequence tag (EST) resource for wheat (*Triticum aestivum* L.): EST generation, unigene analysis, probe selection andbioinformatics for a 16,000-locus bin-delineated map.*Genetics*, 168, 585–593.

Li, G. Q., Li, Z. F. Yang, W. Y., Zhang, Y. He, Z.H. et al. (2006.)Molecular mapping of stripe rust resistance gene *YrCH42* in Chinese wheat cultivar Chuanmai 42 and its allelism with *Yr24* and *Yr26.Theor. Appl. Genet.* 112: 1434–1440.

Li, S., Jia, J., Wei, X., Zhang, X., Li, L., Che H., Fan, Y., Sun, H., Zhao, X., Lei, T., Xu, Y., Jiang, F., Wang, H. & Li, L. (2007). A intervarietal genetic map and QTL analysis for yield traits in wheat. *Molecular Breeding*, 20, 167–178.

Li, Y., Niu, Y.C. & Chen, M. (2009). Mapping a stripe rust resistance gene *YrC591* in wheat variety C591 with SSR and AFLP markers. *Theoretical and Applied Genetics*, 118, 339–346.

Lin, F. & Chen, Y. (2007). Genetics and molecular mapping of genes for race-specific all-stage resistance and non-race-specific high-temperature adult-plant resistance to stripe rust in spring wheat cultivar Alpowa. *Theoretical and Applied Genetics*, 114, 1277–1287.

Maccaferri, M., Sanguineti, M.C., Mantovani, P., Demontis, A., Massi, A., Ammar, K. Kolmer, J.A., Czembor, J.H., Ezrati, S. & Tuberosa, R. (2009). Association mapping of leaf rust response in durum wheat. *Molecular Breeding*, 26, 189-228.

Mackintosh, C.A., Lewis, J., Radmer, L.E., Shin, S., Heinen, S.J., Smith, L.A., Wyckoff, M.N., Dill-Macky, R., Evans, C.K., Kravchenko, S., Baldridge, G.D. Zeyen, R.J. &Muehlbauer, G.J. (2007). Overexpression of defense response genes in transgenic wheat enhances resistance to *Fusarium* head blight. *Plant Cell Reports*, 26, 479–488.

Makandar, R., Essig, J.S., Schapaugh, M.A., Trick, H.N. & Shah, J. (2006).Genetically Engineered resistance to *Fusarium* head blight in wheat by expression of *ArabidopsisNPR1.Molecular Plant Microbe Interactions*, 19, 123–129.

Martin, E.M., Eastwood, R.F., Ogbonnaya, F.C. (2004). Identification of microsatellite markers associated with the cereal cyst nematode resistance gene *Cre3* in wheat. *Australian Journal of Agricultural Research*, 55, 1205–1211.

Meyer, F.D., Talbert, L.E..Martin, J.M., Lanning, S.P., Greene, T.W. & Giroux, M.J. (2007).Field evaluation of transgenic wheat expressing a modified ADP-glucose pyrophosphorylase large subunit.*Crop Science*, 47, 336–342.

Milford, M.H. & Runge, E.C.A. (2007). Beachell and Borlaug, two giants of the American Society of Agronomy's first century. *Agronomy Journal*, 99, 595–598.

National Association of Wheat Growers, U.S. Wheat Associates, North American Millers' Association, Independent Bakers Association & Wheat Foods Council.(2009). How the introduction of modern genetic technology in wheat can help address a competitiveness crisis.8 pp.

Nematollahi, G., Mohler, V. Wenzel, G., Zeller, F.J. &. Hsam, S.L.K. (2007). Microsatellite mapping of powdery mildew resistance allele *Pm5d* from common wheat line IGVI-455. *Euphytica*, 159, 307–313.

Nocente, F., Gazza, L., Pasquini, M. (2007).Evaluation of leaf rust resistance genes *Lr1*, *Lr9*, *Lr24*, *Lr47* and their introgression into common wheat cultivars by marker-assisted selection. *Euphytica*, 155, 329–336.

Ogbonnaya, F.C., Imtiaz, M., Bariana, H.S., McLean, M., Shankar, M.M. Hollaway, G.J., Trethowan, R.M., Lagudah, E.S. & van Ginkel, M. (2008a).Mining synthetic hexaploids for multiple disease resistance to improve bread wheat.*Australian Journal of Agricultural Research*, 59, 421–431.

Ogbonnaya, F. C., Imtiaz, M., Ye, G., Hearnden, P.R., Hernandez, E., Eastwood, R.F., van Ginkel, M., Shorter, S.C. & Winchester, J.M. (2008b). Genetic and QTL analyses of seed dormancy and preharvestsprouting resistance in the wheat germplasm CN10955.*Theoretical and Applied Genetics*, 116, 891-902.

Olivera, P. D., Kolmer, J.A., Anikster, Y. & Steffenson, B.J. (2007).Resistance of Sharon goatgrass (*Aegilops sharonensis*) to fungal diseases of wheat.*Plant Disease*, 91, 942–950.

Ortiz, R. and 28 co-authors (2008a).Wheat genetic resources enhancement by the International Maize and Wheat Improvement Center (CIMMYT). *Genetic Resources and Crop Evolution*, 55, 1095–1140.

Ortiz, R., Iwanaga, M., Reynolds, M.P., Wu, H. & Crouch, J.H. (2007a).Overview on crop genetic engineering for drought-prone environments.SAT eJournal, 4, http://www.icrisat.org/journal/SpecialProject/sp3.pdf.

Ortiz, R., Mowbray, D., Dowswell, C. & Rajaram, S. (2007b). Norman E. Borlaug: The humanitarian plant scientist who changed the world. *Plant Breeding Reviews*, 28, 1–37.

Ortiz, R., Sayre, K.D., Govaerts, B., Gupta, R., Subbarao, G.V., Ban, T., Hodson, D., Dixon, J.M., Ortiz-Monasterio, J.I. & Reynolds, M. (2008b). Climate change: can wheat beat the heat? *Agriculture, Ecosystems & Environment*, 126, 45–58

Ortiz, R., Trethowan, R., Ortiz Ferrara, G., Iwanaga, M., Dodds, J.H., Crouch, J.H., Crossa, J. & Braun, H.J. (2007c). High yield potential, shuttle breeding, genetic diversity, and a new international wheat improvement strategy.*Euphytica*, 157, 365–384.

Ozkan, H., Levy, A.A. & Feldman, M. (2001). Allopolyploidy-induced rapid genome evolution in the wheat(*Aegilops–Triticum*) group. *Plant Cell*, 13, 1735–1747.

Peleg, Z., Saranga, Y., Suprunova, T., Ronin, Y., Röder, M.S., Kilian, A., Korol, A.B. & Fahima, T. High-density genetic map of durum wheat x wild emmer wheat based on SSR and DArT markers. *Theoretical and Applied Genetics*, 117, 103–115.

Pellegrineschi, A., Reynolds, M., Pacheco, M., Brito, R.M., Almeraya, R ,Yamaguchi-Shinozaki, K. & Hoisington, D. (2004). Stress-induced expression in wheat of the *Arabidopsis thaliana DREB1A* gene delays water stress symptoms under greenhouse conditions. *Genome*, 47, 493–500.

Prabhu, K.V., Gupta, S.K., Charpe, A. & Koul, S. (2007). SCAR marker tagged to the alien leaf rust resistance gene *Lr19* uniquely marking the *Agropyron elongatum*-derived gene Lr24 in wheat: a revision. *Plant Breeding*, 417–420.

Pumphrey, M., Bai, J., Laudencia-Chingcuanco, D., Anderson, O. & Gill, B.S, (2009). Nonadditive expression of homoeologous genes is established upon polyploidization in hexaploid wheat. *Genetics*, 181, 1147–1157.

Quarrie, S.A., Steed, A., Calestani, C., Semikhodskii, A., Lebreton, C., Chinoy, C., Steele, D., Pljevljakusic, D., Waterman, E., Weyen, J., Schondelmaier, J., Habash, D.Z., Farmer, P., Saker, L., Clarkson, D.T., Abugalieva, A., Yessimbekova, M., Turuspekov, Y., Abugalieva, S., Tuberosa, R., Sanguineti, M.-C., Hollington, P.A., Aragués, R., Royo, A. & Dodig, D. (2005). A high-density genetic map of hexaploid wheat (*Triticum aestivum* L.) from the cross Chinese Spring x SQ1 and its use to compare QTLs for grain yield across a range of environments. *Theoretical and Applied Genetics*, 110, 865–880.

Ravel, C., Praud, S., Canaguier, A., Dufour, P., Giancola, S., Balfourier, F., Chalhoub, B., Brunel, D., Linossier, L., Dardevet, M., Beckert, M., Rousset,M., Murigneux, A. & Charmet, G. (2006). DNA sequence polymorphisms and their application to bread wheat quality. *Euphytica*, 158, 331–336.

Reynolds, M.P. & Borlaug, N.E. (2006a).Impacts of breeding on international collaborative wheat improvement.*Journal of Agricultural Sciences* (Cambridge). 144, 3–17.

Reynolds, M.P. & Borlaug, N.E. (2006b).Applying innovations and new technologies for international collaborative wheat improvement.*Journal of Agricultural Sciences* (Cambridge).144, 95–110.

Reynolds, M., Dreccer, F. & Trethowan, R. (2007). Drought-adaptive traits derived from wheat wild relatives and landraces. *Journal of Experimental Botany*, 58, 177–186.

Rhone, B., Raquin, A.-L.& Goldringer, I. (2007).Strong linkage disequilibrium near the selected *Yr17* resistance gene in a wheat experimental population.*Theoretical and Applied Genetics*, 114, 787–802.

Rosewarne, GM., Singh, R.P, Huerta-Espino, J., William, H.M., Bouchet, S., Cloutier, S., McFadden, H & Lagudah, E.S. (2006). Leaf tip necrosis, molecular markers and *b*1-proteasome subunits associated with the slow rusting resistance genes *Lr46/Yr29*. *Theoretical and Applied Genetics*, 112, 500-508.

Schoenenberger, N., Guadagnuolo, R., Savova-Bianchi, D., Küpfer, P. & Felber, F. (2006). Introgression lines from GM-wheat to *Aegilops*: Molecular analysis, cytogenetics and fertility of introgression lines from transgenic wheat to *Aegilops cylindrical* Host. *Genetics*, 174, 2061–2070.

Schnurbusch, T., Bossolini, E., Messmer, M. & Keller, B. (2004). Tagging and validation of a major QTL for leaf rust resistance and leaf tip necrosis in winter wheat cultivar Forno. *Genome*, 94, 1040–1041.

Shen, Y.G., Zhang, W.K., He, S.H., Zhang, J.S., Liu, Q. & Chen, S.Y. (2003). An EREBP/AP2-type protein in *Triticum aestivum* was a DRE-binding transcription factor induced by cold, dehydration and ABA stress. *Theoretical and Applied Genetics*, 106, 923–930.

Shewry, P.R., Baudo, M., Lovegrove, A., Powers, S., Napier, J.A., Ward, J.L., Baker, J.M. & Beale, M.H. (2007). Are GM and conventionally bred cereals really different? *Trends in Food Science & Technology*, 18, 201–209.

Shewry, P.R. & Jones, H.D. (2005). Transgenic wheat: where do we stand after the first 12 years? *Annals of Applied Biology*, 147, 1–14.

Slade, A.J., Fuerstenberg, S.I., Loeffler, D., Steine, M.N. & Facciotti, D. (2005).A reverse genetic, nontransgenic approach to wheat crop improvement by TILLING.*Nature Biotechnology*, 23, 75–81.

Snape, J.W., Foulkes, M.J., Simmonds, J., Leverington, M., Fish, L.J., Wang, Y. & Ciavarrella, M. (2006). Dissecting gene x environmental effects on wheat yields via QTL and physiological analysis. *Euphytica*,154, 401–408.

Sorrels, M. (2007).Application of new knowledge, technologies, and strategies to wheat improvement.*Euphytica*, 157, 299–306.

Subbarao, G.V., Ban, T., Kishii, M., Ito, O., Samejima, H., Wang, H.Y., Pearse, S.J., Gopalakrishnan, S., Nakahara, K., Zakir Hossain, A.K.M., Tsujimoto, H. & Berry, W.L. (2007). Can biological nitrification inhibition (BNI) genes from perennial *Leymus racemosus* (Triticeae) combat nitrification in wheat farming? *Plant and Soil*, 299, 55–64.

Tamás, C., Kisgyörgy, B.N., Rakszegi, M., Wilkinson, M.D., Yang, M.-S., Láng, L., Tamás, L. & Bed, Z. (2009).Transgenic approach to improve wheat (*Triticum aestivum* L.) nutritional quality.*Plant Cell Reports*, 28, 1085–1094.

Thompson, J.P. (2008). Resistance to root-lesion nematodes (*Pratylenchus thornei* and *P. neglectus*) in synthetic hexaploid wheats and theirdurum and *Aegilops tauschii* parents.*Australian Journal of Agricultural Research*, 59, 432–446.

Trethowan, R.M. &. Mujeeb-Kazi, A. (2008). Novel germplasm resources for improving environmental stress tolerance of hexaploid wheat. *Crop Science*, 48, 1255–1265.

Trethowan, R., Reynolds, M.P., Ortiz-Monasterio, J.I. & Ortiz, R. (2007).The genetic basis of the Green Revolution.*Plant Breeding Reviews*, 28, 39–58.

Tsilo, T.J., Jin. Y. & J.A. Anderson. (2007). Microsatellite markers linked to stem rust resistance allele *Sr9a* in wheat. *Crop Science*, 47, 2013–2020.

Uauy, C., Paraiso, F., Colasuonno, P., Tran, R.K., Tsai, H., Berardi, S., Comai, L. & Dubcovsky, J. (2009). A modified TILLING approach to detect induced mutations in tetraploid and hexaploid wheat.*BMC Plant Biology*,9, 115 doi:10.1186/1471-2229-9-115.

van Ginkel, M. & Ogbonnaya, F. (2007). Novel genetic diversity from synthetic wheats in breeding cultivars for changing production conditions.*Field Crops Research*, 104, 86–94.

Varshney, R.K., Langridge, P. & Graner, A. (2007).Application of genomics to molecular breeding of wheat and barley.*Advances in Genetics*, 58, 121–155.

Vasil, I.K. (2007). Molecular genetic improvement of cereals: transgenic wheat (*Triticum aestivum* L.). *Plant Cell Reports*, 26, 1133–1154.

Verma, V., Foulkes, M.J., Worland, A.J., Sylvester-Bradley, R., Caligari, P.D.S. & Snape, J.W. (2004).Mapping quantitative trait loci for flag leaf senescence as a yield determinant in winter wheat under optimal and drought-stressed environments.*Euphytica*,135, 255–263.

von Burga, S., Müllera, C.B. & Romeis, J. (2010). Transgenic disease-resistant wheat does not affect the clonal performance of the aphid *Metopolophium dirhodum* Walker. *Basic and Applied Ecology*, 11, 257–263.

Wang, C., Zhang, Y., Han. D., Kang, Z., Li, G., Cao, A. & Chen, P. (2008). SSR and STS markers for wheat stripe rust resistance gene *Yr26*. *Euphytica*, 159, 359–366.

Warburton, M.L., Crossa, J., Franco, J., Kazi, M., Trethowan, R., Rajaram, S., Pfeiffer, W., Zhang, P., Dreisigacker, S. & van Ginkel, M. (2006). Bringing wild relatives back into the family: recovering genetic diversity in CIMMYT improved wheat germplasm. *Euphytica*, 149, 289–301.

Wei, B., Jing, R., Wang, C., Chen, J., Mao, X., Chang, X. & Jia, J. (2009). *Dreb1* genes in wheat (*Triticum aestivum* L.): development of functional markers and gene mapping based on SNPs. *Molecular Breeding,* 23, 13–22.

Wilde, F., Schön, C.C., Korzun, V., Ebmeyer, E., Schmolke, M., Hartl, L. & Miedaner, T. (2008).Marker-based introduction of three quantitative-trait loci conferring resistance to *Fusarium* head blight into an independent elite winter wheat breeding population.*Theoretical and Applied Genetics*, 117, 29–35.

William, H.M., Trethowan, R. & Crosby-Galvan, E.M. (2007). Wheat breeding assisted by markers: CIMMYT's experience. Euphytica, 157, 307–319.

Williams, K.J., Willsmore, K.L., Olson, S., Matic, M. & Kuchel, H. (2006).Mapping of a novel QTL for resistance to cereal cyst nematode in wheat.*Theoretical and Applied Genetics*, 112, 1480-1486.

Wu, H., Doherty, A. & Jones, H.D. (2008).Efficient and rapid *Agrobacterium*- mediated genetic transformation of durum wheat. *Transgenic Research*, 17, 425–436.

Wu, H., Pratley, J., Ma, W. & Haig, T. (2003). Quantitative trait loci and molecular markers associated with wheat allelopathy. *Theoretical and Applied Genetics*, 107, 1477–1481.

Xie, W. & Nevo, E. (2008). Wild emmer: genetic resources, gene mapping and potential for wheat improvement. *Euphytica*, 164, 603–614.

Xu, X.-Y., Bai, G.-H., Carver, B.F., Shaner, G.E. & Hunger, R.M. (2005).Mapping of QTLs prolonging the latent period of *Puccinia triticina* infection in wheat.*Theoretical and Applied Genetics*, 110, 244–251.

Xu, Y. & Crouch, J.H. (2008). Marker-assisted selection in plant breeding: from publications to practice. *Crop Science*, 48, 391–407.

Xue, S., Zhang, Z., Lin, F.,Kong, Z., Cao, Y., Li, C., Yi, H., Mei, M., Zhu, H., Wu, J., Zhao, D., Tian, D., Zhang, C. & Ma, Z.(2008). A high-density intervarietal map of the wheat genome enriched with markers derived from expressed sequence tags. *Theoretical and Applied Genetics*, 177, 181–189.

Yu, J.-B., Bai, G.-H., Cai, S..B. & Ban, T. (2006).Marker-assisted characterization of Asian wheat lines for resistance to *Fusarium* head blight. *Theoretical Applied Genetics*, 113, 308–320.

Zhang, D. & 42 co-authors.(2004). Construction and evaluation of cDNA Libraries for large-scale expressed sequence tag sequencing in wheat (*Triticum aestivum* L.).*Genetics*, 168, 595–608.

Zhang, P., Dreisigacker, S., Melchinger, A.E., Reif, J.C., Mujeeb Kazi, A., Van Ginkel, M., Hoisington D. & Warburton, M.L. (2005).Quantifying novel sequence variation and selective advantage in synthetic hexaploid wheats and their backcross-derived lines using SSR markers. Molecular Breeding, 15, 1–10.

Zhao, T.-J., Zhao, S.-Y., Chen, H.-M., Zhao, Q.-Z., Hu, Z.-M., Hou, B.-K. & Xia, G.-M. (2006). Transgenic wheat progeny resistant to powdery mildew generated by *Agrobacterium* inoculum to the basal portion of wheat seedling.*Plant Cell Reports*, 25, 1199–1204.

Zhao, X.L., Ma, W., Gale, K.R., Lei, Z.S., He, Z.H., Sun, Q.K & Xia, C. (2007).Identification of SNPs and development of functional markers for *LMW-GS* genes at *Glu-D3* and *Glu-B3* loci in bread wheat (*Triticum aestivum* L.).*Molecular Breeding*, 20, 223-231.

Zhou, W., Kolb, F.L., Yu, J., Bai, G., Boze, L.K., & Domier, L.L. (2004). Molecular characterization of *Fusarium* head blight resistance in Wangshuibai with simple sequence repeat and amplified fragment length polymorphism markers. *Genome*, 47, 1137–1143.

In: Wheat: Genetics, Crops and Food Production
Editor: Maria T. Almeida, pp. 179-208
ISBN 978-1-61209-307-9

Chapter 5

UTILIZATION OF WHEAT SPECIES AS CULTURAL FOOD

***Saeed Rauf*[1], *Saeed Akhtar*[2], *Ali Asghar*,[3] *Shahid Nazir*[4], *Sultan Ali Tariq*[5] *and Aziz-ur-Rahman*[4]**
[1]University College of Agriculture, Sargodha, Pakistan
[2]Department of Food and Horticultural Sciences, University College of Agriculture, Bahauddin Zakariya University, Multan, Pakistan
[3]National Institute of Food Sciences, University of Agriculture Faisalabad, Pakistan
[4]Wheat Research Institute, AARI, Faisalabad-Pakistan
[5]Social Science Division, National Agriculture Research Council Islamabad

ABSTRACT

Wheat is the world's most popular crop. It is grown over a large area and under a wide range of conditions and provides more nutrition to humans than any other species. Wheat is the leading source of vegetable protein in human food, having higher protein content than either maize (corn) or rice, the other major starchy crops. Wheat is the primary food staple in North Africa, south Asia and the Middle East, and is growing in popularity in Asia. In world different types of species are grown according to their geographical conditions. Some species are Einkorn (*T. monococcum*), Emmer (*T. dicoccon*), Durum (*T. durum*), Kamut (*T. polonicum), Bread* wheat (*T. aestivum*) and Spelt (*T. spelta*). These species are used for production of different cultural foods based on religion or customs including different breads. Various types of breads are made in different cultures that are consumed as staple or at special festivals. In addition breeding wheat for quality was discussed and possibilities of increasing yield through breeding products such as synthetics, hybrids and biotechnology product i.e. transgenics wheats was reviewed.

Keywords: Wheat species, cultural foods, genetics.

INTRODUCTION

Wheat (various species of the genus *Triticum*) is a grass having significant economic and edible value world wide. It is the principle staple food of South Asia, Middle East and many other nations. It seeds are rich in vegetable proteins and protein percentage in seed is higher than that of other crop species such as maize and rice. Furthermore, wheat occupies the highest cultivated area on global basis i.e. 22.5 billion hectares followed by rice (16.14) and maize (15.95) (FAO, statistics 2009). The history of domestication of wheat is directly related with the emergence of civil society.

Domestication of the wheat provided sustainability to the human societies since its harvest was storable for longer time. Its ability to self pollinate its self readily uncovered the rare, favorable spontaneous mutations which resulted in the origin of significant variation within species. Selection of mutations for traits such as larger seed size, non-shattering type (in ability of the seed to shed upon maturity) and seed dormancy (inability to germinate upon maturity) gave an ease of for harvesting.

Furthermore, harvest index (ratio of seed weight to the total biomass per plant) also show considerable improvement during the course of human selection making excellent harvestable species.

Wheat shows the highest impact of natural and human selection resulting in the emergence of various kinds of genomes, cultivated wheat species, land races and varieties in the genus. These wheat genotypes differ considerably for wheat quality traits. Therefore variable type of wheat products and recipies emerged from these species. All wheat quality traits or parameters are under the genetic control and are also affected by the environment. Therefore a chapter was written to under take the differences among the species for quality traits, variation within cultivated germplasm of species aestivum for quality, progress of improvement in these traits.

WHEAT GLOBAL DISTRIBUTION

In 2005, global wheat consumption was 101 kg per capita, with the highest consumption (623 kg per capita) in Denmark, but most of this (81%) was for animal feed. Wheat is the primary food staple in North Africa Middle East, and the South Asia and is growing in popularity in Asia. Summary of wheat global distribution in major wheat producing nations is given in Table 1. Top ten countries accounted 66% of the total world area and 70% of the total world production. Average yield of the top ten countries stands at 96.92 mounds per hectare and their yield progress at the rate 246% in 48 years (1961-2008). India has the largest area devoted to the wheat while China harvested the highest wheat production (Table 1). In addition China showed the highest progress in yield per hectare over years (1961-2008). Highest yield per unit area was obtained in Germany among the top ten wheat producing countries (Table 1).

Table 1. Wheat statistics among major wheat producing nations arranged in descending order on the basis of their share to the total production

Country	Area Harvested (Million hectares)	Production (million metric tonnes)	% share of the total production	Yield Mounds per hectare	Progress in yield (1961-2008)
China	23.62	112	16.23	119.06	751%
India	28.03	79	11.45	70.05	229%
USA	22.54	68	9.86	75.44	88%
Russia	26.07	64	9.28	61.15	-
France	5.49	39	5.65	177.52	197%
Canada	10.03	29	4.20	71.30	279%
Germany	3.21	26	3.77	202.18	183%
Ukraine	7.05	26	3.77	91.75	-
Australia	13.52	21	3.04	39.47	40%
Pakistan	8.55	21	3.04	61.28	198%
Total	148.11	485	70%	96.92	246%
World total	223.56	690	100%	77.15	183%

FAO, 2008.

Different demographical factors contribute in cultivation of the wheat crop in any region of the world. Some countries have favorable environmental conditions for winter wheat and some for hard type of wheat. So according to geographical locations there are different types of wheat mentioned here in table 2 & 3.

Table 2. Wheat species grown in Asia, North America, and Western Europe

Asia			
Country	Pakistan	India	China
Major classes	Hard Spring	Hard Spring	Hard Spring
North America			
Country	Canada	United States	Mexico
Major classes	Hard Red Spring	Hard Red Winter	Durum
Western Europe			
Country	France	Germany	United Kingdom
Major classes	Semi Hard Winter	Semi Hard Winter	Semi Hard Winter

Foreign Agricultural Service, USDA, 2001; FAO 2001.

Table 3. Wheat species grown in Eastern Europe, Western Asia, South America, and Oceania

Eastern Europe and western Asia					
Country	Russia	Kazakhstan	Ukraine	Turkey	Iran
Major classes	Winter Wheat	Spring Wheat	Winter Wheat	Winter Wheat	Winter Wheat
South America					
Country	Argentina		Brazil		
Major classes	Hard Spring		Semi Hard Spring		
Oceania					
Country	Australia		New Zealand		
Major classes	Hard Spring		Semi Hard Spring		

Foreign Agricultural Service, USDA, 2001; FAO 2001.

History and Origin

It has been shown from ancient archeological remains of einkorn, emmer and durum wheats that it originated about to 7500 to 7800 years BCE at Abu Hureyra near karacadag region of southeastern Turkey. Wheat was the first cereal to be domesticated and its cultivation began in fertile-crescent and the Nile Delta after about 7,000 BCE. Originally domesticated site included the vast areas such as southeastern parts of Turkey, Syria, the Levant, Israel, and Egypt. However recent studies have narrow down the areas to southeastern Turkey. Afterward, its cultivation spread to other parts of the world i.e. Greece, Cyprus and India by 6500 BCE, in Egypt after 6000 BCE, Germany and Spain by 5000 BCE. The Egytian used the wheat flour for making variable baking product. During 3,000 BCE wheat cultivation spread to various European countries (Wrigley, 2004). Ozkan et al. (2002) cited that origin of the cultivated wheat is supported by the two evidences i.e. geographical distribution and archeological records at particular site. However it was also shown that molecular studies have also provided further evidences of wheat origin. These studies have been cited in Ozkan et al. (2002). Ozkan et al. (2002) also studied the relationship between the domesticated and wild tetraploid species using 204 AFLP loci. The data indicated that domesticated types were more closely related to the wild population collected from south eastern Turkey. Furthermore, T. *dicoccoides* wild samples from Izmir, Turkey and was also closely related to cultivated emmer.

Relative Important Traits during Domestication

During the process of domestication, two plant mutations resulting in large grain size and non-shattering were important which latter laid the foundation of modern wheat. Selection by pre-historic farmers for genotype with large grain size improved seed quality and germination while selection for non-shattering types halted the wheat natural phenomenon to disperse seed upon ripening making wheat harvestable at maturity and avoiding seed losses (Eckardt, 2010).

MOLECULAR BASIS OF WHEAT QUALITY TRAITS

Wheat grain quality is dependet on several molecules i.e. proteins (Gluten or non gluten proteins), enzymes, starch, lipids. These molecules would also define the quality charchteristics of dough and bakery end product. Sissons (2008) made good review of all these molecules and may be used for further reading. Among these molecules protein composition have the highest impact on bakery end products (Sissons, 2008).Composition of protein is determined by two types of proteins i.e. glutenin and gliadin. They are made from polymers of very high molecular mass. These polymers may be reduced to several sub units. Glutein sub units may be regared on the basis of the molecular mass as 1>2**>2>6>8>18>12. Gene loci Glu-1 code for several high molecular weight glutenin sub units (HMW-GS) while Glu-A3, Glu-B3 and Glu-B2 encode for LMW-GS (30,000-50,000). Gliadin proteins are coded by the loci such as Gli-1, Gli-2. The products from these loci can be categorized into four types i.e. α,β,γ, ώ. Impact of high molecular weight subunit on gluten strength has been shown to vary and dependent on specfic molecule. For instance HMW-G 20 have been shown to reduce the gluten strength due to lower intra molecule disulpide bonds (Brites and Carillo, 2001; Sissons et al. 2005; Edward 2007). LMW-GS has been shown to contribute in greater denisty for cross linking the molecule, thus increasing strength (Edward et a. 2003). Contrastingly Sissons et al. (2007) showed that induction of HMW-GS increased the dough strength but non signficant impact on pasta firmness. Higher glutenin to gliadin ratio also dough strength (Edward et a. 2003 ; Sissons et al. 2005).

WHEAT: AS CULTURAL FOOD

Thousands of years of cultivation aided by natural and human selection have resulted in the evolution of an immense diversity of crop species and varieties. Throughout history, crops have been shaped and molded by people to meet diverse end-uses, preferences, cultural practices, and to allow continuous adjustment to changing growing conditions. Food diversity is a product of the natural environment and human settlement. Indigenous and traditional foods and food systems are disappearing, however, and this may be a significant loss and threat to future personal health and security at the regional and international level. Heterogeneity in socio-cultural conditions, food traditions, and agro-eco-logical settings, therefore, has favored the cultivation and utilization of diverse crop genetic resources (Bellon, 2004). Each crop species and/or variety has particular significance in the food culture, as a source of daily diet and of food and drink for special occasions (e.g., traditional, religious or other social functions) (Katz and Solomon, 2003).

WHEAT BASED CULTURAL FOODS OF SOME COUNTRIES

Wheat Breads in Different Countries

Because the evolution of wheat has occurred in different ways in different parts of the world, the nature of bread evolved differently in various countries and regions.

1) Those with high specific volume, such as the pan breads of the UK, North America and other Western countries;
2) Those with medium specific volume, such as French bread and many rye breads; those with low specific volume, for example flat breads of the Middle East and Eastern countries.

Table 4. Popular types of bread in Asian, Middle Eastern and North African countries

Country	Bread type
Afghanistan	Naan, Chapati and European bread
Algeria	Matlowa, French bread, Khobz El-daar, European bread
Bahrain	Tanoor, Arabic, Chapati, European bread, Samoon
Djibouti	Kisra
Egypt	Baladi, Shami, Samoon, French, Fatier, Shamsi, Bataw, European
Germany	Spelt
India	Chapati, Phulka, Roti, Parantha, Poori
Iran	Barbari, Tanoor (Taftoon), Lavash, Teeri, Suage, Sangak, European,
Iraq	Tanoor (Khobz), Arabic, Samoon, European bread
Ethiopia	Injera, Genfo, Anebabero, Chechebsa, Mulmul, Dabo-kollo, Nifro
Israel	Sadj, Tarboon, Kimaj (Arabic), European bread
Jordan	Armani, Mafrood, Sauj, European bread
Kuwait	Tanoor, Arabic, European bread
Lebanon	Lebanese, Suaj, European bread
Libya	French bread, Arabic
Morocco	Moroccan Khobz El-daar, French, European
Oman	Arabic, Tanoor, Chapatti, European
Pakistan	Roti, Parantha, Chapati, Naan
Saudi Arabia	Samouli, Mafoud, Tannouri, Burr, Tamees, Korsan European bread
Somalia	Injera
Sudan	Injera, Shamsi, Baladi
Syria	Mafrood (Arabic), Armani, Samoon, Suaj, European
Turkey	Balzuma, Gomme, Yafka, French and other European bread
United Arab	Tanoor, Chapati, Arabic, Samoon, European Emirates
Yemen	Roti, Malouge, French, European bread

Source: (Qarooni, 1999), (Tsegaye and Berg, 2007), (Nagao, 1995), (Abourezk and Sanaa, 2000).

WHEAT CLASSIFICATION

Cyto-Genetic Resources of Wheat

Many taxonomic classification systems are used for wheat species. Chromosome number (degree of polyploidy), and which parental grass genome is present (signified as the A, B, or D genomes) are the two major classification criteria that are used. However the basic set of chromosome remain the same (x=7) in all type of genomes and ploidy levels. Genomes i.e. A, B and D are homeologous to each other i.e. can partially pair with each other. However presence of various type of genes limits their ability to pair with each other. For instance Luo *et al.* (1996) identified a gene *Ph1* which was shown to suppress the pairing between the chromosome 1 of A and B genome.

A) Diploid Species

1) Einkorn (*T. monococcum*)

A diploid (AA) species with wild and cultivated variants; one of the earliest wheats to be cultivated. Suited to cooler climes, it has survived as a cultivated crop in the Vaucluse region of France and the Southern Alps for use in a local porridge. The einkorn flours were characterized by high protein, high ash, a very high carotene content, fiber quality and small flour particle size when compared to the modern bread wheats (Borghi *et al.* 1996; Flippo *et al.* 1999; Løje *et al.* 2003; Hidalgo *et al.* 2006). Borghi *et al.* (1996) showed that protein contents of this wheat ranged from (13.2 to 22.8%), in comparison to bread wheat which contained (10.8-13.3%). They also showed that some other quality traits were comparable to bread wheat such as 16% of the 1,099 accessions have SDS sedimentation values above 60 ml which is considered potential for making bread. Hidalgo *et al.* (2006) showed that carotenoids (lutein), averaged 8.41 µg/g dm, 2–4 times more than non-einkorn wheats (*Turgidum and aestivum*), with a maximum of 13.4 µg/g dm. Tocols averaged 77.96 µg/g dm, exceeding to the non-einkorn wheats, with a maximum of 115.85 µg/g dm. The most abundant tocol was β-tocotrienol (48.22 µg/g dm), followed by α-tocotrienol (12.77 µg/g dm), α-tocopherol (12.18 µg/g dm) and β-tocopherol (4.79 µg/g dm); the mean tocotrienol/tocopherol ratio was 3.68.

According to the Løje *et al.* (2003) the dehulled samples of einkorn had small grain size (21·2–37·4 mg/kernel), very soft endosperm texture (hardness indexes −7·3 to 27·2), high ash content (2·3–2·8% dm), variable protein content (10·3–19·5% dm), low content of (1→3;1→4)-β-glucan (0·29–0·71% dm), the total dietary fibre in einkorn (selected samples) was low (7·6–9·9% dm) compared to common wheat, the lysine content in ranged from (1·51–3·15 g lysine/100 g protein), high falling number (mean 362 sec) and high amylograph viscosities (mean 1185 BU). Degidio *et al.* (1993) compared this species with bread wheat and indicated that this species had similar flour yield than bread wheat but higher yield than durum wheat but this species had poor gluten content which has less extensibility and greater stickiness and low water retention capacity. Similarly, Corbellini *et al.* (1999) compared Einkorn flour with bread wheat and indicated that one of the einkorn lines (ID529) had higher farinograph stability and degree of softening than bread wheat. Furthermore, cookies produced with einkorn flour were larger in diameter and thinner than those produced with soft wheat flour.

B) Allo-Polyploid Species

Allopolyploidy is a condition in which multiple copies of various genomes are concentrated in a single species. Alloploid species are formed when two diverse species able to mate each other followed by genome doubling. Diagrammatically it can be shown as follows:

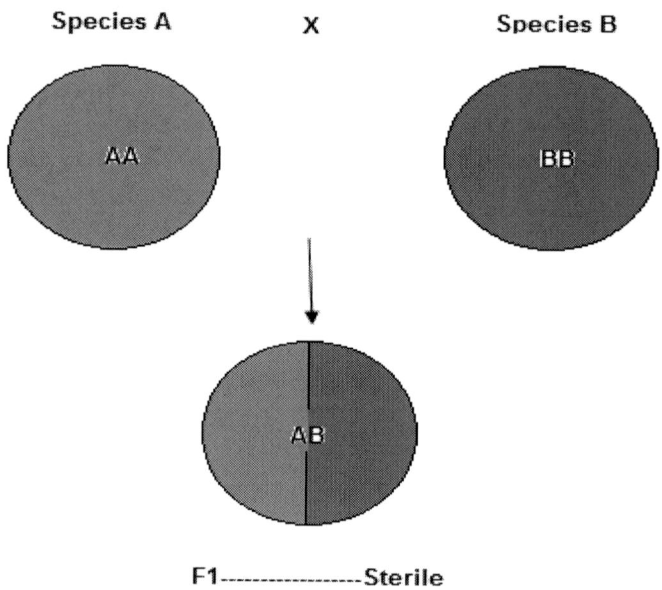

Figure 1. Diagramtic origin of allo-diploid species as aresult of hybridization between two diverse wheat species.

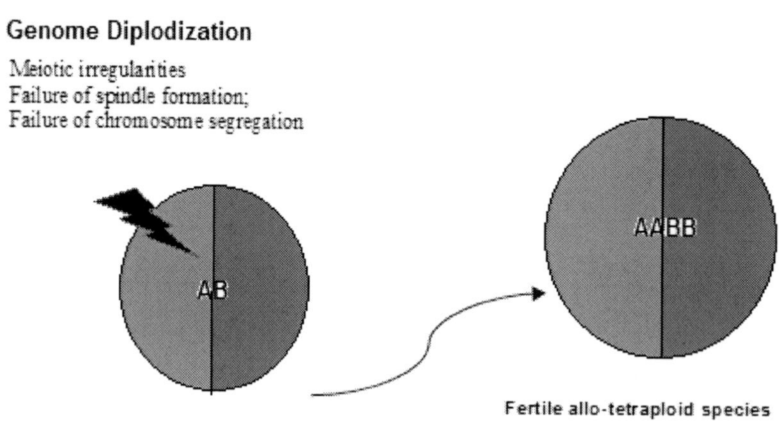

Figure 2. Origin of allo-tetraploid species due to genome duplications.

Allopolidization was of two phases, first it occurred between two diploid species of wheat which resulted in the evolution of tetraploid species which upon further hybridized with another diploid species gave rise to allo-hexaploid species (Figure 1 & 2). Thus evolution of hexaploid species resulted in the concentration of three genomes in single species (AABBDD). Comai (2006) mentioned various advantages of alloploid species such as bringing two diverse genomes in specie was accompanied by more hybrid vigor, growth, masking of lethal alleles. Similarly, Rauf *et al.* (2010) also mentioned that hexaploid species were more tolerant to abiotic stresses in comparison to diploids. Furthermore, allo-ploid species provided the fuel for the million years evolution in wheat and resulted in various differentional and epigenetic changes in gene expression.

Tetraploid Wheats

1) Emmer (*T. dicoccon*)

A tetraploid (AABB) hulled species, cultivated in ancient times where it grew in warmer climates than einkorn. It is considered to be the progenitor of modern tetraploid and hexaploid wheat species (Peng *et al.* 2003). Emmer flour was satisfactory to produce a good loaf of bread; the quality was not equal to breads made from common wheat. No longer in common use, except in Europe where it is sold under its Italian name farro. However, emmer has been extensively used in understanding the pattern of variations, evolution and allele controlling seed storage protein sub units such as HMW glutenin, LMW-glutenin subunits and omega- and gamma-gliadins. Many studies have concluded that emmer wheat have the potential to improve the quality of durum or bread wheat (Nevo and Payne, 1987; Ciaffi *et al.* 1993). Nevo *et al.* (1987) showed that two high molecular weight (HMW) glutenin loci (Glu-A1 and Glu-B1) are rich in allelic diversity i.e. they were shown to be controlled by 11 and 15 alleles in A and B genomes respectively. Ciaffi *et al.* (1993) identified 44 pattern of HMW glutenin sub unit in *dicoccoides* which were found to be controlled by 15 alleles in A genome and 14 alleles in B genome, out of which 27 allelic variants were not present in bread or durum wheat. Similarly, Pflüger *et al.* (2001) also identified four allelic variant at loci *Glu-A1* in Spanish accessions, out which one was not previously mentioned before. In addition high level of diversity was also found for low molecular weight glutenin sub units. 23 pattern were isolated on gel.

2) Durum (*T. durum*)

The only tetraploid (AABB) form of wheat in wide use today, and the second most widely cultivated wheat species. It has a similar wild tetraploid grass progenitor to emmer, but is more suited to semi-arid regions. It is mainly produced for several specific end-products; such as local pasta, couscous and burghul in Mediterarian regions (Elouafi and Nachit 2004; Zhang *et al.* 2008). Bright yellow color, firmness and low cooking losses are important quality attributes for pasta (Zhang *et al.* 2008).

3) Kamut (T. polonicum or turanicum)

A trademarked but ancient durum relative, originally from the middle-east, with a large grain size and high protein content. It is grown in small quantities that are extensively marketed in the USA; more suited for pasta than for bread making. Kamut products include whole grain flour, breads, hot and cold cereals, pastas (Gauthier *et al.* 2006). Piergiovanni *et al.* (2009) noted that distinct feature of kamuts were high selenium and pigments. Grausgruber *et al.* (2005) compared the performance of *turanicum* accessions with *turgidum* and found this species inferior to yield and its resistance against biotic and abiotic stresses. However, it was noted that species had very high 1000-grain weight (50-60g), due to this valuable trait species was easily marketable and this trait may be introgressed in durum wheat.

C) Hexaploid Wheats

Hexaploid wheats arise by hybridization between tetraploid wheat and a diploid native grass (DD), with chromosome formula AABBDD. The evolution of hexaploid wheat can be diagrammatically shown as follows:

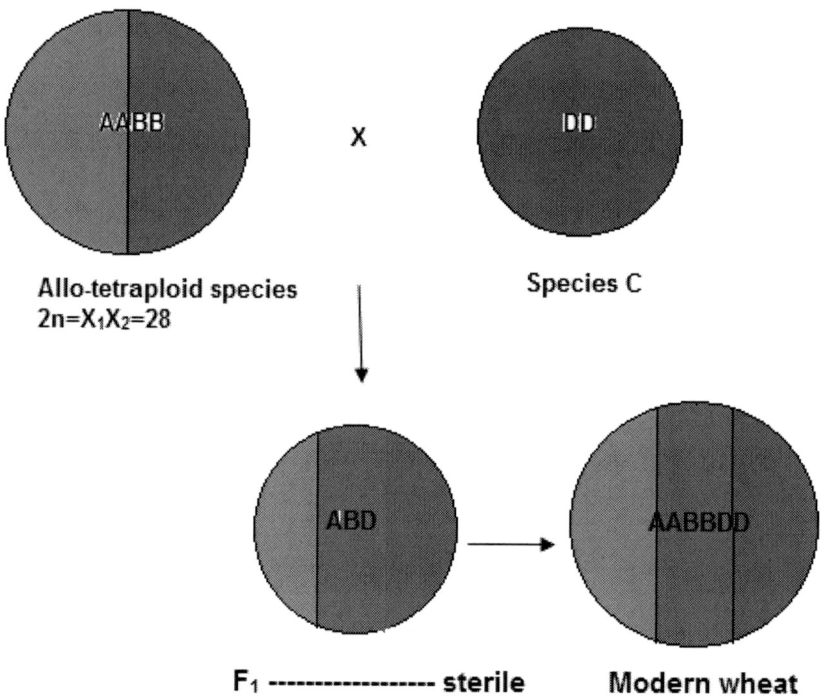

They are all high in the protein gluten, which gives fluffy leavened bread.

1) Bread wheat (*T. aestivum*)

A hexaploid (six chromosome sets, AABBDD) species that is the most widely cultivated wheat, originally arising in antiquity from a *Triticum dicoccoides* (AABB) which crossed with a diploid native grass *Aegilops tauschii*, (DD) near the Caspian sea, introducing improved tolerance of colder climatic conditions. The AA genome of bread wheat has the origin from diploid species *Triticum urartu* (Dvorak et al. 1993) while donar of B genome is unknown or controversial. Kilian et al. (2007) noted that B genome has its origin from *Aegilops speltoides* after analysing fifty nine representative of this species with primers specific to B genome.

2) Spelt (*T. spelta*, Known as Dinkel in Southern Germany and Often Confused with Emmer (Farro))

Another hexaploid (AABBDD) species used in central Europe to make breads and soups. Marconi *et al.* (1999) showed that protein and fat contents in spelt were high in the grains (15.7 and 4.4% db, mean value, respectively), total fiber varied from 10.5 to 14.9% db, the average β-glucan was 1.2% db, uniform milling performance (determined by yield, damaged starch, ash, and particle-size distribution in the flour), gluten strength was low, sensory and chemical evaluations of the pastas indicated that spelt is suitable for obtaining good-quality pasta. Bosanjaska and Francakova (2002) showed that all the spelts cultivars had high wet gluten contents (37.12%) with lower swelling values (9.3 ml) and lower sedimentation values (37.4 ml) and the specific volume (under 310 ml.100 g–1).

Ranhotra *et al.* (1995) compared the performance of spelt with hard red wheat and found that spelt was inferior for milling yield and differed non-significantly for traits such as protein, fiberes and minerals. Yan *et al.* (2003) compared the allelic composition for gluten among European spelt and its related hexaploid and tetraploid species. Seven new glutenin alleles (designated A1a*, B1d*, B1g*, B1f*, B1j*, D1a* at Glu-1 and A3h at the Glu-3 loci, respectively) in European spelt wheat were identified. Two HMW-GS alleles, Glu-B1d* and Glu-B1f* were present in cultivated emmer (*Triticum dicoccum*), but they were not observed in bread wheat (*Triticum aestivum* L.). The allele Glu-B1g* was detected in club wheat (Triticum compactum L.) but absent in bread wheat. Bonafaccia *et al.* (2000) compared the flour product of spelt and bread wheat and indicated that spelt wheat breads contained higher rapidly digested starch (RDS) and a higher starch digestion index (SDI) in comparison to wheat bread. Furthermore whole spelt wheat flour contained more protein, ash and dietary fibre than white spelt wheat flour.

WHEAT BREEDING

Genetic Resources of Wheat

It is generally perceived that rapid urbanization and modern intensified agriculture have threatened wheat natural gene sanctuaries while advanced plant breeding techniques have narrowed down the genetic diversity in cultivated germplasm (Rauf *et al.* 2010). The situation requires the conservation of diverse wheat germplasm in order to conserve the genetic diversity in the crop species. Wheat gremplasm may be categorized as wild species; landraces and cultivated germplasm which may further comprised of cultivated, obsolete cultivars and advanced lines. The genetic diversity was found to be the highest in wild species and the lowest in inbred lines (Rauf *et al.* 2010).

In order to conserve the genetic resources, large wheat exitu collections have been conserved at various international and regional plant genetic resource centers. The collectors include food agriculture organization, international plant genetic resource institute and CIMMYT. According to an estimate, 0.65 million accessions have been conserved in these institutes (FAO, 2006). However it is also important to utilize these genetic resources for improvement the condition of mankind. In the near past, there was trend to utilize only few superior parents in generating trangressive segregates. New better yielding cultivars continue to develop based on the new recombination of alleles but little attention were paid to improve the genetic base of the cultivar. Resultantly narrow genetic base cultivars with high yield potential were developed.

The efforts for the improvement of the species taking place at the CIMMYT are worth full to mention here. Over the years, CIMMYT have been the sole supplier of the improved germplasm to poor or developing nations. The semi dwarf plant material supplied by the CIMMYT has caused green revolution in the countries like Pakistan, India and Mexico in the late 1960s. Theory to decrease the plant height in order to introgress lodging resistance and increase the harvest index in wheat was given by Dr. Norman E. Borlaug and his team. Dr. N.E. Borlaug was awarded Nobel Prize on his efforts to develop semi dwarf wheat and emerged as one of the greatest wheat breeder of the previous century.

Trends have now been changed at CIMMYT; efforts are now been to utilize the wild relative to produce genetic diverse plant material with high yield and better resistance against diseases and insect. Dreisigacker *et al.* (2008) mentioned that more than 1000 synthetic wheat line were developed at CIMMYT. The generated material was superior in term of genetic diversity, resistance to diseases but comparable in term of yield with elite cultivars.

Breeding History of Wheat in Pakistan

In Pakistan, largest crop area is devoted to wheat and quantity produced is more than that of any other crop (Farooq *et al.*, 2010). It contributes 14.4 percent to the value added in agriculture and 3.1 percent to GDP. Wheat was cultivated on an area of 9.04 million hectares with the size of wheat crop is provisionally estimated at 23864 million tons (Pakistan Economic Survey 2009-10).

Wheat breeding in Pakistan comprised of two phases i.e. pre green revolution or post green revolution period.

Research Organizations

In Pakistan significant efforts have been devoted for the evolution of new crop varieties in various organizations. In contrast to the other countries of the world, agricultural research, education and extension are fragmented under the influence of Food Agriculture Organization. Universities are the centre of the post graduate level thesis based research only and faculty members have full work load of teaching. Therefore, little attention is paid to the evolution of new crop varieties or plant material in Agriculture Universities. Evolution of high yielding varieties takes place at the provincial research organization. The efforts at the provincial organization are supplemented and cemented by the federal research organization i.e. Pakistan Agriculture Research Council (PARC).Moreover it provide a national flavor to the research activities carried out at provincial level. PARC do not directly control or influence the research activities of provincial research organizations. However, it may support the research activities by giving funding to the productive research projects on competitive basis. Moreover, it provides a neutral platform for conducting national yield trials of various field crops and yet again it hardly influence the decision of provincial varietal approval committee. Public research organizations are the heart of all research activities while a little contribution is also been provided by the national or multinational private companies.

However their contribution is limited to the introductions and multiplication of advanced breeding material provided by public breeders.

Essence of the plant breeding in Pakistan lies in the welfare of the human being where farmer can grow, multiply or sell the seed of public crop varieties for infinite period unless particular crop variety become outdated or banned. Public breeders cannot demand any right or economic share over the evolved variety from farmer or from any other organization.

Wheat Research Institute Faisalabad, Pakistan

Wheat Research Institute, Faisalabad, has a bright history for developing new wheat varieties. This institute has released more than 70 promising wheat varieties grown very successfully over the years nationwide. The development of Mexi-Pak, Pak81 and Inqilab-91 was among the landmark achievements and attained world fame. Mexi-Pak not only revolutionized the wheat research work in Pakistan but it also spread to many other countries of the world. Among these existing wheat varieties Inqilab-91 covers about 60-70% area under wheat in Pakistan (Hussain et al., 2010). Major emphasis of wheat breeding in the institute revolve around the development of high yield potential, better rust resistance, good quality, tolerance to biotic and abiotic stresses and wider adaptability. Synthetic hexaploid germplasm have been very useful sources for resistance/ tolerance for leaf rust, stripe rust, mineral toxicity, drought, salinity, heat, cold, sprouting, waterlogging (Villareal *et. al.* 2001) and many other traits. Hence a separate segment in wheat breeding program at Wheat Research Institute, Faisalabad has been inducted for the introgression of these novel genes in commercial wheat varieties which hopefully will yield excellent results in future.

Dissection of Pakistani Wheat Germplasm for quality

Evolution of Pakistani wheat germplasm occurred into phases i.e. Pre green revolution or post green revolution. Varieties released before green revolution were evolved during the period of 1933-1964. These varieties were evolved from indigenous land races Type 8A, 8B or Type 9D which belong to *T. aestivum* sub specie *sphaerococum*. Varieties from this species were tall and lodge at the time of maturity. Therefore these varieties were poor yielder and economically less profitable. However, these varieties have excellent seed quality. At this juncture CIMMYT came to rescue from this situation. CIMMYT supplied dwarf wheat germplasm for the evolution of high yielding germplasm. Thus several dwarf wheat cultivars i.e. Mexi-Pak (Figure 3), were released for general cultivation which dramatically increased yield. Therefore, this era was marked as Green Revolution period starting from 1965 onward. During green revolution period and onward, cultivars were continuously released using the CIMMYT breeding lines as source material. Comparison of wheat quality traits of germplasm released during pre-green revolution period with that of post green revolution period will be interesting and show the change in wheat quality due to introduction and adaptation of dwarf material from CIMMYT. Pasha (2006) evaluated the Pakistani wheat germplasm for various quality traits for two years. Mean of varieties were cateogerized into various temporal periods i.e. 1933-1958; 1965-1980; 1981-1990; 1991-2000. These mean were subjected to various biometrical techniques i.e. genotypic coefficient of variation (GCV that is index of genetic variation), heritability. Year effect was used to determine the environmental effects on these traits. Analyses of the data showed differential impact of pre or post green revolution era over the change in the mean, GCV and heritability. Traits such as 1000-seed weight (g), particle size index, grain length, SDS sedimentation were increased. Among these traits SDS sedimentation showed the highest magnitude of increase. On the other hand traits such as test weight peraling value (%), NIR hardness, dry gluten, zeleny value (mL) and water absorption (mL) showing decreasing trend.Zeleny value showed the highest decrease (12.52%), followed by test weight (6%) and dry glutein (5%). Wet gluten and grain width showed no change. Although pearling value showed decreased in its mean value during (1965-1980), yet it showed increasing trend in further decades.However, peraling value of pre green revolution era was never recovered. Wet gluten increased during 1965-1980 and afterward it showed decreasing trend. Mean values of dry gluten and NIR hardness showed continouse decrease in all decades of post green revolution.Wet gluten showed the highest magnitude of GCV% (98.08%) which dried upto 27% in varieties released during the post green revolution period. Post green revolution varieties showed the highest genetic variation for 1000-seed weight (37%). Introduction of dwarfing material generated genetic variation of 32% (post green revolution) as original value of 7% (pre green revolution) in pearling value %. There was no variation or very low magnitude of variation in traits such as NIR hardness, grain length, grain width, water absorption. For traits such as SDS sedimentation there was an increase in genetic variation. In addition dry gluten also showed some minor increase in GCV%. Traits such as 1000-grain weight, test weight, pearling value (%), wet gluten and SDS sedimentation showed very high magnitude of heritability in both eras.

Dry gluten and grain weight showed significant impact of environment as indicated from low heritability. On average heritability estimates of most of the quality traits decreased in post green revolution period thus showing that varieties released during this period were sensitive to the environment. In summarizing the above results it may be concluded that green revolution occurred at the cost of several quality traits such as dry gluten, pearling value, zeleny value and grain hardness and no considerable effort has been made to improve these traits.

A significant genetic diversity has been reported in the land races of the wheat collected from various regions of Pakistan. Tahir et al. (1996) evaluated sixty hexaploid wheat landraces collected from five regions of Pakistan for high molecular weight (HMW) glutenin subunits. The germplasm showed high level of diversity on the basis of HMW glutenin subunit compositions. The Twenty four alleles were distribution in three locus i.e. four belonged to *Glu-Al,* 12 to *Glu-B1* and eight to *Glu-D1* locus. The number of novel HMW glutenin subunits detected were 1,4 and 6 at the three loci *(Glu-A1, Glu-Bl, Glu-D1),* respectively.

On the basis of results it was shown that maximum variation was observed in the Baluchistan and Gilgit regions of Pakistan in terms of distribution of novel *Glu-1* alleles. Similarly, Masood et al. (2004) and sultana et al. (2007) also reported high genetic variation for high molecular weight glutenin subunits in pre green revolution germplasm. Masood et al. (2004) detected 8 alleles in 77 land races collected from various areas of Pakitan at Glu-1 loci out of which three belong to each Glu-A1 and Glu-B1 and 2 belong to Glu-D1 locus. The most frequent subunits were 1 and 2 at Glu-A1 locus, 7+8 at Glu-B1 locus and 2+12 at Glu-D1 locus. Sultana et al. (2007) also detected of 14 alleles at Glu-1 loci, three belonged to Glu-A1, nine to Glu-B1 and two to Glu-D1 locus.

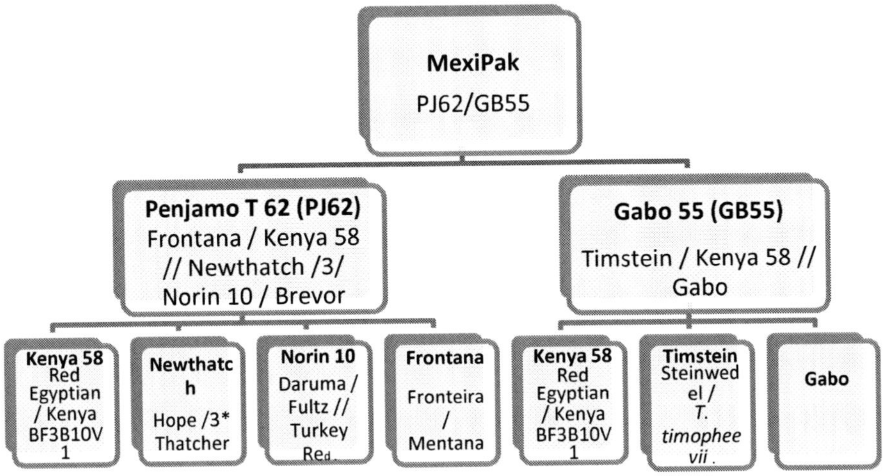

Figure 3. Pedigree of Pakistani wheat cultivar Mexi-Pak.

Wheat Breeding Products

1) Synthetic Hexaploid Wheat

Understanding the pattern of evolution of heaxaploid wheat species led the plant breeder to re-synthesize the hexaploid wheat. Sometime chromosome number was changed to introduce more valuable chromsomes in the species. Zhang *et al.* (2008) crossed *Triticum turgidum* × *Aegilop tauschii* followed by subsequent genome doubling with colchicine. A fertile plant with 56 chromosomes was identified and selected. The synthetic wheat species lines were constituted with the aim of increasing stress resistance and genetic diversity in cultivated wheats (Kent and Evers, 1998). For instance, Qazi *et al.* (2001 a, b) constituted synthetic hexaploid wheat in order to introgress karnal bunt resistance (Tilletia indica, S.tritici and C.Sativus) in the species; wheat powdery mildew, stripe rust (Tan *et al.* 2009); Russian wheat aphid resistance (Lage *et al.* 2004); leaf spot (Iftikhar *et al.* 2009). In addition to the resistance against disease synthetics have also been shown to be superior in terms of yield and quality. Lage *et al.* (2006) showed that synthetics were better in term of proteins (15.5%) over control (13.1%). In addition synthetics with high higher grain weight was also identified.

In addition to this, wheat substitution lines were also developed by replacing one of the chromosomes with other species. For instance, chromosome 1 of B genome was replaced with the chromosome one of R genome of Rye through back crossing. As each chromosome exits in pair, therefore substituted line will have 1B/IR chromosomes. Dhaliwal *et al.* (1987) indicated that these substituted lines are now been widely used in wheat breeding programmes as chromosome 1R from Rye carries resistance genes for wheat four major diseases i.e. powdery mildew, stripe rust, leaf rust and stem rust. However, these substitutions have deleterious effects on wheat quality. Contrasting reports are also present showing non-significant impact of chromosome substitution. For instance Dhalwal *et al.* (1987) showed non-significant impact of chromosome 1 substitution with IR on wheat quality traits such as 1000-grain weight, test weight, grain protein, flour protein, grain hardness, and milling quality. Similarly, Pena *et al.* (1990) also unable to correlate the deterioration of wheat quality attributes such as alveograph characteristics, dough mixing and baking properties with translocation of 1B/IR.

2) Hybrid Wheat

Another possibility to increase the yield potential is to exploit the phenomenon of heterosis in the species. Hybrid wheat has great potential to increase the per capita yield and may provide another breakthrough for yield. Bruns and Peterson (1998) evaluated the hybrid wheat in Great Plains of USA and found that hybrid wheat was 13.5% superior in term of yield over the wheat developed through pure line selection. Increased yield was due to stability, responsiveness to agronomic practices and better heat tolerance of hybrid wheat. Nazir *et al.* (2004) also estimated heterosis (% increase over mid parent value) in various wheat crosses and mentioned a maximum of 36% heterosis in wheat. Efforts to produce hybrid wheat were limited as wheat was naturally self pollinated with hermaphrodite inflorescence. Therefore extra efforts were required for emasculation and pollination (Strieff *et al.* 1997). Hybrid seed in wheat is mainly produced by CMS system or chemical hybridizing agents. Cytoplasmic male sterile, nuclear restoration system derived from *T. taemophaevi* was discovered by Wilson and Ross (1962). Monsanto Company registered a chemical Genesis in USA as a chemical hybridizing agent. France also granted a provisional registration for the use of Genesis.

In addition to the hybrid vigor for the grain yield author(s) have also mentioned some benefits of producing hybrid wheat because of its high quality (Borghi and Perenzin *et al.* 1994). Borghi and Perenzin (1994) mentioned that heterosis (3.3%) breeding potential in wheat was limited but it may double the benefits of growing hybrid due to heterosis manifestation in quality traits such as bread making quality. Summary of other studies conducted to estimate the magnitude of heterosis for various yield and quality traits is given in Table 5.

3) Transgenic Wheat

Modern genetic engineering techniques have allowed the wheat biotechnologist to insert gene from any other species or genera to produce transgenic wheat. Several studies have shown that transgenic plants in wheat have high degree of transgene expression (Rooke *et al.* 1999; Shwery and Jones, 2005). Shwery and Jones (2005) reviewed the work on transgenes in wheat. They concluded that there were some difficulties in tranfroming wheat and transformation was less efficient than other crops such as maize and rice. However recent development in agro-bacterium mediated transformation and new method of selection, removal of unnecessary DNA sequences and invitro mutagenesis will make process more efficient.

Rooke *et al.* (1999) evaluated the transgenic wheat (B73-6-1) containing an additional genes encoding a gluten protein called HMW subunit 1Dx5 which resulted in the 4x increase in the seed proteins and total glutenin sub units. The gene has also pleiotropic effect on dough strength. Barro *et al.* (2003) made a comparison between the transgenic lines containing different transgenes i.e. HMW-GS 1Ax1 or 1Dx5. Both genes were known to control the gluten subunits production. Genes 1Ax1 produced 29-48% while gene 1Dx5 62 to 64% of the total high molecular weight gluten subunit in various lines.

Table 5. Summary of various studies to estimate the magnitude of heterosis

References	Crossing/Plant Material	Region	Heterosis
Xu *et al.* (1984)	Hand Pollination; Emasculation	China	Crude protein (6.41%), lysine (6.75%), tryptophan content (9.10) heterobeltiosis over superior parent
Change *et al.* (1985)	Cytoplasmic male sterility and nuclear gene restoration system	China	Significant heterosis for grain protein contents, grain yield and grain weight per spike
Borghi *et al.* 1988	Chemical hybridizing agent	Italy	17% heterosis over mid parent value
Bruns and Peterson (1998)	Commercial Wheat Hybrids	USA	10.5% yield advantage over cultivars
Perenzin *et al.* 1998	Chemical hybridizing agent	Italy	15% more grain yield than best standard
Abdullah *et al.* (2002)	Hand Pollination; Emasculation	Pakistan	Grain yield (6.67%), number of grain per spike (25.23%), number of spike let per spike (2.53%), plant height (-8.53%) heterosis over mid parent
Oettler eta l. (2008)	Winter Triticale (F_1 and F_2)	Germany	Grain yield (10.5, 5%), falling number (-4, -7%) in F_1 and F_2.

Expression of gene 1Ax1 led to the imparted positive effect of rheological properties (tenacity, extensibility, deformation work, and sedimentation value) of dough while gene 1Dx5 produced undesirable effect on bread making characteristics. Other studies have also confirmed the similar above mentioned results in different environment for transgenic line (B73-6-1) containing 1Dx5 gene or 1Ax1 (Alvarez *et al.* 2003; Rakszegi *et al.* 2005; He *et al.* 2005).

Breeding Wheat for Quality

Genetic Analysis of Wheat Quality Traits

Improving grain quality is a major objective in wheat. Wheat quality is determined by various grain or flour characteristics such as protein contents, sedimentation value or loaf volume (Mladenov *et al.* 2001; Barnard *et al.* 2002; Hristov *et al.* 2010). These traits are itself been effected by environment and thus may be considered as quantitative (Peterson *et al.* 1998; Mladenov *et al.* 2001; Barnard *et al.* 2002; Hristov *et al.* 2010). Hristov *et al.* (2010) estimated the effect of environment on wheat quality traits and found significant variation due to genotype × environment for wheat quality traits.

They were more affected by location than year and protein or sedimentation values were less affected by environment than loaf volume. Park *et al.* (1997) also identified that activity polyphenol oxidase which cause browning in the wheat flour was affected by location and genotypes.

Estimation of total and type of genetic variability associated with quality traits is also important for plant breeders which allow the estimation of selection response or available genetic variability for selection. Genetic variability is composed of three types i.e. additive (cumulative effect of minor genes), dominance (inter allelic interaction) or epistatic (inter genic interactions).

Only additive portion of the variability is available for selection by the plant breeders. Ratio of additive genetic variability to the total variability (Heritability in narrow sense) would provide the information regarding the value of a trait for selection (Barnard *et al.* 2002).

Genetic components estimates have been found to vary in various studies for quality traits, this may be due to genotypes used, mating design adopted and environments in which studies were carried out (Goldringer et a1. 1997).

For instance, Barnard *et al.* (2002) found high narrow sense heritability associated with quality traits such as thousand kernel mass, flour extraction, breakflour yield, flour colour and hectolitre mass. In contrast, Zanetti *et al.* (2008) found that non-additive components (dominance or epistasis) were important in wheat quality traits such as sedimentation values, protein, kernel hardness and 1000-kernel weight. Zonghzhi (1990) showed partial dominance, and high heritability associated with sedimentation value. Thus the value may be readily improved through selection.

Advent of molecular techniques led to the identification and development of molecular markers for the quality traits and large number of studies are now available indicating the effect of various quantitative trait loci (QTL) on the wheat quality traits (Zanetti *et al.* 2008). Crepieux *et al.* (2005) indicated that reliability of the QTLs depends on the careful selection of the parents and inference drawn from the QTL is limited to the specific populations.

They proposed an alternative approach of analyzing breeding populations for QTLs by developing 374 F_6 populations from 80 different parents rather than from only two parents. QTLs identified were confirmatory to the already known studies and hence results were applicable across breeding programs rather than a limited inference.

Zanetti *et al.* (2008) identified molecular markers linked to the QTLs of wheat quality traits using 226 recombinant inbred lines (RILs). They found nine QTLs for sedimentations and protein contents which explained 47% and 51% of the total phenotypic variance respectively. Perretant *et al.* (2000) found major QTL for grain hardness on chromosome 5D, and minor QTLs on chromosome 1A and 6D.

Similarly, for protein contents two QTLs were identified on chromosome 1A and 6A and for dough three QTLs were present on chromosomes of 1A and 5D and 3B in population of 187 double haploid lines. Breseghello and Sorrells (2006) identified markers of Kernel size on three chromosomes (2A, 5A, 5B). Some other studies regarding QTLs mapping on various chromosome of wheat are summarized in Table 6.

Table 6. QTL mapping for wheat quality traits on various chromosome of wheat

Traits	QTLs
Protein contents	1 QTL (explaining 18.73% of phenotypic variation) on large arm of chromosome 2D (Parasad *et al.* 1999); 'Stable' QTLs (i.e. detected in at least four of the six locations exp. 10% variation each) on chromosomes 2A, 3A, 4D and 7D (Gross *et al.* 2003); 5 QTLs identified in more than one location or by more than one method of detection (Parasad *et al.* (2003); QTL (70% of the variation) x subunit of *Glu-B1* and *storage protein activator* located on the chromosome 1BL (Guillaumie *et al.* 2004). Significant QTL on 2B, 6B and 7A (Turner *et al.* 2004); 3QTLs with major effects on chromosome arms 2AS, 6AS and 7BL, identified by the markers *Xcfa2164*, *XP39M37* $_{(250)}$ and *Xgwm577* , respectively (Blanco *et al.* 2006); Flour protein QTL was mapped on chromosome 6A (Kuchel *et al.* 2006); 1QTL for grain and flour protein content on chromosome 2BS (MacCarteny *et al.* 2006); 14 QTLs (19.6%) on chromosome 4B (Kunert *et al.* 2007). Few loci were also found to associate with protein contents and mapped on various chromosomes. For instance, Nelson *et al.* (2006) showed that grain protein was related with *Gli-D2* gliadin locus on chromosome arm 6DS while photoperiod-sensitivity locus *Ppd1* (accounted for 25% protein variation in the population) . *ppd1*-coupled allele was found to increase the protein contents with higher (1.1%) protein concentration. Parasad *et al.* (1999) identified loci wmc41 and a QTL on large arm of 2D which accounted 18.73% of phenotypic variation.
Sedimentation volume	Kunert *et al.* (2007) identified QTLs for sedimentations on chromosome 6D which contributed significantly (21.7%) for this trait. Environment insensitive QTL for SDS sedimentation volume was identified on small arm of chromosome IA in *Triticum monococum* lines (Taenzeler *et al.* 2002).
Grain hardness	4 regions on chromosomes 2A, 2D, 5B, and 6D had single-factor effects on hardness, and other regions on chromosomes 5A, 6D and 7A had interaction effects (Sourdille eta l. 1996); 1 major QTL (explaining 63% of variation) on the short arm of chromosome 5D (Lgrejas *et al.* 2002); 2 QTLs on chromosomes 1A (close to candidate locus *GluA3*) and 5D (close to candidate locus *Ha*) (Arbelbide and Bernardo, 2006); Ha locus on small arm of chromosome 5D (Nelson *et al.* 2006). 3 loci on chromosomes 5D, 6A, and 3A (Pshenichnikova *et al.* 2008); major QTL (contributing 27.7%) on chromosome 1B (Li *et al.* 2009)
Loaf Volume	QTL associated with loaf volume were identified on chromosome 2A (Kuchel *et al.* 2006).

Table 6. (Continued)

Traits	QTLs
Flour yield	3 regions (accounted 22%, 19% and 19% variation) on chromosomes 3A, 5A and 7D (Parker *et al.* 1999); 1QTL on chromosome 6A (Kuchel *et al.* 2006). In addition to the grain hardness, Ha locus on chromosome 5DS related with also related with flour yield by Nelson *et al.* (2006). Broad-sense heritability 0.48 (Parker *et al.* 1999).
Grain/endosperm color	Heritability was shown to moderate (0.67) (Parker *et al.* 1998). 7 major QTLs on chromosome 7A; chromosomes regions 3A and 7A (accounted for 13 and 60% variation) (Nelson *et al.* 1998); 3 QTLs on long arm 3 A, 3B and 3D and 1 on short arm of 5A (Gross *et al.* 2002). Polyphenol oxidase (PPO) in grain was related to undesirable brown discoloration. STS marker *PPO18* for this trait was mapped to chromosome 2AL (Sun *et al.* 2005). PPO gene co-segregated with the STS marker *PPO18* (Sun *et al.* 2005). 2STS marker linked to *Xwmc41* on chromosome 2DL(9.6 and 24.4% variation) across three environments (He *et al.* 2007). Alleles Psy1-1, Psy1-2 and Psy2-1 and Psy2-2 were related with endosperm colour. 4 QTLs on chromosomes 2A, 4B, 6B, and 7B (Pozniak *et al.* 2007).
Pre-harvest sprouting resistance	3 QTLs on long arm of 3 A, 3B and 3D and 1 on short arm of 5A (Gross *et al.* 2002). 5 QTLs by single locus analysis; 2 QTLs detected in more than one environment on 3BL and 3DL near the dormancy gene *TaVp1* (Kulwal *et al.* 2004). Two-locus analysis resolved 14 QTLs including 8 main effect QTLs (accounted 47.95% phenotypic variation), 8 epistatic QTLs (accounted 28.73% phenotypic variation) and 5 QTLs involved in QTL × environment (QE) or QTL × QTL × environment (QQE) interactions (accounted 3.24% phenotypic variation ((Kulwal *et al.* 2004). 1 QTL located on 3B in close proximity to the *R* gene for red grain colour (Kluwals *et al.* 2004); major QTL across all 6 environment (24.68% to 35.21% variation) (*QPhs.ccsu-3A.1*) on 3AL (Kulwal, 2005). 2 QTLs for seed dormancy (QPhs.ocs-3A.1 and QPhs.ocs-3A.2) on the short and long arms of 3AL (Osa *et al.* 2003); QPhs.ocs-3A.1 at the end of chromosome 3AS. QPhs.ocs-3A.1 was dominant QTL (explained 11.6–44.8% variation) on chromosome 4AL, QPhs.ocs-4B.1, on the centromeric region of the long arm of Chr 4B (Mori *et al.* (2005); QPhs.ocs-1 (explained 23-38% variation), QPhs.ocs-2a minor effect (Osa *et al.* 2003); major QTL on chromosome 4A (4AL) via simple sequence repeat (SSR) (Torado *et al.* (2005).

Traits	QTLs
1000grain weight/Grain size	3QTLs for thousand grain weight, detected on chromosomes 2B, 5B and 7A in all environments (Gross *et al.* 2003); 2QTLs (explaining 30% variation) with epistatic effect on chromosomes 7AS and 6BS (Elouafi and Nachit, 2004); 24 QTLs (explaining 42.1% variation) (Haung *et al.* 2004). 5 QTLs (explaining 32% variation), major QTLs on the centromere region of 6B (Elouafi and Nachit, 2004); prominent QTLs on chromosomes 3D and 4A (McCartney *et al.* 2005). 4 markers (with contributions of 8%, 4.66%, 2.92% and 2.61%) linked with high seed size, together contributing 31% of the phenotypic variance. Nulli-tetrasomic and di-telosomic analysis revealed the presence of 3 low seed size QTL-associated markers on three chromosomes, 6BL, 2DL, and 1DS respectively (Ammiraju *et al.* 2001);8 chromosomes (1A, 1D, 2B, 4B, 5B, 6B, 7A and 7D) carried QTLs controlling grain weight, with only 3 of these (1A, 2B, 7A) carrying alleles for high GW; Association of the marker locus *Xwmc333* with grain weight (Varshaney *et al.* 2000).
Dough strength	2QTL identified on chromosomes 2A, 3A (Kuchel *et al.* 2006); 1 QTL on chromosome 3A (Kuchel *et al.* 2006); 4QTLs on chromosomes 1A, 1B, 1D, and 5B. Candidate gene *GluA1* only candidate locus (Arbelbide and Bernardo, 2006).

Conclusions from Table 6

i) Each study uniquely mapped different QTL position on chromosomes, thus confirmatory studies are absent, indicating that (i) QTLs may be randomly distributed throughout the genome for specific trait (ii) QTLs marker presence subjected to specific populations or environment. Generalized picture or there use across the breeding program in different environment is difficult.

ii) QTLs may be categorized into different types on the bases of their size i.e. major QTL (explaining significant large phenotypic variation) or minor QTLs with minor effect on phenotypic variations.

iii) QTLs may show various type of interaction i.e. QTL × QTL (epistatic QTLs), QTL × environment or QTL × QTL × environment. This showed that QTLs may be affected by the environment or due to other QTL of same trait or other trait. In addition QTLs effect may also be dominant or additive.

iv) Stable QTLs may be marked across all environment.

v) Mapping populations such recombinant inbred lines, double haploids and aneuploids are extensively used in identification of QTLs linked to quality traits in wheat.

vi) Parental selection is important to derive these populations and parents are selected on the basis of significant differences for the mean values of the trait to which QTL is to map.

vii) Synthetic species have been proved to be excellent parents and have provided additional QTLs for increasing quality traits values in wheat, indicating that synthetic wheat may contribute their useful alleles while breeding for quality traits.

viii) Transgressive segregants have been recovered in various population for high quality traits.

Correlation Analyses of Wheat Quality Traits

Correlation studies have also been under taken to study the strength of relationship between the quality traits. Such types of studies are helpful in determining the effect of selection for one trait over other. Cornish *et al.* (2001) correlated the glutenin alleles with dough strength and found that high molecular weight (HMW) glutenin producing alleles made major contribution to the dough strength while low molecular weight glutenin (LMW) producing alleles made insignificant contribution. On the basis of their results they devised a glutenin strength score (GSS) which may allow plant breeders to score genotypes on the basis dough strength. LMW alleles significantly contributed for the dough extensibility. Payne *et al.* (2006) related the quality scores of the wheat genotypes with (HMW) glutenin subunits and found that 47-60% of the wheat quality scores were explained by the HMW sub units of the glutenin. Liu *et al.* (2003) estimated relationship between the quality parameters of cooked dry white Chinese noodles (DWCN) and found negative relationship of hardness, farinograph mixing tolerance index, softening and water absorption with DWCN color, appearance, smoothness and taste while flour whiteness and RVA peak viscosity were positively correlated with DWCN quality parameters. In another study, Mladenov *et al.* (2001) found positive relationship of wet gluten with test weight, grain protein contents, sedimentation value, loaf volume and baking score. A significant relationship between starch swelling power and starch swelling volume with starch paste peak viscosity was obtained using the thirteen genotypes planted over two locations by the study carried out by Crosbie (1991). Blanco *et al.* (2006) identified the causes of negative relationship between the grain protein contents and yield and showed that negative protein-yield response maybe due to co-location of grain yield per spike QTLs with opposite pleiotropic gene effects due to the major bio-energetic requirements for synthesis of protein then carbohydrates. Contrary to the above explanation, Gross *et al.* (2003) found no negative relationships between QTLs for yield and grain protein contents; depicting that molecular breeding have the potential to break down this undesirable relationship.

Genes to control the dough quality has also been isolated. Glutenin genes controls dough resistance, dough extensibility (extensibility), and dough development time while puroindoline genes Pina-D1a and Pinb-D1b were shown to control grain hardness, flour water absorption (water absorption), and milling yield (Cane *et al.* 2004; Eagles *et al.* 2008). Genotypes with Pina-D1a/Pina-D1a were shown to produced soft grain, Pina-D1a/Pinb-D1b produced moderately hard grain and Pinb-D1b/PinaD1a produced very hard grain (Cane *et al.* 2004). Combined effect of glutenin genes and puroindoline genes was studied and showed to control the 50% of the total genotypic variance for these traits but larger part of genotypic variance was left unexplained indicating that some other genes also control the wheat dough quality. The Pina-D1b/Pinb-D1a genotype had a significantly higher water absorption and significantly lower milling yield than the Pina-D1a/Pinb-D1b genotype (Eagles *et al.* 2008).

Ahmad *et al.* (2000) indicated that genes that control the HMW-glutenin subunits are located on long arm of chromosome 1. Furthermore, gene complex *Glu-D1* designated as 1Dx5–1Dy10 and 1Dx2–1Dy12 controls the good and poor quality bread. The alleles *Glu-B1*, Bx7 were associated with subunits, By8 or By9. Both allelic pairs contributed to good bread making quality by increasing dough elasticity.

Path analyses technique is based on partial regression method and is employed to partition the direct and indirect effects of independent variable. Path analysis showed that protein grain content had the highest direct effect on loaf volume.

REFERENCES

Abdullah G.M., Khan A.S. and Zulfiqar A. (2002). Heterosis study of certain important traits in wheat International *Journal of Agriculture & Biology*, 04(3): 326–328.

Abourezk and Sanaa. (2000). Secrets of Healthy Middle Eastern Cuisine. New York: Interlink Publishing Group.

Ahmad M. (2000). Molecular marker-assisted selection of HMW glutenin alleles related to wheat bread quality by PCR-generated DNA markers *Theoretical and Applied Genetics*, 101(5-6): 892-896.

Alvarez M.L., Gómez M., Carrillo J.M. and Vallejos R.H. 2001. Analysis of dough functionality of flours from transgenic wheat. *Molecular Breeding*, 8(1):103-108.

Ammiraju J.S.S., Dholakia B.B., Santra D.K., Singh H., Lagu M.D., Tamhankar S.A., Dhaliwal H. S., Rao V.S., Gupta V.S. and Ranjekar P.K. (2001). Identification of inter simple sequence repeat (ISSR) markers associated with seed size in wheat. *Theoretical and Applied Genetics*, 102(5):726-732.

Antje K., Naz A.A., Dedeck O., Pillen K., and Léon J. (2007). AB-QTL analysis in winter wheat: I. Synthetic hexaploid wheat (*T. turgidum* ssp. *dicoccoides* × *T. tauschii*) as a source of favourable alleles for milling and baking quality traits. *Theoretical and Applied Genetics*, 115(5):683-695.

Arbelbide M. and Bernardo R. (2006). Mixed-model QTL mapping for kernel hardness and dough strength in bread wheat. *Theoretical and Applied Genetics*, 112 (5): 885-890.

Atsushi T., Shojiro I. and Michiya K. (2005). Mapping and validation of PCR-based markers associated with a major QTL for seed dormancy in wheat *Euphytica*, 143(3):251-255.

Barnard D. Labuschagne, M.T. and van Niekerk H.A. (2002). Heritability estimates of bread wheat quality traits in the Western Cape province of South Africa. *Euphtica*, 127(1): 115-122.

Barro F., Barceló P., Lazzeri P.A., Shewry P.R. , Ballesteros J., and Martín A. (2003). Functional properties of flours from field grown transgenic wheat lines expressing the HMW glutenin subunit 1Ax1 and 1Dx5 genes. *Molecular Breeding*, 12(3):223-229.

Bellon, M.R. (2004). Conceptualizing interventions to support on-farm genetic resource conservation. *World Development*. 32(1):159–172.

Blanco A., Simeone R. and Gadaleta A. (2006). Detection of QTLs for grain protein content in durum wheat. *Theoretical and Applied Genetics*, 112(7):1195-1204.

Bonafaccia G., Galli V., Francisci R., Mair V., Skrabanja V. and Kreft I. (2000). Characteristics of spelt wheat products and nutritional value of spelt wheat-based bread. *Food Chemistry*, 68(4): 437-441.

Borghi B. and Perenzin M. (1994). Diallel analysis to predict heterosis and combining ability for grain yield, yield components and bread-making quality in bread wheat (*T. aestivum*). *Theoretical and Applied Genetics*, 89(7-8):975-981.

Borghi B., Perenzin M. and Nash R. J. (1988). Agronomic and qualitative characteristics of ten bread wheat hybrids produced using a chemical hybridizing *Euphytica*, 39(2):185-194.

Borghi B., Castagna R., Corbellini M., Heun M., Salamini F. (1996). Breadmaking quality of Einkorn wheat (*Triticum monococcum* ssp. monococcum). *Cereal chemistry*, 73(2): 208-214.

Breseghello F. and Sorrells M.E. (2007). QTL analysis of kernel size and shape in two hexaploid wheat mapping populations *Field Crops Research*, 101(2): 172-179.

Breseghello F., and Sorrells M.E. (2006). Association mapping of kernel size and milling quality in wheat (*Triticum aestivum* L.) cultivars. *Genetics*, 172: 1165-1177.

Bruns R. and C. J. Peterson. (1998). Yield and stability factors associated with hybrid wheat. *Euphytica,* 100: 1–5.

Bruns R. and Peterson C. J. (1998). Yield and stability factors associated with hybrid wheat. *Euphtyica*, 100(1-3): 1-5.

Ciaffi M., Lafiandra D., Porceddu E. and Benedettelli S. (1993). Storage-protein variation in wild emmer wheat (*Triticum turgidum* ssp.*dicoccoides*) from Jordan and Turkey. I. Electrophoretic characterization of genotypes. *Theoretical and Applied Genetics*, 86(4):474-480.

Corbellini M., Empilli S., Vaccino P., Brandolini[1] A., Borghi B., Heun M., and Salamini F. 1999. Einkorn characterization for bread and cookie production in relation to protein subunit composition. *Cereal chemistry*, 76(5):727-733.

Cornish G.B., Békés F., Allen H.M. and Martin D.J. (2001). Flour proteins linked to quality traits in an Australian doubled haploid wheat population. *Australian Journal of Agricultural Research*, 52(12): 1339 – 1348.

Crepieux S., Lebreton C., Flament P. and Charmet G. (2005). Application of a new IBD-based QTL mapping method to common wheat breeding population: analysis of kernel hardness and dough strength *Theoretical and Applied Genetics*, 111(7): 1409-1419.

Crosbie G.B. (1991). The relationship between starch swelling properties, paste viscosity and boiled noodle quality in wheat flours. *Journal of Cereal Science*, 13(2): 145-150

D'Egidio M.G., Nardi S. and Vallega V. (1998). Grain, Flour and Dough Characteristics of selected strains of diploid wheat, *Triticum monococcum* L. *Cereal Chemistry*, 70(3):298-303.

Dahaliwal A.S., Mares D.J., Marshall D.R. (1987). Effect of 1B/IR chromosome translocation on milling and quality characteristics of bread wheat. *Cereal Chemistry*, 64(2):72-76.

Dvorak J., Diterlizzi P., Zhang H.B., Resta P. (1993). The evolution of polyploid wheats: identification of the A genome donor species. *Genome* 36:21-31.

Dosti, R. (1999). Mideast and Mediterranean Cuisines. Tucson, A.Z. Fisher Books. 784- 791.

Dreisigacker S., Kishii M., Lage J. and Warhurton M. (2008). Use of synthetic hexaploid wheat to increase diversity for CIMMYT bread wheat improvement. Synthetic Wheat Symposium No1, Horsham, Victoria, Australie , vol. 59,5 (103 p.).

Eagles H.A., Karen C., Eastwood R.F., Hollamby G.J., Haydn K., Martin P.J. and Cornish G. B. (2008). Contributions of glutenin and puroindoline genes to grain quality traits in southern Australian wheat breeding programs. Crop and Pasture science.

Edwards N.M. Mulvaney S.J. Scanlon M.J., Dexter M.J. (2003). Role of gluten and its component in determining durum semolina dough viscosity properties. *Cereal chemistry* 80: 755-763.

Edwars N.M., Gianibelli M.C., McCaig T.N. Clarke J.M., Anes N.P., Larroque C.R., Dexter J.E. (2007). Relationship between dough strength, polymeric protein quantity and composition for diverse durum wheat genotype. *Journal of Cereal Sciences.* 45: 140-149.

Eckardt, N.A. (2010) Evolution of Domesticated Bread Wheat. *The Plant Cell* 22:993.

Fan Y., Sun H., Zhao J., Ma Y., Li R. and Li S. (2009). QTL mapping for quality traits of northern-style hand-made Chinese steamed bread. *Journal of Cereal Science*, 49(2): 225-229.

Filippo L., Antuono D., Galletti G.C and Paola B. (1998). Fiber quality of emmer (*Triticum dicoccum*Schubler) and einkorn wheat (*T monococcum* L) landraces as determined by analytical pyrolysis. Journal of the Science of Food and Agriculture 78(2): 213–219.

Food and Agriculture Organization(FAO) Rome, 2008.

Gauthier J., Gelinas P., and Beauchemin, R. (2006). Effect of stone-milled semolina granulation on the quality of bran-rich pasta made from khorasan (Kamut®) and durum wheat. *International Journal of Food Science & Technology*, 41(5): 596–599.

Goldringer I., Brabant, P. and Gallais A. 1997. Estimation of additive and epistatic genetic variances for agronomic traits in a population of doubled-haploid lines of wheat. *Heredity*, (1997) 79: 60–71.

Grausgruber H., Oberforster M., Ghambashidze G., and Ruckenbauer P. (2005). Yield and agronomic traits of Khorasan wheat (*Triticum turanicum* Jakubz.). *Field Crop Research*, 91(2-3), 319-327.

Groos C., Robert N., Bervas E. and Charmet G. (2003). Genetic analysis of grain protein content, grain yield and thousand-kernel weight in bread wheat. *Theoretical and Applied Genetics*, 106(6):1032-1040.

Groos C., Gay G., Perretant M.R., Gervais L., Bernard M., Dedryver F. and Charmet G. (2002). Study of the relationship between pre-harvest sprouting and grain color by quantitative trait loci analysis in a white×red grain bread-wheat cross. *Theoretical and Applied Genetics*, 104(1): 39-47.

Guillaumie S., Charmet G., Linossier L., Torney V., Robert N. and Ravel C. Colocation between a gene encoding the bZip factor SPA and an eQTL for a high-molecular-weight glutenin subunit in wheat (*Triticum aestivum*) *Genome*, 47(4): 705–713.

Hanson, H., Borlaug, N.E. and Anderson, R.G. (2001). Wheat in the third world. West view Press, Boulder Co

He X.Y., He Z.H., Zhang L.P., Sun D.J., Morris C.F., Fuerst E.P. and Xia X.C. (2007). Allelic variation of polyphenol oxidase (PPO) genes located on chromosomes 2A and 2D and development of functional markers for the PPO genes in common wheat. *Theoretical and Applied Genetics*, 115(1): 47-58.

He X.Y., Zhang Y.L., He Z.H., Wu Y.P., Xiao Y.G., Ma C.X. and Xia X.C. (2008). Characterization of phytoene synthase 1 gene (*Psy1*) located on common wheat chromosome 7A and development of a functional marker. *Theoretical and Applied Genetics*, 116(2):213-221.

Hea G.Y., Jonesb H.D., D'Ovidioc R., Stefania M., Chena M., Westb J., Butowd B., Andersone O.D., Lazzerif P., Fidob R., Shewryb P.R. (2005). Expression of an extended HMW subunit in transgenic_wheat and the effect on dough mixing properties. *Journal of Cereal Science*, 42:225–231.

Hidalgo A., Brandolini A., Pompei C. and Piscozzi R. (2006). Carotenoids and tocols of einkorn wheat (*Triticum monococcum* ssp. *monococcum* L.) *Journal of Cereal Science*, 44(20):182-193.

Hristov N., Mladenov N., Djuric V., Kondic-Spika A., Marjanovic-Jeromela A. [1] and Simic D. (2010). Genotype by environment interactions in wheat quality breeding programs in southeast Europe. *Eupytica*, 174(3):315-324.

Huang X.Q., Cloutier S., Lycar L., Radovanovic N., Humphreys D.G., Noll J.S., Somers D. J. and Brown P. D. (2006). Molecular detection of QTLs for agronomic and quality traits in a doubled haploid population derived from two Canadian wheats (*Triticum aestivum* L.). *Theoretical and applied genetics*, 113(4): 753-766.

Igrejas G., Leroy P., Charmet G., Gaborit T., Marion D. and Branlard G. (2002). Mapping QTLs for grain hardness and puroindoline content in wheat (Triticum aestivum L.) *Theoretical and Applied Genetics*, 106(1):19-27.

Jaradat, A. A. (1998). Triticeae III. Enfield, N.H.: Science Publishers.

Karen C., Spackman M., and Eagles H.A. (2004). Puroindoline genes and their effects on grain quality traits in southern Australian wheat cultivars. *Australian Journal of Agricultural Research*, 55(1) 89 – 95.

Kathy S., André B. and Armand G. (1997). Hybrid wheat seed production potential using the chemical hybridizing agent SC2053. *Plant Growth regulators*, 21(2):103-108.

Katz and Solomon, H. (2003). Encyclopedia of food and culture. New York: Charles Scribner. 543-552.

Kazi A.M., Davila G.F., Villareal R.L., Cortes A., Roasas V. and Delgado R. (2001). Registration of 10 synthetic hexaploid wheat and six bread wheat germplasms resistant to karnal Bunt *Crop Science,* 41: 1652-1653.

Kent, N.L. and Evers, A. D. (1998). Technology of the Cereals. Fifth Edition. 79-82.

Kilian B., Özkan H, Deusch O., Effgen S., Brandolini A., Kohl J., Martin W. and Salamini F. (2007). Independent wheat B and G genome origins in outcrossing *Aegilops* Progenitor Haplotypes. *Molecular Biology and Evolution* 24 (1): 217-227.

Kuchel H., Langridge P., Mosionek L., Williams K. and Jefferies S.P. (2006). The genetic control of milling yield, dough rheology and baking quality of wheat. *Theoretical and Applied Genetics*, 112(8): 1487-1495.

Kulwal P.L., Kumar N., Gaur A., Khurana P., Khurana J.P., Tyagi A.K., Balyan H.S. and Gupta P. K.(2004). Mapping of a major QTL for pre-harvest sprouting tolerance on chromosome 3A in bread wheat. *Theoretical and Applied Genetics*, 111(6): 1052-1059.

Kulwal P.L., R. Singh, H.S. Balyan and P.K. Gupta. (2004). Genetic basis of pre-harvest sprouting tolerance using single-locus and two-locus QTL analyses in bread wheat. *Functional & Integrative Genomics*, 4(2): 94-101.

Lage J., Skovmand B., Andersen S.B. (2004). Field Evaluation of Emmer Wheat-Derived Synthetic Hexaploid Wheat for Resistance to Russian Wheat Aphid (Homoptera: Aphididae). *Journal of Economic Entomology*, 97(3):1065-1070.

Lage J., Skovmand B., Peña R.J. and Andersen[3] S.B. (2006). Grain quality of emmer wheat derived synthetic hexaploid wheats genetics resource and crop evolution 53(5): 955-962.

Li, Y., Song, Y., Zhou R., Branlard G., Jia, J. (2009). Detection of QTLs for bread-making quality in wheat using a recombinant inbred line population *Plant Breeding*, 128(3): 235-243.

Liu J.J., He Z.H., Zhao Z.D., Peña R.J. and Rajaram S. (2003). Wheat quality traits and quality parameters of cooked dry white Chinese noodles. *Euphytica*, 131(2),147-154.

Løje H., Møller B., Laustsen A.M. and Hanse Å. (2003). Chemical Composition, Functional Properties and Sensory Profiling of Einkorn (*Triticum monococcum* L. *Journal of Cereal Science*, 37(2):231-240.Luo, M.C., Dubcovsky J. and Dvorak J. (1996). Recognition of Homeology by the Wheat Ph1 Locus. *Genetics*, 144: 1195-1203.

Masood M.S., Asghar M. ANWAR R. (2004). Genetic diversity in wheat landraces from Pakistan based on polymorphism for high molecular weight glutenin subunits (HMW-GS). *Pakistan. Journal of Bot*any 36(4): 835-843.

Marconi E., Carcea M., Graziano M., and Cubadda R. (1999). Kernel properties and pasta-making quality of five European spelt wheat (*Triticum spelta* L.) cultivars. *Cereal chemistry*, 76(1): 25-29.

McCartney A., Somers D.J., Humphreys D.G., Lukow O., Ames N., Noll J., Cloutier S., and McCallum. B.D. (2005). Mapping quantitative trait loci controlling agronomic traits in the spring wheat cross RL4452 × 'AC Domain'**C.** *Genome*, 48(5): 870–883.

McCartney C.A., Somers D.J. Lukow O., Ames N., Noll J., Cloutier S., Humphreys D.G., McCallum B.D. (2006). QTL analysis of quality traits in the spring wheat cross RL4452 × 'AC Domain'. *Plant Breeding*, 125(6):565-575.

Mingli W., Chang A., Tiecheng H., Enchung L., Wenhua S. (1985). Studies on quality and other agronomic characteristics of hybrid wheat with *T. timopheevii* cytoplasm-1. Heterosis and combining ability. Acta Agronomica Sinica.

Mladenov N., Przulj N., Hristov N., Djuric V., and Milovanovic M. (2003). Cultivar by environment interactions for wheat quality traits in semiarid conditions. 78(3):363-367.

Mori, M., Uchino N., Chono M., Kato K. and Miura H. Mapping QTLs for grain dormancy on wheat chromosome 3A and the group 4 chromosomes, and their combined effect. *Theoretical and Applied Genetics*, 110(7): 1315-1323.

Nagao, S. (1995). Wheat usage in East Asia, in Wheat End Uses Around the World, American Association of Cereal Chemists, St. Paul, Minnesota, USA. 170- 176.

Nelson J.C., Andreescu C., Breseghello F., Patrick L.F., Daisy G.G., Christine J.B., Roberto J.P., Marie R.P., Philippe L. and Calvin O.Q., et al. Quantitative trait locus analysis of wheat quality traits. *Euphytica*, 149 (1-2): 145-159.

Nevo E. and Payne P.I. (1987). Wheat storage proteins: diversity of HMW glutenin subunits in wild emmer from Israel. *Theoretical and Applied Genetics*, 74(6):827-836.

Oeittler G., Becker H.C., and Hoppe G. (2008). Heterosis for yield and other agronomic traits of winter triticale F_1 and F_2 hybrids. *Plant Breeding*, 120(4), 351-353.

Osa M., Kato K., Mori M., Shindo C., Torada A. and Miura H. (2003). Mapping QTLs for seed dormancy and the Vp1 homologue on chromosome 3A in wheat. *Theoretical and Applied Genetics*, 106(8):1491-1496.

Owens, G. (2001). Cereals processing technology. Woodhead Publishing Ltd., and CRC Press, LLC. 62-64.

Ozkan H.O, Brandolini, A., Schafer-Pregl R., Salamini F. (2002). Hard Wheat Domestication in Southeast Turkey. *Molecular Biology and Evolution* 19(10): 1797–1801.

Park W. J., Shelton D. R., Peterson C. J., Martin T.J., Kachman S.D., and Wehling R.L. (1997). Variation in polyphenol oxidase activity and quality characteristics among hard white wheat and hard red winter wheat samples *Cereal chemistry*, 74(1), Pages 7-11.

Parker G. D., Chalmers K. J., Rathjen A. J. and Langridge P. (1998). Mapping loci associated with flour colour in wheat (*Triticum aestivum* L.). *Theoretical and Applied Genetics*, 97(1-2): 238-245.

Pasha, Imran (2006) Biochemical characterization of Pakistani wheats in relation to grain hardness. *PhD thesis, University of Agriculture, Faisalabad*

Pena R.J., Amaya A., Rajaram S. and Mujeeb-Kazi A. (1990). Variation in quality characteristics associated with some spring 1B/1R translocation wheats. *Journal of Cereal Science*, 12,_(2):105-112.

Peng J., Ronin Y., Fahima T.*, Roder M.S., Li Y., Nevo E. and Korol A. (2003). Domestication quantitative trait loci in Triticum dicoccoides, the progenitor of wheat. *Proceeding national academy of science*, 100(5): 2489-2949.

Perenzin M., Corbellini M., Accerbi M., Vaccino P. and Borghi B. (1998). Bread wheat: F_1 hybrid performance and parental diversity estimates using molecular markers *Euphytica*, 100(1-3): 273-279.

Perretant M.R., Cadalen T., Charmet G., Sourdille P., Nicolas P., Boeuf C., Tixier M.H., Branlard G. and Bernard S. (2000). QTL analysis of bread-making quality in wheat using a doubled haploid population. *Theoretical and applied genetics*, 100(8):1167-1175.

Peter I.P., Mark A., Nightingale, Anatole F.K., Linda M.H. (2008). The relationship between HMW glutenin subunit composition and the bread-making quality of British-grown wheat varieties. Journal of the science of food quality and agriculture 40(1): 51-65.

Peterson C.J., Graybosch R.A., Shelton D.R. and Baenziger P.S. (1998). Baking quality of hard winter wheat: Response of cultivars to environment in the Great Plains. *Euphytica*, 100:157-162.

Pflüger L.A., Martín L.M. and Alvarez J.B. (2001). Variation in the HMW and LMW glutenin subunits from Spanish accessions of emmer wheat (*Triticum turgidum ssp. dicoccum Schrank*). *Theoretical and Applied Genetics*, 102: 767-772.

Pozniak C.J., Knox R.E., Clarke F.R. and Clarke J.M. (2007). Identification of QTL and association of a phytoene synthase gene with endosperm colour in durum wheat *Theoretical and Applied Genetics*, 114(3): 525-537.

Prasad M., Kumar N., Kulwal, P. Röder M., Balyan H., Dhaliwal H. and Gupta P. (2003). QTL analysis for grain protein content using SSR markers and validation studies using NILs in bread wheat. *Theoretical and Applied Genetics*, 106 (4): 659-667.

Pshenichnikova T.A., Ermakova M.F., Chistiakova A.K., Shchukina L.V., Berezovskaia E.V., Lohwasser U., Roder M., Borner A. (2008). Mapping of the quantitative trait loci (QTL) associated with grain quality characteristics of the common wheat grown under different environmental conditions. *Genetika*, 44(1):90-101.

Qarooni. J. (1999). Flat bread Technology. Chapman and Hall, 115. Fifth Ave., New York, NY. 206 -213.

Rakszegi M., Békés F., Láng[a] L., Tamás[c] L., Shewry[d] P.R. and Bedő Z. (2005). Technological quality of transgenic wheat expressing an increased amount of a HMW glutenin subunit. *Journal of Cereal Science*, 42(1): 15-23.

Ranhotra G.S., Gelroth J.A., Glaser B.K. and Lorenz K.J.(1995).Baking and nutritional qualities of a spelt wheat sample. LWT - *Food Science and Technology*, 28(1):118-122.

Ravindra M.P., Manoj D.O., Shubhada A.T., Pierre S. and Veliventi S.R. (2008). Mapping and validation of a major QTL for yellow pigment content on 7AL in durum wheat (*Triticum turgidum* L. ssp. *durum*) *Molecular Breeding*, 21(4):485-496.

Rooke L., Békés F., Fido R., Barro F., Gras P., Tatham A.S., Barcelo P., Lazzeri P. and Shewry P. R. (1999). Overexpression of a gluten protein in transgenic wheat results in greatly increased dough strength. *Journal of Cereal Science*, 30(2): 115-120.

Shamim I., Shahzad A., Kazi A.M. and Iftikhar A. (2009). In vitro screening of synthetic hexaploid wheat lines against bipolaris sorokiniana in pakistan. *Pak. J. Bot.*, 41(4): 1989-2001.

Shewry P.R. and Jones H.D. (2005). Transgenic wheat: where do we stand after the first 12 years? *Annals of Applied Biology*, 147(1): 1 – 14.

Sissons, M.J., Egan, M.E., Gianibelli, M.C. (2005). New insight into the role of gluten on durum pasta quality using reconsititution method. *Cereal chemistry* 82: 601-608.

Sissons, M.J., Soh. H.N., Turner, M.A. (2007). Role of gluten and its components in influencing durum wheat dough properties and spaghetti cooking quality. *Journal of Science of Food and Agriculture* 87: 1874-1885.

Sissons, M.J. (2008). Role of durum wheat composition on quality of pasta and bread food. Food 2(2), 75-90.

Sun D.J., He Z.H., Xia X.C., Zhang L.P., Morris C.F., Appels R., Ma W.J. and Wang H. (2005). A novel STS marker for polyphenol oxidase activity in bread wheat. *Molecular Breeding*, 16 3, 209-218.

Sultana T. Ghafoor A. Ashraf M. (2007). Genetic variability in bread wheat (Triticum aestivum L.) of Pakistan based on polymorphism for high molecular weight glutenin subunits. *Genetic Resource and Crop Evolution* 54:1159–1165.

Taenzler B., Esposti R.F., Vaccino P., Brandolini A., Effgen S. , Heun M., Schäfer-Pregl R., Borghi B. and Salamini F. (2002). Molecular linkage map of Einkorn wheat: mapping of storage-protein and soft-glume genes and bread-making quality QTLs. *Genetical Research*, 80(2): 131-143.

Tahir M., Turchetta T., Anwar R. Lafiandra D. (1996). Assessment of genetic variability in hexaploid wheat landraces of Pakistan based on polymorphism for HMW glutenin subunits. *Genetic Resources and Crop Evolution* 43:211-220.

Tan F., Zhou J., Yang Z., Zhang Y., Pan Li., and Zhenglong R. (2009) Characterization of a new synthetic wheat – *Aegilops biuncialis* partial amphiploid. *African Journal of Biotechnology*, 8 (14), 3215-3218.

Tsegaye, B. and Berg, T. (2007). Utilization of durum wheat landraces in East Shewa, central Ethiopia: Are home uses an incentive for on-farm conservation? *Agriculture and Human Values*. 24:219–230.

Turner S., Bradburne R.P., Fish L. and Snape J.W. New quantitative trait loci influencing grain texture and protein content in bread wheat . *Journal of Cereal Science*, 40 (1): 51-60.

Varshney R.K., Prasad M., Roy J.K., Kumar N., Harjit-Singh, Dhaliwal H.S., Balyan H.S. and Gupta P.K. (2000). Identification of eight chromosomes and a microsatellite marker on 1AS associated with QTL for grain weight in bread wheat. *Theoretical and Applied Genetics*, 100(8): 1290-1294.

Villareal, R.L., K.Sayre, O. Banuelos, A. Mujeeb-Kazi, (2001). Registration of four synthetic hexaploid wheat (Triticum turgidum/ Aegilops taushii) germplasm lines tolerant to water logging. *Crop science*,.41: 274-279.

Wrigley, C. (2004). Encyclopedia of grain science. Food science Australia and wheat CRC, North Ryde, NSW, Australia. 337- 339.

Yan Y., Hsam S.L.K., Yu J.Z., Jiang Y., Ohtsuka I. and F.J. Zeller. (2003). HMW and LMW glutenin alleles among putative tetraploid and hexaploid European spelt wheat (*Triticum spelta* L.) progenitors. *Theoretical and Applied Genetics*, 107(7): 1321-1330.

Zanetti S., Winzeler M., Feuillet C., Keller, B., Messmer M. (2001). Genetic analysis of bread-making quality in wheat and spelt. *Plant Breeding*, 120: 13-19.

Zhang L., Liu D., Lan X., Zheng Y, Yan Z (2008). A synthetic wheat with 56 chromosomes derived from *Triticum turgidum* and *Aegilops tauschii J Appl Genet.* 49(1): 41–44.

Zhang W., Chao S., Manthey F., Chicaiza O., Brevis J.C., Echenique V. and Dubcovsky J. (2008). QTL analysis of pasta quality using a composite microsatellite and SNP map of durum wheat. *Theoretical and Applied Genetics*, 117(8):1361-1377.

Zhu M.X., Hongping A.P., Yu Z. (1984). The analysis of heterosis and combining ability in six wheat cultivars for grain protein,lysine and tryptophan content (1984). *Acta Agronomica Sinica.*

Zongzhi Li Studies on inheritance and correlation of some quality characteristics in winter wheat. *Acta Agronomica Sinica.*

In: Wheat: Genetics, Crops and Food Production
Editor: Maria T. Almeida, pp. 209-229
ISBN 978-1-61209-307-9

Chapter 6

IMPACT OF NITROGEN AND SULFUR FERTILIZATION ON GLUTEN, COMPOSITION, AND BAKING QUALITY OF WHEAT

Dorothee Steinfurth, Karl H. Mühling and Christian Zörb*
Christian Albrechts University Kiel, Institute of Plant Nutrition and Soil Science, Hermann-Rodewald-Str. 2, D-24118 Kiel, Germany

ABSTRACT

Wheat has a broad genetic potential for the expression of various gluten proteins and is strongly influenced by the environment, e. g. with respect to the fertilization of nitrogen (N) and sulfur (S) during wheat plant development. Nitrogen and S availability markedly change wheat flour and baking quality. However, without adequate management of N and S fertilization, the genetic potential of wheat cannot be exploited. Nitrogen fertilization primarily affects the concentration of gliadines and glutenines as it is source-regulated. Storage protein concentration, particularly of gliadine and glutenine, is highly correlated with loaf volume. The use of three instead of one dressing of fertilizer, especially as a late dressing at ear emergence, the gliadine and glutenine concentration of the grain and, furthermore, loaf volume can be enhanced. However, not only quantity, but also the quality of gluten proteins is important for their composition and characteristics. Despite protein amount being only slightly influenced by increasing S fertilization levels, the composition of gluten proteins is highly affected. Increasing S availability for plants enhances S-containing amino acids, namely cysteine and methionine, which are foremost in S-rich gluten proteins. For an improvement of baking quality, a high concentration of S-rich proteins is of particular importance in order to form an appropriate gluten network within the dough. Additionally, a late S dressing can further improve baking quality, since such a dressing increases the synthesis of high molecular weight (HMW) glutenins that considerably affect baking quality in wheat.

S-rich components such as sulfate and glutathione are, moreover, important molecules for the transport and supply of S to the grain. Both these molecules represent important S transport forms from source to sink organs. The potential to synthesize large

* Corresponding author: Christian Zörb czoerb@plantnutrition.uni-kiel.de

amounts of storage protein depends on the ability of the plant to take up and to transport S into sink organs such as ears and kernels. Sulfur and glutathione are therefore important molecules functioning as actuators during plant growth and grain development. Furthermore, glutathione can function as internal plant signal for the S status of the plant. In terms of baking quality, glutathione is able to interact with gluten proteins responding in a change of existing disulfide bonds between gluten proteins and resulting in a different rheological property and baking quality. Concerning the interaction of N and S in wheat, a well-balanced N/S ratio is exceedingly important for suitable gluten and baking quality. If high N fertilization is accompanied by inadequate S fertilization, S deficiency is provoked, resulting in changed gluten protein composition and a loss of nutritional quality of the grains.

1. Introduction

Gluten proteins in cereals belong to the family of plant storage proteins and primarily provide nitrogen (N) and sulfur (S) stores for the growth of seedlings. In terms of human and animal nutrition, these storage proteins are not only a source of essential amino acids, but also of particular importance for product quality. Especially in wheat, gluten proteins are highly important, since they have special characteristics, appearing exclusively in wheat, for dough preparation and baking quality (Zhao et al., 1999b).

Fertilization of wheat plants with N and especially with S has been extensively discussed, since environmental S depositions from the atmosphere have decreased during the last decades (Vestreng et al., 2007). Sulfur dioxide (SO_2) from the atmosphere thus no longer acts as a sufficient source for S. Hence, S application to the soil in any form and its management during plant development needs to be further investigated. Although wheat is not such a high-S-demanding crop plant as canola, S deficiency becomes obvious as soon as S fertilization is inadequate. Sulfur deficieny in wheat results in a change of color from dark green to light-green in young leaves (Bell et al., 1995), whereas N deficiency is diagnosed by the lightening of old leaves. Consequently, an inadequate nutrient supply results in reduced yield (Zhao et al., 1999a), the lower production of S-containing amino acids and S-rich storage proteins (Wieser et al., 2004), and low baking quality. These aspects, when taken together, lead to lower profits for farmers and the baking industry.

Furthermore, as a reinforcing situation, wheat breeding during the last few decades has focused mainly on increasing yield and resistance against plant pathogens and only to a lesser extent on baking quality. As a consequence, yield has increased at the expense of protein concentration in grain (Feil, 1997; Statistisches Bundesamt, 2010).

Each wheat cultivar has its characteristic, genetically determined pattern of gluten proteins, which have a major influence on baking quality. Irrespective of the genetic background, N and S fertilization further highly influence the composition of storage proteins. Thus, fertilization is a means to influence not only grain yield, but also grain quality. As early as the middle of the last century, Finney and Barmore (1948) showed that wheat cultivars exhibit different slopes with respect to loaf volume, and thus baking quality increases with protein concentration independently of the genetic background of the cultivars. Is it therefore possible to increase both yield and baking quality by adequate N and S fertilizer application?

The main focus of this chapter is to demonstrate the impact of N and S fertilization on the modification of gluten proteins and their rheological characteristics and baking quality parameters. Nitrogen and S fertilization can be managed in several ways: (1) the amount of N and S fertilizer applied, (2) the application of soil or foliar fertilizers, and (3) the application at different developmental stages of plant growth. Moreover, an elucidation of the way that N and S interact during the formation of gluten proteins is of interest.

2. Gluten Protein Composition and Synthesis

Since T. J. Osborne's (1924) publication concerning the extractability of storage proteins, they have been classified into four groups in wheat: the albumins and globulins, which are mainly metabolic and structural proteins (Goesaert et al., 2005), are extractable with water and NaCl solution, respectively. Other storage proteins such as the prolamin or gluten proteins are divided into gliadins, which are extractable with ethanol, and glutenins, which need to be extracted with reducing agents because of covalent cross-links such as disulfide bonds (Osborne, 1924; Shewry, 1995). Gliadins and glutenins account for about 75 % of mature wheat grain proteins (Shewry and Halford, 2002) and are the most important factors for the remarkable breadmaking properties of wheat.

Since the role of globulins in baking quality is still not fully understood (Shewry and Halford, 2002), this chapter will focus on gliadin and glutenin proteins. Gluten proteins further consist of subunits that contain no (S-poor), low (high molecular weight glutenin subunits, HMW-GS), or high (S-rich) S amounts in the form of cysteine (Shewry and Halford, 2002)

Gliadin proteins are monomeric structures that can only form intra-molecular subunits, whereas glutenins are able to form both intra- and inter-molecular subunits. B- and C-type low molecular weight (LMW) subunits of glutenin, and γ- and α-gliadins are S-rich molecules and contain about 2-3% cysteine. D-type LMW subunits of glutenins and ω-gliadins are S-poor and contain only 0 to <0.5% cysteine. Cysteine possesses a free thiol-group that is able to form disulfid cross-links with other cysteine-containing molecules.

HMW glutenins contain repetitive domains that may be intrinsically elastic because of ß-turns in the protein structure that are bordered by α-helices on the N- and C-terminal domain of the molecule. Additionally, HMW-GS are notably divided into two types according to their genetic background: y-type HMW-GS that are of low molecular weight and x-type HMW-GS subunits that are of high molecular weight. The y-type HMW-GS contain more SH-groups at the N-terminal domain than the x-type HMW-GS (Shewry et al., 1992). Further information concerning glutenin composition can be obtained by reference to the above-mentioned reviews on HMW glutenin subunits by Shewry et al. (1992) and Shewry and Halford (2002).

Gluten protein expression is highly tissue-specific and does exclusively occur in seeds during grain development. The expression of gluten proteins is further regulated by available N and S metabolites that are synthesized during plant growth (Tabe et al., 2002). Nitrogen is on first instance important for amino acid and protein synthesis. Therefore, the next chapter will discuss the role of N fertilization on gluten composition and its effect on baking qualiy.

3. The Role of Nitrogen Fertilization in Gluten Formation and the Resulting Baking Quality

3.1. Assimilation and Translocation of Nitrogen within the Wheat Plant

In wheat plant roots, nitrate transporters are mainly responsible for the uptake of nitrate (NO_3) into the root symplast (Yin et al., 2007). Nitrogen is also taken up by wheat roots in the form of ammonium (NH_4^+) (Cox and Reisenauer, 1973). Ammonium, however, cannot be stored in vacuoles in the same manner as nitrate, and furthermore, ammonia (NH_3) is a cell-toxic compound that needs to be converted into amino acids or amides as rapidly as possible. Most plant species prefer to take up N in the form of nitrate, although some species do indeed take up N in the form of ammonium. This can further be observed under special soil conditions, such as low soil pH and low redox potential (Ismunadji and Dijkshoorn, 1971). For an adequate growth rate and yield of wheat, both forms of N (nitrate and ammonium) have to be applied, since the application of only one N form has adverse effects on the uptake of other nutrients and further on pH regulation (van Beusichem et al., 1988).

In brief, N is assimilated as follows. NO_3 is taken up into the cytosol of root cells and further converted to nitrite by nitrate reductase. In the next step, nitrite is transported to chloroplasts or plastids and is converted to NH_4 by nitrite reductase. NH_4 is the cation that is used to form the first amino acids in plant metabolism: glutamine and glutamate. In the plastids (roots) and chloroplasts (leaves), glutamine synthetase and glutamate synthase (GOGAT) convert NH_4 into glutamine and glutamate (Miflin and Habash, 2002). These amino acids are now available to synthesize all the other amino acids and to build up metabolic and structural proteins that are needed during the vegetative growth of wheat plants. Nitrogen assimilation takes place in both root and leaf cells. NO_3 is transported from roots to leaves via the xylem. If NO_3 is present to excess in leaves, it is stored in vacuoles and can be regenerated as soon as NO_3 concentrations decrease in the cytosol. During generative growth, NO_3 concentrations in the soil without additional fertilization is not sufficiently supplied, and the high demands of sink tissues cannot be met. Therefore, plant organs, especially the flag leaf, regenerate N from protein degradation to meet the N demands of the developing grain (Gooding et al., 2007). Mainly glutamine and asparagine are transported to the grains via the phloem to synthesize storage proteins (Macnicol, 1977; Anderson and Fitzgerald, 2001).

The wheat grain has different potentialities for attaining N. First, as discussed above, N that has been stored during vegetative growth can be mobilized, or second, further N can be taken up directly from the soil (deRuiter and Brooking, 1996). If N is not limiting, the primary source is N from vegetative tissues (60-92%) that has previously been assimilated (Simpson et al., 1983). Furthermore, Martre et al. (2003) have found N in grain to be source-regulated, i. e. total grain N is regulated by N in vegetative tissues as incorporated prior to anthesis (Barneix and Guitman, 1993; Fuertes-Mendizábal et al., 2010) but not by the activity of the grain namely the sink tissue. However, increasing N fertilization enhances the N concentration in the leaves of wheat and further provides the potential to obtain higher amounts of N for translocation to the grain and therefore the synthesis of gluten proteins.

3.2. Nitrogen Fertilization Increases the Amount of Gluten Proteins: Is this All We Know?

Nitrogen fertilization increases the protein concentration in all wheat plant tissues, and especially in wheat grain (Finney and Barmore, 1948). First and foremost, an increase in N fertilization enhances grain yield. However, Gauer et al. (1992) and Garrido-Lestache et al. (2005) have observed stagnating grain yield after N rates of 120 kg ha^{-1} and 100 kg ha^{-1}, respectively, despite protein concentration still increasing. The enhancement of grain protein concentration by incremental additions of N fertilizer is limited in the same way (Barneix et al., 1992; Triboi and Triboi-Blondel, 2002). After reaching a maximum N concentration in grain, only the N concentrations within the vegetative plant parts increase on additional N application. Mechanisms mediated by ammonium and glutamine reduce the expression of nitrate transporters in roots by negative feedback reactions. As long as storage protein synthesis is in progress, ammonium and glutamine concentrations will remain low and result in a constant N uptake (Foulkes et al., 2009). Furthermore, application at different stages of plant development influence grain yield and N concentration. Early N fertilization during vegetative growth enhances grain yield by influencing kernels per ear and ears per plant. Nitrogen fertilization at anthesis predominantly benefits grain N and therefore grain protein concentration (Jahn-Deesbach and Weipert, 1964; Wuest and Cassman, 1992).

Investigations of gluten protein fractions have shown that Osborne fractions are differently affected by N fertilization. Unlike gliadin and glutenin proteins, albumins and globulins are hardly influenced (Wieser and Seilmeier, 1998; Fuertes-Mendizábal et al., 2010). Wieser and Seilmeier (1998) have provided further insight into the gluten protein composition of various wheat cultivars that were raised on different N regimens. Interestingly, the quantity of gliadins increases more than the quantity of glutenins. These findings have been confirmed by Triboi et al. (2000) in field experiments.

With respect to the pattern of gliadin and glutenin subunits, the quantity of S-poor gluten proteins such as ω-gliadins and HMW-glutenins is mainly enhanced after increasing N fertilization. Wieser and Seilmeier (1998) have further investigated the HMW subunits, and observed that x-type glutenins are more influenced by N fertilization than y-type glutenins. X-type glutenins contain fewer thiol-groups than do y-type glutenins (Shewry et al., 1992). As a consequence, N fertilization enhances primarily S-poor subunits by increasing the expression of S-poor gluten subunit genes. Prieto et al. (1992) have observed increasing ω-gliadin concentrations with increasing N fertilization in agreement with the findings by Wieser and Seilmeier (1998).

Recent investigations concerning late N fertilization have provided new insights concerning gluten protein modification. Late fertilization has the following advantages: late N fertilization at ear emergence enhances protein concentration and, furthermore, recovers deficiency symptoms, since fast nutrient translocation mechanisms exist in plants. For example, Foulkes et al. (2009) have observed that N deficiency symptoms at anthesis can be even recovered by a later N fertilization. Tea et al. (2007) have examined the translocation of N and S after a late foliar application and found high recovery rates in leaves and grain with a beneficial impact on N/S ratios and protein concentration. Therefore, under inconvenient weather conditions with high or no precipitation, a split application of nutrients prevents leaching or helps to enable nutrient uptake by foliar application, respectively. Furthermore, the application of N at a time point when demand for N is high makes sense. This high demand occurs predominantly during the first few days of grain development, when proteins have started to be synthesized (Emes et al., 2003; Shewry et al., 2009).

Unlike increasing N fertilization treatments that predominantly increase gliadin concentration, a late application of N, when the flag leaf is just visible, mainly influences the concentration of glutenins (Fuertes-Mendizábal et al., 2010). With respect to the subunits of gluten proteins, each HMW-GS evidently increases with increasing N fertilization. Although HMW glutenins are minor components of gluten proteins, compromising approximately 12% of all gluten proteins, these glutenin subunits notably account for 45-70% of variation in baking quality (Shewry et al., 2000). This aspect largely explains the variation in baking quality after different N fertilization treatments.

3.3. Nitrogen Fertilization Increases Baking Quality, But the Question is How?

As early as 60 years ago, Finney and Barmore (1948) showed convincingly that the protein concentration in wheat flour of individual cultivars is highly correlated with loaf volume. However, the parameter that is responsible for this high correlation remains unknown. Parameters such as the gliadin/glutenin ratio (MacRitchie, 1980; Reinbold et al., 2008), amino acid composition (Granvogl et al., 2007), and thiol and disulfide status (Graveland et al., 1978) have been considered as possible explanations for the correlation but have failed to account for the correlation completely. The interdependency seems to be more complex than can be explained by only one parameter. However, can single gluten subunits indeed explain the impact on baking quality?

Many direct and indirect methods are available to examine baking quality on dough or on gluten proteins. The standardized baking test, for example the rapid-mix-test (RMT) (Pelshenke et al., 1978) is a direct method for examining loaf volume, dough consistency during preparation, and sensory parameters. Other methods such as sedimentation value, extensigram, alveogram, and mixograph-test are only capable of predicting loaf volume by the correlation of the flocculation of gliadins and glutenins, elasticity, resistance, and mixing properties, respectively, with loaf volume. Despite the standardized baking test being the only direct method for obtaining information about loaf volume, particularly in small-scale investigations such as pot experiments, indirect methods are appropriate for providing information about the impact of single gluten proteins on rheological parameters.

Several investigations have examined both the impact of N fertilization on baking quality parameters and the change of individual gluten proteins (Luo et al., 2000; Flaete et al., 2005; Fuertes-Mendizábal et al., 2010). Flæte et al. (2005) have observed increasing sedimentation values and resistence of dough (R_{max}) in gluten extensigraph experiments when comparing high N with low N fertilization levels. A diminished S concentration in gluten proteins as a result of inadequate S fertilization has been assumed to explain the increased resistance of dough. To underline the obvious impact of S, Popineau et al. (1994) used balanced N and S levels and concluded that this combination did not result in increased R_{max}.

Furthermore, two-dimensional (2D)-gel electrophoresis has been used to explain the effect of increased N fertilization on protein composition. Flæte et al. (2005) have found two proteins to be changed by increased N fertilization. However, these proteins have been identified to be non-gluten proteins such as glyceraldehyde-3-phosphate dehydrogenase and one serpin protein. With respect to the sedimentation value, Luo et al. (2000) have also compared early N fertilization at booting with late N fertilization at flowering and have observed the same positive impact of N fertilization on the sedimentation value as that reported by Flæte et al. (2005), irrespective of whether either early or late N application was conducted. Unlike Flæte et al. (2005), who used extensibility tests on dough to elucidate the impact of single gluten proteins on baking properties, Luo et al. (2000) used a mixograph test. The mixograph test revealed that the mid-line peak value, but not the mid-line peak time, was enhanced with N fertilization. To conclude, both groups of authors found an increase of dough resistance and a decrease of extensibility attributable to N fertilization. These quality parameters indicate reduced baking quality, since higher resistance and lower extensiblity account for a dough that is firm and therefore proves less well on dough fermentation.

Are there any hints in gluten composition that can explain the increased resistance? Ayoub et al. (1994), for example, have investigated both N rate and timing on loaf volume and found that an increasing N rate and late N fertilization increases both, the protein concentration and the loaf volume. Evidently, increasing N fertilization produces wheat doughs with lower extensibility and higher resistance, since those gluten proteins, which are poor in S, increase. Fuertes-Mendizabal et al. (2010) have further elucidated the effect of a late N application on baking properties and additionally found explanations in the gluten composition. Baking quality parameters examined on an alveograph increase with both an increasing N rate and a late N application. Dough strength increases by 50% with increasing N rate, whereas splitting enhances dough strength by about 80%. A late N application obviously has pivotal effects on both total protein concentration and baking quality characteristics. Investigations into the effects of late N fertilization on gluten composition by Fuertes-Mendizabal et al. (2010) have further enlightened the impact of individual proteins on dough characteristics. The x-type HMW-GS, which contain a smaller amount of SH-groups, are more affected by N fertilization than are y-type HMW-GS. Furthermore, late N fertilization seems to enhance all individual HMW-GS and thus increases dough strength and extensibility (Fuertes-Mendizábal et al., 2010). However, an evaluation of the effect of N fertilization on baking quality parameters remains difficult, since the S effect cannot be excluded from the evaluation.

4. Influence of S Fertilization on Gluten Composition and Baking Quality

4.1. Assimilation and Translocation of Sulfur within the Wheat Plant

Since S depositions from the atmosphere have decreased over the last few decades, the most important source of S for plants is the sulfate in soil, which is taken up by roots via sulfate transporters (Clarkson et al., 1993; Hawkesford et al., 1993; Buchner et al., 2004). Instead of assimilation in root cells, sulfate is, to a higher extent, transported to leaves for storage or direct assimilation (Larsson et al., 1991). Comparable to N assimilation, sulfate is stored in vacuoles as soon as sulfate uptake exceeds the rates of S assimilation. The first step of S assimilation is the activation of sulfate by ATP resulting in the formation of adenosine phosphosulfate. In the various subsequent steps, adenosine phosphosulfate is bound to thiol groups and is then reduced to sulfite (SO_3^{2-}) and further to sulfide (S^{2-}). All these assimilation steps are located in the chloroplasts of wheat leaf cells. In the next step, H_2S is incorporated into O-acetyle-serine to form the first S-containing amino acid, cysteine (Schiff and Saidha, 1987). Cysteine is further transported from chloroplasts to the cytosol to form methionine or is combined with glycine and glutamine to form the S-containing tripeptide glutathione (Ravanel et al., 1998; Hesse and Hoefgen, 2003). Glutathione fulfills several functions in plant metabolism, such as the reduction of oxidative stress within the glutathione-ascorbate cycle (Noctor and Foyer, 1998) and the detoxification of xenobiotics (Schröder et al., 2007) and heavy metals (Cobbett and Goldsbrough, 2002). Additional functions concern S translocation in wheat plants. Glutathione is one component that is transported from vegetative tissues, predominantly the flag leaf, to the developing grain (Rennenberg et al., 1979). However, the component that is primarily transported to the grain is not yet completely known. The S-containing compounds sulfate, glutathione, and S-methyl-methionine have been found in phloem sap during grain development (Bonas et al., 1982; Larsson et al., 1991). However, in wheat, S-methyl-methionine is presumed to be the main transport form of S to developing grains (Bourgis et al., 1999). Grains seem to prefer S-containing metabolites for the formation of S-rich gluten proteins (Steinfurth et al., submitted). Another option involves the direct assimilation of sulfate that is taken up from the soil and then directly transported to the grain (Steinfurth et al., submitted). Nevertheless, grains are able to take up sulfate directly from the xylem stream which has been revealed by investigations on sulfate transporter expression in wheat plant tissues (Buchner et al., 2004). Recently, Buchner et al. (2010) have further elucidated the impact of vegetative plant tissues on the translocation of S within the plant during grain development. If the duration of grain development and storage protein synthesis is prolonged, an increasing number of leaves is responsible for maintaining S transport toward the grain. Young leaves and primarily the flag leaf are not able to meet the S demand of the grain, and therefore, older leaves have to mobilize S to supply the sink organ, viz., the grain.

Furthermore, the uptake and translocation of S is highly regulated. Endless absorption of S by the roots during vegetative growth is impossible, since sulfate transporters undergo a negative feedback mechanism (Saito, 2000). Nevertheless, the compound in plant metabolism that is responsible for the negative feedback on sulfate transporters is not yet known. Anderson and McMahon (2001) have assumed sulfate or glutathione to down-regulate the transporter expression. Additional investigations have revealed that glutathione alone is not able to repress the transporter expression. Thus, the ratio of sulfate to glutathione is suggested to act more probably as an internal signal (Rennenberg, 2001).

Grain usually only contains small amounts of sulfate (Steinfurth et al. submitted). Unlike vegetative tissues that contain about 50% of total S as sulfate, grains only contain about 1-5% of the total S as sulfate (Zhao et al., 1999a). Nevertheless, the way that wheat grain forms amino acids and polypeptides with a special function in bread making is of interest.

4.2. Does Sulfur Fertilization Increase Gluten Proteins as it Has Been Reported for Nitrogen Fertilization?

Increasing S fertilization is accompanied by a higher S concentration in grain (Zhao et al., 1999a; Zörb et al., 2009). High S amounts are incorporated into the S-containing amino acids cysteine and methionine, which accumulate in grain during development (Zhao et al., 1999a). Unlike replete S-supplied plants that predominantly store glutamate as a main amino acid in grain, S-starved plants accumulate the amino acid asparagine. This amino acid has devastating effects on the nutritional quality of bakery products. As the authors Granvogl et al. (2007) have shown acrylamide formation increases during the baking process since concentrations of free asparagine and sugars are enhanced because of S deficiency.

However, next to the negative effects of product processing, the protein quality of wheat is also highly important for human and animal nutrition since cereal products cover major amounts of daily protein intake (Max Rubner-Institute, 2008). To examine the biological value of each essential amino acid in wheat flour the protein digestibility corrected amino acid score (PDCAAS) was calculated (Schaafsma, 2000).

For this purpose a pot experiment with increasing S fertilization (0 g S, 0.1 g S, and 0.2 g S at sowing) and, in particular, a late S fertilization (0.1 g S at sowing and additionally 0.1 g S at ear emergence) was conducted. This score has the advantage of evaluating the biological quality of protein by considering the characteristic digestibility and requirements for a 5-year-old child. The PDCAAS was not affected by S fertilization (Figure 1).

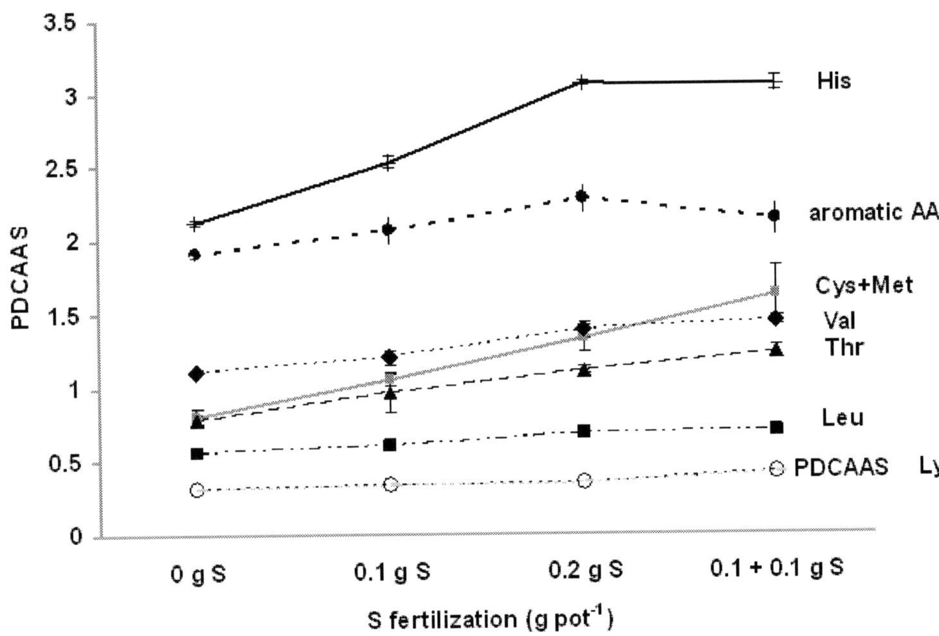

Figure 1. Impact of S fertilization on the biological value of wholemeal flour as calculated by protein digestibility corrected amino acid score (PDCAAS). Plant cultivation of cultivar Türkis was conducted according to Zörb et al. (2009). The cultivar indicated on the x-axis including a late S dressing (0.1 + 0.1 g S). PDCAAS was calculated according to Schaafsma (2000). His: histidine, aromat. AA: aromatic amino acids Tyr+Phe, Cys+Met: cysteine+methionine, Val: valine, Thr: threonine, Leu: leucine, Lys: lysine.

Unlike S fertilization, N fertilization decreases lysine concentrations (Shewry, 2007) because higher gluten concentrations are accompanied with a low lysine abundance (Kasarda et al., 1978). Nevertheless, the PDCAAS of cysteine and methionine increases with S fertilization (Figure 1). Methionine is, in contrast to lysine, not a limiting amino acid in wheat but is essential for human and animal nutrition. The facilitation of an adequately S-fertilized wheat might represent a chance to enhance the amount of methionine in the human diet and, hence, the protein quality consumed. Since humans and animals consume not only wheat-based products, high methionine levels in wheat can compensate for other low methionine sources.

Unfortunately, a further negative parameter concerns the correlation of the lysine concentration in wheat protein with baking quality. Nitrogen fertilization predominantly enhances the gluten protein concentration (Wieser and Seilmeier, 1998; Fuertes-Mendizábal et al., 2010), but does not affect the albumin and globuline fractions. Since these fractions contain higher lysine amounts than gluten fractions (Shewry, 2009), increasing baking quality is negatively correlated with nutritional protein quality. However, the question arises as to whether S fertilization changes gluten composition, and whether these changes influence baking quality.

In freshly prepared flour, 95% of cysteine is present as inter- or intramolecular disulfides and only 5% as the thiol form (Shewry and Tatham, 1997; Grosch and Wieser, 1999). These disulfide bonds are predominantly formed between LMW and HMW glutenin subunits resulting in gluten-protein aggregates. Therefore, S fertilization changes the protein composition and baking quality mainly by increasing the cysteine concentration in gluten proteins. Unlike N fertilization that predominantly affects the quantity of protein in grain, S fertilization is known to "fine-tune" gluten proteins (Wieser et al., 2004; Zörb et al., 2009).

With respect to the Osborne fractions, the quantity of total gliadin and glutenin proteins is not affected by increasing S fertilization (Zörb et al., 2010). Unlike plants that receive no or little S, well-fertilized plants synthesize large amounts of S-rich γ-gliadin and LMW-glutenin proteins (Wieser et al., 2004; Zörb et al., 2009). Accordingly, S-poor gluten proteins such as ω-gliadins and HMW-glutenins are reduced. However, how can the reduced synthesis of S-rich gluten proteins be explained by plant metabolism?

An adequate S supply during wheat plant development enhances the S pool in leaves and provides large amounts of S available for translocation to the grain. Furthermore, large amounts of glutathione are synthesized in the leaves (Steinfurth et al., submitted), and proteins might contain higher amounts of cysteine and methionine. Therefore, high concentrations of glutathione (Rennenberg, 1982) and S-methylmethionine (Bourgis et al., 1999) are translocated to the grain via the phloem. Exclusively under S deficiency, protein degradation in leaves is necessary to meet the high S sink capacity of grains (Fitzgerald et al., 1999). Because of high amounts of glutathione available in grain, S-rich proteins are primarily synthesized (Shewry et al., 2001). As long as large amounts of S are supplied to the grain, S-rich gluten proteins such as LMW glutenin subunits and γ-gliadins are predominantly synthesized (Wieser et al., 2004; Zörb et al., 2009). The higher the S concentration of the grain, the higher the LMW glutenin concentration, and the lower the HMW glutenin concentration (Wieser et al., 2004). As a result, the HMW/LMW ratio increases. HMW glutenin subunits are thought to interact with other HMW or LMW subunits to form glutenin aggregates (Shewry et al., 1992). Several investigations concerning polymer formation have concluded that the associations between glutenin subunits are non-random. Glutenin subunits encoded by chromosome 1D are extensively involved in oligomer formation, and variants of genome 1B subunits differ in their ability to combine with 1A and 1D subunits (Lawrence and Payne, 1983). Furthermore, x-type subunits are able to form inter-chain bonds with two cysteine molecules, whereas y-type subunits use six thiol-groups for the formation of aggregates (Moonen et al., 1985). However, these aggregation models are still under discussion and need to be examined further in detail.

Higher S fertilization evidently leads to greater amounts of S being translocated to the grain, and therefore, a higher potential for aggregates in the dough. Without the aggregation of gluten proteins, wheat dough has no ability to gain volume during dough fermentation, since the matrix formation containing predominantly starch and gluten proteins is restricted (Gan et al., 1995). However, how does a late S fertilization or foliar application of S influence gluten composition? The impact of a late S fertilization on gluten composition has recently been investigated. The effects of late S fertilization on baking quality characteristics are highly controversial. Zörb et al. (2009) have observed no effects of late S fertilization on the protein and S concentration in wheat grain compared with grain obtained following fertilization with the same amount of S applied at sowing. Even more controversially, Luo et al. (2000) have concluded that a late S application is not necessary to optimize baking quality, but glutenin proteins respond to both late S and combined late N and S fertilization with quality reduction.

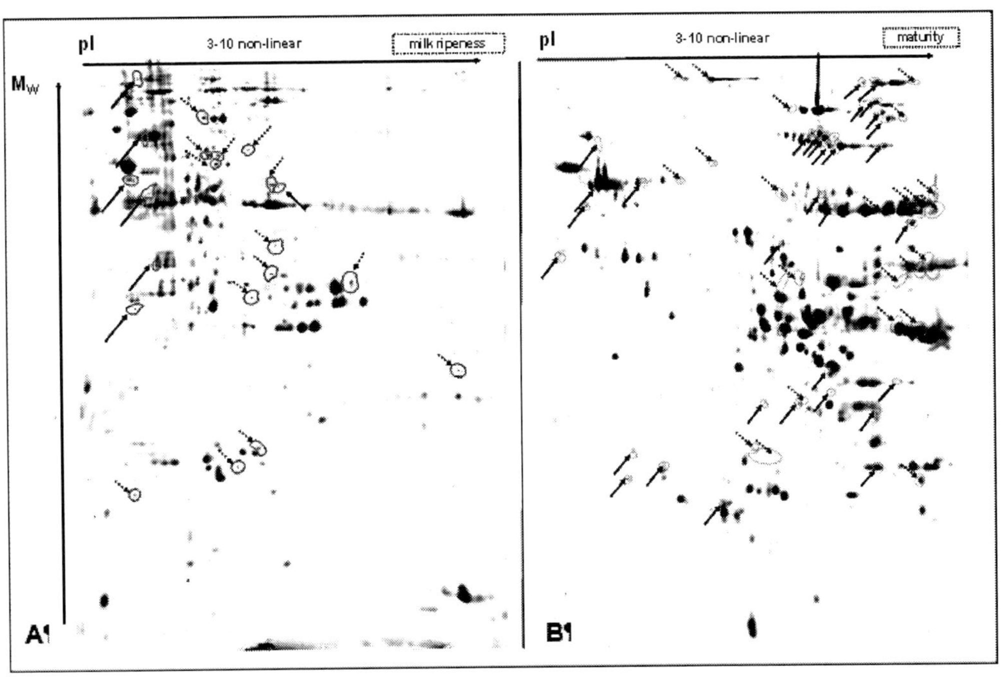

Figure 2. Impact of S fertilization on protein synthesis in various developmental stages of wheat grain. Overlays of wheat flour proteins comparing high with late S fertilization. A) Overlay of images of two-dimensional (2D)-gel electrophoresis of immature milk ripe grain and B) overlay of flour of mature wheat grain. Continuous (rising) arrows illustrate the higher quantity of protein with fertilization levels after late S application, dotted (falling) arrows indicate higher protein quantity in plants with 0.2 g S at sowing. Wheat plants were grown under high (0.2 g S at sowing) and late S fertilization (0.1 g S at sowing + 0.1 g S at ear emergence). Kernels were harvested at milk ripeness and maturity. Pot experiment with cultivar Türkis and overlays of 2D gels comparing high S with late S fertilization were conducted according to Zörb et al. (2009).

The experiment of Luo et al. (2000) has nevertheless a shortcoming concerning the N and S application time. Only the impact of N and S fertilization and its combination either at booting or at flowering has been examined. Hence, wheat plants might be grown on N- and S-deficient soil until booting and flowering, which has an important influence on plant metabolism in vegetative and generative tissues. To prevent quality losses, Haneklaus et al. (1995) concluded from S-starvation studies that S has to be supplied before S deficiency symptoms become apparent. Furthermore, Zörb et al. (2009) have demonstrated that a late S fertilization at ear emergence prevents S deficiency attributable to low S application at sowing.

By comparing S fertilization at sowing with the split application of S, *viz.*, at sowing and at ear emergence, a late S application effectively results in an increased loaf volume (Seling et al., 2006; Zörb et al., 2009). Reversed phase-high performance liquid chromatography (RP-HPLC) and the 2D-gel electrophoresis of flour proteins might provide further insight into the alteration of gluten protein composition after late S fertilization. Zörb et al. (2009) have found HMW glutenin subunits of type 1By9 and 1Bx7 to be enhanced after late S fertilization. However, Grove et al. (2009) have identified LMW glutenins and γ-gliadins as being enhanced after late S fertilization at early heading. Unfortunately, glutenins of high molecular weight were excluded from their analysis, because HMWs were negatively stained.

In addition, late S fertilization changes the protein profile of wheat grain during whole grain development time. Figure 2 illustrates the changes in protein pattern of wheat grain with respect to milk ripe kernels (13 days post anthesis) and mature wheat grain. Grain material of milk ripe and mature kernels from a pot experiment with the winter wheat cultivar Türkis was harvested, and 2D-gel overlays of high S fertilization (0.2 g S at sowing) with late S fertilization (0.1 g S at sowing and 0.1 g S at ear emergence) were compared (Figure 2). However, notably the protein profile predominantly changes in mature but not in milk ripe grain. In mature grain, more proteins are up-regulated in comparison with those in milk ripe grain. To conclude, S fertilization has an severe impact on the protein profile during the whole grain development and appears to change protein synthesis mainly during later stages of grain development. Shewry et al. (2009) have further illustrated protein synthesis during grain development by using 1D-gels and have demonstrated that LMW and HMW-GS are synthesized continuously until maturity, whereas other gluten proteins, such as ω-gliadins do not eventually attain further concentration. Since protein synthesis occurs until grain maturity and continues during the dessication phase, changes in S fertilization appear to regulate gluten protein gene expression until maturity.

To date, further investigations, especially those under field conditions with diverse fertilization techniques involving soil or foliar application are necessary to elucidate completely the impact of a late S fertilization on gluten composition and baking quality characteristics. Foliar application of S, for example, is a strategy to prevent S deficiency in the late growth stages of wheat, since foliar applied S is still incorporated in grain gluten proteins (Tea et al., 2003). In addition to the impact of foliar S application on S translocation, Tea et al. (2005) have elucidated the impact of foliar S application on the thiol-disulfide status and polymeric protein formation in wheat grain. Glutathione is an interacting molecule in gluten formation during grain development (Huttner and Wieser, 2001; Tea et al., 2005). High levels of reduced glutathione (GSH) decrease the maximum resistance of dough and concomitantly increase extensibility (Kieffer et al., 1990). Glutathione subsequently interacts with thiol-groups of gluten proteins and changes their properties. In particular, during the desiccation phase of grain development, the content of polymeric proteins increases. Furthermore, the impact of S fertilization on glutathione concentration and the polymerization of gluten proteins becomes noticeable during grain development (Tea et al., 2005). With low S fertilization, low polymeric protein-bound-glutathione is available and further low aggregated proteins emerge. Glutathione binds exclusively to cysteine residues that form inter-molecular disulfide bonds and are consequently responsible for the formation of LMW-GS aggregates. However, the effect of glutathione on gluten composition and its results in baking quality are not yet completely understood. Further investigations are needed. Reduced glutathione is known to affect the rheological properties of dough by the depolymerization of glutenin polymers via the specific cleavage of inter-molecular disulfide bonds. Reinbold et al. (2008) have elucidated the effect of increasing S fertilization on the thiol and glutathione status of S-deficient wheat. Since the dough made from S-deficient flour is highly resistant to expansion, it is of interest to determine the way that S fertilization changes SH-groups in wheat flour. With increasing S fertilization, cysteine and glutathione levels are enhanced. These authors have revealed that, in S-deficient wheat cysteine and glutathione correlated with bread volume in the same extent as the protein concentration. So far, S fertilization is known not only to enhance glutathione concentration in wheat flour, but concomitantly also the amount of S-containing gluten proteins. There is a need to elucidate further whether the impact of glutathione on gluten proteins is different between low S plants and high S plants.

4.3. Is There a Potential to Increase Baking Quality Further by Adequate Sulfur Fertilization?

Whereas an examination of the individual impact of N fertilization on baking quality parameters is difficult, the impact of S fertilization on baking quality has been demonstrated in several investigations. So far, S concentration in grain alone is not able to explain all the impact on loaf volume. An increase of the S concentration in grains, accompanied by adequate N fertilization, increases the extensibility of dough while doughs become less short and resistance decreases (Kettlewell et al., 1998; Zhao et al., 1999b; Flaete et al., 2005). Accordingly, S deficiency leads to high resistant doughs with low extensibility (Wrigley et al., 1984). These quality characteristics account for the higher potential of doughs to prove during dough fermentation because of the disulfide bonds in the gluten structure (Shewry et al., 1992), resulting in higher loaf volume (Moss et al., 1981; Flaete et al., 2005) among adequate S fertilization. Zhao et al. (1999b) have shown a positive but low correlation of S concentration in grain, with loaf volume. Therefore, grain S concentration more clearly explains an increasing dough extensibility. Furthermore, the gel-protein also called glutenin macro polymer, which is difficult to extract from wheat flour because of its high aggregation behavior, has a higher correlation with dough extensibility than the S concentration by itself. Haneklaus et al. (1992) have also been able to explain 40% of the variability in loaf volume based on grain S concentration. Reinbold et al. (2008) have further found cysteine and glutathione concentrations in S-deficient wheat to be positively correlated with R_{max}, extensibility, and extension area. Thus, these data confirm the way in which thiol groups and glutathione interact with S-rich gluten proteins and explain the changes in baking properties. As another important aspect, late S fertilization is considered to affect baking quality positively. Seling et al. (2006) have enhanced RMT loaf volume of winter wheat with a late S fertilization by 117 mL 100 g^{-1} flour. The enhanced baking quality may be refered to an increase of HMW subunits 1Bx7 and 1By9 glutenins found to be increased after late S fertilization (Zörb et al., 2009). This combination of proteins was formerly demonstrated to have a strong positive relationship with the Zeleny sedimentation value and dough resistance and therefore with high baking quality (Branlard and Dardevet, 1985; Seilmeier et al., 1991).

Tea et al. (2007) have examined the effect of foliar N and S application as a late application at anthesis after soil N treatment with ammonium nitrate during vegetative growth. Flour protein content was increased each with N and S treatments. However, combined N and S foliar applications resulted in the highest protein concentration. Alveograph measurements confirmed previous studies with soil-applied fertilizers. Nitrogen or combined N and S treatments increased dough strength, whereas S and combined N and S treatments increased dough swelling and extensibility. Hence, foliar N and S application at anthesis is an alternative to soil application during late wheat development, with applied nutrients still being assimilated and incorporated into gluten proteins.

An increase in the quantity of gluten proteins leads to a higher potential of dough to expand during dough fermentation, since more gas cells develop during mixing surrounded by the characteristic starch-protein-matrix (Gan et al., 1995). Furthermore, cysteine residues within this protein-matrix are able to asure extensibility and lower resistance. A balance between extensibility and resistance is of particular importance, since exessive resistance of dough limits expansion, and excessive elasticity leads to a failure of the dough to retain CO_2 during fermentation (Zhao et al., 1997). An adequate relationship of N and S has therefore to be maintained for adequate baking results, and further increases in baking quality are possible by the addition of adequate S fertilization concomitantly with adequate N fertilization.

5. The Way that Nitrogen and Sulfur Interact in Gluten Composition

A range of investigations highlights the importance of the adequate management of concomitant N and S fertilization. For example, Zörb et al. (2010) have found total protein and as well gliadin and glutenin concentrations to increase in flour following the N fertilization of wheat. In addition, a concomitant increase in S fertilization has no effect on total protein concentrations. However, increasing N fertilization without adequate S fertilization can provoke S-deficiency symptoms, despite an increase in yield and protein concentration. Moreover, Eriksen and Mortensen (2002) have examined the effect of S fertilization on barley; S-containing amino acids cysteine and methionine increase following S fertilization, but N concentration decreases because of a dilution effect caused by increased yield. Hence, baking quality can only be enhanced by increasing N fertilization levels providing that the S abundance in proteins is adequate. The question is: what exactly is meant by "adequate"? Randall et al. (1981) and Zhao et al. (1999a) have stated that N/S ratios of 17:1 and higher in wheat grain indicate S deficiency. Loaf volume can be enhanced by N fertilization at first, but as soon as the N/S ratio increases to inadequate levels, S fertilization needs to be adjusted. On the contrary, the S concentration in grain depends on the protein concentration, since the abundance of N and S in gluten protein subunits is genetically manifested. A high S concentration in grain consequently occurs when both high N and high

S supply is applied (Moss et al., 1981). Throughout plant development the abundance of metabolites following N and S assimilation determines the expression of individual gluten protein genes (Tabe et al., 2002). Therefore, splitting the rates of N and S or of different fertilizer forms, such as soil or foliar applications, are alternatives to guarantee nutrient availability throughout plant and grain development. Fast mobilization and retranslocation rates within the plant enable the reliable assembly of late-applied N and S into storage proteins and lead to high quality gluten protein properties during breadmaking.

ACKNOWLEDGMENTS

The authors thank Stephanie thor Straten, Fabian Braukmann and Bärbel Biegler for technical assistance. Financial support by the German Research Foundation (DFG Ge 1009/4-1) is gratefully acknowledged.

REFERENCES

Anderson J. W. & Fitzgerald M. A. (2001) Physiological and metabolic origin of sulphur for the synthesis of seed storage proteins. *Journal of Plant Physiology*, *158*, 447-456.

Anderson J. W. & McMahon P. J. (2001) The Role of Glutathione in the Uptake and Metabolism of Sulfur and Selenium. In D. Grill, M. Tausz and L. J. de Kok (Eds.), *Significance of Glutathione in Plant Adaptation to the Environment* (pp. 57-99). Dordrecht: Kluwer Academic Publishers.

Ayoub M., Guertin S., Fregeaureid J., Smith D. L. (1994) Nitrogen-fertilizer effect on breadmaking quality of hard red spring wheat in Eastern Canada. *Crop Science*, *34*, 1346-1352.

Barneix A. J., Arnozis P. A., Guitman M. R. (1992) The regulation of nitrogen accumulation in the grain of wheat plants (*Triticum aestivum*). *Physiologia Plantarum*, *86*, 609-615.

Barneix A. J. & Guitman M. R. (1993) Leaf Regulation of the Nitrogen Concentration in the Grain of Wheat Plants. *Journal of Experimental Botany*, *44*, 1607-1612.

Bell C., Fordham M., Richardson P., Cram J., Tomos D. (1995) Cellular and subcellular compartmentation of sulfate in leaves in relation to low-sulfur mobility. *Journal of Plant Nutrition and Soil Science*, *158*, 63-65.

Bonas U., Schmitz K., Rennenberg H., Bergmann L. (1982) Phloem Transport of Sulfur in *Ricinus*. *Planta*, *155*, 82-88.

Bourgis F., Roje S., Nuccio M. L., Fisher D. B., Tarczynski M. C., Li C. J., Herschbach C., Rennenberg H., Pimenta M. J., Shen T. L., Gage D. A., Hanson A. D. (1999) S-methylmethionine plays a major role in phloem sulfur transport and is synthesized by a novel type of methyltransferase. *Plant Cell*, *11*, 1485-1497.

Branlard G. & Dardevet M. (1985) Diversity of grain protein and bread wheat quality. 2. Correlation between high molecular-weight subunits of glutenin and flour quality characteristics. *Journal of Cereal Science*, *3*, 345-354.

Buchner P., Parmar S., Kriegel A., Carpentier M., Hawkesford M. J. (2010) The sulfate transporter family in wheat: tissue-specific gene expression in relation to nutrition. *Molecular Plant, 3*, 374-389.

Buchner P., Takahashi H., Hawkesford M. J. (2004) Plant sulphate transporters: co-ordination of uptake, intracellular and long-distance transport. *Journal of Experimental Botany, 55*, 1765-1773.

Clarkson D. T., Hawkesford M. J. and Davidian J.-C. (1993) Membrane and long-distance transport of sulfate. In L. J. de Kok, I. Stulen, H. Rennenberg, C. Brunold and W. E. Rauser (Eds.), *Sulphur Nutrition and Assimilation in Higher Plants* (pp. 3-19). The Hague: SPB Academic Publishing.

Cobbett C. & Goldsbrough P. (2002) Phytochelatins and metallothioneins: roles in heavy metal detoxification and homeostasis. *Annual Review of Plant Biology, 53*, 159-182.

Cox W. J. & Reisenauer H. M. (1973) Growth and ion uptake by wheat supplied nitrogen as nitrate, or ammonium, or both. *Plant and Soil, 38*, 363-380.

deRuiter J. M. & Brooking I. R. (1996) Effect of sowing date and nitrogen on dry matter and nitrogen partitioning in malting barley. *New Zealand Journal of Crop and Horticultural Science, 24*, 65-76.

Emes M. J., Bowsher C. G., Hedley C., Burrell M. M., Scrase-Field E. S. F., Tetlow I. J. (2003) Starch synthesis and carbon partitioning in developing endosperm. *Journal of Experimental Botany, 54*, 569-575.

Eriksen J. & Mortensen J. V. (2002) Effects of timing of sulphur application on yield, S-uptake and quality of barley. *Plant and Soil, 242*, 283-289.

Feil B. (1997) The inverse yield-protein relationship in cereals: possibilities and limitations for genetically improving the grain protein yield. *Trends in Agronomy, 1*, 103-119.

Finney K. F. & Barmore M. A. (1948) Loaf volume and protein content of hard winter and spring wheats. *Cereal Chemistry, 25*, 291-312.

Fitzgerald M. A., Ugalde T. D., Anderson J. W. (1999) S nutrition affects the pools of S available to developing grains of wheat. *Journal of Experimental Botany, 50*, 1587-1592.

Flæte N. E. S., Hollung K., Ruud L., Sogn T., Færgestad E. M., Skarpeid H. J., Magnus E. M., Uhlen A. K. (2005) Combined nitrogen and sulphur fertilisation and its effect on wheat quality and protein composition measured by SE-FPLC and proteomics. *Journal of Cereal Science, 41*, 357-369.

Foulkes M. J., Hawkesford M. J., Barraclough P. B., Holdsworth M. J., Kerr S., Kightley S., Shewry P. R. (2009) Identifying traits to improve the nitrogen economy of wheat: Recent advances and future prospects. *Field Crops Research, 114*, 329-342.

Fuertes-Mendizábal T., Aizpurua A., Gonzalez-Moro M. B., Estavillo J. M. (2010) Improving wheat breadmaking quality by splitting the N fertilizer rate. *European Journal of Agronomy, 33*, 52-61.

Gan Z., Ellis P. R., Schofield J. D. (1995) Gas cell stabilization and gas retention in wheat bread dough. *Journal of Cereal Science, 21*, 215-230.

Garrido-Lestache E., Lopez-Bellido R. J., Lopez-Bellido L. (2005) Durum wheat quality under Mediterranean conditions as affected by N rate, timing and splitting, N form and S fertilization. *European Journal of Agronomy, 23*, 265-278.

Gauer L. E., Grant C. A., Gehl D. T., Bailey L. D. (1992) Effects of nitrogen fertilization on grain protein content, nitrogen uptake, and nitrogen use efficiency of 6 spring wheat

(*Triticum aestivum* L) cultivars, in relation to estimated moisture supply. *Canadian Journal of Plant Science*, *72*, 235-241.

Goesaert H., Brijs K., Veraverbeke W. S., Courtin C. M., Gebruers K., Delcour J. A. (2005) Wheat flour constituents: how they impact bread quality, and how to impact their functionality. *Trends in Food Science & Technology*, *16*, 12-30.

Gooding M. J., Gregory P. J., Ford K. E., Ruske R. E. (2007) Recovery of nitrogen from different sources following applications to winter wheat at and after anthesis. *Field Crops Research*, *100*, 143-154.

Granvogl M., Wieser H., Koehler P., von Tucher S., Schieberle P. (2007) Influence of sulfur fertilization on the amounts of free amino acids in wheat. Correlation with baking properties as well as with 3-aminopropionamide and acrylamide generation during baking. *Journal of Agricultural and Food Chemistry*, *55*, 4271-4277.

Graveland A., Bosveld P., Marseille J. P. (1978) Determination of thiol-groups and disulfide bonds in sheat-flour and dough. *Journal of the Science of Food and Agriculture*, *29*, 53-61.

Grosch W. & Wieser H. (1999) Redox reactions in wheat dough as affected by ascorbic acid. *Journal of Cereal Science*, *29*, 1-16.

Grove H., Hollung K., Moldestad A., Færgestad E. M., Uhlen A. K. (2009) Proteome changes in wheat subjected to different nitrogen and sulfur fertilizations. *Journal of Agricultural and Food Chemistry*, *57*, 4250-4258.

Haneklaus S., Evans E., Schnug E. (1992) Baking quality and sulphur content of wheat. I. Influence of sulphur and protein concentration on loaf volume. *Sulphur in Agriculture*, *16*, 31-36.

Haneklaus S., Murphy D. P. L., Nowak G., Schnug E. (1995) Effects of the timing of sulfur application on grain yield and yield components of wheat. *Zeitschrift für Pflanzenernährung und Bodenkunde*, *158*, 83-85.

Hawkesford M. J., Davidian J. C., Grignon C. (1993) Sulfate proton cotransport in plasma-membrane vesicles isolated from roots of *Brassica napus* 1 - Increased transport in membranes isolated from sulfur-starved plants. *Planta*, *190*, 297-304.

Hesse H. & Hoefgen R. (2003) Molecular aspects of methionine biosynthesis. *Trends in Plant Science*, *8*, 259-262.

Huttner S. & Wieser H. (2001) Studies on the distribution and binding of endogenous glutathione in wheat dough and gluten. II. Binding sites of endogenous glutathione in glutenins. *European Food Research and Technology*, *213*, 460-464.

Ismunadji M. & Dijkshoorn W. (1971) Nitrogen nutrition of rice plants measured by growth and nutrient content in pot experiments. Ionic balance and selective uptake. *Netherlands Journal of Agricultural Science*, *19*, 223-236.

Jahn-Deesbach W. & Weipert D. (1964) Untersuchungen über den Einfluß der Stickstoffdüngung auf Ertrag und backereitechnologische Qualitätseigenschaften des Weizens. *Landwirtschaftliche Forschung*, *18*, 132-145.

Kasarda D. D., Nimmo C. C. and Kohler G. O. (1978) Proteins and the Amino Acid Compositon of Wheat Fractions. In Ed. Y. Pomeranz, *Wheat Chemistry and Technology* (p 227). St.Paul, USA: American Association of Cereal Chemists, Inc.

Kettlewell P. S., Griffiths M. W., Hocking T. J., Wallington D. J. (1998) Dependence of wheat dough extensibility on flour sulphur and nitrogen concentrations and the influence of foliar-applied sulphur and nitrogen fertilisers. *Journal of Cereal Science*, *28*, 15-23.

Kieffer R., Kim J. J., Walther C., Laskawy G., Grosch W. (1990) Influence of glutathione and cysteine on the improver effect of ascorbic acid stereoisomers. *Journal of Cereal Science, 11*, 143-152.

Larsson C. M., Larsson M., Purves J. V., Clarkson D. T. (1991) Translocation and cycling through roots of recently absorbed nitrogen and sulfur in wheat (*Triticum aestivum*) during vegetative and generative growth. *Physiologia Plantarum, 82*, 345-352.

Lawrence G. J. & Payne P. I. (1983) Detection by gel-electrophoresis of oligomers formed by the association of high molecular-weight glutenin protein subunits of wheat endosperm. *Journal of Experimental Botany, 34*, 254-267.

Luo C., Branlard G., Griffin W. B., McNeil D. L. (2000) The effect of nitrogen and sulphur fertilisation and their interaction with genotype on wheat glutenins and quality parameters. *Journal of Cereal Science, 31*, 185-194.

Macnicol P. K. (1977) Synthesis and interconversion of amino acids in developing cotyledons of pea (*Pisum-Sativum* L.). *Plant Physiology, 60*, 344-348.

MacRitchie F. (1980) Studies of gluten protein from wheat flours. *Cereal Foods World, 25*, 382-385.

Martre P., Porter J. R., Jamieson P. D., Triboi E. (2003) Modeling grain nitrogen accumulation and protein composition to understand the sink/source regulations of nitrogen remobilization for wheat. *Plant Physiology, 133*, 1959-1967.

Max Rubner-Institute (2008) Nationale Verzehrsstudie II. Ergebnisbericht, Teil 2; http://www.was-esse-ich.de/.

Miflin B. J. & Habash D. Z. (2002) The role of glutamine synthetase and glutamate dehydrogenase in nitrogen assimilation and possibilities for improvement in the nitrogen utilization of crops. *Journal of Experimental Botany, 53*, 979-987.

Moonen J. H. E., Scheepstra A., Graveland A. (1985) Biochemical-properties of some high molecular-weight subunits of wheat glutenin. *Journal of Cereal Science, 3*, 17-27.

Moss H. J., Wrigley C. W., MacRitchie F., Randall P. J. (1981) Sulfur and nitrogen fertilizer effects on wheat. 2. Influence on grain quality. *Australian Journal of Agricultural Research, 32*, 213-226.

Noctor G. & Foyer C. H. (1998) Ascorbate and glutathione: keeping active oxygen under control. *Annual Review of Plant Physiology and Plant Molecular Biology, 49*, 249-279.

Osborne T.B. (1924) The vegetable proteins. Longmans, Breen & Co, London.

Pelshenke P. F., Schulz A. and Stephan H. (1978) Rapid-Mix-Test, Standard-Backmethode für die Weizen- und Weizenmehlbeurteilung. In Ed. Arbeitsgemeinschaft Getreideforschung e.V., *Merkblatt der Arbeitsgemeinschaft Getreideforschung e.V.* (. Detmold: Granum-Verlag.

Popineau Y., Cornec M., Lefebvre J., Marchylo B. (1994) Influence of high M_r glutenin subunits on glutenin polymers and rheological properties of glutens and gluten subfractions of near-isogenic lines of wheat Sicco. *Journal of Cereal Science, 19*, 231-241.

Prieto J. A., Kelfkens M., Weegels P. L., Hamer R. J. (1992) Variations in the gliadin pattern of flour and isolated gluten on nitrogen application - implications for baking potential and rheological properties. *Zeitschrift fur Lebensmittel-Untersuchung und Forschung, 194*, 337-343.

Randall P. J., Spencer K., Freney J. R. (1981) Sulfur and nitrogen fertilizer effects on wheat. 1. Concentrations of sulfur and nitrogen and the nitrogen to sulfur ratio in grain, in relation to the yield response. *Australian Journal of Agricultural Research*, *32*, 203-212.

Ravanel S., Gakiere B., Job D., Douce R. (1998) The specific features of methionine biosynthesis and metabolism in plants. *Proceedings of the National Academy of Sciences of the United States of America*, *95*, 7805-7812.

Reinbold J., Rychlik M., Asam S., Wieser H., Koehler P. (2008) Concentrations of total glutathione and cysteine in wheat flour as affected by sulfur deficiency and correlation to quality parameters. *Journal of Agricultural and Food Chemistry*, *56*, 6844-6850.

Rennenberg H. (1982) Glutathione metabolism and possible biological roles in higher-plants. *Phytochemistry*, *21*, 2771-2781.

Rennenberg H. (2001) Glutathione - An Ancient Metabolite with Modern Tasks. In D. Grill, M. Tausz and L. J. De Kok (Eds.), *Significance of Glutathione in Plant Adaptation to the Environment* (pp. 1-11). Dordrecht, The Netherlands: Kluwer Academic Publishers.

Rennenberg H., Schmitz K., Bergmann L. (1979) Long-distance transport of sulfur in *Nicotiana tabacum*. *Planta*, *147*, 57-62.

Saito K. (2000) Regulation of sulfate transport and synthesis of sulfur-containing amino acids. *Current Opinion in Plant Biology*, *3*, 188-195.

Schaafsma G. (2000) The protein digestibility-corrected amino acid score. *Journal of Nutrition*, *130*, 1865S-1867S.

Schiff J. A. & Saidha T. (1987) Overview - Inorganic sulfur and sulfate activation. *Methods in Enzymology*, *143*, 329-334.

Schröder P., Scheer C. E., Diekmann F., Stampfl A. (2007) How plants cope with foreign compounds - translocation of xenobiotic glutathione conjugates in roots of barley (*Hordeum vulgare*). *Environmental Science and Pollution Research*, *14*, 114-122.

Seilmeier W., Belitz H. D., Wieser H. (1991) Separation and quantitative-determination of high-molecular-weight subunits of glutenin from different wheat-varieties and genetic-variants of the variety Sicco. *Zeitschrift fur Lebensmittel-Untersuchung und Forschung*, *192*, 124-129.

Seling S., Weigelt W., Wissemeier A. H. (2006) Bedeutung der Schwefeldüngung für Ertrag und Qualität von Weizen. *Getreidetechnologie*, *60*, 148-152.

Shewry P. R. (1995) Plant storage proteins. *Biological Reviews of the Cambridge Philosophical Society*, *70*, 375-426.

Shewry P. R. (2007) Improving the protein content and composition of cereal grain. *Journal of Cereal Science*, *46*, 239-250.

Shewry P. R. (2009) Wheat. *Journal of Experimental Botany*, *60*, 1537-1553.

Shewry P. R. & Halford N. G. (2002) Cereal seed storage proteins: structures, properties and role in grain utilization. *Journal of Experimental Botany*, *53*, 947-958.

Shewry P. R., Halford N. G., Tatham A. S. (1992) High-molecular-weight subunits of wheat glutenin. *Journal of Cereal Science*, *15*, 105-120.

Shewry P. R., Popineau Y., Lafiandra D., Belton P. (2000) Wheat glutenin subunits and dough elasticity: findings of the EUROWHEAT project. *Trends in Food Science & Technology*, *11*, 433-441.

Shewry P. R. & Tatham A. S. (1997) Disulphide bonds in wheat gluten proteins. *Journal of Cereal Science*, *25*, 207-227.

Shewry P. R., Tatham A. S., Halford N. G. (2001) Nutritional control of storage protein synthesis in developing grain of wheat and barley. *Plant Growth Regulation*, *34*, 105-111.

Shewry P. R., Underwood C., Wan Y. F., Lovegrove A., Bhandari D., Toole G., Mills E. N. C., Denyer K., Mitchell R. A. C. (2009) Storage product synthesis and accumulation in developing grains of wheat. *Journal of Cereal Science*, *50*, 106-112.

Simpson R. J., Lambers H., Dalling M. J. (1983) Nitrogen redistribution during grain-growth in wheat (*Triticum aestivum* L). 4. Development of a quantitative model of the translocation of nitrogen to the grain. *Plant Physiology*, *71*, 7-14.

Statistisches Bundesamt (2010) https://www-genesis.destatis.de/genesis/online;jsessionid=BAA09C17B0BC21CF3C72E94F81454E77.tomcat_GO_1_2?operation=abruftabelleAbrufen&selectionname=41241-0003&levelindex=1&levelid=1282214538094&index=2. (18-9-2010)

Steinfurth D., Zörb C., Braukmann F., Mühling K.H. (submitted) Time-dependent distribution of sulphur, sulphate and glutathione in wheat tissues and grain as affected by late sulphur fertilization.

Tabe L., Hagan N., Higgins T. J. V. (2002) Plasticity of seed protein composition in response to nitrogen and sulfur availability. *Current Opinion in Plant Biology*, *5*, 212-217.

Tea I., Genter T., Naulet N., Marie L. M., Kleiber D. (2007) Interaction between nitrogen and sulfur by foliar application and its effects on flour bread-making quality. *Journal of the Science of Food and Agriculture*, *87*, 2853-2859.

Tea I., Genter T., Naulet N., Morvan E., Kleiber D. (2003) Isotopic study of post-anthesis foliar incorporation of sulphur and nitrogen in wheat. *Isotopes in Environmental and Health Studies*, *39*, 289-300.

Tea I., Genter T., Violleau F., Kleiber D. (2005) Changes in the glutathione thiol-disulfide status in wheat grain by foliar sulphur fertilization: consequences for the rheological properties of dough. *Journal of Cereal Science*, *41*, 305-315.

Triboi E., Abad A., Michelena A., Lloveras J., Ollier J. L., Daniel C. (2000) Environmental effects on the quality of two wheat genotypes: 1. quantitative and qualitative variation of storage proteins. *European Journal of Agronomy*, *13*, 47-64.

Triboi E. & Triboi-Blondel A. M. (2002) Productivity and grain or seed composition: a new approach to an old problem - invited paper. *European Journal of Agronomy*, *16*, 163-186.

van Beusichem M. L., Kirkby E. A., Baas R. (1988) Influence of nitrate and ammonium nutrition on the uptake, assimilation, and distribution of nutrients in *Ricinus communis*. *Plant Physiology*, *86*, 914-921.

Vestreng V., Myhre G., Fagerli H., Reis S., Tarrason L. (2007) Twenty-five years of continuous sulphur dioxide emission reduction in Europe. *Atmospheric Chemistry and Physics*, *7*, 3663-3681.

Wieser H., Gutser R., von Tucher S. (2004) Influence of sulphur fertilisation on quantities and proportions of gluten protein types in wheat flour. *Journal of Cereal Science*, *40*, 239-244.

Wieser H. & Seilmeier W. (1998) The influence of nitrogen fertilisation on quantities and proportions of different protein types in wheat flour. *Journal of the Science of Food and Agriculture*, *76*, 49-55.

Wrigley C. W., Ducros D. L., Fullington J. G., Kasarda D. D. (1984) Changes in polypeptide composition and grain quality due to sulfur deficiency in wheat. *Journal of Cereal Science*, *2*, 15-24.

Wuest S. B. & Cassman K. G. (1992) Fertilizer-nitrogen use efficiency of irrigated wheat .1. Uptake efficiency of preplant versus late-season application. *Agronomy Journal*, *84*, 682-688.

Yin L. P., Li P., Wen B., Taylor D., Berry J. O. (2007) Characterization and expression of a high-affinity nitrate system transporter gene (TaNRT2.1) from wheat roots, and its evolutionary relationship to other NTR2 genes. *Plant Science*, *172*, 621-631.

Zhao F. J., Hawkesford M. J., McGrath S. P. (1999a) Sulphur assimilation and effects on yield and quality of wheat. *Journal of Cereal Science*, *30*, 1-17.

Zhao F. J., Salmon S. E., Withers P. J. A., Evans E. J., Monaghan J. M., Shewry P. R., McGrath S. P. (1999b) Responses of breadmaking quality to sulphur in three wheat varieties. *Journal of the Science of Food and Agriculture*, *79*, 1865-1874.

Zhao F. J., Withers P. J. A., Evans E. J., Monaghan J., Salmon S. E., Shewry P. R., McGrath S. P. (1997) Sulphur nutrition: an important factor for the quality of wheat and rapeseed (Reprinted from Plant nutrition for sustainable food production and environment, 1997). *Soil Science and Plant Nutrition*, *43*, 1137-1142.

Zörb C., Grover C., Steinfurth D., Mühling K. H. (2010) Quantitative proteome analysis of wheat gluten as influenced by N and S nutrition. *Plant and Soil*, *327*, 225-234.

Zörb C., Steinfurth D., Seling S., Langenkämper G., Koehler P., Wieser H., Lindhauer M. G., Mühling K. H. (2009) Quantitative protein composition and baking quality of winter wheat as affected by late sulfur fertilization. *Journal of Agricultural and Food Chemistry*, *57*, 3877-3885.

In: Wheat: Genetics, Crops and Food Production
Editor: Maria T. Almeida, pp. 231-251
ISBN 978-1-61209-307-9

Chapter 7

IMPROVING DROUGHT AND SALINITY TOLERANCE IN WHEAT

Faïçal Brini, Inès Yakoubi, Habib Khoudi, Moez Hanin and Khaled Masmoudi*

Plant Protection and Improvement Laboratory, Centre of Biotechnology of Sfax (CBS)/University of Sfax, B.P ''1177'' 3018 Sfax –Tunisia

ABSTRACT

Drought and salinity are major constraints on crop production and food security, and adversely affect entire countries over several years and result in serious social, economic, and environmental cost. Wheat production in the Mediterranean region is limited mainly by the availability of water resources. Water-deficit stress caused by drought and soil salinization adversely affects plant growth and crop productivity. This situation is compounded firstly by the predicted change in climate with increased temperatures and decreased precipitation as a result of global warming and secondly by the narrowness of genetic diversity of modern wheat cultivars due to domestication, selection and conventional breeding programs that minimizes plant tolerance to such stresses. The challenge of the majority of breeding programs, which are in continuous profit from the development of biotechnology, genetic engineering and genomic–based approaches, is the release of improved wheat cultivars able to withstand these adverse conditions. The identification of genes that control discrete agronomic traits, the mapping of QTLs (Quantitative Trait Loci) affecting tolerance to drought and salinity in model species and crops are powerful support for the development of functional markers, which will assist selection in the breeding process.

Keywords: wheat, drought, salinity, functional genomics, marker-assisted selection, QTL, genetic engineering.

---

* For correspondence: Khaled Masmoudi, Plant Protection and Improvement Laboratory, Centre of Biotechnology of Sfax (CBS), B.P'1177', 3018 Sfax, Tunisia. Tél/Fax: 216-74 872 091 ; Email: khaled.masmoudi@cbs.rnrt.tn

1. INTRODUCTION

One of the major challenges facing agriculture today is the global water shortage caused by the increasing world population and worldwide climate change. Stresses such as drought and salinity affect the productivity of most field crops to variable degrees, depending on the onset time, duration, and intensity of the stress. Current and predicted climatic conditions, such as prolonged drought and heat episodes, pose a serious challenge for agricultural production worldwide, affecting plant growth and yield, and causing annual losses estimated at billions of dollars [1]. The acclimation of plants to abiotic stress conditions is a complex and coordinated response involving hundreds of genes. These responses are also affected by interactions between different environmental factors and the developmental stage of the plant and could result in shortened life cycle, reduced or aborted seed production, or accelerated senescence. Traditional breeding for yield stability under water-deficit conditions is labour-intensive and time-consuming, primarily because of the complicated genetics that control yield and the difficulty in controlling moisture levels in the field. Furthermore, the long-time practice of breeding and selecting for higher yield under well watered conditions has narrowed the genetic spectrum of current cultivars with respect to their ability to tolerate drought stress. The genetic characterization of segregating populations of various crop species facilitated the identification of QTLs (Quantitative Trait Loci) associated with root growth [2], early flowering [3], drought tolerance [4, 5], etc. Even so, few of these QTLs were successfully used for breeding programs. The most important limitation of stress-related QTLs is that they are dependent on the environmental conditions to which they were characterized (high G × E interaction) [6]. Functional analysis of the genes can be significantly aided through the application of reverse genetics approaches such as RNA interference (RNAi) and by screening TILLING populations in order to characterize the individual gene function(s) [7]. Emphasis should be given to forward genetics studies where the identified genes can be expressed in genotypes that have been already selected for their adaptation to stressful environmental conditions.

The genomic tools and methods that are available will provide new opportunities to characterize the gene networks involved and to gain a more holistic view of abiotic stress responses. Expressed sequence tags (ESTs) from abiotic stress-treated libraries of various crop plants, complete genome sequence information for rice and *Arabidopsis* and the development of new bioinformatic tools allow us to identify the key stress-responsive gene-pools. Furthermore, use of multi-parallel techniques such as expression profiling by microarrays, random and targeted mutagenesis, complementation and promoter-trapping strategies provide important clues for functional characterization of stress responsive genes and stress tolerance mechanisms [8]. Genomic studies showed considerable overlap of plant responses to cold, drought and salinity stresses underlining the complexity and provides opportunities to engineer new stress-resistant crop varieties [9 -13]. Genetic engineering is thus being intensively explored to enhance plants' tolerance to abiotic stresses, and many engineered plants have manifested improved stress resistance phenotypes [14 - 17]. Although such genetic engineering has been successfully applied to some crop species, such as rice, canola (*Brassica napus*), and maize, the majority of published results were based on the analysis of transgenic *Arabidopsis*, and stress tolerance was often evaluated in potted soil with rapid stress imposition under controlled growth conditions.

With the advent of newly developed genomics, two major approaches could be used in exploiting the gene-pool for imparting abiotic stress tolerance: (1) identification and introduction of genes imparting stress-tolerance into crops of interest, and (2) development and identification of molecular markers associated with genes or QTLs conferring tolerance to stress in germplasm collections and their use in marker-assisted breeding programs. Here, we review the results of functional genomic approaches to analyze and manipulate salt and drought stress tolerance in wheat.

2. Functional Genomics and Stress Response in Wheat

Genome analysis has been mostly limited to model plants that fulfil some specific requirements such as: (1) small genome size, (2) short generation time, (3) small size to enable growth in limited space, and (4) availability of gene manipulation technologies [18]. In particular, two of the most important model species are *Arabidopsis thaliana* and rice (*Oryza sativa* L.) for dicotyledonous and monocotyledonous plant species, respectively. Besides its importance as a crop, rice has a high degree of synteny with genomes of other cereals, such as maize, wheat, barley and other grasses [19 - 21]. Great advances in the comparison of genomes and transcriptomes of different organisms have contributed to the development of comparative genomics as one of the most promising fields in the area [19, 22]. In this way, finding variations in the genome or the transcriptome from the current model species related to interesting agronomic traits is of the highest importance for crop biotechnology [23].

2.1. Role of Model Species

Many of the recent advances in genomics research have been developed in model species. *Arabidopsis* has led the way but, more recently, resources and information on rice and maize have gained importance. The complete sequencing of the *Arabidopsis* and rice genomes has been crucial to the success of these models. For *Arabidopsis*, the mutant and tagged populations and extensive microarray databases [24] have provided powerful resources for gene discovery and functional analysis. The microarray information covers a diverse series of stresses including cold, salt, heat, water deficit and UV treatments [24]. Through these and related studies used in *Arabidopsis* many genes involved in the control of abiotic stress tolerance have been identified and, in some cases, the information has been transferable to wheat. A clear example can be seen with the Dehydration-responsive element-binding (DREB) transcription factors. These were first identified in *Arabidopsis* as being involved in the control of drought and cold responses [25]. Subsequent experiments have identified orthologous genes in cereals (maize [26], rice [27] and wheat [28]). In the case of wheat, the DREB-1A transcription factor has been used to generate transgenic lines that show increased drought tolerance in preliminary trials [29]. A further example is the ornithine amino transferase gene that was found to confer salt tolerance when overexpressed in *Arabidopsis* [30]. This gene has been used to transform wheat and the transgenic lines are currently being evaluated in field trials in Australia [31]. Although these two examples indicate that genes discovered in *Arabidopsis* can be used directly or via orthologues to engineer abiotic stress tolerance in cereals, the key validation of this approach will only be effective when the field trials, currently underway, are completed. Many of the crucial stress responses in crop plants relate to their behaviour as a crop, when growing as part of a plant community and when faced with multiple stresses. Therefore, the preliminary results with the *Arabidopsis* genes are still well away from real application.

2.2. Genome Similarity in Wheat, Barley and Other Cereals

Wheat, barley and rye are all closely related. Indeed, stable hybrids can be produced between wheat and barley (*Trithordeum*) and wheat and rye (*Triticale*). The high level of genome similarity between wheat, rye and barley has been recognized for some time. Not only are the sequences of individual genes similar between the species, but gene order on the chromosomes is highly conserved (synteny). Indeed it is possible to replace many wheat chromosomes with related chromosome from rye or barley. The gene order across the grasses is generally well maintained and clear relationships can be drawn across the grass genomes [32, 33]. However, at the single gene level the order does start to break down. For example, a detailed comparison of a 20 cM region of chromosome 3 DS of wheat with rice chromosome 1 showed about a 20% loss of colinearity [34]. There are also certain classes of genes, such as the race-specific disease resistance genes, that are of recent evolutionary origin and these are not usually located at syntenous regions. However, this problem does not seem to apply to many of the key loci controlling abiotic stress tolerance, and the rice genome sequence [35, 36] has proved to be a powerful tool for positional cloning of genes from wheat and barley [37, 38].

2.3. Mutant Populations

Modern genomics makes reverse genetics possible as large amounts of genomic and expressed sequence information become available. One reverse genetics method involves the use of chemical mutagenesis coupled with high-throughput screening to detect sequence variations in genes of interest, so called TILLING (Targeting Induced Local Lesion IN Genome). In contrast to typical reverse genetics techniques such as RNA interference and insertional mutagenesis, TILLING is a non transgenic approach and generates allelic series of mutations, including knockouts, in the desired gene [39].

The TILLING method was first applied in plants [7] and has the advantage that seeds of mutant populations can be banked for later use. The TILLING method is useful for both functional genomics as demonstrated in *Arabidopsis* [40] and crop improvement as demonstrated in wheat (*Triticum aestivum* L.) [41] and applied successfully to maize (*Zea mays* L.) [42]; rice (*Oryza sativa* L.) [43, 44]; barley (*Hordeum vulgare* L.) [45]; sugar beet (*Beta vulgaris* L.) [46]; *Lotus japonicus* (Regel) K. Larsen [47]; and soybean [*Glycine max* (L.) Merr.] [48].

Wheat appears to be particularly suitable for mutant screening since it is able to carry a very high mutation load, presumably due to polyploidy. Slade et al. [41] identified 246 alleles after screening for each homologue in only 1920 mutated individuals. A large wheat mutant population constructed for mutation screening was used to identify a series of mutations in the granule-bound starch synthase I (GBSSI) of wheat [41]. There are currently several projects underway around the world to develop mutated populations of both wheat and barley, and diploid progenitors of wheat, and many of these will be available as public resources.

2.4. Transcript Profiling

A fundamental step in any functional genomics study is the analysis of gene expression. One of the greatest strengths of genomics compared to other disciplines is the prospect of analyzing the expression of thousands of genes simultaneously; resulting in a more comprehensive picture of changes occurring in the transcriptome across different conditions [49]. The technology available for the analysis of gene expression can be divided into two categories: closed and open systems. Closed systems are characterized by a finite number of genes that can be assessed by virtue of their inclusion by selection. Therefore, the coverage of genes will be related to the completeness of the knowledge of the genome being studied, limiting this kind of analysis to the most well characterized species or systems [49]. Typically, closed systems such as microarrays and real-time polymerase chain reaction (PCR) have been extensively used in gene expression analysis in plants [50 - 57]. On the other hand, with open systems there is no need for previous knowledge of the genome or transcriptome of the organism. cDNA-AFLP (cDNA-Amplified Fragment Length Polymorphism), MPSS (Massively Parallel Signature Sequencing), and specially SAGE (Serial Analysis of Gene Expression) have been successfully used to quantify transcript abundance and generate expression data across different tissue types or developmental stages in higher plants [58 - 65]. Worth mentioning are 454-sequencing technology and digital gene expression (DGE) that have recently been used to study the transcriptome of different organisms and promise to become an efficient and cost-effective alternative with high potential in crop research [66 - 70]. Open and closed systems should not be considered as competitors, but rather as complementary technologies to be used depending on the subject to be analyzed and the objectives of the research.

3. Gene Expression Profiling during Drought and Salt Stress in Wheat

Transcriptome analysis using microarray technology offers a powerful platform to find out the expression of genes during abiotic stress at the global level. This led to a cascade of challenges that occur at molecular, metabolic, physiological, cellular and morphological levels. This perhaps would not have been possible without technical advancement in (a) gene discovery, (b) high-throughput gene expression, (c) altering gene expression by transformation technologies and (d) functional characterization of genes of interest *via* high-throughput gene inactivation techniques.

Analysis of gene expression in wheat using EST databases showed that homologous genes can be expressed in one genome, but are silent in one or both of the remaining genomes [71]. The tissue specificity of expression from homologous genes may also change. Consequently, a gene in one genome may be expressed in roots while the homologues are expressed in leaf tissue. There are now well over 1 million cereal EST sequences in the public databases with wheat and barley dominating. The large number of ESTs and the diversity of cDNA libraries that have been used to generate the sequences have made 'electronic Northerns' a useful method for assessing gene expression and this provides a good first measure of transcript abundance. Several microarray and macroarray platforms have been generated for cereals. For wheat and barley, there are a number of proprietal arrays such as the 10000 cDNA array reported by [72]. Recently, a NimbleGen Sequence Capture protocol adapted to crops containing more complex genomes including polyploidy crops such as wheat and rapeseed (canola) was pioneered by Roche NimbelGen [73].

Abiotic stress tolerance involves similar transcription factors in both dicotyledonous and monocotyledonous plants and some molecular mechanisms of drought tolerance have been extensively described [74]. It includes signal transduction cascade and activation/regulation of transcription, functional protection of proteins by late-embryogenesis abundant proteins (e.g. dehydrins) and chaperone proteins (e.g. heat shock proteins), accumulation of osmolytes (proline, glycine betaine, trehalose, mannitol, myo-inositol), induction of chemical antioxidants (ascorbic acid and glutathione), and enzymes reducing the toxicity of reactive oxygen species (superoxide dismutase, glutathione S-transferase). Homologous genes of these different classes were also identified in transcriptomic experiments comparing wheat lines grown under well-watered and water-stressed conditions [75 – 80].

Despite the existence of common regulatory mechanisms across species, the conservation of the molecular response to dehydration across experiments is low due to variation in stress dynamics, stage of development and tissue analysed [76, 78]. Interestingly, microarray assays revealed unexpected results such as a decrease in the expression of glutathione-related genes following withholding of water in tolerant synthetic wheat line [76], or the accumulation of proline in drought-sensitive emmer wheat line [79]. Thus, suggesting that some pathways/mechanisms are dependent upon genotype and the duration, intensity, and type of stress applied. Comparison of transcript profiles between tolerant and susceptible lines under water stress has revealed differences in regulatory pathways. A drought-tolerant emmer wheat genotype showed induction in bZIP and HD-ZIP gene expression (transcription factors known to be related to the ABA regulatory pathway) in response to shock-like drought stress, whereas the sensitive genotype induced some genes encoding transcription factors that bind to ethylene responsive elements. The two genotypes also showed differences in expression of the phospholipase C gene, involved in 1, 4, 5-triphosphate (IP3) signalling, and MAPK cascade elements [80]. A direct link between gene expression profiles and specific physiological mechanisms has not been demonstrated, probably because the drought stress scenario and the germplasm used for microarrays and physiological experiments are seldom the same. Mohammadi et al. [76] identified drought-responsive genes in the root of a synthetic hexaploid wheat line considered as a drought avoider. This line retained a higher net photosynthetic rate and relative water content compared with a sensitive line, Opata. However, these researchers could not establish a causal relationship between the transcripts and the maintenance of relative water content.

4. QTL Analysis and Breeding

Drought and salinity are two of the most complex stress tolerance to breed for as the type (combinations of heat and drought or salinity and sodicity), timing in relation to plant growth stage and intensity of stress can all vary considerably [81]. Tolerance to these stresses, like the majority of other abiotic stresses, is quantitative trait, with complex phenotype, mechanism and genetic control (multigenic, low heritability and high GxE interactions) [82, 83]. The goal of the QTL approach is to dissect the complex inheritance of quantitative traits into "mendelian-like" factors amenable to marker-assisted selection (MAS) and eventually their cloning [84]. QTLs mapping approach allows to assess the locations, numbers, magnitude of phenotypic effects, and pattern of gene action [85]. It entails the phenotypic evaluation and the profiling with molecular markers (RFLPs, SSRs, AFLPs, SNPs etc) of a mapping population coupled with statistical analysis to pinpoint the loci controlling a trait [86]. Genetic maps are prerequisite for QTLs identification. The availability of high density wheat consensus maps allows loci controlling traits related to drought resistance to be identified and mapped in the genome. In practice, for wheat (bread and durum), many genetic maps with different resolution, incorporating different categories of molecular markers (RFLPs, RAPDs, SSRs, AFLPs, STS, EST) obtained with different crosses are now available [87 – 92].

4.1. Traits Associated with Drought and Salinity Tolerance

One of the major difficulties in drought QTLs identification in crops in general and in wheat specially is the identification of the key physiological and morphological determinants of drought tolerance. The complexity of the drought phenomenon and the number of putative traits of unknown value has hampered the improvement of drought tolerance through QTL approach. Many reviewers have analysed and discussed the mechanisms and the traits underlying tolerance to drought [93 – 95] and salinity [96]. The response to drought is determined by component traits that interact and differ in their individual responses to the intensity and duration of water deficit and temperature. Most QTLs for drought tolerance in wheat have been identified through yield and yield measurement under water-limited conditions [97]. A grain yield QTL on the proximal region of chromosome 4AL was found. This region was associated with QTL for grain yield, grain fill rate, spike density, grains m^{-2}, biomass production, biomass production rate and drought susceptibility index [98]. Because of low heritability of yield itself, the breeders are always looking for secondary traits associated with drought resistance that are determinant more simply or having high heritability. A complementary strategy is also to target traits closely correlated with yield and yield potential. Crop physiology studies on cereals under drought environments have identified several indirect traits: radiation, water use efficiency, green leaf duration, harvest index, rate of senescence, grain fill duration, leaf area index, deep roots, vigorous crop establishment, stem–reserve utilization and maintaining cellular hydratation [99, 100]. Other constitutive type traits have also been selected and have proven to be very useful in escaping drought such as time of flowering [101]. Loci for yield, yield components, heading date, plant height and physiological and development traits under drought condition have been established in mapping populations [97].

Major genomic region controlling productivity and related traits were identified on chromosome 2B, 4A, 5A and 7B using a 152 recombinant inbreed lines (RILs), derived from a cross between durum wheat (*cv.* langdon) and wild emmer (acc- G18-16), under contrasting water availability. QTLs for productivity :grain yields, spike dry matter, total dry matter, harvest index were demonstrated to be associated with QTLs for drought–adaptative traits, carbon isotope ratio, osmotic potential at heading, chlorophyll content at heading, flag leaf rolling index [102].

The influence of root architecture on yield and other agronomic traits, especially under stress conditions, has been widely reported in all major crops [103, 104]. However, in wheat it had attracted little attention in drought QTLs identification. QTLs for elongation of axial and lateral roots in response to water potential were identified in maize [105 – 107]. A major effect on root architecture in durum wheat was detected on chromosomes arms 2AL (at Xgwm294), 7AL (at *Xcfa2257* and *Xgwm332*) and 7BL (at *Xgwm577* and *Xcfa2040) [108]*.

Salt tolerance is a complex and multigenic trait [109, 110]. In salt tolerance, the control of salt uptake by the roots and transport within the plant are critical factors. Many wild relatives of wheat crop show greater tolerance than their domesticated descendants and can be used as a source of superior alleles in tetraploid wheat, lines were identified that have low Na^+ accumulation and high K^+/Na^+ discrimination ratio and they were used to improve the tolerance of durum wheat [111]. The sodium exclusion trait was mapped on chromosome 2AL using QTL approach on a population derived from a cross between the low Na^+ landrace and the cultivar Tamaroi [112]. The sodium exclusion in durum wheat line 149 is conferred by two QTLs *Nax1* and *Nax2* (for Na^+ exclusion) derived from *T. monococcum* [113]. *Nax* genes are transferred from the durum wheat line 149 into durum wheat cultivars assisted by molecular markers. A much linked SSR marker, *gwm312*, for *Nax1* is being used routinely to select low Na^+ accumulator's progeny in the durum wheat breeding program [113]. SSR markers, *gpw2181* might be used for selection of the *Nax2* gene [114]. A doubled-haploid bread wheat population (Berkut/krichauf) are used to identify QTL associated with salinity tolerance traits (leaf symptoms, tiller number, seedling biomass, chlorophyll content, and shoot Na^+ and K^+ concentration). Among a total of 40 QTLs detected, two QTLs identified for Na^+ exclusion co-located with seedling biomass, the 2A QTLs appears to coincide with the previous reported Na^+ exclusion locus in durum wheat that hosts one active HKT1;4 (*Nax1*) and one inactive HKT1;4 gene [115].

4.2. Methods for QTLs Identification

Identification of QTLs can be done mostly in two ways: (1) using the experimental populations (also referred as "biparental" mapping populations) that is called QTL mapping and (2) using the diverse lines from the natural populations or germaplasm collections that is called LD-mapping or "association mapping" [116]. To perform linkage analysis, experimental populations such as F2, back cross (BC), double haploid (DH), recombinant inbred lines (RIL), and near isogenic lines (NIL), need to be developed. In wheat, drought related QTLs were mapped using these two strategies. The QTLs for grain yield (GY), heading date (HD), and plant height (PH) were identified in a durum wheat population of 249 recombinant inbred lines in 16 environments characterized by a broad range of water availability [97]. Although traditional QTL mapping remains to be an important tool in gene tagging of crops, it is now considered as a "classical approach". The major limitation, hampering the fine mapping, is associated with the availability of only a few meiotic events to be used and that occurs in recent past experimental hybridization. However, the natural populations are the products of many cycles of recombination. The basic objective of association mapping (AM) studies is to detect the correlation between genotypes in a sample of individuals on the basis of linkage disequilibrium (LD) [117]. QTLs for root and shoot traits at the seedling stage were mapped using mapping association on a panel of 57 elite durum wheat accessions [108]. The association mapping can also enhance the QTL information and achieves higher resolution with short LD extend [118].

4.3. Role of QTLs in Wheat Breeding Programs

Breeding for drought and /or salinity tolerance in wheat have proven to be very difficult. Crop improvement under the adverse conditions will increasingly benefit from the manipulation of QTLs through MAS and followed QTLs cloning. The most important step towards the application of molecular markers is the cloning of the identified drought related QTLs; reports on these topics are currently scanty despite high success rate of cloning QTLs or genes for other traits such as yellow pigment content [119]; pre-harvest sprouting resistance [120] and *Fusarium* head blight resistance [121].

Identification of QTLs for drought and salinity in wheat is not sufficient to improve tolerance in wheat but is a prerequisite for MAS which can be used to pyramid several different QTLs. Identification of functional markers that could be used for identification of superior alleles to enhance tolerance to drought that benefit from the identification and manipulation of QTLs are very scars. The plethora of loci (QTLs) related to drought have showed a different effect in many stress scenario. The majorities of genomic regions associated with drought QTL are still very large and are usually unsuitable for screening in breeding program [122]. Most drought QTLs identified are adaptative QTLs, detected only on specific environmental conditions or increase in expression with level of an environmental factor. The accurate characterization and validation of a QTL, requires its isogenisation (i.e production of congenic strains differing only for chromosomes region flanking QTLs). MAS for component traits of salinity tolerance in wheat have produced some positive results and the pyramiding of stable QTLs may provide a solution [113, 114]. There is a great potential of breeding genetic for drought and salinity tolerance through the contribution of wild relatives that will benefit from identification of drought and salinity QTLs and functional markers. *Triticum dicccoides* which is the progenitor of durum wheat is fully compatible with tetraploid durum wheat and can be crossed with hexaploid bread wheat; offers an allelic repertoire required for wheat improvement of agronomically important trait such as drought tolerance [123, 124].

5. Genetic Transformation of Wheat

Wheat (*T. aestivum and T. Durum*) are major component of human diet thus food security in the future can not be achieved without significant increases in wheat production. Although progress has been made in improving wheat productivity using conventional breeding method, this procedure is very tedious and labour intensive. In addition, it is complicated by the multigenic nature of yield, the large size of wheat genome and the limited gene pool available for breeders [125]. In recent years, genetic transformation of wheat has become a powerful tool. It allows specific trait to be transferred into selected genotypes without adversely affecting their desirable genetic background. Genetic engineering was successfully used to improve agronomically important traits, such as resistance to biotic (pests, pathogens) and abiotic (drought, soil salinity, heat or cold) stresses [126]. The improvement of drought stress in wheat was achieved through the overexpression of genes such as *HVA1* gene for drought tolerance [127], oat arginine decarboxylase cDNA [128], and gene for glycine betaine synthase [129]. Salinity tolerance was improved using other sets of genes such as the vacuolar Na^{+}/H^{+} antiporter gene *AtNHX1* from *Arabidopsis thaliana* [130]. Heat tolerance was also improved in wheat using genetic engineering [131].

Wheat improvement by genetic engineering requires the delivery, integration and expression of defined foreign genes into suitable regenerable explants. There are two main methods widely used for wheat transformation (1) *Agrobacterium*-based systems which exploits the unique ability of this bacterium to introduce the transgene into plant cells and (2) direct gene transfer via microprojectile bombardment developed by Sanford [132] and which relies on the acceleration of gold or tungsten particles, coated with plasmid DNA, into plant cells as a method of directly introducing the DNA. Both of these methods involve delivery of the transgene to callus tissue, followed by selection of transformed cells and regeneration of plantlets carrying the gene of interest. Each method has its advantages and limitations: microprojectile bombardment facilitates a broad variety of transformation strategies with a wide range of gene expression, has no biological constraints or host limitations, and diverse cell types can be targeted efficiently for foreign DNA delivery [133]. The main limitations of the bombardment approach include fragmentation of the DNA during bombardment, the insertion of backbone vector DNA and the insertion of multiple copies [134]. The multiple copies of the transgene at single loci and rearrangements of the transgenes have been frequently reported in cereal transformation studies using the microprojectile bombardment approach [135]. Numerous downstream breeding cycles are required to select out those transgenic plants with the best insertions and then to generate the homozygous lines required for the downstream breeding program and development of a commercial product. Whereas *Agrobacterium*-mediated method is simple, low cost alternative to particle bombardment. In addition, the *Agrobacterium*-mediated transformation system facilitates the precise integration of a small number of gene copies into the plant genome and shows a greater degree of stability for the transgene. Unlike microprojectile bombardment, *Agrobacterium* method seems to induce less transgene silencing since introduced genes remain transcriptionnaly active and has higher transformation efficiency than the microprojectile bombardment method. For all these reasons *Agrobacterium*-mediated transformation has been adopted as the method of choice for wheat.

The first success in cereal transformation using *Agrobacterium* was reported by Hiei et al. [136] for stable transformation of Japonica rice. A super binary vector in the *Agrobacterium* strain LBA4404 was demonstrated to be the most effective for transformation of all three japonica cultivars tested. *Agrobacterium*-mediated transformation of rice has now emerged as a reliable and highly reproducible method for transferring genes of interest into the rice genome. The success of Hiei prompted other researchers to try *Agrobacterium* transformation on other agronomically important cereal crop species, such as maize, barley, and wheat. In fact, maize transformation was accomplished with freshly isolated immature embryos followed by barley and wheat [137]. Compared to rice and maize, progress with wheat has been slower. *Agrobacterium*-mediated genetic transformation of wheat has been largely confined to particular genotypes that combine the amenability to gene transfer by *Agrobacterium* with adequate regeneration potential. The most suitable line used as models in routine transformation of wheat was the breeding line "Bobwhite 26" [137]. Lately, some other genotypes such as: Fielder, Cadenza, Verry5 [138 – 141] and the winter type Florida [138, 141] have turned out to be useful for *Agrobacterium*-mediated transformation, albeit with significantly lower efficiency [138, 142 - 144].

Most of the protocols efficiently used for wheat transformation generally rely on the use of hypervirulent *Agrobacterium* strains such as, AGL-0 and AGL-1 in barley and wheat [141, 143 - 146] as well as hypervirulent derivatives of LBA4404 in maize, barley and wheat [139, 143, 147]. Hypervirulence can be mediated by accessory *Vir* genes that are either contained on particular Ti-plasmids [148], on so-called superbinary vectors [149], or on an additional plasmid present in the *Agrobacterium* clone employed [147].

The recovery of stable plant cells after *Agrobacterium*-mediated transformation remains however influenced by many factors such as: genotype, explant, *Agrobacterium* strain, binary vector, selectable marker gene and promoter, inoculation and co-culture conditions, inoculation and co-culture medium, osmotic treatment, dessication, *Agrobacterium* density and surfactants, tissue culture, and regeneration medium [150, 151]. Some of these factors represent a drawback in extending the *Agrobacterium*-mediated transformation system to elite cultivars of economically important cereals. These limitations inspired some investigators to search for new alternative transformation procedures such as in *planta* transformation which involves no in *vitro* culture of plant cells or tissue [152, 153] or the flower dipping method originally developed for *Arabidopsis thaliana* transformation [154]. Although these alternative methods seem simple and straightforward, they are technically challenging and the results are not always convincing.

CONCLUSION

Our knowledge of the mechanism of drought and salt tolerance has been enhanced by research programmes targeting specific physiological, genetic or molecular aspects of the drought and salt response. However, in wheat, these approaches have not led to an increase in tolerance over that already achieved by breeders using empirical selection. Although the idea of linking physiology, 'omics and quantitative genetics have already been proposed, only a few research programmes have taken this integrative approach. The great strength of genetic and genomics analysis in wheat has been the ability to generate large populations and well-developed field phenotyping capabilities. However, genetic studies have often been too ambitious and have not permitted effective dissection of the drought and salt mechanisms or focused on specific drought and salt scenarios or regimes. Analysis of the response to drought and salinity has been further complicated by the absence of a genome sequence and the generally poor genomic resources. New developments in sequencing, marker development, and genome analysis have created the opportunity to reconsider the way in which we structure populations for analysis and tackle specific components of drought and salt tolerance. Recently, Scientists at the University of Liverpool have decoded the genome of wheat, the largest genome to be sequenced to date, using two Roche 454 FLX instruments and three SOLiD next generation genome analysers [155]. Gene expression profiling constitutes an exciting tool to unveil mechanisms involved in the response of plants to environmental stress. Its application in crop research is just starting as technologies are becoming more accessible and cost-effective and are expected to fuel huge advances in agriculture in the coming decades.

REFERENCES

[1] Mittler, R. Abiotic stress, the field environment and stress combination. *Trends Plant Sci.*, 2006 11, 15-19.

[2] Zhu, J; Kaeppler, SM; Lynch, JP. Mapping of QTLs for lateral root branching and length in maize (Zea mays L.) under differential phosphorus supply. *Theor. Appl. Genet.*, 2005 111, 688-95.

[3] Thomson, MJ; Edwards, JD; Septiningsih, EM; Harrington, SE; McCouch, SR. Substitution mapping of dth1.1, a flowering-time quantitative trait locus (QTL) associated with transgressive variation in rice reveals multiple sub-QTL. *Genetics*, 2006 172, 2501-14.

[4] Edgerton, MD. Increasing crop productivity to meet global needs for feed, food, and fuel. *Plant Physiol*, 2009 149, 7 – 13.

[5] Tuberosa, R; Salvi S. Genomics approaches to improve drought tolerance in crops, *Trends Plant Sci.*, 2006 11, 415–412.

[6] Collins, NC; Tardieu, F; Tuberosa, R. Quantitative trait loci and crop performance under abiotic stress: where do we stand? *Plant Physiol*, 2008 147, 469-86.

[7] McCallum, CM; Comai, L; Greene, EA; Henikoff, S. Targeted screening for induced mutations. *Nat. Biotechnol.*, 2000a 18, 455–457.

[8] Bohnert, HJ; Ayoubi, P; Borchert, C; Bressan, RA; Burnap, RL; Cushman, JC; Cushman, MA; Deyholos, M; Fischer, R; Galbraith, DW; et al. A genomics approach towards salt stress tolerance. *Plant Physiol. Biochem.*, 2001 39, 295-311.

[9] Knight, H; Knight, MR. Abiotic stress signalling pathways: specificity and cross-talk. *Trends Plant Sci.*, 2001 6, 262-267.

[10] Kreps, JA; Wu, YJ; Chang, HS; Zhu, T; Wang, X; Harper, JF. Transcriptome changes for *Arabidopsis* in response to salt, osmotic, and cold stress. *Plant Physiol*, 2002 130, 2129-2141.

[11] Chen, WQ; Provart, NJ; Glazebrook, J; Katagiri, F; Chang, HS; Eulgem, T; Mauch, F; Luan, S; Zou, GZ; Whitham, SA; et al. Expression profile matrix of *Arabidopsis*

transcription factor genes suggests their putative functions in response to environmental stresses. *Plant Cell*, 2002 14, 559-574.
[12] Seki, M; Narusaka, M; Ishida, J; Nanjo, T; Fujita, M; Oono, Y; Kamiya, A; Nakajima, M; Enju, A; Sakurai, T; et al. Monitoring the expression profiles of 7000 *Arabidopsis* genes under drought, cold and high-salinity stresses using a full-length cDNA microarray. *Plant J.*, 2002 31, 279-292.
[13] Abe, H; Urao T, Ito, T; Seki, M; Shinozaki, K; Yamaguchi-Shinozaki, K. *Arabidopsis* AtMYC2 (bHLH) and AtMYB2 (MYB) function as transcriptional activators in abscisic acid signaling. *Plant Cell*, 2003 15, 63-78.
[14] Vinocur, B; Altman A. Recent advances in engineering plant tolerance to abiotic stress: achievements and limitations. *Curr. Opin. Biotechnol.*, 2005 16, 123-132.
[15] Umezawa, T; Fujita, M; Fujita, Y; Yamaguchi-Shinozaki, K; Shinozaki, K. Engineering drought tolerance in plants: discovering and tailoring genes to unlock the future. *Curr Opin Biotechnol.*, 2006 17, 1–10.
[16] Pennisi, E. Genetics. Hopping to a better protein. *Science*, 2008 322, 1454-5.
[17] Wan, H; Zhao, Z; Qian, C; Sui, Y; Malik, AA; Chen, J. Selection of appropriate reference genes for gene expression studies by quantitative real-time polymerase chain reaction in cucumber. *Anal Biochem.*, 2010 399, 257-61.
[18] Tabata, S. Impact of genomics approaches on plant genetics and physiology. *J. Plant Res.*, 2002 115, 271-275.
[19] Gale, MD; Devos, KM. Comparative genetics in the grasses. *Proc. Natl. Acad. Sci. USA.*, 1998 95, 1971-1974.
[20] Bowers, JE; Arias, MA; Asher, R; Avise, JA; Ball, RT; Brewer, GA; et al. Comparative physical mapping links conservation of microsynteny to chromosome structure and recombination in grasses. *Proc. Natl. Acad. Sci. USA.*, 2005 102, 13206-13211.
[21] Paterson, AH; Freeling, M; Sasaki, T. Grains of knowledge: Genomics of model cereals. *Genome Res.*, 2005 15, 1643-1650.
[22] Caicedo, AL; Purugganan, MD. Comparative plant genomics. Frontiers and prospects. *Plant Physiol*, 2005 138, 545-547.
[23] van de Mortel, JE; Aarts, MGM. Comparative transcriptomics - model species lead the way. *New Phytol*, 2006 170, 199-201.
[24] The *Arabidopsis* Information Resource, http://www.*arabidopsis*.org/Carnegie, Inst, Washington [25 October 2005, date last accessed].
[25] Liu, Q; Kasuga, M; Sakuma, Y; et al. Two transcription factors, DREB1 and DREB2, with an EREBP/AP2 DNA binding domain separate two cellular signal transduction pathways in drought- and low-temperature-responsive gene expression, respectively, in *Arabidopsis. Plant Cell*, 1998 10, 1391–406.
[26] Qin, F; Sakuma, Y; Li, J; et al. Cloning and functional analysis of a novel DREB1/CBF transcription factor involved in cold-responsive gene expression in *Zea mays* L. *Plant Cell Physiol*, 2004 45, 1042–52.
[27] Dubouzet, JG; Sakuma, Y; Ito, Y; et al. OsDREB genes in rice, *Oryza sativa* L., encode transcription activators that function in drought-, high-salt- and cold-responsive gene expression. *Plant J*, 2003 33, 751–63.
[28] Shen YG, Zhang WK, He SJ, et al. An EREBP/AP2-type protein in *Triticum aestivum* was a DRE-binding transcription factor induced by cold, dehydration and ABA stress. *Theor Appl Genet.*, 2003 106, 923–30.

[29] Pellegrineschi, A; Hoisington, D. 2005. Results of Transgenic Wheat Trial Look Promising, http://www.cimmyt.org/ english/wps/news/dreb.htm.
[30] Roosens, NH; Thu, TT; Iskandar, HM; et al. Isolation of the ornithine-delta-aminotransferase cDNA and effect of salt stress on its expression in *Arabidopsis thaliana*. *Plant Physiol*, 1998 117, 263–71.
[31] DIR 053/2004 – Field trial of genetically modified salt tolerant wheat on saline land, http://www.ogtr.gov.au/ir/ dir053.htm, Office of the Gene Technology Regulator [30 January 2006, date last accessed].
[32] Devos, KM. Updating the 'Crop circle'. *Current Opin Plant Biol.*, 2005 8, 155–62.
[33] Ware, D; Stein, L. Comparison of genes among cereals. *Current Opin Plant Biol.*, 2003 6, 121–7.
[34] Sutton, T; Whitford, R; Baumann, U; et al. The Ph2 pairing homoeologous locus of wheat (*Triticum aestivum*), identification of candidate meiotic genes using a comparative genetics approach. *Plant J.*, 2003 36, 443–56.
[35] Yu, J; Hu, SN; Wang, J; et al. A draft sequence of the rice genome (*Oryza sativa* L. ssp indica). *Science*, 2002 296, 79–92.
[36] Goff, SA; Ricke, D; Lan, TH; et al. A draft sequence of the rice genome (*Oryza sativa* L. ssp japonica). *Science*, 2002 296, 92–100.
[37] Feuillet, C ; Travella, S ; Stein, N ; et al. Map-based isolation of the leaf rust disease resistance gene Lr10 from the hexaploid wheat (*Triticum aestivum* L.) genome. *Proc Natl Acad Sci USA.*, 2003 100, 15253–8.
[38] Yan, L; Loukoianov, A; Tranquilli, G; et al. Positional cloning of the wheat vernalization gene VRN1. *Proc. Natl. Acad. Sci. USA.*, 2003 100, 6263–8.
[39] Henikoff, S; Comai, L. Single-nucleotide mutations for plant functional genomics. *Annu. Rev. Plant Biol.*, 2003 54, 375–401.
[40] McCallum, CM; Comai, L; Greene, EA; Henikoff, S. Targeting Induced Local Lesions IN Genomes (TILLING) for plant functional genomics. *Plant Physiol.*, 2000b 123, 439–442.
[41] Slade, AJ; Fuerstenberg, SI; Loeffler, D; et al. A reverse genetic, non transgenic approach to wheat crop improvement by TILLING. *Nature Biotech.*, 2005 23, 75–81.
[42] Till, BJ; Reynolds, SH; Weil, C; Springer, N; Burtner, C; Young, K; Bowers, E; Codomo, CA; Enns, LC; Odden, AR; Greene, EA; Comai, L; Henikoff, S. Discovery of induced point mutations in maize genes by TILLING. *BMC Plant Biol.*, 2004 4, 12.
[43] Wu, JL; Wu, C; Lei, C; Baraoidan, M; Bordeos, A; Madamba, MRS; Ramos-Pamplona, M; Mauleon, R; Portugal, A; Ulat, VJ; Bruskiewich, R; Wang, G; Leach, J; Khush, G; Leung, H. Chemicaland irradiation-induced mutants of indica rice IR64 for forward and reverse genetics. *Plant Mol. Biol.*, 2005 59, 85–97.
[44] Till, BJ; Cooper, J; Tai, TH; Colowit, P; Greene, EA; Henikoff, S; Comai, L. Discovery of chemically induced mutations in rice by TILLING. *BMC Plant Biol.*, 2007 7, 19.
[45] Caldwell, DG; McCallum, N; Shaw, PP; et al. A structured mutant population for forward and reverse genetics in Barley (*Hordeum vulgare* L.). *Plant J.*, 2004 40, 143–50.
[46] Hohmann, U; Jacobs, G; Jung, C. An EMS mutagenesis protocol for sugar beet and isolation of non-bolting mutants. *Plant Breeding*, 2005 124, 317–321.

[47] Perry, JA; Wang, TL; Welham, TJ; Gardner, S; Pike, JM; Yoshida, S; Parniske, M. A TILLING reverse genetics tool and a web accessible collection of mutants of the legume *Lotus japonicus*. *Plant Physiol.*, 2003 131, 866–871.

[48] Cooper, JL; Till, BJ; Laport, RG; Darlow, MC; Kleffner, JM; Jamai, A; El-Mellouki, T; Liu, S; Ritchie, R; Nielsen, N; Bilyeu, KD; Meksem, K; Comai, L; Henikoff, S. TILLING to detect induced mutations in soybean. *BMC Plant Biol.*, 2008 8, 9.

[49] Green, CD; Simons, JF; Taillon, BE; Lewin, DA. Open systems: panoramic views of gene expression. *J. Immunol. Methods*, 2001 250, 67-79.

[50] Ma, L ; Chen, C ; Liu, X ; Jiao, Y ; Su, N ; Li, L ; et al. A microarray analysis of the rice transcriptome and its comparison to *Arabidopsis*. *Genome Res.*, 2005 15, 1274-1283.

[51] Rensink, W; Iobst, S; Hart, A; Stegalkina, S; Liu, J; Buell, C. Gene expression profiling of potato responses to cold, heat, and salt stress. *Funct. Integr. Genomics*, 2005 5, 201-207.

[52] Oono, Y; Seki, M; Satou, M; Iida, K; Akiyama, K; Sakurai, T; et al. Monitoring expression profiles of *Arabidopsis* genes during cold acclimation and de-acclimation using DNA microarrays. *Funct. Integr. Genomics*, 2006 6, 212-234.

[53] Xu, WF; Shi, WM. Expression profiling of the 14-3-3 gene family in response to salt stress and potassium and iron deficiencies in young tomato (*Solanum lycopersicum*) roots: Analysis by real-time RT-PCR. *Ann. Bot.*, 2006 98, 965-974.

[54] Mantri, N; Ford, R; Coram, T; Pang, E. Transcriptional profiling of chickpea genes differentially regulated in response to high-salinity, cold and drought. *BMC Genomics*, 2007 8, 303.

[55] Monroy, A; Dryanova, A; Malette, B; Oren, D; Ridha Faramalla, M; Liu, W; et al. RegulatoryRegulatory gene candidates and gene expression analysis of cold acclimation in winter and spring wheat. *Plant Mol. Biol.*, 2007 64, 409-423.

[56] Fernandez, P; Rienzo, JD; Fernandez, L; Hopp, HE; Paniego, N; Heinz, RA. Transcriptomic identification of candidate genes involved in sunflower responses to chilling and salt stresses based on cDNA microarray analysis. *BMC Plant Biol.*, 2008 8, 11.

[57] Remans, T; Smeets, K; Opdenakker, K; Mathijsen, D; Vangronsveld, J; Cuypers, A. Normalisation of real-time RT-PCR gene expression measurements in *Arabidopsis thaliana* exposed to increased metal concentrations. *Planta*, 2008 227, 1343-1349.

[58] Fizames, C; Munos, S; Cazettes, C; Nacry, P; Boucherez, J; Gaymard, F; et al. The Arabidopsis root transcriptome by serial analysis of gene expression. Gene identification using the genome sequence. *Plant Physiol*, 2004 134, 67-80.

[59] Meyers, BC; Tej, SS; Vu, TH; Haudenschild, CD; Agrawal, V; Edberg, SB; et al. The use of MPSS for whole-genome transcriptional analysis in *Arabidopsis*. *Genome Res.*, 2004 14, 1641-1653.

[60] Calsa, T; Figueira, A. Serial analysis of gene expression in sugarcane (*Saccharum* spp.) leaves revealed alternative C4 metabolism and putative antisense transcripts. *Plant Mol. Biol.*, 2007 63, 745-762.

[61] Chen, J; Agrawal, V; Rattray, M; West, MW; St Clair, D; Michelmore, R; et al. A comparison of microarray and MPSS technology platforms for expression analysis of *Arabidopsis*. *BMC Genomics*, 2007 8, 414.

[62] Leymarie, J; Bruneaux, E ; Gibot-Leclerc, S ; Corbineau, F. Identification of transcripts potentially involved in barley seed germination and dormancy using cDNA-AFLP. *J. Exp. Bot.*, 2007 58, 425-437.

[63] McIntosh, S; Watson, L; Bundock, P; Crawford, A; White, J; Cordeiro, G; Barbary, D; Rooke, L; Henry, R. SAGE of the developing wheat caryopsis. *Plant Biotechnol. J.*, 2007 5, 69-83.

[64] Song, S; Qu, H; Chen, C; Hu, S; Yu, J. Differential gene expression in an elite hybrid rice cultivar (*Oryza sativa*, L.) and its parental lines based on SAGE data. *BMC Plant Biol.*, 2007 7, 49.

[65] Ritter, E; Ruiz de Galarreta, J; van Eck, H; Sánchez, I. Construction of a potato transcriptome map based on the cDNA–AFLP technique. *Theor. Appl. Genet.*, 2008 116, 1003-1013.

[66] Mikkilineni, V; Mitra, RD; Merritt, J; DiTonno, JR; Church, GM; Ogunnaike, B; Edwards, JS. Digital quantitative measurements of gene expression. *Biotechnol. Bioeng.*, 2004 86, 117-124.

[67] Margulies, M; Egholm, M; Altman, E; Attiya, S; Bader, JS; Bemben, LA; et al. Genome sequencing in microfabricated high-density picolitre reactors. *Nature,* 2005 437, 376-380.

[68] Velculescu, VE; Kinzler, K. Gene expression analysis goes digital. *Nat. Biotech.*, 2007 25, 878-880.

[69] Weber, APM; Weber, KL; Carr, K; Wilkerson, C; Ohlrogge, JB. Sampling the Arabidopsis transcriptome with massively parallel pyrosequencing. *Plant Physiol.*, 2007 144, 32-42.

[70] Torres, TT; Metta, M; Ottenwälder, B; Schlötterer, C. Gene expression profiling by massively parallel sequencing. *Genome Res.*, 2008 18, 172-177.

[71] Mochida, K; Yamazaki, Y; Ogihara, Y. Discrimination of homologous gene expression in hexaploid wheat by SNP analysis of contigs grouped from a large number of expressed sequence tags. *Mol. Genet. Genom.*, 2004 270, 371–7.

[72] Leader, DJ. Transcriptional analysis and functional genomics in wheat. *J. Cereal Sci.*, 2005 41, 149–63.

[73] (http://www. highbeam.com/doc/1G1-225681293.html).

[74] Yamaguchi-Shinozaki, K; Shinozaki, K. Transcriptional regulatory networks in cellular responses and tolerance to dehydration and cold stresses. *Annual Review of Plant Biol.*, 2006 57, 781–803.

[75] Xue, GP; McIntyre, CL; Chapman, S; Bower, NI; Way, H; Reverter, A; Clarke, B; Shorter, R. Differential gene expression of wheat progeny with contrasting levels of transpiration efficiency. *Plant Mol. Biol.*, 2006 61, 863–881.

[76] Mohammadi, M; Kav, NN; Deyholos, MK. Transcriptional profiling of hexaploid wheat (*Triticum aestivum* L.) roots identifies novel, dehydration-responsive genes. *Plant Cell and Environment*, 2007 30, 630–645.

[77] Mohammadi, M; Kav, NN; Deyholos, MK. Transcript expression profile of water-limited roots of hexaploid wheat (*Triticum aestivum* 'Opata'). *Genome*, 2008 51, 357–367.

[78] Aprile, A; Mastrangelo, AM; De Leonardis, AM; Galiba, G; Roncaglia, E; Ferrari, F; De Bellis, L; Turchi, L; Giuliano, G; Cattivelli, L. Transcriptional profiling in response

to terminal drought stress reveals differential responses along the wheat genome. *BioMed Central Genomics*, 2009 10, 279.

[79] Ergen, NZ; Budak, H. Sequencing over 13000 expressed sequence tags from six subtractive cDNA libraries of wild and modern wheat's following slow drought stress. *Plant Cell and Environ.*, 2009 32, 220–236.

[80] Ergen, NZ; Thimmapuram, J; Bohnert, HJ; Budak, H. Transcriptome pathways unique to dehydration tolerant relatives of modern wheat. *Funct. Integr. Genomics.*, 2009 9, 377–396.

[81] Witcombe, JR; Hollington, PA; Howarth, CJ; Reader, S; Steele, KA. Breeding for abiotic stresses for sustainable agriculture. *Philos Trans R Soc Lond B Biol Sci.*, 2008 363, 703-716.

[82] Mc William, RG. The dimensions of drought. In: Baker, F.W. (ed.), *Drought Resistance in Cereal*, 1989 1–11.

[83] Mohammadi, M; Taleei, A; Zeinali, H; Naghavi, MR; Ceccarelli, S; Grando, Baum, M. QTL Analysis for Phenologic Traits in Doubled Haploid Population of Barley. *Int. J. Agri. Biol.*, 2005 7, 820-823.

[84] Tuberosa,, R; Salvi, S; Giuliani, S; Sanguineti, MC; Bellotti, M; Conti, S; Landi, P. Genome-wide approaches to investigate and improve maize response to drought. *Crop Sci.*, 2007 47, 120-141.

[85] Vinh, NT; Paterson, AH. Genome mapping and its implication for stress resistance in plants. In: Asraf M, Harris PJC, editors. *Abiotic stresses: plant resistance through breeding and molecular approaches.* 2005.

[86] Tuberosa, R; Salvi, S. Dissecting Qtls For Tolerance to Drought and Salinity. In: *Advances in Molecular Breeding Toward Drought and Salt Tolerant Crops*, 2007 381-411.

[87] Röder, MS; Korzun, V; Wendehake, K; Plaschke, J; Tixier, MH; Leroy, P; Ganal, MW. A microsatellite map of wheat. *Genetics*, 1998 149, 2007-2023.

[88] Somers, DJ; Isaac, P; Edwards, K. A high-density microsatellite consensus map for bread wheat (*Triticum aestivum* L.). *Theor Appl Genet.*, 2004 109, 1105-1114.

[89] Sourdille, P; Singh, S; Cadalen, T; Brown-Guedira, G; Gay, G; Qi, L; Gill, BS; Dufour, P; Murigneux, A; Bernard, M. Microsatellite-based deletion bin system for the establishment of genetic-physical map relationships in wheat (*Triticum aestivum* L.). *Funct Integr Genomics.*, 2004 4, 12-25.

[90] Goyal, A; Bandopadhyay, R; Sourdille, P; Endo, TR; Balyann, HS; Gupta, PK. Physical molecular maps of wheat chromosomes. *Funct. Integr. Genomics.*, 2005 5, 260-263.

[91] Peleg, Z; Saranga, Y; Suprunova, T; Ronin, Y; Röder, MS; Kilian, A; Korol, AB; Fahima, T. High-density genetic map of durum wheat x wild emmer wheat based on SSR and DArT markers. *Theor Appl Genet.*, 2008a 117, 103-15.

[92] Mantovani, P; Maccaferri, M; Sanguineti, MC; Tuberosa, R; Catizone, L; Wenzl, P; Thomson, B; Carling, J; Huttner, E; DeAmbrogio, E; Kilian, A. An integrated DArT-SSR linkage map of durum wheat. *Mol. Breeding*, 2008 22, 629–648.

[93] Passioura, JB. Environmental biology and crop improvement *Funct. Plant biol.*, 2002 29, 537-546.

[94] Richards, RA; Rebetzke, GJ; ondon, AG; van Herwaarden, AF. Breeding Opportunities for Increasing the Efficiency of Water Use and Crop Yield in Temperate Cereals. *Crop Sci.*, 2002 42, 111-121.

[95] Tuberosa, R; Salvi, S. QTLs and Genes for Tolerance to Abiotic Stress in Cereals. *Cereal Genomics*, 2005 253-315
[96] Munns, R. Genes and salt tolerance: bringing them together. *New Phytol.*, 2005 167, 645-663.
[97] Maccaferri, M; Sanguineti, MC; Corneti, S; Ortega, JL; Salem, MB; Bort, J; DeAmbrogio, E; del Moral, LF; Demontis, A; El-Ahmed, A; Maalouf, F; Machlab, H; Martos, V; Moragues, M; Motawaj, J; Nachit, M; Nserallah, N; Ouabbou, H; Royo, C; Slama, A; Tuberosa, R. Quantitative trait loci for grain yield and adaptation of durum wheat (*Triticum durum* Desf.) across a wide range of water availability. *Genetics*, 2008 178, 489-511
[98] Kirigwi, FM; Van Ginkel, M; Brown-Guedira, G; Gill, BS; Paulsen, GM; Fritz, AK. Markers associated with a QTL for grain yield in wheat under drought. *Mol. Breeding*, 2007 20, 401–413.
[99] Araus, JL; Slafer, GA; Royo, C; Serret, MD. Breeding for yield potential and stress adaptation in cereals. *Critical Reviews in Plant Sci.*, 2008 27, 377-412.
[100] Reynolds, M; Tuberosa, R. Translational research impacting on crop productivity in drought-prone environments. *Curr. Opin. Plant Biol.*, 2008 11, 171-179.
[101] Habash, DZ; Kehe, lZ; Nachit, M. Genomic approaches for designing durum wheat ready for climate change with a focus on drought. *J. Exp. Bot.*, 2009 60, 2805-2815.
[102] Peleg, Z; Fahima, T; Krugman, T; Abbo, S; Yakir, D; Korol, AB; Saranga, Y. Genomic dissection of drought resistance in durum wheat x wild emmer wheat recombinant inbreed line population. *Plant Cell Environ.*, 2009 32, 758-779.
[103] Price, AH; Cairns, JE; Horton, P; Jones, HG; Griffiths, H. Linking drought-resistance mechanisms to drought avoidance in upland rice using a QTL approach: progress and new opportunities to integrate stomatal and mesophyll responses. *J. Exp. Bot.*, 2002 53, 989-1004.
[104] Tuberosa, R; Sanguineti, MC; Landi, P; Giuliani, MM; Salvi, S; Conti, S. Identification of QTLs for root characteristics in maize grown in hydroponics and analysis of their overlap with QTLs for grain yield in the field at two water regimes. *Plant Mol. Biol.*, 2002 48, 697-712.
[105] Ruta, N; Liedgens, M; Fracheboud, Y; Stamp, P; Hund, A. QTLs for the elongation of axile and lateral roots of maize in response to low water potential. *Theor. Appl. Genet.*, 2010 120, 621-31.
[106] Trachsel, S; Messmer, R; Stamp, P; Hund, A. Mapping of QTLs for lateral and axial root growth of tropical maize. *Theor. Appl. Genet.*, 2009 119, 1413-24.
[107] Landi, P; Giuliani, S; Salvi, S; Ferri, M; Tuberosa, R; Sanguineti, MC. Characterization of root-yield-1.06, a major constitutive QTL for root and agronomic traits in maize across water regimes. *J. Exp. Bot.*, 2010 61 3553-3562.
[108] Sanguineti, MC; Li, S; Maccaferri, M; Corneti, S; Rotondo, F; Chiari, T; Tuberosa, R. Genetic dissection of seminal root architecture in elite durum wheat germplasm. *Ann. Appl. Biol.*, 2007 151, 291–305.
[109] Shannon, MC. Adaptation of plants to salinity. *Advances in Agronomy*, 1997 60, 75-120.
[110] Flowers, TJ. Improving crop salt tolerance. *J. Exp. Bot*, 2004 55, 307–319.
[111] Munns, R; Hare, RA; James, RA; Rebetzke, GJ. Genetic variation for improving the salt tolerance of durum wheat. *Aust. J. of Agr. Res.*, 2000 51, 69–74.

[112] Lindsay, MP, Lagudah, ES; R. Hare, A; Munns, R. A locus for sodium exclusion (Nax1), a trait for salt tolerance, mapped in durum wheat. *Funct. Plant Biol.*, 2004 31, 1105–1114.

[113] James, RA; Davenport, R; Munns, R. Physiological characterisation of two genes for Na^+ exclusion in wheat: Nax1 and Nax2. *Plant Physiol.*, 2006 142, 1537-1547.

[114] Byrt, CS; Platten, JD; Spielmeyer, W; James, RA; Lagudah, ES; Dennis, ES; Tester, M; Munns, R. HKT1;5-like cation transporters linked to Na^+ exclusion loci in wheat, Nax2 and Kna1. *Plant Physiol.*, 2007 143, 1918-1928.

[115] Genc, Y; Verbyla, A; Lott, G; Hassan, M; Tester, M; Wallwork, H; McDonald, GK. Sodium exclusion QTL associated with improved seedling growth in bread wheat under salinity stress. *Theor.l Appl. Genet.*, 2010 121, 877-94.

[116] Abdurakhmonov, IY; Abdukarimov, A. Application of association mapping to understanding the genetic diversity of plant germplasm resources. *Int. J. Plant Genomics*, 2008 2008, 574927.

[117] Zondervan, KT; Cardon, LR. The complex interplay among factors that influence allelic association. *Nat. Rev. Genet.*, 2004 5, 89-100.

[118] Yao, J; Wang, L; Liu, L; Zhao, C; Zheng, Y. Association mapping of agronomic traits on chromosome 2A of wheat. *Genetica*, 2009 137, 67-75.

[119] Patil, RM; Oak, MD; Tamhankar, SA; Sourdille, P; Rao, VS. Mapping and validation of a major QTL for yellow pigment content on 7AL in durum wheat (*Triticum turgidum* L. ssp durum). *Mol. Breed.*, 2008 21, 485–496.

[120] Liu, S; Bai, G. Dissection and fine mapping of a major QTL for pre-harvest sprouting resistance in white wheat Rio Blanco. *Theor Appl Genet*, 2010 DOI 10.1007/s00122-010-1396-4.

[121] Xue, S; Lin, G; Jia, H; Xu, F; Lin, F; Tang, M; Wang, Y; An, X; Xu, H; Zhang, L; Kong, Z; Ma, Z. Fine mapping Fhb4, a major QTL conditioning resistance to Fusarium infection in bread wheat (*Triticum aestivum* L.). *Theor. Appl. Genet.*, 2010 121, 147-156.

[122] Fleury, D; Jefferies, S; Kuchel, H; Langridge, P. Genetic and genomic tools to improve drought tolerance in wheat. *J. Exp. Bot.*, 2010 61, 3211-22.

[123] Peleg, Z; Abbo, S; Fahima, T; Krugman, T; Nevo, E; Yakir, D; Saranga, Y. Genetic diversity for drought resistance in wild emmer wheat and its eco-geographical associations. *Plant, Cell & Environ.*, 2005 28, 176-191.

[124] Peleg, Z; Saranga, Y; Abbo, S; Krugman, T; Nevo, E; Fahima, T. Allelic diversity associated with aridity gradient in wild emmer wheat populations. *Plant, Cell & Environ.*, 2008b 31, 39-49.

[125] Malik, ST; Hamid, R; Tayyaba, Y; Minhas, NM. Effects of 2, 4-D on callus induction from mature wheat (*Triticum aestivum* L.). *Seeds Int J Agri Biol.*, 2003 6, 156-159.

[126] Vasil, IK. Molecular genetic improvement of cereals: transgenic wheat (*Triticum aestivum* L.). *Plant Cell Reports*, 2007 26, 1133–1154.

[127] Bahieldin, A; Mahfouz, H; Eissa, H; Saleh, OM; Ramadan, AM; Ahmed, IA; Dyer, W; El-Itriby Madkour, MA. Field evaluation of transgenic wheat plants stably expressing the *HVA1* gene for drought tolerance. *Physiol. Plantarum*, 2005 123, 421–427

[128] Bassie, L; Zhu, C; Romagosa, I; Christou, P; Capell, T. Transgenic wheat plants expressing an oat arginine decarboxylase cDNA exhibit increases in polyamine content in vegetative tissue and seeds. *Mol. Breed.*, 2008 22, 39-50.

[129] He, C; Yang, A; Zhang, W; Gao, Q; Zhang, J. Improved salt tolerance of transgenic wheat by introducing beta gene for glycine betaine synthesis. *Plant Cell Tissue Organ Cult.*, 2010 101, 65–78.

[130] Xue, ZH; Zhi, DY; Xue, GP; Zhang, H; Zhao, YX; Xia, GM. Enhanced salt tolerance of transgenic wheat (*Tritivum aestivum* L.) expressing a vacuolar Na^+/H^+ antiporter gene with improved grain yields in saline soils in the field and a reduced level of leaf Na^+. *Plant Sci.*, 2004 167, 849-859.

[131] Fu, J; Momcilovic, I; Clemente, TE; Nersesian, N; Trick, HN; Ristic, Z. Heterologous expression of a plastid EF-Tu reduces protein thermal aggregation and enhances CO2 fixation in wheat (*Triticum aestivum*) following heat stress. *Plant Mol. Biol.*, 2008 68, 277–288.

[132] Sanford, JC. The biolistic process. *Trends in Biotechnology*, 1988 6, 299–302.

[133] Altpeter, F; Baisakh, N; Beachy, R; et al. Particle bombardment and the genetic enhancement of crops: myths and realities. *Mol. Breed.*, 2005 15, 305– 327.

[134] Hu, T; Metz, S; Chay, C; Zhou, HP; Biest, N; Chen, G; Cheng, M; Feng, X; Radionenko, M; Lu, F; Fry, J. *Agrobacterium*-mediated large-scale transformation of wheat (*Triticum aestivum* L.) using glyphosate selection. *Plant Cell Reports*, 2003 21, 1010–1019.

[135] Janakiraman, V; Steinau, M; Mccoy, SB; Trick, HN. Recent advances in wheat transformation. *In Vitro Cellular & Developmental Biology – Plant*, 2002 38, 404–414.

[136] Hiei, Y; Ohta, S; Komari, T; Kumashiro, T. Efficient transformation of rice (*Oryza sativa* L) mediated by *Agrobacterium* and sequence-analysis of the boundaries of the T-DNA. *Plant J.*, 1994 6, 271–282.

[137] Cheng, M; Fry, JE; Pang, SZ; Zhou, HP; Hironaka, CM; Duncan, DR; Conner, TW; Wan, YC. Genetic transformation of wheat mediated by *Agrobacterium tumefaciens*. *Plant Physiol.*, 1997 115, 971–980.

[138] Jones, HD. Wheat transformation: current technology and applications to grain development and composition. *J. Cereal Sci.*, 2005 41, 137–147.

[139] Khanna, HK; Daggard, GE. *Agrobacterium tumefaciens* mediated transformation of wheat using a superbinary vector and a polyamine-supplemented regeneration medium. *Plant Cell Reports*, 2003 21, 429–436.

[140] Weir, B; Gu, X; Wang, MB; Upadhyaya, N; Elliott, AR; Brettell, RIS. *Agrobacterium tumefaciens*-mediated transformation of wheat using suspension cells as a model system and green fluorescent protein as a visual marker. *Aust J. of Plant Physiol.*, 2001 28, 807–818.

[141] Wu, H; Sparks, C; Amoah, B; Jones, HD. Factors influencing successful *Agrobacterium*-mediated genetic transformation of wheat. *Plant Cell Report*, 2003 21, 659-668.

[142] Huang, X; Wei, Z. Successful *Agrobacterium*-mediated genetic transformation of maize elite inbred lines. *Plant Cell, Tissue and Organ Culture*, 2005 83, 187-200.

[143] Hensel, G; Valkov, V; Middlefell-Williams, J; Kumlehn J. Efficient generation of transgenic barley: the way forward to modulate plant-microbe interactions. *J. Plant Physiol.*, 2008 165, 71–82.

[144] He, Y; Jones, HD; Chen, S; Chen, XM; Wang, DW; Li, KX; Wang, DS; Xia, LQ. *Agrobacterium*-mediated transformation of durum wheat (*Triticum turgidum* L. var. *durum cv* stewart) with improved efficiency. *J. Exp. Bot.*, 2010 61, 1567-1581.

[145] Tingay, S; McElroy, D; Kalla, R; Fieg, S; Wang, MB; Thornton, S; Brettell, R. *Agrobacterium* tumefaciens-mediated barley transformation. *Plant J.*, 1997 11, 1369–1376.

[146] Matthews, PR; Wang, MB; Waterhouse, PM. Marker gene elimination from transgenic barley, using co-transformation with adjacent 'twin T-DNAs' on a standard *Agrobacterium* transformation vector. *Mol. Breed.*, 2001 7, 195–202.

[147] Kumlehn, J; Serazetdinova, L; Hensel, G; Becker, D; Loerz, H. Genetic transformation of barley (*Hordeum vulgare* L.) *via* infection of androgenetic pollen cultures with *Agrobacterium tumefaciens. Plant Biotech. J.*, 2006 4, 251–261.

[148] Jin, SG; Komari, T; Gordon, MP; Nester, EW. Genes responsible for the super virulence phenotype of *Agrobacterium tumefaciens* A281. *J. Bacteriology*, 1987 169, 4417-25.

[149] Komori, T; Imayama, T; Kato, N; Ishida, Y; Ueki, J; Komari, T. Current status of binary vectors and superbinary vectors. *Plant Physiol*, 2007 145, 1155-60.

[150] Shrawat, AK; Lorz, H. *Agrobacterium*-mediated transformation of cereals: a promising approach crossing barriers. *Plant Biotech. J.*, 2006 4, 575–603.

[151] Cheng, M; Hu, TC; Layton, J; Liu, CN; Fry, JE. Desiccation of plant tissues post *Agrobacterium* infection enhances T-DNA delivery and increases stable transformation efficiency in wheat. *In Vitro Cellular & Developmental Biology – Plant*, 2003 39, 595–604.

[152] Supartana, P; Shimizu, T; Shioiri, H; Nogawa, M; Nozue, M; Kojima, M. Development of simple and efficient in planta transformation method for rice (*Oryza sativa* L.) using *Agrobacterium tumefaciens. J. Biosci. Bioeng.*, 2005 100, 391–397.

[153] Lin, J; Zhou, B; Yang, Y; Mei, J; Zhao, X; Guo, X; Huang, X; Tang, D; Liu, X. Piercing and vacuum infiltration of the mature embryo: a simplified method for *Agrobacterium*-mediated transformation of indica rice. *Plant Cell Reports*, 2009 28, 1065–1074.

[154] Zale, JM; Agarwal, S; Loar, S; Steber, M. Evidence for stable transformation of wheat by floral dip in *Agrobacterium tumefaciens. Plant Cell Reports*, 2009 28, 903–913.

[155] (http://www.liv.ac.uk/news/press_releases/2010/08/decoding_of_wheat_genome_will_help_address_global_food_short.htm).

In: Wheat: Genetics, Crops and Food Production
Editor: Maria T. Almeida, pp. 253-278
ISBN 978-1-61209-307-9

Chapter 8

WHEAT STARCH

Djurdjica Ackar**, Drago Subaric and Jurislav Babic
Faculty of Food Technology, F. Kuhaca 20, 31000 Osijek, Croatia

ABSTRACT

The main component of wheat is starch, which constitutes 63 – 72 % of wheat grain (d.m.). Wheat flour, produced by wheat grain milling, is major ingredient in most European breads and pastries. Since part of the grain is removed during production of flour, starch content in flour rises to app. 80 % d.m., and in bread it averages around 50 % of total bread mass. Starch is main source of energy in human diet. Therefore, it is very important to understand its properties and changes it undergoes during processing, as well as its physiological effects and influence of botanical source on these characteristics.

It is well known that starch constitutes of two polysaccharides: linear amylose and branched amylopectin, organised in granules of different sizes and shapes, depending on botanical source. The cluster model of organisation of amylose and amylopectin is widely accepted. However, its chemistry, biochemistry, genetics, processing and modifications, as well as behaviour of starch and starchy materials in different thermodynamic conditions, although revealed to some extent, are still object of many researches and are not yet fully understood.

This chapter will deal with isolation of starch from wheat and its characterisation. The first part of book chapter will give literature review of different processes of starch isolation from wheat and its properties, while in the second part of chapter laboratory isolation of starch from two Croatian wheat varieties (*V1* and *V2*) and results gained by its characterisation will be presented. Properties of starch: gelatinisation, retrogradation, pasting properties, swelling capacity, gel texture, resistant starch content etc. will be discussed.

---

* Corresponding author: E-mail adress: dackar@ptfos.hr (D. Ackar); tel: 00385 31 224 300; fax: 00385 31 207 115

1. Introduction

1.1. Starch Structure

Starch is polysaccharide with molecular formula $(C_6H_{10}O_5)_n$. Virtually all green plants accumulate it in different tissues and organs (grain, fruit, tuber…). During photosynthesis plants convert CO_2 to glucose which condensates to primary starch to prevent raise of osmotic pressure in the cell. During night starch is hydrolysed and transported to "storage" organs, where starch is synthesized in amyloplasts. During grain or tuber germination and fruit ripening the "storage" starch is hydrolyzed and hydrolization products are used as sources of energy.

Starch content in potato tubers averages around 30 %, while in grain it constitutes app. 70 % of wheat (Eliasson, 2004).

Shape and size of starch granules are characteristic for plant species, and can be used for determination of starch origin. Grain starches are usually small polygons, while potato starch granules are large ellipsoids. Starch is virtually insoluble in cold water, and consists of two polymers: amylose and amylopectin (Eliasson, 2004; BeMiller & Whistler, 2009).

Amylose

Amylose is linear polymer made up of α-(1→4) bound glucose molecules (Figure 1). However, amylose of most starches contains very low proportions of branched chains, with α-(1→6) glycosidic bonds at branching points (Eliasson, 2004). Molar mass of amylose is 50 – 500 × 10^6 (BeMiller & Whistler, 1996), and degree of polymerization (DP) 0.51 – 6.34 × 10^3 (Eliasson, 2004).

Axial – equatorial position of (1→4) links in amylose chain results in chain twisting to a right-handed helix. Helix interior is lipophilic because it contains only hydrogen atoms. All hydroxyl groups are positioned on the exterior of the helix (BeMiller & Whistler, 1996).

X-ray diffraction showed existence of two types of amylose: A and B. Structure elements of B-type are double helices, antiparalel packed into hexagonal units. Central channel is enclosed with 6 double helices and filled with water (36 mol H_2O/structure unit). Type A is enclosed with parallel double helix and has more compact structure. Internal channel is filled with 8 mol H_2O/unit (Sajilata et al., 2006).

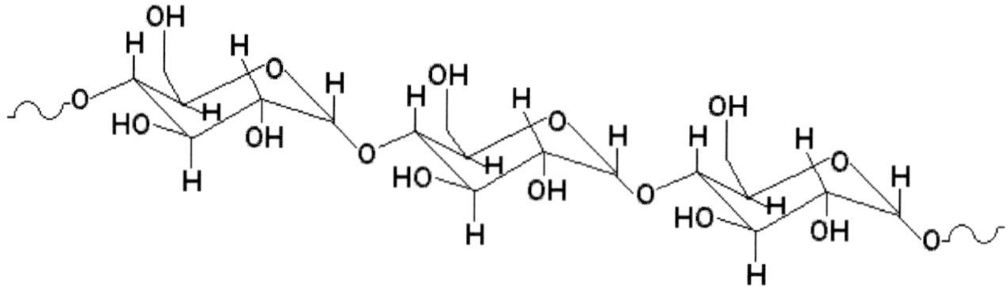

Figure 1. Structure formula of amylose.

Due to hydrophobic internal part of helix, amylose forms blue complex with iodine, which is commonly used for determination of starch presence in food (BeMiller & Whistler, 1996).

In dilute solutions amylose tends to retrograde, while in concentrated solutions it forms gel. Retrogradation of amylose is influenced by degree of polymerization. Molecules with DP ~ 80 have highest tendency to retrograde, while molecules with higher and lower DP are more stable. DP has similar influence on solubility. Molecules with DP 80 – 100 are poorly soluble, and solubility increases with chain shortening or lengthening (Walter, 1998).

Amilopectin

Amilopectin is highly branched molecule where glucose units are bounded with α-(1→4) glycosidic bonds into linear chain and with α-(1→6) bonds on branching sites (Figure 2). Amylopectin DP is around 2 million, and molar mass 10^7 – 10^9. Amylopectin molecule consists of 3 chain types: A-, B- and C-type. A-chains are straight fragments in which glucose molecules are bounded with α-(1→4) links, they are short (DP 12-16), may form double helix and are ordered into clusters. B-chains are linked to one ore more A-chains, and glucose molecules are linked by α-(1→4) and α-(1→6) linkages. DP of these chains is 40 – 110 (Oates, 1997). C-chain is the only one with reducing end of amylopectin molecule.

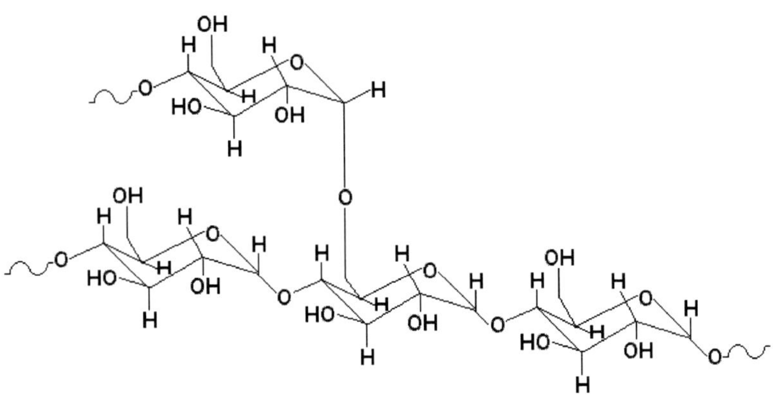

Figure 2. Structure formula of amylopectin.

Side chains of amylopectin twine towards the main chain and form tridimensional structure with parallel sided chains (Biliaderis, 1998; BeMiller & Whistler, 2009). Due to conformations of glycosidic bonds on branching points double helixes can be cross-linked by α-(1→6) linkages with minimal torsions. This model responds to Hizukuri's cluster model, where amylopectin chains exist as single cluster or as linkages between two or more clusters (Biliaderis, 1998; BeMiller & Whistler, 2009).

Amylopectin is partially crystalline polymer of crystalline double helices, amorphous compact branched chains and long chain segments which link clusters. Due to branched structure, amylopectin does not form stabile complexes with iodine. Only side chains react with iodine forming redish – brown complex (Babić, 2007).

Starch Granule

Amylose and amylopectin are radially organized into alternating amorphous and crystalline shells which form "growth rings". In addition to starch, granules contain non-starch components, mainly lipids, proteins and phosphorous.

Lipids are most important non-starch components of starch granule. The amount of lipids in granule averages between 5 – 10 g/kg, with free fatty acids and phospholipids as main components, which are complexed with amylose. Lipids are mostly bounded to the surface of starch granule. Lipid:starch complexes weaken enzyme:supstrate contact and therefore reduce digestibility of starch. In addition, these complexes reduce swelling of the granule due to increased hydrophobicity. Smaller granules contain larger amounts of lipids than larger ones. This can be result of larger total surface area of granules (Svihus, Uhlen & Harstad, 2005).

Starch granules contain up to 3 g/kg protein. The amount of proteins increases from the center of the granule to the surface (Svihus, Uhlen & Harstad, 2005).

Wheat Starch

Wheat starch consists of two granule types: A-granules, which are disk or lentil shaped, with diameter 10 μm or larger, and B-granules, which are spherical or polygonal, with diameter < 10 μm. During grain filling C-granules, smaller than 5 μm, can also be formed (Park, Wilson & Seabourn, 2009).

A-granules develop asymmetrically and have characteristic equatorial groove. Their number and size are increasing during early stages of grain development, during cell division. They are formed from initial spherical nucleus, by forming thin plate in the equatorial lane. This plate then enlarges above and under equatorial lane until granule becomes lentil-shaped. After cell division the number of A-granules remains constant, but they continue to enlarge during grain filling (BeMiller & Whistler, 2009).

B-granules are developed symmetrically and their number is continuously growing until grain maturity, when they account for 90 % of total granule number, but only 20 – 30 % of total volume. B-granules don't have equatorial groove, but voids and crevices which penetrate to the surface (BeMiller & Whistler, 2009).

Kim & Huber (2008) investigated so-called "channels" in starch granules. They identified channel structures both in A- and B-type granules. In A-granules main channels were large and mostly situated in area of equatorial groove, with smaller channels at other parts of granule. On the other hand, channels in B-granules resembled more to small crevices and voids which were blocked or filled with proteins. It is believed that channels in both granule types have function in transportation of chemical compounds inside the granule. Hydration and removal of proteins from the channels enhanced this effect. This leads to conclusion that channels facilitate migration of chemical compounds into the granule and influence chemical reactions during starch modification. Since channels are large enough to receive enzymes, they may have important role in starch hydrolysis.

Granule size distribution (A:B type ratio) is influenced by wheat sort and variety, and growing conditions. Li et al. (2008) compared size of starch granules between soft and hard wheat varieties grown in China. They found that hard wheat starch contained more B-granules and less A-granules compared to soft wheat.

Dai et al. (2008) investigated distribution of starch granule size of wheat grown under irrigated and rainfed conditions. Compared to irrigated treatment, rainfed treatment resulted in increase of percent volume and percent surface of starch granule smaller than 9.8 μm, and decrease of those in granules larger than 18.8 μm. Lack of water in the soil caused decrease of both amylose and starch content, and increase of protein content, peak and final viscosities.

In addition to differences in size and shape, A- and B-granules differ both in chemical composition (amylose, lipid and phosphorous content), molecular structure, functional properties, gelatinization, and reactivity during chemical modification etc. (Kim & Huber, 2008).

Generally, A-granules contain higher proportions of amylose and lipids (BeMiller & Whistler, 2009). Dai et al. (2008) reported that smaller granules were low in contents of amylose and starch, and high in protein content, while larger granule were high both in contents of amylose and starch.

Yoo & Jane (2002) investigated structural and physical properties of starches isolated from waxy wheat, amylose-reduced wheat and normal hard red winter wheat. Amylopectin of waxy wheat had the highest molar mass and no detectable extra-long side chains. Molar mass of amylopectin was in negative correlation with amylose content, while proportion of extra-long side chains of amylopectin was in positive correlation with amylose content. Onset gelatinization temperature increased, while gelatinization enthalpy decreased with increase of amylose content. Differences in gelatinization temperature and maximum viscosity between normal and waxy wheat starch were more pronounced than differences between maize counterparts. Lack of amylose in waxy starch didn't influence crystal polymorphism, size and morphology of granules, nor gelatinization temperatures, but it significantly influenced degree of crystalinity.

Van Hung, Maeda & Morita (2006) described properties, functionality and application of waxy and high-amylose wheat starches and flours. They reported that amylopectin structure in waxy varieties resembled structure in non-waxy varieties. No significant difference was determined for chain length and DP of amylopectin, which indicated that waxy character had very little impact on chain length of amylopectin. X-ray diffraction showed that waxy starch contained A-type granules and had higher degree of cristallinity (due to high content of amylopectin) than non-waxy starches. Structure of amylopectin in high-amylose starch is significantly different. Proportion of short chains (DP 6 – 10) in amylopectin increases, while proportion of chains with DP 11 – 25 decreases.

Park, Wilson & Seaburn (2009) investigated distribution of starch granule size in hard winter and spring wheat and their influence on bread quality. B-granules occupied larger volumes in spring wheat varieties in comparison to winter varieties. Protein content was inversely proportional to parameters of B-granules. Granule size influenced physico-chemical properties of flour, wheat and bread quality.

B-granules are used mostly as fat replacers, while, for instance, A-granules are used in production of biodegradable plastics (Xie et al., 2008).

Wheat starch has unique properties which are very important for its use in food industry. It is primarily its exceptional white color, resulting from absence of pigments. In addition, protein, ash and fiber contents are very low, and sulfites are not present at all.

Protein content in wheat starch averages between 0.06 – 0.22 %. The main proteins are starch synthase bounded to granule and friabilin on the granule surface. Lysophospholypids entrapped in starch granule contain nitrogen and affect detection of protein content. Well purified starch from soft wheat contains 0.014 – 0.019 % lysophospholipid nitrogen, 0.0034 – 0.0086 % friabilin nitrogen and 0.013 – 0.014 % starch synthase nitrogen. Wheat starch contains around 1 % lipids, mainly lysophospholipids.

1.2. Physico-Chemical Properties of Starch

Intact starch granules are insoluble in cold water. Nevertheless, they reversibly absorb small proportion of water. If starch suspension in water is heated, starch granules will gelatinize and dissolve.

Gelatinisation

Gelatinization is process which starch granules undergo when subjected to high temperatures in presence of water, although same result can be achieved in different solvent, under high pressure, shearing or milling (Sopade et al., 2004). During this process structure in the granule is disrupted, granule is irreversibly dissolved, birefringence and crystal structure are lost and digestibility is increased. Gelatinization process takes place within specific temperature interval, depending on starch origin and influence of different factors.

While tapioca and wheat starch start to gelatinize around 52 °C, maize starch does not gelatinize until 62 °C. Gelatinization temperature is increased by addition of sugars and other polyols (Perry & Donald, 2002).

Although processes that occur during gelatinization are not fully understood, following model is adopted:

Water first penetrates into amorphous regions of granules, which have glass structure and granules swell. When granules are heated in substantial amount of water (minimum 60 %), at specific temperature (t_g, glass transition temperature) glass phase transforms to rubbery. Energy absorbed during this transition cannot be detected by differential scanning calorimetry (DSC), since right after glass transition crystal parts of amylopectin, covalently bounded to amorphous regions, are dissolved. Since enthalpy of glass transition is significantly lower than enthalpy of initial crystal dissolving, it can't be detected.

Consequently, granule swells increasingly and amylose and amylopectin molecules leach from granule into solution. Shearing enhances disruption of granule and amylopectin and amylose dissolving. As more molecules dissolve, viscosity of system is higher.

Ratnayake & Jackson (2007) investigated changes occurring during gelatinization of seven starch types: maize, waxy maize, amylomaize, wheat, rice, tapioca and potato. They reported that, during heating, structural changes in granules occur even before visible morphological changes occurring during gelatinization can be detected. The nature of these structural changes depended on starch type. The researchers concluded that starch gelatinization was process consisting of many consecutive processes and not rapid or sudden process occurring in narrow temperature range. It started at low temperatures and proceeded until complete degradation of granules.

During first phase of change water absorbed by granules increases mobility of starch polymers, primarily amylose in amorphous regions and new intermolecular bonds are formed. Increase of enthalpy during heating starch below gelatinization temperature (determined by DSC) confirms this phenomenon. Changes of relative cristallinity at lower temperatures show that structural changes include changes in crystal parts of granule which are dependent on starch origin.

Further heating makes starch polymers more mobile, interactions between them weaken or completely disappear and starch granules breakdown. Onset and progress of this phase transition can be detected by DSC or microscopic techniques.

One can conclude that gelatinization is process consisted of three steps: 1) absorption of water in starch granules and increase of chain mobility in amorphous regions of the granule; 2) reorganization of polymers in amorphous regions and new intermolecular bonding; 3) increase of mobility due to new hydrothermal conditions, loss of intermolecular bonds and loss of granule structure.

Since gelatinization is endothermic process, it can be determined by Differential Scanning Calorimetry (DSC). The result of measurement is curve (Figure 3) from which four parameters can be determined: onset temperature (t_o), peak temperature (t_p), endset temperature (t_e) and gelatinization enthalpy (ΔH).

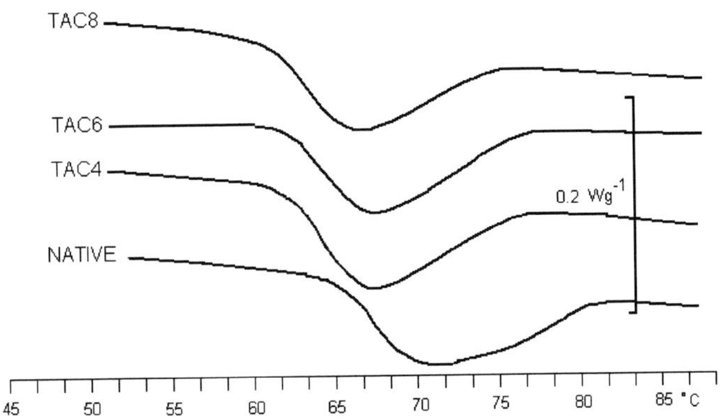

Figure 3. DSC curves of starch gelatinization: (NATIVE) native tapioca starch; (TAC4, TAC6 and TAC8) acetylated tapioca starch with 4, 6 and 8 % of acetic anhydride (Babic, 2007).

DSC has been extensively used for determination of parameters of starch gelatinization (Table 1) and different values have been obtained, depending on applied conditions.

Table 1. Gelatinization parameters of different starch types determined by DSC

Starch	Starch:water ratio	t_o [°C]	t_p [°C]	t_e [°C]	ΔH_{gel} [J/g]
Maize[a]	1:1.5	62.3	67.7	84.3	14.0
Maize[b]	1:9	65.7	71.0	-	12.0
Wheat[a]	1:1.5	51.2	56.0	76.0	9.0
Wheat[c]	1:2:3	51.2	56.0	76.0	9.0
Rice[a]	1:1.5	62.0	67.4	97.5	11.0
Rice[d]	1:9	57.7	65.1	-	11.5

T_o = onset temperature; t_p = peak temperature; t_e = endset temperature; ΔH_{gel} = gelatinization enthalpy; [a]Jenkis i Donald, 1998.; [b]Li i Yeh, 2001.; [c]Sasaki i sur., 2000.; [d]Sodhi i Singh, 2002.

Sodhi & Singh (2005) studied gelatinization parameters of five rice starches and influence of acetylation on them. Singh et al. (2004) studied influence of acetylation on gelatinization properties of maize and potato starches, and Babic et al. (2006) investigated influence of lactose and whey powder addition on gelatinization and retrogradation of native and acetylated tapioca starch. Ortega & Eliasson (2001) studied gelatinization and retrogradation properties of maize, potato and barley starch, and their mixtures.

Thermal properties of waxy and high-amylose starches, determined by DSC (Van Hung et al., 2006) are significantly different and specific for each cultivar. High-amylose starches have two endothermic peaks. The first one is starch gelatinization, while the second represents melting of amylose-lipid complexes. Waxy starches have only one peak – starch gelatinization. Temperatures and enthalpies of gelatinization of high-amylose wheat starches are significantly lower than waxy and common wheat starches. This indicates that amylopectin, due to high degree of cristallinity, requires more energy.

Blazek et al. (2009) determined thermo-physical properties of wheat starches with different amylose content. They stated that gelatinization of wheat starch occurred in temperature range 61.5 – 65.4 °C. Waxy starch had higher gelatinization temperature (66.8 °C) due to higher degree of cristallinity.

Retrogradation

Cooling gelatinized starch leads to retrogradation, process in which starch molecules are re-crystallized. Amylose molecules are bounded into double helices and amylopectin molecules are linked by short side chaines (Singh et al., 2003). Due to highly branched structure, amylopectin molecules recrystallize significantly slower.

Retrogradation is influenced by several factors: starch origin, concentration of starch in solution, gelatinization temperature, pH, presence of other substances (lipids, electrolytes, sugars) and storage conditions (Walter, 1998). It is highest at pH 5 – 7, retards at $pH < 2$ and at $pH > 10$ it stops. Degree of retrogradation increases with temperature decrease and starch content increase, and decreases in presence of monovalent salts, $Ca(NO_3)_2$ and urea.

Liu & Thompson (1998) studied retrogradation of starch after different thermal treatments. Results showed that increase of gelatinization temperature increases rate of retrogradation during storage. Rate of retrogradation was also increased by increase of water content and storage period (Jouppila & Ross, 1997; Liu & Thompson, 1998).

Retrogradation of starch paste results in: viscosity increase, turbidity, starch settling and syneresis. It can deteriorate organoleptic properties of food products (bread staling, loss of viscosity, settling of gravies and soups). Addition of polar lipids, such as monoglycerides and their derivatives, extends shelf-life of bakery products (Eliasson, 2004).

Starch Paste Viscosity

Continuous heating and shearing of starch suspension induces number of changes in starch granules, such as swelling, exudation of amylose and amylopectin, granule disruption, lose of birefringence, increased viscosity and formation of starch paste. Starch paste is two-phase system consisted of dispersed phase of swollen starch granules and continuous phase of dissolved amylose and/or amylopectin. It can also be described as complex polymer system in which swollen granules are incorporated into amylose matrix (Cui, 2005).

Rheological properties are important both in food product quality and control of production processes. Although numerous techniques are developed for determination of rheological properties of starch, Brabender viscoamylograph and Rapid ViscoAnalyzer are mostly used.

Brabender Viscoamylograph is used for characterization of viscosity change during heating starch suspensions as well as rheological properties of starch pastes after cooling (BeMiller & Whistler, 1996). The result of measurement is the curve (Figure 4) from which 6 parameters are determined: pasting temperature, peak viscosity, hot paste viscosity, viscosity after holding at 95 °C, cold paste viscosity and viscosity after holding at 50 °C.

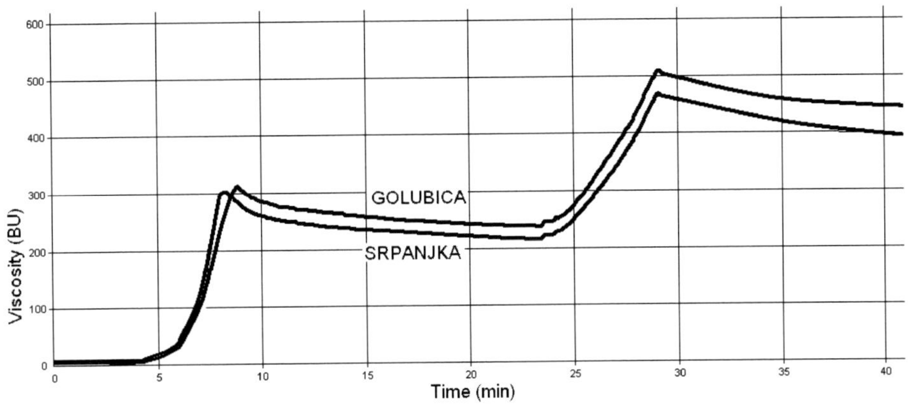

Golubica and Srpanjka .

Figure 4. RVA viscoamylogram of starches from two wheat varieties.

RVA provides characterization of starch similar to Brabender viscoamylograph, with additional versatility of testing parameters. Advantages of this technique are smaller sample amounts, shorter test period and ability to modify testing conditions (Cui, 2005).

Based on paste properties, starches can be divided in four categories:

1) starches with high swelling power, high peak viscosity and high decrease of viscosity after holding period at 95 °C (potato, tapioca, waxy cereal starches);
2) starches with medium swelling power, lower peak viscosity and lower decrease of viscosity after holding period at 95 °C (cereal starches);
3) starches with low swelling power, slight peak viscosity (or no peak viscosity) and high viscosity which changes very little or not at all during heating (cross-linked starches);
4) starches with exceptional low swelling power which do not swell enough to make viscous suspension (high-amylose maize starch) (Cui, 2005).

Viscosity of wheat starch paste is highly influenced by wheat variety, amylose content, cultivar and genetics, heating rate, pH, granule size, starch concentration, modification, interaction with other substances in food etc. (BeMiller & Whistler, 2009).

Since paste viscosity of wheat starch is lower than maize starch, wheat starches have advances in application in soups, gravies and dressings. Difference in viscosity is explained through higher swelling power, higher molar mass of amylopectin and lower phospholipids' content in maize starch.

Paste viscosity is in negative correlation with amylose content in range 23 – 43 %, due to negative impact of amylose on swelling capacity. However, due to high breakdown values, end viscosity of waxy wheat starch is lower than amylose starches. Non-ionic surfactants increase pasting temperature and decrease viscosity of non-waxy wheat starches, but have very low impact on pasting properties of waxy starches.

Gel Texture

Cooling of paste increases viscosity and, at sufficient starch concentration, results in formation of gel. Gel hardness is influenced by water amount and presence of different ingredients, such as lipids, proteins, sugars and acids (Eliasson, 2004).

Gel texture and mechanical properties are mostly determined by rheological properties of continuous phase, volume proportion of swollen granules, their deformation and interactions between continuous and dispersed phase (Choi, 2001).

Heating at high temperatures (> 125 °C) with/or shearing causes degradation of starch granule structure, changes in rheological and structural properties of starch gel, which result in decrease of gel hardness (Choi, 2001).

Water is very important in development of texture and mechanical properties of starch gels and foods based on starch. Inaba et al. (1994) investigated influence of water content on starch gel texture. They reported that compression force and work increase with increase of starch concentration, while elasticity and compressibility aren't depended on it (Choi, 2001).

Chung & Lee (1991) studied relation between texture of starch based products and swelling capacity. They concluded that bound water in surimi gels with added potato starch significantly influence compression force.

Gel hardness of wheat starch is higher than maize starch at starch concentrations above 6 %. At lower concentrations, maize starch gives harder gel (BeMiller & Whistler, 2009).

1.3. Isolation of Starch from Wheat

In addition to flour, bread, pasta and other wheat based products, wheat is important raw material in starch production. Approximattely 33 % of starch produced in EU is wheat starch (http://www.zuckerforschung.at) (Figure 5).

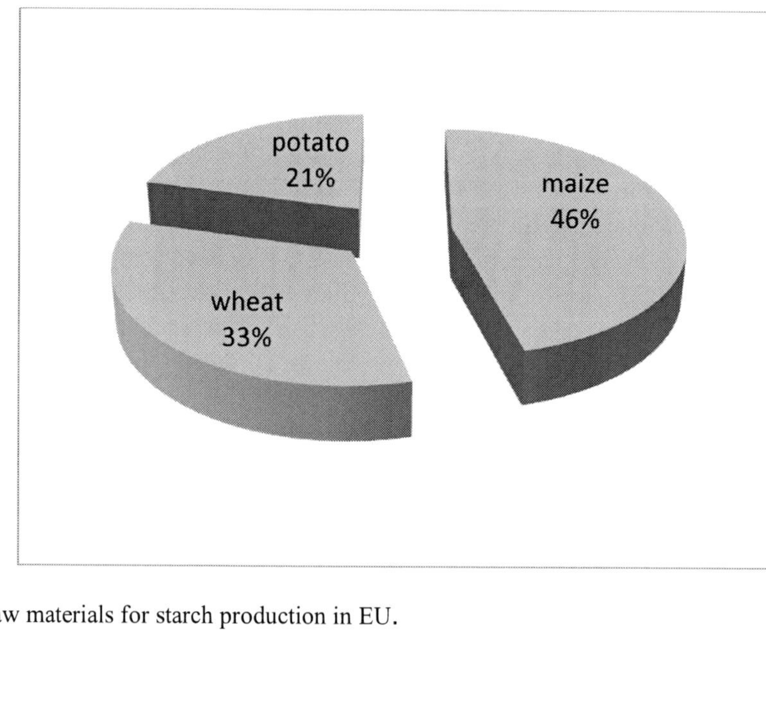

Figure 5. Raw materials for starch production in EU.

During isolation of starch from wheat grain, starch and gluten can be damaged. Therefore, starch fractionation from gluten is generally performed from flour. Wheat flour for commercial production of starch has to comply following demands: high protein content (> 11 %); low content of damaged starch, low ash content and not contain α-amylase (low falling number) (BeMiller & Whistler, 2009).

In majority of industrial processes starch is separated from gluten by centrifugation, decanting or in hydrocyclones due to different sizes of starch and gluten particles (Van der Borght et al., 2005). According to some authors, starch and gluten can be separated by 15 different processes, but only few of them are used in industrial scale. Van der Borght et al. (2005) gave a review of methods for starch isolation from wheat and wheat flour.

Martin Process

This is one of the oldest processes for production of wheat starch. It is based on five basic steps: mixing flour with water, formation of dough, washing out the starch, refining and drying starch (BeMiller & Whistler, 2009). Flour and water are mixed in 10:6 – 7 or 2:1 ratio to give dough with formed gluten network. Hard wheat forms stiff elastic dough and needs more water than soft wheat, which forms crumbling dough. Water must contain mineral salts; otherwise it will cause formation of slimy gluten. Separation of starch from gluten is performed by washing out the dough in drums, mixers or similar type of equipment. Starch slurry passes over the sieves, where remaining gluten particles are removed, starch is then purified and dried. Isolated starch contains approximately 0.3 % proteins and 10 – 12 % water (Knight & Olson, 1984).

Dough-Batter Process

Stiff dough is formed and gluten matrix is dispersed by mixing with water. Formed batter is either sieved or centrifugated. Starch yield is 69 – 79 % (Van Der Borght et al., 2005; Knight & Olson, 1984).

Batter Process

Approximately same amounts of water and flour are mixed into soft dough which is mechanically broken up by addition of water. Suspended gluten curd is formed and separated on the sieves, and starch slurry is centrifuged and dried. Starch yield is 68 – 77 %, based on starting flour (Van Der Borght et al., 2005).

Fesca Process

Wheat flour is quickly mixed with water to form thin batter. Agglomeration of gluten is prevented by shearing and lower temperatures compared to batter process. Starch is separated by centrifugation, by which three layers are formed: primary starch, squeegee starch and supernatant containing proteins and water-soluble components. It yields 63 – 87 % starch (Van Der Borght et al., 2005; Knight & Olson, 1984).

Raisio Process

This is modification of Fesca process, in which gluten is not agglomerated. Instead, primary starch is separated by continuous centrifugation, and gluten is recovered by purification of supernatant (BeMiller & Whistler, 2009; Van Der Borght et al., 2005).

High-Pressure Disintegration Process and Tricanter Process

Wheat flour is mixed with water to form smooth dough, which is transported to high-pressure pump and homogenization valve. High shear causes agglomeration of gluten in micrometer particles and liberation of starch from endosperm. Water is added to dough to achieve 30 % d. m. and mixture is separated in centrifugal decanter in three fractions: 1) A-starch, 2) B-starch with gluten and 3) viscous pentosans with solubles. B-starch is separated from gluten in perforated drums and on vibrating sieves. A-starch is refined in hydrocyclones. Advantages of this process are lesser consumption of water, higher yield of A-starch, quality of flour doesn't influence the process, and it can be used for isolation of starch from potato, beans and corn (BeMiller & Whistler, 2009).

1.4. Isolation of Starch From Wheat Grain

Halle Fermentation

Wheat grain is soaked in water at 25 °C during one week, milled and meal is fermented. During fermentation proteins are dissolved and hydrolised. Starch is washed out in rotating drums, centrifuged, refined and dried (Van Der Borght et al., 2005).

Alsatian Process

Wheat grain is soaked 1 – 2 days at 30 – 35 °C, coarsely grinned and washed out in perforated trough with rotating arms. Starch slurry is centrifuged and starch yield averages 45 – 70 % (Van Der Borght et al., 2005).

2. Materials and Methods

2.1. Wheat Varieties for Starch Isolation

Starch was isolated from two Croatian wheat varieties: *Golubica* (V1) and *Srpanjka* (V2). V1 is winter high-quality bread wheat. Its farinograph and extensograph properties are given in Table 2.

Table 2. Farinograh and extensograph properties of wheat variety *Golubica* (V1)

Farinograph		Extensograph	
Water absorption	63,0 %	Time	135 min
Development time	5.5 min	Energy	85 cm^2
Stability	2.6 min	Resistance	226 EJ
Resistance	8.1 min	Extensibility	183 mm
Degree of softening	36 FJ	Maximum	360 EJ
Quality group	A1	Ratio number	1.2

Kovačević, 2009.

V2 is most cultivated winter wheat variety in Croatia. It has high yields, high quality and is resistant to diseases. Its farinograph and extensograph properties are given in Table 3.

Table 3. Farinograh and extensograph properties of wheat variety *Srpanjka* (V2)

Farinograph		Extensograph	
Water absorption	57.4 %	Time	135 min
Development time	1.9 min	Energy	88 cm^2
Stability	0.4 min	Resistance	304 EJ
Resistance	2.3 min	Extensibility	145 mm
Degree of softening	28 FJ	Maximum	478 EJ
Quality group	A2	Ratio number	2.1

Kovačević, 2009.

Analysis of wheat properties by Infratec Grain Analyzer showed differences in wet gluten, sedimentation and flour strength between the two varieties (Table 4).

Table 4. Wheat properties determined by Infratec Grain Analyzer

	V1	V2
Moisture content / %	12.27	12.20
Protein / % d. m.	14.37	12.57
Starch / % d.m.	67.00	68.73
Wet gluten / % d. m.	37.70	33.73
Sedimentation / mL	58.67	33.93
W from alveograph / J	336.00	279.33
Hardness	85.90	52.60
Hectoliter mass / kg/hL	84.03	83.90

2.2. Isolation of Starch from Wheat

Starch was isolated from wheat by following procedure: Wheat was fully milled in laboratory miller (IKA Werke M20), sifted and flour was suspended in 0.25% NaOH (1:3 w/w) by stirring with propeller stirrer (Tehtnica UM-405) at 1500 rpm for 1 hr at room temperature.

Suspension was centrifuged at 3000 rpm for 5 min in benchtop centrifuge IEC Centra MP4R and upper greyish layer was discarded. Sediment containing starch was washed with distilled water over cheese cloth and filter cloth.

Washings were centrifuged (3000 rpm for 5 min) and upper grey layer of proteins was removed with spatula. White starch sediment was suspended in water and centrifugation and re-suspension of starch in water were repeated until grey layer disappeared. Washed starch was neutralised with 1M HCl, centrifuged once more and air-dried. Dry matter content of starch was determined by oven-drying (130°C/90 min).

2.3. Starch Purity

To determine starch purity, protein content was determined by Kjeldahl method, ash by annealing at 550 °C, cellulose by Harrer-Kürschner method and fat content by Soxhlet. Moisture content was determined by drying at 130 °C for 90 min.

2.4. Amylose Content

Amylose content was determined according to the Megazyme method. Starch samples were completely dispersed in dimethyl sulphoxide (DMSO), precipitated in ethanol and dissolved in acetate/salt solution. Amylopectin was specifically precipitated by the addition of Con A and removed by centrifugation. The amylose in the aliquot of the supernatant was enzymatically hydrolysed to D-glucose, which was analyzed using glucose oxidase/peroxidase reagent. Total starch in separate aliquot of acetate/salt solution was also enzymatically hydrolysed to D-glucose, which was analyzed using glucose oxidase/peroxidase reagent. Concentration of amylose in starch sample was calculated from Eq.1.:

$$Amylose = \frac{Absorbance\ (ConA)}{Absorbance\ (Total\ Starch\ Aliquot)} \times 66.8\ \ [\%] \tag{1}$$

where 66.8 was dilution factor for Con A and Total Starch extracts.

2.5. Gelatinization and Retrogradation Properties of Starch

Gelatinization and retrogradation properties of native and modified starches were determined my method described by Babić et al. (2009), using differential scanning calorimeter DSC 822^e (Mettler Toledo) equipped with STAR^e software. Starch samples were weighed into standard aluminium pan (40μL) and distilled water was added by Hamilton microsyringe to achieve suspension containing 65% water. Samples were hermetically sealed and equilibrated for 24 hr at room temperature before heat treatment in the DSC apparatus. The starch samples were heated at a rate of 10°C/min from 25 to 95°C.

After heat treatment, samples were cooled to 25°C and removed from DSC. The starch gels were aged at 4°C and monitored for retrogradation after 7 and 14 days. The retrogradation experiments were conducted at a heating rate of 10°C/min from 25 to 95°C. The changes in enthalpy (ΔH in J/g of dry starch), onset temperature (T_o), peak temperature (T_p) and conclusion temperature (T_c) for gelatinisation and retrogradation were obtained from the exotherm DSC curves. Experiments were run in triplicates.

2.6. Pasting Properties of Starch

Pasting properties of starches (7% d.w.b., 100 g total weight) were determined using a Rapid Visco-Analyser (Model 803202, Brabender Gmbh & Co KG, Duisburg, Germany). The starch suspensions were equilibrated at 50°C for 1 min, heated at 7.5°C/min to 92°C, held at 92°C for 20 min, cooled at 7.5°C/min to 50°C, and held at 50°C for 20 min. Experiments were run in triplicates.

2.7. Gel Texture

Gel texture was determined using texture analyzer TA-XT Plus (Stable Microsystem). A starch suspension (11% w/w) was prepared in a flask and then heated at 95 °C for 30 min in a temperature controlled shaking water bath with constant shaking. The cooked paste was cooled to 25 °C and allowed to gel at 25 °C and 85 % RH. The gel formed in the can (45 mm in height and 35 mm in diameter) was compressed at the speed of 2.0 mm/sec to the distance of 20 mm with a flat cylinder probe having 20 mm diameter. The peak height at 20 mm compression was termed hardness, and the negative area of the curve during retraction of the probe was termed adhesiveness (Saartrat et al 2005). Five repeated measurements were performed for each sample.

2.8. Swelling Power and Solubility of Starch

Swelling power (SP) and solubility (SOL) were determined by method described by Babić (2007). Briefly, 1% starch suspensions (d.w.b.) were heated in shaking water bath for 30 min at 65, 75, 85 and 95°C. This temperature range is commonly used in SP and SOL measurements. After heating suspensions were cooled and centrifuged at 4000 rpm for 30 min. Supernatant was decanted, and gel was weighed. Dry matter of supernatant was determined by oven drying at 105 °C until constant mass was reached. Experiments were run in triplicates. SP and SOL were calculated from Eq.2. and 3.:

$$SP = \frac{weight of\ gel}{weight of\ dry matter in\ gel} \left[\frac{g}{g}\right] \quad (2)$$

$$SOL = \frac{weight\ of\ dry\ matter\ in\ supernatant}{weight\ of\ dry\ matter\ in\ suspension} [\%] \quad (3)$$

2.9. Paste Clarity

Paste clarity (in triplicates) was determined by method described by Raina et al. (2006). 1% (d.w.b.) starch suspensions were heated for 30 min in boiling water bath with occasional shaking. After 1 hr holding at room temperature % transmittance was read at 650 nm against distilled water as blank.

2.10. Resistant Starch Content

Resistant starch content was determined by official AOAC 2002.02 method, using RSTAR Megazyme kit. In principle, samples are incubated in a shaking water bath with pancreatic α-amylase and amyloglucosidase (AMG) for 16 hr at 37°, during which time nonresistant starch is solubililzed and hydrolyzed to glucose by the combined action of the two enzymes. The reaction is terminated by the addition of an equal volume of aqueous ethanol or industrial methylated spirits (IMS), and the RS is recovered as a pellet on centrifugation. This is then washed twice by suspension in aqueous ethanol or IMS (50% v/v), followed by centrifugation. Free liquid is removed by decantation. RS in the pellet is dissolved in 2M KOH by vigorously stirring in an ice-water bath over a magnetic stirrer. This solution is neutralized with acetate buffer, and the starch is quantitatively hydrolyzed to glucose with AMG. Glucose is measured with glucose oxidase/peroxidase reagent (GOPOD), and this is a direct measure of the RS content of the sample (Eq. 4).

$$RS\,/\,\%\text{ d. m.} = \Delta E \times F \times \frac{10{,}3}{0{,}1} \times \frac{1}{1000} \times \frac{100}{W} \times \frac{162}{180} = \Delta E \times \frac{F}{W} \times 9{,}27 \qquad (4)$$

where:

ΔE Absorbance read against reagent blank

F Conversion from absorbance to micrograms (absorbance for 100 μg of glucose is determined and: F = 100 (μg glucose) / absorbance).

100/0,1 Volume correction (0,1 mL taken from 100 mL)

1/1000 Conversion from micrograms to milligrams

W Dry weight of sample

162/180 Factor to convert free glucose to anhydro-glucose

3. Results and Discussion

Starch isolated from wheat variety V1 had moisture content 15.43 %, while starch isolated from variety V2 had moisture content 15.45 % (Table 5). Proteins weren't detected by applied method and ash content was 0.29 % d.m. in V1 starch and 0.20 % d.m. in V2 starch. Fiber content was 0.07 % in V1 starch and 0.05 % in V2 starch. Fat content was 0.49 % d.m. for V1 and 0.42 % for V2 starch.

Starch isolated from soft wheat in research of Sung & Stone (2003) contained 0.8 % protein, 0.2 % fat and 0.2 % fibers, while starch from hard wheat (in the same research) contained 0.6 % proteins, 0.2 % fat and 0.1 % fibers. Mohan & Maleshi (2006) reported 0.3 % protein, 0.2 % fat and 0.06 % ash in wheat starch.

Amylose content was 20.27 % in V1 starch and 22.49 % in V2 starch. These results are in accordance with researches of Wasserman et al. (2007) and Yoo & Jane (2002).

Following gelatinization parameters were determined for V1 starch: onset temperature t_o 59.49 °C, peak temperature t_p 62.68 °C and endset temperature t_c 66.66 °C (Table 6). For V2 starch these parameters were: t_o 59.94 °C, t_p 63.72 °C and t_c 67.82 °C. These temperatures are in accordance with research of Mohan & Malleshi (2006), according to which wheat starch gelatinizes at t_o 59.7 °C, t_p 64.5 °C and t_c 70.0 °C. Yoo i Jane (2002) determined that wheat starch gelatinised at temperature range t_o 54.9 – 55.7 °C, t_p 58.9 – 62.1 °C and t_c 63.1 – 67.6 °C, depending on wheat variety.

Table 5. Chemical composition of starch isolated from wheat varieties V1 and V2

	V1	V2
Moisture / %	15.43	15.45
Protein / % d. m.	n.d.*	n.d.*
Ash / % d. m.	0.29	0.20
Fiber / % d. m.	0.07	0.05
Fat / % d. m.	0.49	0.42
Amylose / % starch	20.27	22.49

*n.d. Not detected.

Table 6. DSC gelatinization and retrogradation properties of starch isolated from wheat varieties V1 and V2

	T_o [°C]	T_p [°C]	T_c [°C]	$\Delta H_{g/r}$ [J/g]
		Gelatinization (n=3)		
V1	59.49±0.07	62.68±0.03	66.66±0.06	8.87±0.07
V2	59.94±0.17	63.72±0.02	67.82±0.09	8.47±0.01
		Retrogradation after 7 days at 4°C (n=3)		
V1	42.57±0.29	52.10±0.26	60.45±0.05	3.42±0.01
V2	43.64±0.42	51.67±0.27	60.44±0.49	3.23±0.03
		Retrogradation after 14 days at 4°C (n=3)		
V1	42.30±0.36	51.40±0.70	60.74±0.41	3.81±0.04
V2	43.14±0.32	51.55±0.24	60.64±0.06	3.70±0.04

T_o, onset temperature; T_p, peak temperature; T_c, conclusion temperature; $\Delta H_{g/r}$, gelatinization/ retrogradation enthalpy. Values are means ±SD of triplicate.

Wickramasinghe et al. (2005) reported that starch from Japanese soft wheat cultivars had onset gelatinization temperature t_o 55.6 °C and peak temperature t_p 61.2 °C, while for hard cultivars these parameters were t_o 52.0 – 54.4 °C and t_p 58.8 – 60.5 °C. Van Hung & Morita (2005) investigated differences in gelatinization of small and large starch granules. They reported that small granules gelatinized at t_o 55.8 °C, t_p 61.6 °C and t_c 66.9 °C, while large granules had somewhat lower gelatinization temperatures (t_o 55.6 °C, t_p 60.5 °C, t_c 64.2 °C). Commercial wheat starch gelatinized at t_o 55.62 °C, t_p 58.50 °C, t_c 69.19 °C (Majzoobi et al., 2009).

Gelatinization enthalpy ΔH was 8.87 J/g for V1 starch and 8.47 J/g for V2 starch (Table 6). These values are lower than those determined in research of Yoo & Jane (2002), who determined ΔH 10.6 – 13.6 J/g and Wickramasinghe et al. (2005) who determined ΔH 10.3 – 11.4 J/g for hard wheat and 11.7 J/g for soft wheat. However, they are higher than enthalpies determined by Van Hung & Morita (2005) (4.70 J/g for small and 6.20 J/g for large granules) and Majzoobi et al. (2009) (5.16 J/g).

Retrogradation enthalpy after 7 days of storage at 4 °C was 3.42 J/g for V1 starch and 3.23 J/g for V2 starch (Table 6). These results show that V1 and V2 starch were less prone to retrogradation than starches in research of Yoo and Jane (2002), which had retrogradation enthalpy 3.8 – 4.6 J/g.

Pasting properties, tested by Brabender microviscoanalyzer are shown in Table 7. V1 starch started to paste at 64.87 °C, and V2 starch at 65.6 °C. These values are similar to those determined by Mohan & Maleshi (2006) and Yoo & Jane (2002).

Table 7. Pasting properties of starch isolated from wheat varieties: V1 and V2. Pastes contained 7% starch, w/w (d.w.b.)

	V1	V2
Pasting temperature [°C]	64.87±0.25	65.60±0.36
V_{max} [BU]	306.33±5.51	291.00±3.61
Viscosity at 92°C [BU]	256.67±3.79	272.33±4.16
Hold 20 min at 92°C [BU]	240.67±4.16	214.00±3.00
Viscosity at 50°C [BU]	501.67±4.04	457.67±3.79
Hold 20 min at 50°C [BU]	435.00±4.58	392.33±6.11
Breakdown	65.67±2.08	77.00±2.65
Setback	261.00±1.73	243.67±1.53

V_{max}, maximum viscosity; Breakdown= V_{max}-viscosity at 92°C after 20 min; Setback=viscosity at 92°C after 20 min - viscosity at 50°C before holding. Values are means ±SD of triplicate.

Pasting of starch becomes prominent after gelatinization, when starch granules absorb sufficient amounts of water and swell. Swollen granules brush each other and amylose leaches to the solution. This causes rise in viscosity. Fragile starch granules contribute to the further rise of viscosity by forming homogenous mass (Mohan & Malleshi, 2006).

Maximum viscosity reflects the ability of the granules to swell freely prior to their physical breakdown (Singh et al., 2007). V1 starch had somewhat higher peak viscosity (306 BU) than V2 starch (291 BU). Both these values are significantly higher than peak viscosities reported by Mohan & Maleshi (2006) (140 RVU) and Yoo & Jane (2002) (96 – 122 RVU). Peak viscosity is highly influenced by amylose content – higher the amylose content, hence the lower peak viscosity (Yoo & Jane, 2002). Results shown in this chapter support this claim, since V1 starch contained less amylose than V2 starch (Table 5).

Viscosity at 92 °C, however, was lower for V1 starch (256.67 BU) than for V2 starch (272.33 BU), as well as breakdown value. Breakdown reflects stability of starch paste during shearing at high temperatures. Lower breakdown indicates more stable paste, since it shows lower drop in viscosity (Mohan & Malleshi, 2006).

According to Hoover and Vasanthan (1992) setback values reflect the extent of water immobilization around the charged centres of starch components and those of free and helically complexed lipid molecules (rather than starch paste retrogradation). The association of water molecules with these charged centres would decrease the effective water concentration in the continuous phase, resulting in a rise in viscosity during the cooling cycle. However, setback values can reflect tendency of starch paste toward gel formation (Adebowale, 2002) or towards retrogradation (Luo et al., 2009).

Setback value was lower for V2 starch (243.67) than for V1 starch (261). This value is consistent with retrogradation entalpies obtained by DSC after 7 and 14 days of storage, which were also lower for V2 starch and confirms higher tendency of V1 starch towards retrogradation.

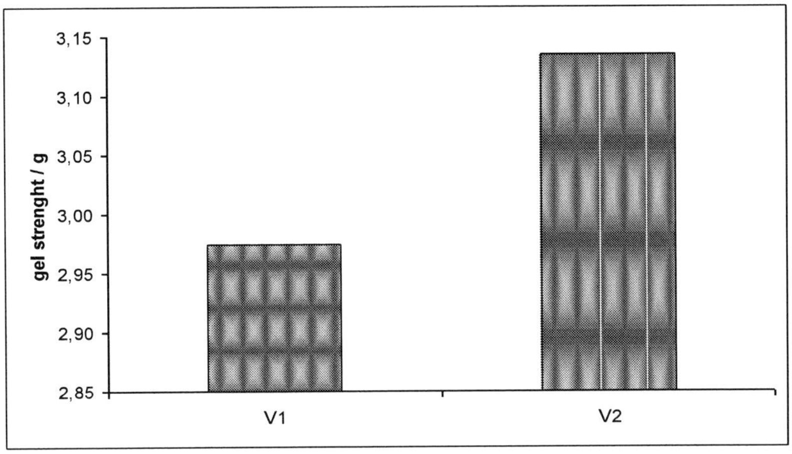

Figure 6. Gel strenght (11 % d. .m.) of starch isolated from wheat varieties: V1 and V2.

Texture properties of starch are very important in food industry. It determines palatability of the product and its acceptability by consumers, and affects mode of production processes.

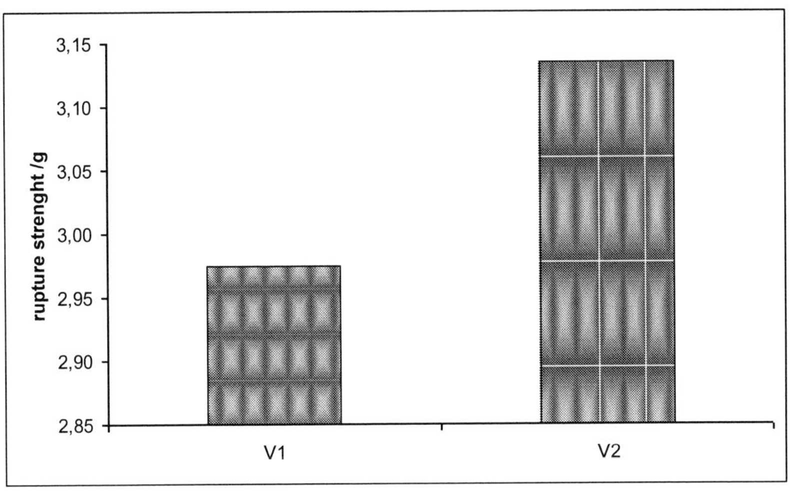

Figure 7. Rupture strenght of gels (11 % d. .m.) prepared from starches isolated from wheat varieties: V1 and V2.

Gel strength is associated with amylose content. In general, starches with higher amylose content form stronger gels (Blazek & Copeland, 2009). Results obtained for V1 and V2 (Figure 6) support this claim, since V1 starch contained less amylose and gave weaker gel than V2 starch. Rupture strength (Figure 7) of V1 starch was also lower than for V2 starch.

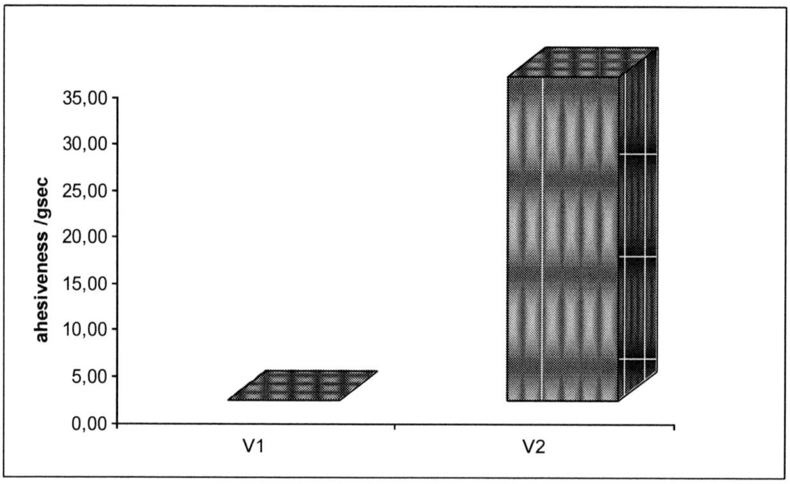

Figure 8. Adhesiveness of gels (11 % d. .m.) prepared from starches isolated from two wheat varieties: V1 and V2.

Adhesiveness is surface characteristic of starch gel, which depends on combined effect of cohesive and adhesive forces, as well as viscosity. Adhesiveness of two starches shown in this chapter (Figure 8) differed significantly. While V1 starch wasn't adhesive, V2 starch showed significant adhesiveness. These results correspond well with pasting properties of starch. Since V2 had lower cold viscosity, cohesive forces in gel were weaker, which resulted in higher adhesiveness.

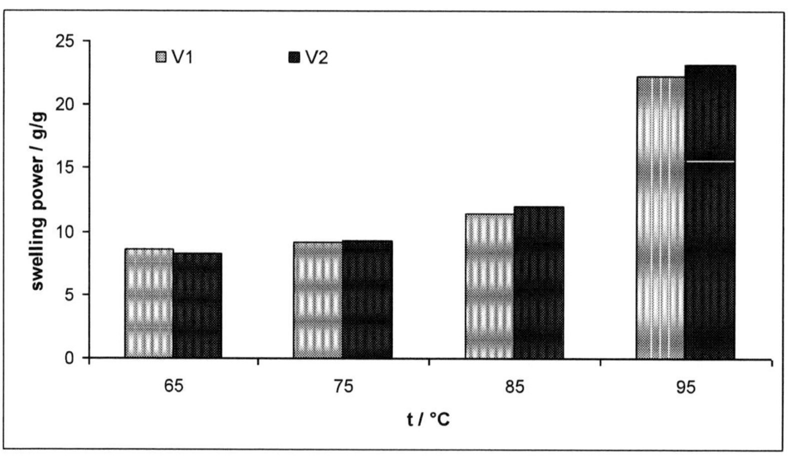

Figure 9. Swelling power of starches isolated from two wheat varieties: V1 and V2.

Although V2 starch had somewhat higher swelling power and solubility than V1 starch at all investigated temperatures, swelling power and solubility of these starches did not differ significantly (Figures 9 & 10).

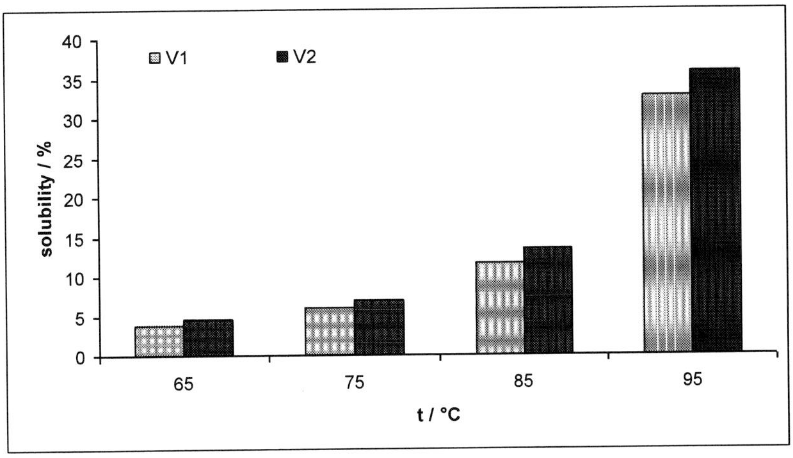

Figure 10. Solubility of starches isolated from two wheat varieties: V1 and V2.

Paste clarity is important characteristic for starch application in food industry, esp. in juice production. It is influenced by amylose content, molecular and granular structure, all of which directly influence swelling of the granule. Paste clarity is increased with increase of swelling (Bhandary & Singhal, 2002; Van Hung & Morita, 2005).

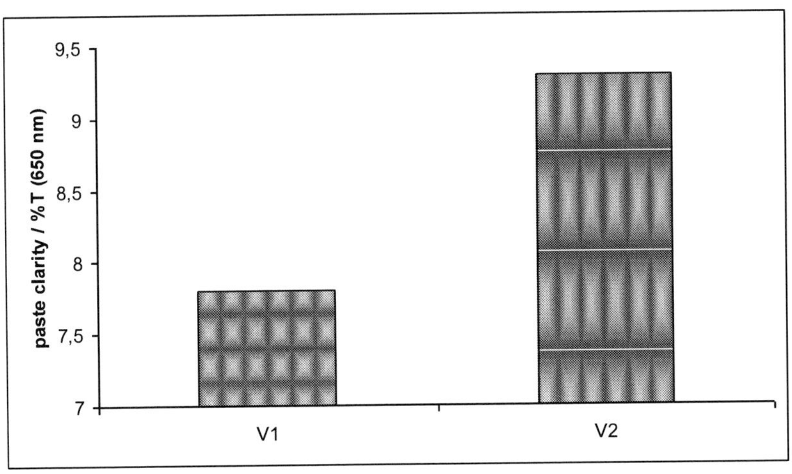

Figure 11. Paste clarity of starches isolated from two wheat varieties: V1 and V2.

Starch isolated from V2 wheat variety had higher paste clarity than V1 starch, which was in accordance with their swelling power (Figure 11).

Singh et al. (2003) reported that paste clarity was influenced by amylose leaching from starch granule during swelling. Leaching of solubles from starch granule is greatly influenced by lipid content and amylose-lipid complexes. Amylose-lipid complexes are water-insoluble and degrade at very high temperatures. Therefore, amylose bounded into these complexes does not leach during granule swelling. Since V1 starch contained higher proportion of lipids, this probably reduced its swelling and paste clarity.

According to results shown in Figure 12, both V1 and V2 starches are highly digestible. Resistant starch content in V1 starch was 0.48 % d. m. and in V2 starch 0.57 % d. m.

However, Garcia-Alonso et al. (1999) reported 14.4 % d. m. of resistant starch in commercial wheat starch, Xie et al. (2008) 2.48 % d. m. and Eerlingen et al. (1994) reported that wheat starch after removal of lipids contained 1 % resistant starch.

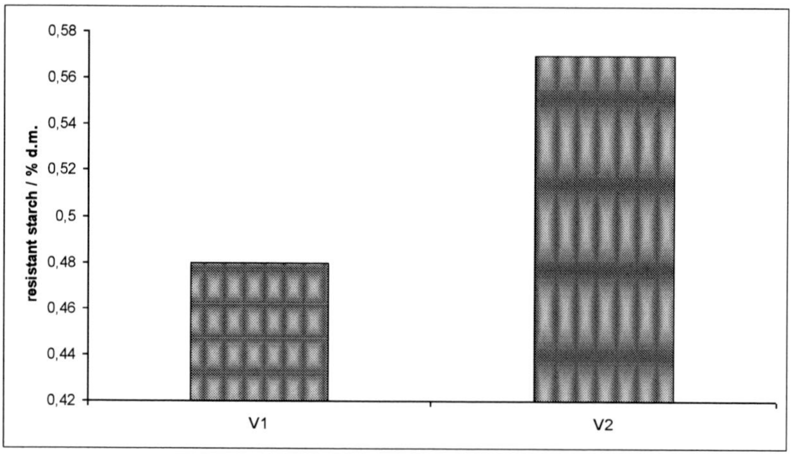

Figure 12. Resistant starch content in starches isolated from two wheat varieties: V1 and V2.

Resistant starch content doesn't have technological importance, since it has all the properties as digestible starch. However, numerous researches (Champ et al., 1998; Cook & Sellin, 1998; De Schrijver et al., 1999; Eerlingen et al., 1994) have shown its great impact on human health. Since it is not digested in small intestine, it undergoes fermentation in the large intestine. Products of fermentation are short-chain fatty acids (SCFA), which have prebiotic, anti-carcinogenic and anti-inflammatory effect (Cook & Sellin, 1998).

Conclusion

Wheat starch is important raw material in the food industry. Properties, such as thermo-physical and pasting properties, swelling power, solubility and paste clarity are very important in its applications, since they play very important role in process manipulation and quality of the products.

Based on the research shown in this chapter, following conclusions may be drawn:

- Starch isolated from Croatian wheat varieties V1 and V2 gelatinizes in the temperature range characteristic for common wheat starches, but is less prone to retrogradation.
- Pasting properties are also in the range for common wheat starches. Starch isolated from Croatian wheat varieties is rather stable during shearing at high temperatures and cooling.
- Although investigated starches didn't differ significantly in thermo-physical and pasting properties, and chemical composition, they have shown large differences in texture. V2 starch had significantly higher gel strength, rupture strength and

adhesiveness than V1 starch. This is very significant, since texture properties may determine different usage of these starches in food industry.

- Swelling power and solubility of both starches increased with temperature increase and weren't significantly different at any temperature of testing.
- V2 starch formed paste with higher clarity than V1 starch. This difference is also important for suitability of utilization of starches in specific food industries (eg. juice & dressing production).
- Resistant starch content is not industrially significant characteristic, but has great importance from nutritional aspect. Both investigated starches had very low resistant starch content which indicated high digestibility.

Acknowledgment

Results shown have outcome from scientific project "Development of new modified starches and their application in food industry" supported by the Ministry of Science, Education and Sports of the Republic of Croatia.

References

http://www.zuckerforschung.at

Adebowale, KO; Adeniyi Afolabi, T; Lawal, OS (2002). Isolation, chemical modification and physico-chemical characterisation of Bambarra groundnut (*Voandzeia subterranean*) starch and flour. *Food Chemistry*, 78, 305-311.

Babić, J; Šubarić, D; Nedić Tiban, N; Kopjar, M (2009). Acetylation and characterization of corn starch. *Journal of Food Science and Technology*, 46(5), 423–426.

Babić, J; Šubarić, D; Nedić Tiban, N; Kopjar, M (2006). The effect of lactose and whey powder on the gelatinization and retrogradation of tapioca starch. *Proceedings of the 4th international symposium on food rheology and structure, ISFRS,* 707-708.

Babić, J (2007). The influence of acetylation and additives on rheological and thermophysical properties of maize and tapioca starch. *Dissertation*. Faculty of Food Technology, Osijek.

BeMiller, J; Whistler, R (2009). *Starch chemistry and technology (3rd Edition)*. Academic Press, New York, SAD, 149 – 188, 441 – 505.

BeMiller, JN; Whistler, RL (1996). Carbohydrates. In: *Food Chemistry*. O.R. Fennema (ed.), MarcelDekker, Inc. New York, 191 – 204.

Bhandari, PN; Singhal, RS (2002). Effect of succinylation on the corn and amaranth starch pastes - review. *Carbohydrate Polymers*, 48, 233-240.

Biliaderis, CG (1998). Structures and phase transitions of starch polymers. In: *Polysaccharide Association Structures in Food*. Marcel Dekker, New York, 57 – 168.

Blazek, J; Salman, H; Lopez Rubio, A; Gilbert, E; Hanley, T; Copeland, L (2009). Structural characterization of wheat starch granules differing in amylose content and functional characteristics. *Carbohydrate Polymers,* 75, 705–711.

Blazek, J; Copeland, L (2009). Effect of monopalmitin on pasting properties of wheat starches with varying amylose content. *Carbohydrate Polymers*, 78, 131-136.

Champ, M; Molis, C; Flourie, B; Bornet, F; Pellier, P; Colonna, P; Galmiche, JP; Rambaud, JC (1998). Small-intestinal digestion of partially resistant cornstarch in healthy subjects. *American Journal of Clinical Nutrition*, 68, 705-710.

Choi, SG (2001). The effect of chemical modification on the molecular mobility, texture, and retrogradation of starch-water suspensions. *Dissertation.* Univeristy of Georgia, Athens, Georgia, SAD.

Chung, KH; Lee, CM (1991). Water binding and ingredient dispersion pattern effects on surimi gel texture. *Journal of Food Science*, 56, 1263-1266.

Cook, SI; Sellin, JH (1998). Review article: short chain fatty acids in health and disease. *Aliment Pharmacology & Therapeutics*, 12, 499-507.

Cui, SW (2005). *Food Carbohydrates*. CRC Press, Boca Raton, SAD.

Dai, ZM; Yin, YP; Zhang, M; Li, WY; Yan, SH; Cai, RG; Wang, ZL (2008). Distribution of Starch Granule Size in Grains of Wheat Grown Under Irrigated and Rainfed ConditionsActa. *Agronomica Sinica* 34(5), 795-802.

De Schrijver, R; Vanhoof, K; Vande Ginste, J (1999). Nutrient utilization in rats and pigs fed Enzyme resistant starch. *Nutrition Research*, 19(9), 1349-1361.

Eerlingen, RC; Cillen, G; Delcour, JA (1994). Enzyme-resistant starch. IV. Effect of endogenous lipids and added sodium dodecyl sulfate on formation of resistant starch. *Cereal Chemistry*, 71(2), 170-177.

Eliasson, AC (2004). *Starch in Food.* Woodhead Publishing Ltd., Cambridge, UK.

Hoover, R; Vasanthan, T (1992). Studies on isolation and characterization of starch from oat (Avena nuda) grains. *Carbohydrate Polymers*, 19, 285–297.

Jouppila, K; Roos, YH (1997). The physical state of amrphous corn starch and its impact on crystallization. *Carbohydrate polymers,* 32, 95-104.

Kim, HS; Huber, KC (2008): Channels within soft wheat starch A- and B-type granules. *Journal of Cereal Science,* 48, 159-172.

Knight, JW; Olson, RM (1984). Wheat starch: production, modification and uses. In: *Starch chemistry and technology (1st Edition).* Academic Press, Orlando, Florida, SAD, 491 – 505.

Kovačević, J. http://www.poljinos.hr/dokumenti/SORTE%20PSENICE%20I%20JECMA.pdf

Li, WY; Yan, SH; Yin, YP; Li, Y; Liang, TB; Gu, F; Dai, ZM; Wang, ZL (2008). Comparison of starch granule size distribution between hard and soft wheat cultivars in Eastern China. *Agricultural Sciences in China,* 7(8), 907-914.

Liu, Q; Thompson, DB (1998). Effects of moisture content and different gelatinization heating temperatures on retrogradation of waxy-type maize starches. *Carbohydrates research,* 314, 221-235.

Luo, FX; Huang, Q; Fu, X; Zhang, LX; Yu, SJ (2009). Preparation and characterisation of cross-linked waxy potato starch. *Food Chemistry,* 115, 563-568.

Majzoobi, M; Radi, M; Farahnaky, A; Jamalian, J; Tongdang, T (2009): Physico-chemical properties of phosphoryl chloride cross-linked wheat starch. *Iranian Polymer Journal,* 18(6), 491-499.

Mohan, BH; Maleshi, NG (2006). Characteristics of native and enzymatically hydrolised common wheat (*Triticum aestivum*) and *diococcum* wheat (*Triticum diococcum*) starch. *Eur Food Res Technol*, 223, 355–361.

Oates, CG (1997). Towards an understanding of starch granule structure and hydrolysis. *Trends in Food Science and Technology,* 8, 375-382.

Ortega, FE; Eliasson, AC (2001). Gelatinisation and retrogradation behaviour of some starch mixtures. *Starch/Starke,* 53, 520-529.

Park, SH; Wilson, JD; Seabourn, BW (2009). Starch granule size distribution of hard red winter and hard red spring wheat: Its effects on mixing and breadmaking quality. *Journal of Cereal Science,* 49, 98-105.

Park, SH; Wilson, JD; Seabourn, BW (2009). Starch granule size distribution of hard red winter and hard red spring wheat: Its effects on mixing and breadmaking quality. *Journal of Cereal Science,* 49, 98-105.

Perry, PA; Donald, AM (2002). The effect of sugars on the gelatinisation of starch. *Carbohydrate polymers,* 49, 155-165.

Raina, CS; Singh, S; Bawa, AS; Saxena, DC (2006). Some characteristics of acetylated, cross-linked and dual-modified Indian rice starches. *Eur Food Res Technol*, 223, 561-570.

Ratnayake, WS; Jackson, DS (2007). A new insight into the gelatinization process of native starches. *Carbohydrate Polymers,* 67, 511-529.

Sajilata, MG; Singhal, RS; Kulkarni, PR (2006). Resistant starch – A Review. *Comprehensive Reviews in Food Science and Food Safety*, 5, 1-17.

Singh, J; Kaur, L; McCarthy, OJ (2007). Factors influencing the physico-chemical, morphological, thermal and rheological properties of some chemically modified starches for food applications - a review. Food Hydrocolloids, 21, 1-22.

Singh, J; Kaur, L; Singh, N (2004). Effect of acetylation on some properties of corn and potato starches. *Starch/Starke,* 56, 586-601.

Singh, N; Singh, J; Kaur, L; Sodhi Singh, N; Singh Gill, B (2003). Morphological, thermal and rheological properties of starches from different botanical sources. *Food Chemistry,* 81, 219–231.

Sodhi, NS; Singh, N: Characteristics of acetylated starches prepared using starches separated from different rice cultivars. *Journal of Food Engineering* 70:117–127, 2005.

Sopade, PA; Halley, PJ; Junming, LL: Gelatinisation of starch in mixtures of sugars II. Application of differential scanning calorimetry. *Carbohydrate Polymers* 58, 311–321, 2004.

Sung, W-C; Stone, M (2003). Characterisation of various wheat starch in pasta development. *Journal of Marine Science and Technology*, 11(2), 61-69.

Svihus, B; Uhlen, AK; Harstad, OM (2005). Effect of starch granule structure, associated components and processing on nutritive value of cereal starch: A review. *Animal Feed Science and Technology,* 122, 303-320.

Van Der Borght, A; Goesaert, H; Veraverbeke, WS; Delcour, JA (2005). Fractionation of wheat flour into starch and gluten: overview of main processes and the factors involved. *Journal of Cereal Science,* 41, 221-237.

Van Hung, P; Maeda, T; Morita, N (2006). Waxy and high-amylose wheat starches and flours—characteristics, functionality and application. *Trends in Food Science and Technology,* 17, 448-456.

Van Hung, P; Morita, N (2005). Physiccochemical properties of hydroxypropylated and cross-linked starches from A-type and B-type wheat starch granules. *Carbohydrate Polymers*, 59, 239-246.

Walter, RH (1998). *Polysaccharide association structures in food.* Marcel Dekker, Inc. New York, SAD.

Wasserman, LA; Signorelli, M; Schiraldi, A; Yuryev, V; Boggini, G; Bertini, S; Fessas, D (2007). Preparation of wheat resistant starch: Treatment of gels and DSC characterization. *Journal of Thermal Analysis and Calorimetry*, 87, 153–157.

Wickramasinghe, HAM; Miura, H; Yamauchi, H; Noda, T (2005). Comparison of the starch properties of Japanese wheat varieties with those of popular commercial wheat classes from the USA, Canada and Australia. *Food Chemistry*, 93, 9-15.

Xie, X; Cui, SW; Liu, Q (2008). Isolation and characterisation of wheat bran starch. *Food Research International*, 41, 882-887.

Yoo, SH; Jane, JL (2002). Structural and physical characteristics of waxy and other wheat starches. *Carbohydrate Polymers,* 49, 297-305.

In: Wheat: Genetics, Crops and Food Production
Editor: Maria T. Almeida, pp. 279-295
ISBN 978-1-61209-307-9

Chapter 9

HANDING WHEAT (*TRITICUM AESTIVUM* L.) UNDER STRESS

Mohammad Miransari*
Professor of Soil Science, Department of Soil Science,
College of Agricultural Sciences, Shahed University, Tehran,
Tehran, Qom Highway, 18151/159, Iran

ABSTRACT

Wheat (Triticum aestivum L.) is an important source of food for people of the world. It is widely planted and is subjected to different stresses. Therefore, it is pertinent to suggest and test methods and/or techniques, which may contribute to enhanced wheat production under stress. For example, a large part of cultivating lands are saline, acidic, and compacted, as a result of using agricultural machinery in the field, or are subjected to drought stress. Methods or techniques, which are tested, must be environmentally and economically friendly and recommendable. There have been biological and molecular methods, which have shown to be promising for wheat cultivation under different stresses. For example, using the symbiotic soil fungi, arbuscular mycorrhiza (AM) or identification and insertion of stress controlling genes in the planted seeds are among such methods. Using AM fungi under compaction stress has been effective on alleviating the stress, or molecular techniques have made possible the use of stress controlling genes for higher production under stress. The most important related findings are reviewed in this article.

Keywords: Wheat (*Triticum aestivum* L.), stress genes, arbuscular mycorrhizal fungi, molecular and biological methods.

* E-mail: Miransari1@gmail.com, Miransari@shahed.ac.ir; Tel: (98)2151212243

INTRODUCTION

Wheat is among the most important crop plants feeding a large number of people in the world. The most important reasons for the popularity of wheat as a source of food are: 1) its morphological and physiological properties, which make it grow in a wide range of soils, and 2) its use as a source of different types of food. Crop production is affected by different parameters including soil and climate properties, plant species, and soil stresses. Soil stresses such as salinity, drought, acidity, and compaction can significantly decrease plant growth and yield production (Auge, 2001; Miransari et al., 2007; 2008; Daei et al., 2009; Miransari, 2010a).

Salinity adversely affects plant growth by influencing osmotic potential in soil and plant, and by increasing the completion between salt ions and the necessary nutrients for plant growth. There is a large area of agricultural soils in the world, which are saline or are subjected to salinity affecting crop production (Munns and Richards, 2007). There are different methods of treating saline soils to enhance their efficiency including soil leaching with irrigation water, use of tolerant plant species and use of biological methods such as soil microbes (Munns and Tester, 2008; Colmer et al., 2006; Jalili et al., 2009; Miransari, 2010a). Plants can use different mechanisms to alleviate salinity stress on their growth including the accumulation of salt ions in their cellular vacuoles, production of solutes to adjust cellular water potential, and salt exudation by leaf or root (James et al., 2006; Cuin et al., 2008; Munns and Tester, 2008).

Soil moisture efficiency is a parameter, highly affecting crop production. The higher the efficiency of soil moisture, the higher the rate of plant growth is. Drought can affect plant growth by influencing soil and plant water potential. Water is necessary for plant functioning because of the followings: 1) uptake of nutrients from soil by plant roots, 2) transfer of soluble compounds in plant, 3) cellular growth and turgor, 4) a medium for biochemical processes, and 5) adjusting plant temperature by evpo-transpiration. Plants can alleviate the adverse effects of drought on their growth by expanding their root growth into soil, production of solutes, decreasing their growth, and reducing evapo-transpiration (Auge, 2001; Blum, 2005).

In humid areas, because of high amounts of precipitation, alkaline cations are leached through the soil profile resulting in the reduction of soil pH and increase in the concentration of elements, such as iron (Fe), manganese (Mn), and aluminum (Al). Under acidity stress, crop plants may show signs of nutrient deficiency resulting in the reduction of crop yield. Low pH can also decrease plant growth through interfering with plant physiological processes as well as plant morphological properties. It is because different plant activities including enzymes activities can be conducted in a certain pH range and other pH's can adversely affect enzyme structure and its activities. The presence of high rate of hydrogen (H^+) as well as Al and Mn can have unfavorable effects on plant growth and morphological properties (Garvin and Carver, 2003; Ernani et al., 2004; Caires et al., 2008).

Soil compaction is resulted by using agricultural machinery in the field, especially at high rate of soil moisture. Soil compaction increases soil bulk density and decreases the rate of soil macropores adversely affecting soil structure and hence plant growth. The exchange of soil gases including O_2, CO_2, nitrogen oxides and methane is influenced by soil compaction indicating its environmental and agricultural significance. Although use of mechanical methods may reduce the unfavorable effects of soil compaction on soil properties and plant growth, they are expensive and laborious and hence not recommendable. Researchers have indicated that use of biological methods such as arbuscular mycorrhizal (AM) fungi can alleviate the stress of soil compaction on plant growth (Miransari et al., 2007; 2008; 2009a,b).

Determination and evaluation of the mechanisms, developed by plant for adaptation and resistance under stress can be very useful to the researchers to find solutions for the alleviation of stress. For example, how it is likely to increase wheat production under different kinds of stresses. Although wheat is not classified among tolerant crop plants to stress, it has some unique morphological and physiological properties, which can help the plant to tolerate stresses up to some extent (Miransari et al., 2008; 2009a; Munns and Tester, 2008).

Wheat, Salinity and Drought

Salinity is common in large areas of the world. It is resulted by low level of precipitation, high rate of evaporation, and the quality of irrigation water as well as soil mineralogical structure. Salt ions can increase soil and plant osmotic tension decreasing the availability of water for plant use. The competition of salt ions and the necessary nutrients for plant growth can also adversely affect plant growth. Due to their electron arrangement, salt ions are absorbed by dipolar water molecules. Such kind of affinity reduces the activity potential of water molecules and hence their availability to plant. Therefore, although there may be enough water, the presence of salt ions can significantly decrease its utilization by plant. There are different parameters, determining plant tolerance to salinity including soil and climate properties, agricultural practices, and plant species (Miransari and Smith, 2007; Munns and Tester, 2008; Daei et al., 2009).

Salinity and drought decrease plant growth by adversely affecting photosynthesis and cell homeostasis and increasing the rate of photorespiration as well as production of reactive oxygen species. Plants define new pathways during stress, which make the plant tolerate the stress. For example, production of reactive oxygen species and their regulation during the stress is among the mechanisms, utilized by plant to tolerate the stress (Miller et al., 2010). There are other mechanisms used by plants under drought or salinity stress to alleviate their adverse effects on plant growth including the adjustment of plant growth stage, water use efficiency, water uptake and light absorption (Reynolds et al., 2005).

Drought stress can adversely affect different plant functioning including water and nutrient uptake. As a result of drought stress, the uptake of water and nutrients by plant decreases resulting in oxidative stress in plant. Excessive absorption of light by drought stressed plants, relative to non stressed plants also increases the rate of evapotranspiration and hence reduces plant growth. Accordingly, plant attempts to decrease the rate of evapotranspiration as a result of stomatal closure negatively affecting the process of photosynthesis. Such controlling affects causes oxidative stress in plant by producing reactive oxygen species, especially hydrogen peroxide (H_2O_2). Hydrogen peroxide is also produced as a result of ABA effects on stomatal closure during drought stress. The process of photorespiration also produces high rates of hydrogen peroxide, through the reaction of glycollate oxidase in peroxisomes (Pei et al., 2000; Noctor et al., 2002; Luna et al., 2005).

Catalase is the enzyme, removing the produced H_2O_2 in proxisomes as a result of photorespiration. However, processes, which adversely affect photorespiration, including the enhanced rate of CO_2, can inhibit catalse activity. In other words, processes inhibiting photosynthesis, such as decreased level of CO_2, high light and temperature, can increase the rate of photorespiration and decreases the activities of enzymes such as catalse. Photorespiration is enhanced under drought, because of stomatal closure and increased level of non-assimilated CO_2 (Luna et al., 2005).

Drought and salinity tolerance are different. Crop tolerance to salinity stress is determined by the regulated uptake of salt and its transport in the roots, which is not the case for drought tolerance (Munns and Richards, 2007). Plant performs mechanisms, which depending on the level of stress can partially or completely control the adverse effects of stress on plant growth. However, such responses may differ with respect to the present conditions (Reynolds et al., 2006). It has been indicated that it is likely to transfer the tolerant genes from the wild types of wheat plant to the domesticated species. For example, there are wild types of wheat and barley, which are able to survive with only 50 mm of precipitation per year. Such wild varieties are extremely tolerant to drought and salt stresses. Hence, identification of tolerant genes, using the new methods, in such plants and their transfer to non- or partially-tolerant species can significantly increase their tolerance to stress (Genc et al., 2007; Nevo and Chen, 2010).

There are two distinct mechanisms by which plant can control salinity stress including rapid and slow phases. In the rapid phase plant control the stress by adjusting the osmotic potential, which results in the growth inhibition of young leaf. In the slow ionic phase, the controlling process is by the senescence of old leaf. In the osmotic phase, there is a threshold level of 40 mM for most plants. During the ionic phase Na^+ is accumulated in the leaf (Munns and Tester, 2008).

There are three different plant responses to salinity stress including the followings. 1) Tolerance to osmotic stress, which regulates cellular growth and expansion under salinity stress by adjusting the osmotic potential, 2) exclusion of Na^+ from the leaf, which is performed by controlling ions movement to the shoots, adjusting the unfavorable concentrations of ions in the chloroplast, accumulating Na^+ into the root vacuoles and altering the movement pathways resulting in the reduction of Na^+ accumulation, and 3) tissue tolerance, which is regulated by controlling ions storage in vacuoles, delayed senescence of old leaf, adjusting the unfavorable concentrations of salt ions in the chloroplast, decreasing the energy for Na^+ exclusion, increasing storage of Na^+ in the leaf vacuole and increasing the concentration of similar solutes in the cytoplasm (Munns and Tester, 2008). The increase in the concentration of K^+ may also be a factor controlling the adverse effects of Na^+ on plant growth under salinity (Zhu, 2002).

There are plants such as soybean (*Glycine max* L.), which are able to accumulate high rates of Cl^- and not Na^+ in their tissues, because in these species Cl^- toxicity is of more importance than Na^+ toxicity. However, it does not imply that Cl^- has more unfavorable effects on plant growth than Na^+, but such species are more efficient in the exclusion of Na^+ than Cl^-. The best way to determine the differences between the toxicity of Na^+ and Cl^- is to investigate the genetic properties of species (Munns and Tester, 2008).

As previously indicated, the genetic differences between for example wheat and soybean make them absorb higher rates of Na^+ and Cl^- under salinity stress, respectively (Munns and Tester, 2008). In addition, different species of a plant can also indicate signs of differences in their response to salinity. For example, bread wheat is more tolerant to salinity than durum wheat, because the former is able to exclude higher rates of Na^+, and hence keep the cytoplasm with the lowest likely concentration of Na^+ (Colmer et al., 2005; James et al., 2006). This is the reason for the more attention of researchers to breeding plants for the ability to exclude higher rates of Na^+ under salinity.

Although Na^+ exclusion is considered important for the alleviation of salt stress on plant growth, the role of K^+ has also been investigated as a likely reason for alleviating salinity stress on plant growth. Cuin et al. (2008) indicated that under salinity flux of K^+ from the roots of wheat seedlings significantly affected plant morphological and physiological properties. Hence, K^+ homeostasis is also another important mechanism by which plant can tolerate salinity stress by adjusting the ratio of K^+/Na^+ (Maathuis and Amtmann, 1999). However, excess absorption of Na^+ in cytoplasm may result in the membrane leakage of K^+, as a result of membrane depolarization by NaCl, adversely affecting the K^+/Na^+ ratio (Shabala et al., 2003).

Under drought, water deficiency can significantly decrease plant growth. Accordingly, water use efficiency can be important to plane growth under drought, because plant with higher rates of water efficiency can grow more efficiently under drought. The mechanism by which plants can survive under drought is dehydration avoidance. Plants can avoid dehydration by the followings. 1) Higher water uptake, 2) adjusting the process of avapotranspiration, and 3) maintenance of cellular hydration under drought stress. Plants can increase their water use efficiency by decreasing their growth (Blum, 2005).

Durum wheat is a tolerant species to drought stress. Among the most cellular components, which are able to assist plant during stress is mitochondria. In combination with chloroplasts and peroxisomes and by regulating the process of reduction, mitochondria prevent the excess production of electron carriers during the photoespiratory cycle and by photosynthesis process (Kromer, 1995). Mitochondria are able to alleviate stresses on plant growth through regulating the homeostasis of cell redox or in other words diminishing the produced reactive oxygen species by using the following mechanisms: 1) the potassium channel in mitochondria, 2) the uncoupling protein mechanism and 3) alternated oxidase (Dutilleul et al., 2003; Pastore et al., 2007).

The enzyme, proline dehydrogenase, which is responsible for the production of proline (osmoprotectant) is localized in the mitochondrion membrane. Respiration can also influence the production of ascrobate, which can effectively detoxify the produced reactive oxygen species (Milar et al., 2003; Pastore et al., 2007). In addition, mitochondria can also regulate cell activities under stress through signaling with chloroplast and by altering the expression of genes located in the nucleus (Millar et al., 2001; Rhoads et al., 2006). Proteomic analyses has indicated that under drought and salinity stress, mitochondria is the place of producing reactive oxygen species and hence the related unfavorable effects (Bartoli et al., 2004; Taylor et al., 2005; Pastore et al., 2007).

Use of Microbes Under Drought and Salinity

As previously indicated, precise investigation of plant response under drought and salinity can be useful for the development of methods, which can be used to alleviate the stress. For example, indication of genes, which are expressed under drought and salinity stress for inbreeding plants, can be useful under stress. In addition, use of soil microbes including arbuscular mycorrhizal (AM) fungi and plant growth promoting rhizobacteria (PGPR) such as N-fixing bacteria, *rhizobium*, and *Azospirillum* spp. has also been proved to be effective on the alleviation of different stresses such as drought and salinity on plant growth (Miransari and Smith, 2007; 2008; 2009; Daei et al., 2009; Arzanesh et al., 2009; 2010).

AM fungi are among the most efficient soil microbes alleviating different stresses including salinity and drought. AM fungi can develop symbiotic association with most terrestrial plants and provide water and nutrients for their host plant in the exchange for carbon. The extensive hyphal network, which is the result of spore germination, can grow into the root cortical cells and absorb water and nutrient from the surrounding soil. AM hypha can also develop arbuscules, the branched hyphal structure, as the interface for the exchange of water and nutrients between plant roots and soil, and vesicles, which are the storage tissues (Smith and Read, 2008).

Most of the experiments regarding the alleviating effects of AM species on plant growth under stress have been performed under greenhouse conditions and there are little data under field conditions. Daei et al. (2009) tested the effects of different AM species on the nutrient uptake, growth and yield of different wheat cultivars under field saline conditions. The soil salinity of 7.41 dS/m and the AM species of *Glomus etunicatum*, *G. mosseae* and *G. intraradices* and wheat genotypes of Roshan, Kavir, and a mutated line of Tabasi were tested. The salinity of the irrigation water was also equal to 13.87 dS/m. Growth and nutrient uptake, of the mutated Tabasi was significantly affected by AM species under salinity, compared with other genotypes particularly Kavir genotype.

The lowest concentration of Na^+ and Cl^- was related to the mycorrhizal Tabasi genotype. *G. etunicatum* was the most effective species under salinity, followed by *G. mosseae* and *G. intraradices*. The highest (42.08%) and the lowest (7.55%) wheat dry weight was resulted by the symbiosis between *G. etunicatum* and *G. intraradices*, respectively. Similarly, *G. etunicatum* and *G. intraradices* produced the highest (38.3%) and the lowest (4.5%) grain dry weight, respectively. Accordingly, it may be mentioned that some symbiotic association between specific AM species and wheat cultivars may be more efficient under stresses such as salinity as *G. etunicatum* was the most efficient species in this experiment, compared with other species.

AM fungi are also able to alleviate the stress of drought on plant growth and yield production. As previously mentioned, AM fungi are able to significantly increase plant water and nutrient uptake under different conditions including stress. Different experiments have indicated the alleviating effects of AM fungi on the growth of different plant species under drought (Miransari, 2010) including corn (Sylvia et al., 1993; Subramanian et al., 1995), soybean (Bethlenfalvay et al., 1988), wheat, onion, lettuce and other crop plants, which has been completely reviewed by Auge (2001).

The most important cause for the alleviating effects of AM fungi on plant growth under drought stress is probably the osmotic adjustment. In comparison with non-mycorrhizal plants, mycorrhizal plants are able to accumulate organic compounds such as proline and carbohydrates as well as mineral ions under drought stress (Azcon et al., 1996; Goichea et al., 1998; Ruiz-Lozano, 2003; Ruiz-Lozano et al., 2006). The improving effects of AM fungi on soil structure can also contribute to the alleviation of AM fungi on drought stress. AM fungi are able to produce a glycoprotein called glomalin, binding soil particles and hence improving soil structure (Auge, 2001; Ruiz-Lozano, 2003; Auge et al., 2004; Rillig, 2004).

In addition, the symbiotic association between AM fungi and the host plant can enhance plant tolerance to drought stress by altering plant physiological properties as well as gene expression (Ruiz-Lozano et al., 2006; Aroca et al., 2008; Boomsma and Vyn, 2008; Miransari et al., 2008). Mycorrhizal plants are also able to produce antioxidant enzymes enhancing their tolerance and hence their growth under stress (Ruiz-Lozano, 2003).

Miransari and Smith (2007; 2008; 2009) tested the process of N-fixation by *Bradyrhizobium japonicum* and soybean (*Glycine max* L.) under different stresses including salinity, acidity and suboptimal root temperature. They hypothesized that under stress the process of signal molecule (genistein) exchange between the bacteria and the host plant is disrupted resulting in the inhibition of N-fixation process between the two symbionts. Accordingly, they preincubated the bacterial culture with different concentrations of genistein at μM level and found that addition of genistein could partially or completely alleviate the effects of different stresses on the process of N-fixation. Genistein, which is produced by the roots of the legume host, is able to activate the *Nod* genes in the bacteria, which in response produces the *Nod* factors. *Nod* factors are able to induce morphological changes in the roots, especially, in hair roots of the host plant including root curling and root bulging. Eventually as a result of infection tread production the bacterium enter the nodules, where they fix atmospheric N.

The effects of *Azospirillum* spp. on wheat growth was also tested by Arzanesh et al. (2009; 2010). Inoculation of wheat plants with *Azospirillum* spp. enhanced wheat growth under drought. In their first experiment they used different strains of *Azopirillum* spp. including 25 isolated ones and four standard strains of *A. halopreaferanse*, *A. brasilense*, *A. irakense* and *A. lipoferum*. In their second experiment they tested three different strains of *A. lipoferum* to inoculate wheat plants under different drought intensities and longevity.

The plant growth promoting effects of different bacterial strains were also measured. Strain B3 with the ability to fix N and produce the highest amount of N and auxin, and with P solublization and ACC deaminase activities was the most efficient strain increasing wheat yield at the highest level of drought by 109%. However, the most resistant strain under drought was strain B9 (siderophore producer). Accordingly, the results of these experiments in combination with the other experiments indicate that *Azospirillum* spp. is able to alleviate the unfavorable effects of drought on wheat growth under different physiological and ecological conditions.

AM fungi and soil microbes interactions in the soil can affect their functionality. This indicates that at the time of inoculating seeds with inoculums, the soil microbes can influence the efficiency of inoculum. Hence, when using inoculum under different conditions including stress, evaluation of soil biological properties (especially in soils with significant microbial activities) or activity of soil microbes can be a determining parameter in effectiveness of inoculum. Accordingly, when isolating soil microbes for use under stress, it is more beneficial that soil microbes be isolated from stressed soils. Under such conditions, the soil microbes that are used for inoculum are more tolerant to the stress, because they are adapted to the stress. Under stress soil microbes develop mechanisms, which can enhance their ability to tolerate the stress (Miransari, 2010 a,b,c).

Wheat and Soil Compaction

Soil compaction is a result of using agricultural machinery in the field, especially at high soil moisture. In a compacted soil, bulk density increases limiting root growth and hence the uptake of water and nutrients and subsequent decreased plant growth (Pereira et al., 2007). Root growth is a function of soil properties affecting the interactions between roots and shoots including the signaling process. Under compaction the signaling between roots and shoots can limit root growth and hence uptake of water and nutrients. Root length is determined by parameters such as turgor pressure, yield threshold and the growth of cell wall (Munns and Cramer, 1996; Passiora, 2002).

Under compaction soil structure is adversely affected influencing different soil chemical, physical and biological properties. The binding of soil particles, soil porosity and hence the flux of water and gases in the soil are among soil physical properties affected by compaction. Soil compaction can also influence soil chemical properties such as soil acidity and the emission of greenhouse gases including carbon dioxide, methane and nitrogen oxides, to the atmosphere. Such a phenomenon is of environmental and agricultural importance. Because the emission of greenhouse gases to the atmosphere can affect the atmospheric climate and the gases combination. In addition, the emission of greenhouse gases, which include nutrients such as C and N, can also affect soil fertility and soil structure (Miransari et al., 2007; 2008; 2009a, b).

Due to the process of N denitrification in a compacted soil, use of N fertilization is necessary to supply adequate N for plant growth and yield production. Under compaction, root growth is less affected by shoot growth because of the followings. 1) Under stress plant allocate more C to the roots and AM fungi are able to intensify such allocation, resulting in enhanced root growth (Miransari and Smith 2007; 2008; 2009; Miransari et al., 2007; 2008), and 2) decreased N uptake under compaction stress reduces the production of cytokinins in the roots and their subsequent transfer to the shoots. When the amounts of cytokinins in the shoots is reduced, the rate of cell division also decreases, however in roots; such a reduction can neutralize the suppressing effects of cytokinins on cell division (Van der Werf and Nagel, 1996).

AM fungi have some unique properties, which can assist the plant to survive under stress. As previously mentioned, they develop a very extensive hyphal network with arbuscules and vesicles. Due to their fine diameter, AM hypha are able to grow even in the finest soil micropores, which is not accessible by root hairs, and absorb water and nutrients (Jakobsen, 1995). This can be beneficial in a compacted soil, where most part of soil porosity is devoted to the soil micropores. Furthermore, AM hypha are able to produce glomalin, which is a glycoprotein and can bind soil particles improving soil structure and contributing to soil C and hence soil productivity (Wright and Upadhyaya, 1996; Rillig, 2004). Under stress there is some kind of signal communications between AM fungi and the host plant, which can contribute to the alleviation of stress.

Roots have the following main functions in the soil: 1) maintaining plant in the soil, 2) nutrient and water uptake, and 3) interaction with soil microbes. However, there also some other functions for the roots in soil including root growth in the soil, fluctuation of soil moisture, rhizodeposition, contributing to soil organic matter by degradation and improving soil structure by binding soil particles (Rillig, 2004). AM may affect such functions by affecting plant growth (root/shoot ratio). Usually under compaction root growth is more affected by AM fungi than shoot growth, because of the enhanced P uptake (Miransari et al., 2007; 2008).

Plant hormonal balance is also important under stress. Through controlling meristem activities, plant hormones including auxin (usually produced in shoot meristems), cytokinins (usually produced in root meristems) and gibberellins (usually produced in differentiating tissues) can regulate plant response under stress (Amzallag, 1999; Dodd, 2005). For example, under P deficient conditions auxin, ethylene and cytokinins are able to alter root morphological properties by increasing the growth of root hairs and production of cluster roots. Auxin is also able to alter root exudates and hence P mobilization in soil (Wittenmayer and Merbach, 2005). AM fungi may also affect such hormonal balance in plant under stress (Miransari et al., 2009a).

Under P deficient conditions, the roots, which are distributed on the soil surface and shallower depth, are more efficient in absorbing P from the soil. According to Zhu et al. (2005) AM fungi were able to change root structure in plant species with deep rooting, make it shallower. This can also be the case under soil compaction (Miransari et al., 2009a). Interestingly, with increasing compaction, soil P and Zn were the most sensitive nutrients to the stress (Barzegar et al., 2006; Miransari et al., 2009a), however, AM species were able to enhance their uptake by plant.

Different experiments have indicated the enhancing effects of AM species on wheat (*Triticum aestivum* L.) growth (Vierheilig and Ocampo, 1991; Schweiger and Jakobsen, 1999). AM fungi are able to influence plant metabolism and physiology (Nemec and Meredith, 1981; Gupta, 2003) by enhancing photosynthetic rate and nutrient uptake and altering the allocation of photosynthates in plant roots and shoots. Such alterations result in structural and biochemical changes in the root cellular structure including membrane permeability and hence the quality and quantity of root exudates (Linderman, 1992; Harrison, 1999). Interestingly, under stress, AM fungi are able to alter plant physiology in a way that host plant can handle the stress (Miransari et al., 2008).

Inoculation of wheat with AM species significantly increased wheat nutrient uptake, growth and yield under compaction, relative to the control treatment. However, with increasing the level of compaction the effectiveness of AM fungi decreases as at high compaction levels the symbiotic association between AM fungi and the host plant may turn from mutualism to parasitism (Miransari et al., 2006; 2008). Brown et al. (2006) indicated that with increasing compaction the expression of phosphate transporter genes decreases.

P is among the most important macro-nutrients and hence plant use different strategies to supply their necessary P for their growth including the followings. 1) Symbiosis with AM fungi, 2) changes in root architecture (Lupez-Bucio et al., 2003), 3) alteration in root exudates (Neumann et al., 2000), and 4) adjustment of P transferring methods (Smith and Gianinazzi-Pearson, 1988; Marschner, 2005).

Comparatively mentioning, Miransari et al. (2009a, b) found that corn is more sensitive to nutrient uptake under compaction stress, relative to wheat. This can be due to the more demanding proprieties of corn and the differences in the root architecture of the two plants. Corn plants develop a more extensive root network, compared with wheat plants. As a result the growth of corn roots in a compacted soil, where most of the porosity is devoted to micropores is more demanding because more energy must be spent by the plant and hence higher rate of nutrients must be absorbed by plant. However, use of AM fungi under compaction stress, can effectively alleviate the stress of soil compaction on the nutrient uptake, growth and yield of corn and wheat (Miransari et al., 2007; 2008; 2009a, b).

WHEAT AND ACIDITY

Similar to salinity, acidity is also common in different areas of the world, especially humid areas. Soil acidity is a result of high rate of precipitation and/or the properties of soil forming materials. High amounts of rain leach away soil alkaline cations such as Ca^+ and Mg^+ increasing the amount of H^+ in soil. Additionally, high amounts of rain speed up the rate of soil weathering producing 1:1 clay minerals as well as Al and Fe hydroxides, which are also sources of H^+ in soil.

In addition to the unfavorable effects of H^+ on soil properties and growth and activity of plants and soil microbes, Al^{3+} toxicity can also adversely affect soil productivity in acidic soils (Sumner et al., 1986; Marsh and Grove, 1992). The adverse effects of low pH on soil productivity are through its unfavorable effects on plant roots (Caires et al., 2008) and on the biological activities including enzyme activities. Soil acidity can also adversely affect plant growth and microbial activities by influencing the process of nutrient uptake. Under low pH, the solubility of elements such as Fe, Mn and Al increases, negatively influencing the uptake of other necessary nutrients for plant growth and microbial activity. As a result of high concentrations of Fe, Mn and Al in plant, signs of toxicity appear in the plant (Marschner, 1995).

To adjust the problem of low pH, usually lime ($CaCO_3$) is applied to the soil. Lime is able to produce hydroxyl ions and Ca^{2+}, when hydrolysed by water in soil, replacing the acidic ions, such as H+, Fe, Mn and Al on the soil colloids. Hydroxyl ions can affect soil pH, and Ca^{2+} ions can affect both soil pH and soil structure. Calcium can significantly affect the formation of soil aggregates and hence soil structure. This is the reason for the unfavorable soil structural properties in humid area or highly alkaline soil, dominated by Na^+. The amount of lime necessary for pH adjustment is determined by soil type, soil pH, climate, plant species, use of chemical fertilization and agricultural practices (Oliveira and Pavan, 1996; Ernani et al., 2004; Caires et al., 2005). Briefly, in humid areas, where soils are acidic, the availability and solubility of nutrients is subjected to processes such as leaching (NO_3^-), fixation (PO_4^{3-}), and denitrification (N_2). High amounts of rain result in NO_3^- leaching from the soil profile and hence contamination of ground water. High concentrations of Fe and Al in soil cause the fixation of soil PO_4^{3-} (Marschner, 1995).

The most wide spread element in soil is Al, released into the soil solution at pH < 5. The uptake of exchangeable Al by plant roots can adversely affect root and hence plant growth. Al toxicity is common in 30 to 40% of crop lands (von Uexküll and Mutert, 1995). The interesting point about Al and wheat is that it is possible to genetically modify wheat to resist Al toxicity (Garvin and Carver, 2003). Accordingly, compared with other methods of ameliorating soil acidity such as liming, use of inbred wheat cultivars with Al resistance under acidic conditions, have proved to be the most efficient method. It is not yet known that how many genetic loci control resistance to Al in wheat.

Use of transgenic plants may be useful to ameliorate the stress of acidity on plant growth and yield production. There are some plant species, which are genetically resistant to the unfavorable effects of acid soils on plant growth. For example, high Al concentration can adversely affect the growth of plant root and hence water and nutrient uptake. Such tolerant species of plants can either exclude Al from plant or tolerate its high concentration in plant. The transfer of such trait to other plant species can increase their tolerance under acidity (von Uexküll and Mutert, 1995; Garvin and Carver, 2003). Accordingly, the genes responsible for plant resistance to Al toxicity have been isolated (Sasaki et al., 2004; Furukawa et al., 2007; Magalhaes et al., 2006; 2007; Wang et al., 2007) and the possibility of overexpressing genes, which increase plant tolerance to Al toxicity, has been indicated (Ezaki et al., 2000; Basu et al., 2001; Tesfaye et al., 2001; Ryan et al., 2007).

The expression of two families of the genes, responsible for plant tolerance to Al toxicity, produce membrane proteins, activated in the presence of Al, and result in the exudation of organic acids by plant roots (Delhaize et al., 2007). Such root exudation can neutralize the toxic effects of Al on plant growth by chelating Al in a wide range of plant species (Matsumoto, 2000; Ryan et al., 2001; Kochian et al., 2004). Delhaize et al. (2009) indicated that expression of *TaALMT1*, which is a a malate transporter gene and causes Al resistance in wheat (Sasaki et al., 2004; Zhou et al., 2007) enhances barley ability to absorb P in acidic soils. Using the hemotaxolin stain method, Zhou et al., (2007) identified two genetic loci (QTL) responsible for Al resistance in wheat. One was located on chromosome 4D and the other on chromosome 3BL accounting for about 57% of variation related to hemotaxolin stain method and 50% related to plant root growth. These loci's can be used to improve Al resistance in wheat.

Conclusion

Different methods have been suggested to handle wheat plant under stress including soil amelioration by for example liming, use of breeding techniques and soil microbes. However, the latter two have been proved to be the most efficient ones agriculturally and economically. Although wheat is not considered as a tolerant crop plant to stress, it has some unique cellular, morphological and physiological properties that makes it suitable for breeding under stress. For example, its genetical modification for Al resistance can make the plant grow under acidic conditions. Root symbiotic or non-symbiotic association with soil microbes including AM fungi, and PGPR such as *rhizobium* and *Azospirillum* spp. can also significantly improve wheat potential to grow under stress. Therefore, future research may concentrate more precisely on breeding methods and the ability of soil microbes to handle wheat under stress.

References

Amzallag, G.N., 1999. Regulation of growth: the meristem network approach. *Plant, Cell and Environment*, 22, 483–493.

Aroca, R., Vernieri, P., Ruiz-Lozano J.M. 2008. Mycorrhizal and non-mycorrhizal *Lactuca sativa* plants exhibit contrasting responses to exogenous ABA during drought stress and recovery. *Journal of Experimental Botany*, 8, 2029–2041.

Arzanesh, M.H., Alikhani, H.A., Khavazi, K. Rahimian, H.A., Miransari M. 2009. In vitro growth of wheat (*Triticum aestivum* L.) seedlings, inoculated with *Azospirillum* sp., under drought stress. *International Journal of Botany*, 5, 244-249.

Arzanesh, M.H., Alikhani, H.A., Khavazi, K., Rahimian, H.A., Miransari, M. 2010. Wheat (*Triticum aestivum* L.) growth enhancement by *Azospirillum* spp. under drought stress. *World Journal of Microbiology and Biotechnology*, in press.

Auge, R.M. 2001. Water relations, drought and vesicular arbuscular mycorrhizal symbiosis. Mycorrhiza 11, 3–42.

Auge, R.M., Sylvia, D.M., Park, S.J., Buttery, B.R., Saxton, A.M., Moore, J.L., Cho, K. 2004. Partitioning mycorrhizal influence on water relations of *Phaseolus vulgaris* into soil and root components. *Canadian Journal of Botany*, 82, 503–514.

Azcon, R., Gomez, M., Tobar, R.M. 1996. Physiological and nutritional responses of *Lactuca sativa* to nitrogen sources and mycorrhizal fungi under drought. *Biology and Fertility of Soils*, 22, 156–161.

Bartoli, C.G., Gomez, F., Martinez, D.E., Guiamet, J.J. 2004. Mitochondria are the main target for oxidative damage in leaves of wheat (*Triticum aestivum* L.). *Journal of Experimental Botany*, 55, 1663–1669.

Barzegar, A.R., Nadian, H., Heidari, F., Herbert, S.J., Hashemi, A.M., 2006. Interaction of soil compaction, phosphorus and zinc on clover growth and accumulation of phosphorus. *Soil and Tillage Research*, 87, 155–162.

Basu, U., Good, A.G. and Taylor, G.J. 2001. Transgenic *Brassica napus* plants overexpressing aluminium-induced mitochondrial manganese superoxide dismutase cDNA are resistant to aluminium. *Plant, Cell and Environment,* 24, 1278–1269.

Bethlenfalvay, G.J., Brown, M.S., Ames, R.N., Thomas, R.E. 1988. Effects of drought on host and endophyte development in mycorrhizal soybeans in relation to water use and phosphate uptake. *Physiologia Plantarum*, 72, 565–571.

Blum, A. 2005. Drought resistance, water-use efficiency, and yield potential—are they compatible, dissonant, or mutually exclusive? Australian Journal of Agricultural Research, 56, 1159–1168.

Boomsma, C., Vyn, T. 2008. Maize drought tolerance: potential improvements through arbuscular mycorrhizal symbiosis? *Field Crops Research*, 108, 14–31.

Brown, D.A., Clark, L.J., Howarth, J.R., Parmar, S., Hawkesford, M., 2006. Mechanical impedance and nutrient acquisition in rice. *Plant and Soil*, 280, 65–76.

Caires, E.F., Alleoni, L.R.F., Cambri, M.A., Barth, G., 2005. Surface application of lime for crop grain production under a no-till system. *Agronomy Journal*, 97, 791–798.

Caires, E., Garbuio, F., Churka, S., Barth, G., Correa, J. 2008. Effects of soil acidity amelioration by surface liming on no-till corn, soybean, and wheat root growth and yield. *European Journal of Agronomy*, 28, 57–64.

Colmer, T.D., Flowers, T.J., Munns, R. 2006. Use of wild relatives to improve salt tolerance in wheat. Journal of Experimental Botany 57, 1059–1078.

Cuin, T., Betts, S., Chalmandrier, R., Shabala, S. 2008. A root's ability to retain K^+ correlates with salt tolerance in wheat. *Journal of Experimental Botany*, 59, 2697–2706.

Daei, G., Ardakani, M., Rejali, F., Teimuri, S., Miransari, M. 2009. Alleviation of salinity stress on wheat yield, yield components, and nutrient uptake using arbuscular mycorrhizal fungi under field conditions. *Journal of Plant Physiology*, 166, 617-625.

Delhaize, E., Gruber, B.D. and Ryan, P.R. 2007. The roles of organic anion permeases in aluminum resistance and mineral nutrition. *FEBS Letter*, 581, 2255–2262.

Delhaize, E., Taylor, P., Hocking, P., Simpson, R., Ryan, P., Richardson, A. 2009. Transgenic barley (*Hordeum vulgare* L.) expressing the wheat aluminium resistance gene (*TaALMT1*) shows enhanced phosphorus nutrition and grain production when grown on an acid soil. *Plant Biotechnology Journal*, 7, 391-400.

Dodd, I.C., 2005. Root-to-shoot signaling: assessing the roles of 'up' in the up and down world of long-distance signaling In *Planta. Plant and Soil*, 274, 251–270.

Dutilleul, C., Garmier, M., Noctor, G., Mathieu, C., Chetrit, P., Foyer, C.H., de Paepe, R. 2003. Leaf mitochondria modulate whole cell redox homeostasis, set antioxidant capacity, and determine stress resistance through altered signaling and diurnal regulation. *The Plant Cell*, 15, 1212–1226.

Ernani, P.R., Ribeiro, M.F.S., Bayer, C., 2004. Chemical modifications caused by liming below the limed layer in a predominantly variable charge acid soil. *Communications in Soil Science and Plant Analysis*, 35, 889–901.

Ezaki, B., Gardner, R.C., Ezaki, Y. and Matsumoto, H. 2000. Expression of aluminum-induced genes in transgenic Arabidopsis plants can ameliorate aluminum stress and/or oxidative stress. *Plant Physiology*, 122, 657–666.

Furukawa, J., Yamaji, N., Wang, H., Mitani, N., Murata, Y., Sato, K., Katsuhara, M., Takeda, K. and Ma, J.F. 2007. An aluminumactivated citrate transporter in barley. *Plant and Cell Physiology,* 48, 1081–1091.

Garvin, D.F. and Carver, B.F. 2003. Role of the genotype in tolerance to acidity and aluminum toxicity. In: Handbook of Soil Acidity (Rengel, Z., ed.), pp. 387–406. New York: Marcel Dekker.

Genc, Y., Mcdonald, G., Tester, M. 2007. Reassessment of tissue Na+ concentration as a criterion for salinity tolerance in bread wheat. *Plant, Cell and Environment*, 30, 1486–1498.

Goicoechea, N., Szalai, G., Antolin, M.C., Sanchez-Diaz, M., Paldi, E. 1998. Influence of arbuscular mycorrhizae and Rhizobium on free polyamines and proline levels in water-stressed alfalfa. *Journal of Plant Physiology*, 153, 706–711.

Gupta, S., 2003. Chemotactic response of plant-growth-promoting bacteria towards roots of vesicular–arbuscular mycorrhizal tomato plants. *FEMS Microbiological Ecology*, 45, 219–227.

Harrison, M.J., 1999. Molecular and cellular aspects of the arbuscular mycorrhizal symbiosis. *Annual Reviews in Physiology and Plant Molecular Biology*, 50, 361–389.

Jalili, F., Khavazi, K., Pazira, E., Nejati, A., Asadi Rahmani, H., Rasuli Sadaghiani, H., Miransari, M. 2009. Isolation and characterization of ACC deaminase producing fluorescent pseudomonads, to alleviate salinity stress on canola (*Brassica napus* L.) growth. *Journal of Plant Physiology*, 166, 667-674.

James, R.A., Munns, R., von Caemmerer, S., Trejo, C., Miller, C., Codon, T. 2006. Photosynthetic capacity is related to the cellular and subcellular partitioning of Na^+, K^+, and Cl^- in saltaffected barley and durum wheat. *Plant, Cell and Environment*, 29, 2185–2197.

Jakobsen, I., 1995. Transport of phosphorous and carbon in VA mycorrhiza. In: Varma, A., Hock, B. (Eds.), Mycorrhiza Structure, Function, Molecular Biology and Biotechnology. Springer-Verlag, Berlin, pp. 297–324.

Kochian, L.V., Hoekenga, O.A. and Piñeros, M.A. 2004. How do crop plants tolerate acid soils? Mechanisms of aluminum tolerance and phosphorus efficiency. *Annual Review of Plant Biology*, 55, 459–493.

Kromer S. 1995. Respiration during photosynthesis. *Annual Review of Plant Physiology and Plant Molecular Biology*, 46, 45–70.

Linderman, R.G., 1992. Vesicular–arbuscular mycorrhizae and soil microbial interactions. In: Bethlenfalvay, G.J., Linderman, R.G. (Eds.), Mycorrhiza in Sustainable Agriculture. American Society of Agronomy, Madison, WI, pp. 45–70.

Lopez-Bucio, J., Cruz-Ramirez, A., Herrera-Estrella, L., 2003. The role of nutrient availability in regulating root architecture. *Current Opinion in Plant Biology*, 6, 280–287.

Luna, C., Pastori, G., Driscoll, S., Groten, K., Bernard, S., Foyer, C. 2005. Drought controls on H2O2 accumulation, catalase (CAT) activity and CAT gene expression in wheat. *Journal of Experimental Botany*, 56, 417–423.

Maathuis, F.J.M., Amtmann, A. 1999. K^+ nutrition and Na^+ toxicity: The basis of cellular K^+/Na^+ ratios. *Annals of Botany*, 84, 123–133.

Magalhaes, J. 2006. Aluminum tolerance genes are conserved between monocots and dicots. *Proceeidng of the National Academy of Sciences, USA,* 103, 9749–9750.

Magalhaes, J.V., Liu, J., Guimaraes, C.T., Lana, U.G.P., Alves, V.M.C., Wang, Y.-H., Schaffert, R.E., Hoekenga, O.A., Pineros, M.A., Shaff, J.E., Klein, P.E., Carneiro, N.P., Coelho, C.M., Trick, H.N. and Kochian, L.V. 2007. A gene in the multidrug and toxic compound extrusion (MATE) family confers aluminum tolerance in sorghum. *Nature Genetics*, 39, 1156–1161.

Marschner, H., 1995. Mineral Nutrition of Higher Plants. Academic Press, London, UK.

Marsh, B.H., Grove, J.H., 1992. Surface and subsurface soil acidity: soybean root response to sulfate-bearing spent lime. *Soil Science Society of America Journal*, 56, 1837–1842.

Matsumoto, H. 2000. Cell biology of aluminum toxicity and tolerance in higher plants. *International Reviews of Cytology*, 200, 1–46.

Millar, H., Considine, M,J,, Day, D.A., Whelan, J. 2001. Unraveling the role of mitochondria during oxidative stress in plants. IUBMB Life 51, 201–205.

Millar, A., Mittova, V., Kiddle, G., Heazlewood, J.L., Bartoli, C.G., Theodoulou, F.L., Foyer, C.H. 2003. Control of ascorbate synthesis by respiration and its implications for stress responses. *Plant Physiology*, 133, 443–447.

Miller, G., Suzuki, Ciftci-Yilmazi, S., Mittler, R. 2010. Reactive oxygen species homeostasis and signaling during drought and salinity stresses..*Plant, Cell and Environment*, 33, 453–467.

Miransari, M., Smith, D.L., 2007. Overcoming the stressful effects of salinity and acidity on soybean [*Glycine max* (L.) Merr.] nodulation and yields using signal molecule genistein under field conditions. *Journal of Plant Nutrition*, 30, 1967-1992.

Miransari, M., H.A. Bahrami, F. Rejali, M. J. Malakouti, H. Torabi. 2007. Using arbuscular mycorrhiza to reduce the stressful effects of soil compaction on corn (*Zea mays* L.) growth. *Soil Biology and Biochemistry*, 39, 2014-2026.

Miransari, M., Smith, D.L., 2008. Using signal molecule genistein to alleviate the stress of suboptimal root zone temperature on soybean-Bradyrhizobium symbiosis under different soil textures. *Journal of Plant Interactions*, 3, 287-295.

Miransari, M., Bahrami, H.A., Rejali, F., Malakouti, M.J., 2008. Using arbuscular mycorrhiza to reduce the stressful effects of soil compaction on wheat (*Triticum aestivum* L.) growth. *Soil Biology and Biochemistry*, 40, 1197-1206.

Miransari M., Smith, D. 2009. Alleviating salt stress on soybean (*Glycine max* (L.) Merr.) - Bradyrhizobium japonicum symbiosis, using signal molecule genistein. *European Journal of Soil Biology*, 45, 146-152.

Miransari M., Rejali F., Bahrami H.A., Malakouti, M.J. 2009a. Effects of soil compaction and arbuscular mycorrhiza on corn (*Zea mays* L.) nutrient uptake. *Soil and Tillage Research*, 103, 282-290.

Miransari M., Rejali F., Bahrami H.A., Malakouti, M.J. 2009. Effects of arbuscular mycorrhiza, soil sterilization, and soil compaction on wheat (*Triticum aestivum* L.) nutrients uptake. *Soil and Tillage Research*, 104, 48-55.

Miransari, M. 2010a. Contribution of arbuscular mycorrhizal symbiosis to plant growth under different types of soil stresses. Review article. *Plant Biology*, 12, 563-569.

Miransari, M., 2010b. Arbuscular Mycorrhiza and Soil Microbes. In: Mycorrhizal Biotechnology, Thangadurai, D., Busso, C.A., Hijri, M. (Eds.), Science Publishers, CRC Press and Taylor and Francis Group, 226p. USA.

Miransari, M. 2010c. Biological Fertilization. In: A. Méndez-Vilas, A. (Ed.) Current Research, Technology and Education Topics in Applied Microbiology and Microbial Biotechnology. Microbiology Book Series – 2010 Edition, Spain.

Munns, R., Cramer, G.R., 1996. Is coordination of leaf and root growth mediated by abscisic acid? *Plant and Soil*, 185, 33–49.

Munns, R., Richards, R.A. 2007. Chapter 22 Recent Advances In Breeding Wheat For Drought and Salt Stresses. In Advances in Molecular Breeding toward Drought and Salt Tolerant Crops (eds. M.A. Jenks, et al.), pp. 565–585. Springer-Verlag, Berlin.

Munns, R., Tester, M., 2008. Mechanisms of salinity tolerance. *Annual Review of Plant Biology*, 59, 651–81.

Nemec, S., Meredith, F.I., 1981. Amino acids content of leaves in mycorrhizal and nonmycorrhizal citrus root stocks. Annals of Botany 51, 351.

Neumann, G.,Massonneau, A., Langlade, N., Dinkelaker, B., Hengeler, C., Romheld, V., Martinoia, E., 2000. Physiological aspects of cluster root function and development in phosphorus-deficient white lupine (*Lupinus albus* L.). *Annals of Botany*, 85, 909–919.

Nevo, E., Chen, G. 2010. Drought and salt tolerances in wild relatives for wheat and barley improvement. *Plant, Cell and Environment,* 33, 670–685.

Noctor, G., Veljovic-Jovanovic, S.D., Driscoll, S., Novitskaya, L., Foyer, C.H. 2002. Drought and oxidative load in wheat leaves. A predominant role for photorespiration? *Annals of Botany*, 89, 841–850.

Oliveira, E.L., Pavan, M.A., 1996. Control of soil acidity in no-tillage system for soybean production. *Soil and Tillage Research*, 38, 47–57.

Passioura, J.B., 2002. Soil conditions and plant growth. *Plant, Cell and Environment*, 25, 311–318.

Pastore, D., Trono, D., Laus, M., Fonzo, N., Flagella1, Z., 2007. Possible plant mitochondria involvement in cell adaptation to drought stress A case study: durum wheat mitochondria. *Journal of Experimental Botany*, 58, 195–210.

Pei, Z.M., Murata, Y., Benning, G., Thomine, S., Klusener, B., Allen, G.J., Grill, E., Schroeder, J.I. 2000. Calcium channels activated by hydrogen peroxide mediate abscisic acid signalling in guard cells. *Nature*, 406, 731–734.

Pereira, J.O., Defossez, P., Richard, G., 2007. Soil susceptibility to compaction by wheeling as a function of some properties of a silty soil as affected by the tillage system. *European Journal of Soil Science*, 58, 34–44.

Reynolds, M., Mujeeb-Kazi, A., Sawkins, K. 2005. Prospects for utilising plant-adaptive mechanisms to improve wheat and other crops in drought- and salinity-prone environments. *Annals of Applied Biology*, 146, 239–259.

Reynolds, M.P., Rebetzke, G., Pellegrinesci, A., Trethowan R. 2006. Chapter 11, Drought Adaptation in Wheat. In Drought Tolerance in Cereals (ed. J.M. Ribaut), pp. 402–436. Haworth's Food Products Press, New York.

Rhoads, D.M., Umbach, A.L., Subbaiah, C.C., Siedow, J.N. 2006. Mitochondrial reactive oxygen species: contribution to oxidative stress and interorganellar signaling. *Plant Physiology*, 141, 357–366.

Rillig, M.C., 2004. Arbuscular mycorrhizae, glomalin and soil quality. *Canadian Journal of Soil Science*, 84, 355–363.

Ruiz-Lozano, J.M. 2003. Arbuscular mycorrhizal symbiosis and alleviation of osmotic stress. New perspectives for molecular studies. Mycorrhiza, 13, 309–317.

Ruiz-Lozano, J.M., Porcel, R., Aroca, R. 2006. Does the enhanced tolerance of arbuscular mycorrhizal plants to water deficit involve modulation of drought-induced plant genes? *New Phytologist*, 171, 693–698.

Ryan, P.R., Delhaize, E. and Jones, D.L. 2001. Function and mechanism of organic anion exudation from plant roots. *Annual Review of Plant Physiology and Plant Molecular Biology*, 52, 527–560.

Ryan, P.R., Liu, Q., Sperling, P., Dong, B., Franke, S. and Delhaize, E. 2007. A higher plant delta-8 sphingolipid desaturase with a preference for (*Z*)-isomer formation confers aluminum tolerance to yeast and plants. *Plant Physiology*, 144, 1968–1977.

Sasaki, T., Yamamoto, Y., Ezaki, B., Katsuhara, M., Ahn, S.J., Ryan, P.R., Delhaize, E. and Matsumoto, H. 2004. A wheat gene encoding an aluminum-activated malate transporter. Plant Journal 37, 645–653.

Shabala, S., Shabala, L., Van Volkenburgh, E. 2003. Effect of calcium on root development and root ion fluxes in salinised barley seedlings. *Functional Plant Biology*, 30, 507–514.

Schweiger, P.F., Jakobsen, I., 1999. Direct measurement of arbuscular mycorrhizal phosphorus uptake into field-grown winter wheat. *Agronomy Journal*, 91, 998–1002.

Smith, S., Gianinazzi-Pearson, V., 1988. Physiological interactions between symbionts in vesicular–arbuscular mycorrhizal plants. *Annual Review of Plant Physiology and Plant Molecular Biology*, 39, 221–244.

Smith, S.E., Read, D.J. 2008. Mycorrhizal Symbiosis, 3rd ed. Academic Press.

Subramanian, K., Charest, C., Dwyer, L.M., Hamilton, R.I. 1995. Arbuscular mycorrhizas and water relations in maize under drought stress at tasselling. *New Phytologist*, 129, 643–650.

Subramanian, K., Santhanakrishnan, P., Balasubramanian, P. 2006. Responses of field grown tomato plants to arbuscular mycorrhizal fungal colonization under varying intensities of drought stress. *Scientia Horticulturae*, 107, 245–253.

Sumner, M.E., Shahandeh, H., Bouton, J., Hammel, J., 1986. Amelioration of an acid soil profile through deep liming and surface application of gypsum. *Soil Science Society of America Journal*, 50, 1254–1258.

Sylvia, D.M., Hammond, L.C., Bennet, J.M., Hass J.H., Linda S.B. 1993. Field response of maize to a VAM fungus and water management. *Agronomy Journal*, 85, 193–198.

Taylor, N.L., Heazlewood, J.L., Day, D.A., Millar, A.H. 2005. Differential impact of environmental stresses on the pea mitochondrial proteome. *Molecular and Cellular Proteomics*, 4, 1122–1123.

Tesfaye, M., Temple, S.J., Allan, D.L., Vance, C.P. and Samac, D.A. 2001. Overexpression of malate dehydrogenase in transgenic alfalfa enhances organic acid synthesis and confers tolerance to aluminum. *Plant Physiology*, 127, 1836–1844.

Van der Werf, A., Nagel, O.W., 1996. Carbon allocation to shoots and roots in relation to nitrogen supply is mediated by cytokinins and sucrose. *Plant Soil*, 185, 21–32.

Vierheilig, H., Ocampo, J.A., 1991. Receptivity of various wheat cultivars to infection by VA-mycorrhizal fungi as influenced by inoculum potential and the relation of VAM—effectiveness to succinic dehydrogenase activity of the mycelium in roots. *Plant and Soil*, 133, 291–296.

von Uexküll, H.R. and Mutert, E. 1995. Global extent, development and economic impact of acid soils. Plant and Soil 171, 1–15.

Wang, J.P., Raman, H., Zhou, M.X., Ryan, P.R., Delhaize, E., Hebb, D.M., Coombes, N. and Mendham, N. 2007. High-resolution mapping of the *Alp* locus and identification of a candidate gene *HvMATE* controlling aluminium tolerance in barley (*Hordeum vulgare* L.). *Theoretical and Applied Genetics,* 115, 265–276.

Wittenmayer, L., Merbach, W., 2005. Plant responses to drought and phosphorus deficiency: contribution of phytohormones in root-related processes. *Journal of Plant Nutrition and Soil Science*, 168, 531–540.

Wright, S.F., Upadhyaya, A., 1996. Extraction of an abundant and unusual protein from soil and comparison with hyphal protein of arbuscular mycorrhizal fungi. *Soil Science*. 161, 575–586.

Zhou, L., Bai, G., Ma, H., Carver, B. 2007. Quantitative trait loci for aluminum resistance in wheat. *Molecular Breeding*, 19, 153–161.

Zhu J-K. 2002. Salt and drought signal transduction in plants. *Annual Review of Plant Biology,* 53:247–73.

Zhu, J., Kaeppler, S.M., Lynch, J.P., 2005. Topsoil foraging and phosphorus acquisition efficiency in maize (*Zea mays* L.). *Functional Plant Biology*, 32, 749–762.

In: Wheat: Genetics, Crops and Food Production
Editor: Maria T. Almeida, pp. 297-314
ISBN 978-1-61209-307-9

Chapter 10

MYCOPARASITES OF *FUSARIUM* PATHOGENS ON WHEAT: FROM TAXONOMY, GENOMICS AND PROTEOMICS TO BIOTECHNOLOGY

Vladimir Vujanovic** and Yit Kheng Goh
Department of Food and Bioproduct Sciences, University of Saskatchewan,
51 Campus Drive, Saskatoon SK, Canada, S7N 5A8

ABSTRACT

Environmental hazards and health problems due to massive application of chemical pesticides in agriculture incite huge public concerns. A solution is introduction of biological control means to manage outbreaks of plant diseases. To date, only small numbers of beneficial microorganisms, belonging to the category of hyperparasitic or mycoparasitic fungi, have proven to protect crops from plant pathogen infection. Wheat crops, in all phases of growth, are subjected to infections by several groups of pathogens or pests. *Fusarium* pathogenic fungi are responsible for many different severe diseases, such as Fusarium head blight (FHB), Fusarium damaged kernels (FDK), Fusarium wilt, Fusarium crown rot, as well as Fusarium root rot. *Fusarium* mycotoxins produced are also important contaminants on food and agricultural products. Control of Fusarium pathogens is, therefore, crucial. *Fusarium*-associated biotrophic mycoparasitic fungi represent important source of novel biological control agents to manage *Fusarium* diseases and promote plant growth. The specificity of biotrophic mycoparasites distinguishes them as the most promising, eco-friendly solutions for suppressing *Fusarium* pathogens on wheat and preventing mycotoxins accumulations in food and feed. Compared to necrotrophic generalists, biotrophic mycoparasites preserve other beneficial micro-organisms in wheat ecosystems.

* Corresponding author: Dr. Vladimir Vujanovic , Department of Food and Bioproduct Science, College of Agriculture, University of Saskatchewan, 51 Campus Dr.Saskatoon, SK, S7N 5A8 Tel+1 (306) 966-5048; E-mail: vladimir.vujanovic@usask.ca

1. Introduction

Wheat is one of the principal cereal grains or crops produced for the world's food supplies and Canada is one of the major wheat producers as well as exporters in the world (FAO/UN 2009). However, wheat and other crops are vulnerable to a huge number of plant pathogens or pests. Wheat crops, at all different phases of growth, can be subjected to infection by several different groups of microbes or pests. These pathogens associated with wheat can be grouped into fungi, bacteria, viruses, insects, and nematodes (Wiese 1977; Wiese 1987; Prescott et al. 1986). Fungal diseases that are commonly found on wheat crops are black kernel smudge, common bunt, root rot, crown root, Fusarium head blight (FHB), rust, powdery mildew, blotch, take-all, tan spot, smut and wilt (Agrios 2005; Bridges 1995; Howard 2003; Prescott et al. 1986; Wiese et al. 2000).

1.1. *Fusarium* Pathogens

Fusarium fungal pathogens are omnipresent and commonly found on wheat. Most *Fusarium* species are pathogenic to plants; these phytopathogens are responsible for major economically important crops diseases (Bai and Shaner 2004; Goh et al. 2009; Sutton 1982). *Fusarium* diseases are not only found in Canadian wheat fields. Pathogenic *Fusarium* spp. can also infect barley, canola, asparagus, and other crops important to the world bioeconomy (Calman et al. 1986; Fernandez et al. 2007b; Fernando et al. 2000; Vujanovic et al. 2002; Yergeau et al. 2005). In cereals, *Fusarium* species are commonly reported to cause crown and root rot diseases, FHB, Fusarium damaged kernels (FDK) and Fusarium wilt (Gilbert et al. 2001; Mavragani 2008). Crown and root rot diseases are frequently caused by *F. graminearum, F. pseudograminearum, F. culmorum, F. acuminatum, F. avenaceum, F. oxysporum, F. crookwellense,* and *F. equiseti* in wheat (Fernandez et al. 2007a; Fernandez and Jefferson 2004; Smiley and Patterson 1996; William et al. 2002). These pathogenic fungi showed high capability in reducing yields of wheat production, inhibiting seed germination, and affecting grain quality (Mavragani 2008). *Fusarium* spp. are main mycotoxin producers in small cereal kernels, affect the quality of food and feeds and threaten animals and humans (Fernando et al. 2000; Miller et al. 2007; Mills 1990).Therefore, it is important to control the outbreaks of *Fusarium* diseases or reduce the effects of *Fusarium* inoculants on crops and increase the plant yields. Biological control has been proposed as promising environmental solution (Pal and McSpadden 2006) and practical option against *Fusarium* pathogens (Vujanovic 2008).

1.2. *Fusarium* Mycotoxins

Mycotoxins are toxic secondary metabolites generated by micro-organisms. These secondary metabolites are reported to cause disease and death in humans or mammals and other animals (Bennett and Klich 2003; Peraica et al. 1999). Accumulation of *Fusarium* mycotoxins in stored grains was also suggested to be affecting the grain quality and reducing the shelf life (Magan et al. 2003). These toxins were found to contaminate domestic and imported cereal grains or cereal products (Chelkowski 1998). *Fusarium* species are able to produce four different types of trichothecenes: types A, B, C, and D (Foroud and Eudes 2009). Type A trichothecenes, such as T-2 and HT-2 toxins, are highly toxigenic to mammals compared to other trichothecenes (Mirocha et al. 2003). However, type B trichothecenes, such as DON, nivalenol (NIV), 3- and 15-Acetyldeoxynivalenol (ADON), are common toxins reported with FHB and also more phytotoxigenic (Foroud and Eudes 2009; Sudakin 2003). Type C and D are less toxicity and not found to associate with FHB (Desjardins and Proctor 2007; Foroud and Eudes 2009; Liddell 2003; Ueno 1983). Some mycotoxins, such as DON produced by *F. graminearum* were observed to increase the level of virulence and pathogenesis ability in pathogenic *Fusarium* strains to infect wheat or other plants (Bai et al. 2001; Harris et al. 1999). In addition, mycotoxigenic and DON-producing *Fusarium* species were demonstrated to suppress the expression of a chitinase gene in *Trichoderma atroviride* P1 (biocontrol agent). This could be reducing the efficiency of this biocontrol agent in controlling *Fusarium* strains (Lutz et al. 2003).

2. Biological Control

Biological control is one of the promising alternatives to chemical pesticides and has received increasing attention over the last 20 years (Paulitz and Bélanger 2001). Environmental hazards due to the effects of excessive doses of chemical pesticides in agricultural sectors over past hundred years have prompted debates. Therefore, more and more pest management and biocontrol scientists/researchers are working on developing new alternatives for managing pests and controlling plant/crop diseases (Pal and McSpadden Gardener 2006). The term biological control (biocontrol) bears several definitions, which cover various fields. The commonly used definition for biological control is "*the use of living organisms to curtail the growth and proliferation of other, undesirable ones*" (Gnanamanickam et. al. 2002). According to Pal and McSpadden Gardener (2006), a narrower definition of biological control is "*the purposeful utilization of introduced or resident living organisms, other than disease resistant host plants, to suppress the activities and populations of one or more plant pathogens*". The living organisms and micro-organisms, which are showing suppression, inhibition, or control of the pathogens, have been named as the biological agents (BCA) (Pal and McSpadden Gardener 2006) (Table 1).

Pal and McSpadden Gardener (2006) and Cortes-Penagos and associates (2007) suggested that the antagonisms between microbes are categorized into three major types: *direct*, *mixed-path*, and *indirect interactions*. Moreover, each type of interaction is sub-divided into various biological control mechanisms. *Mycoparasitism* or *hyperparasitism* and predation mechanisms belong to direct interaction. Mixed-path antagonism included four different mechanisms, such as antibiotics secretion, lytic enzyme production, unregulated by-products, and physical-chemical influences. Lastly, indirect interaction includes competition and host defenses induction mechanisms (Figure 1). According to Pal and McSpadden Gardener (2006), the most efficient biocontrol agents are those with multiple mechanisms to suppress the pathogens. The meaning of multiple mechanisms refers to the engagement of more than one kind of biocontrol mechanisms. For instance, certain bacteria were reported to generate antibiotic compounds, and also involve the induction of host defenses towards pathogens (Iavicoli et al. 2003). Similarly, *Pythium oligandrum*, the mycoparasite of *Fusarium oxysporum*, also has the potential to trigger plant defenses (Benhamou et al. 1997).

Table 1. Biocontrol agents (BCA) agents against plant pathogens

Biocontrol agent (BCA)	Pathogens	Reference
Bacteria	Bacteria	
Nonpathogenic	*Pseudomonas syringae*	Wilson et. al. 2002.
Antagonistic	*P. solanacearum*	Anuratha and Gnanamanickam 1990.
Fungi & Bacteria	Fungi	
Serratia plymuthica	*Verticillium dahlia*	Kurze et al. 2001.
S. plymuthica	*Phytophora cactorum*	Kurze et al. 2001.
Trichoderma lignorum	*Rhizoctonia solani*	Howell 2003; Weindling 1932.
Antagonistic fungi	*Fusarium culmorum*	Knudsen et al. 1995.
T. harzianum	*F. graminearum*	Dal Bello et al. 2002.
Bacillus cereus	*F. graminearum*	
Fungi & Bacteria	Nematodes	
Hirsutella spp. (fungi)	*Heterodera glycines*	Chen et al. 2000.
Rhizosphere antagonistic bacteria	*Meloidogyne incognita*	Kloepper et al. 2004.

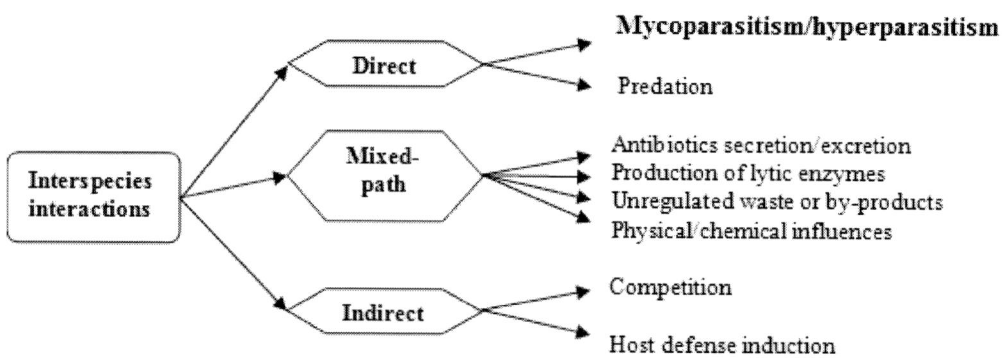

Source: Adapted from Pal and McSpadden Gardener 2006; Cortes-Penagos et al. 2007.

Figure 1. Types of interactions/anatagonisms between organisms/microbes and mechanisms involved in the biological control.

Mycoparasitism or hyperparasitism is a direct biological control mechanism to control plant pathogenic fungi (Howell 2003; Paulitz and Bélanger 2001). Mycoparasitism is the parasitic interaction between one fungus with another fungus (Butler 1957). Hyperparasitism or direct parasitism is considered a synonym for mycoparasitism (Boosalis 1964; Kranz 1981). Indeed, parasites or fungi that parasitize on other fungi (host) are defined as mycoparasites or hyperparasites (Boosalis 1964). The first hyperparasite, *Darluca filum*, was described in 1813 (Kranz 1981), whereas more in-depth study into parasitic interactions between fungi were initiated in 1865 (Barnett 1964). Since then, more and more mycoparasites have been reported and in depth studies were established, such as taxonomy, morphological, cytological, histological, nutritional, interaction studies, molecular, and enzymes characterization and synthesis (Almeida et al. 2007; Assante et al. 2004; Barnett 1964; Fravel et al. 2003; Jeffries and Young 1994; Kranz 1981).

Another mean of biocontrol mechanism is to utilize the indirect methods, which consist of two interactions – competition and host defense induction. In this chapter, we will briefly discuss the host defense induction. Plants are not only showing responses to environmental stresses, however, they also do respond to signals generated by non-pathogenic microbial agents. The signals produced by non-pathogenic microbial agents will trigger the increment of tolerance/resistance towards pathogens. This defense induction is an induced systemic resistance (ISR) response (Fravel et al. 2003; Gnanamanickam et al. 2002; Pal and McSpadden Gardener 2006).

3. Mycoparasitism

According to Boosalis (1964), mycoparasitism interactions are categorized into two groups, biotrophic and necrotrophic parasitism. The classification of the mycoparasitism interactions is based on the fungal parasitic bioactivity and the effects produced on the pathogenic host by the parasitism. Usually, necrotrophic parasites kill the host cells (contact cells or nearby regions) before occupying them, thus deriving nutrients from dead cells. Necrotrophic parasites commonly have broader host ranges compared to biotrophic parasites because necrotrophs produce nonspecific toxic compounds (Barnett 1963; Boosalis 1964). Weindling (1932) pioneered initial reports of *Trichoderma lignorum* (necrotroph) mycoparasitism interactions. Following these first reports on necrotrophic mycoparasitism, further reviews were made by Warren, Boosalis, Backus and Stowell, Howell, and so on (reviewed in Barnett 1963). According to Butler (1957) and Barnett (1963), the level of destruction might be due to several factors such as host resistance, characteristics of the host, the mycoparasite nutritional requirement, and environmental factors. *Trichoderma* species are the famous necrotrophic mycoparasites with their rapid growth capacities, which play a vital role in biological control. Howell (2003) described mechanisms that were utilized by *Trichoderma* spp. to control plant pathogens (mycoparasitism, antibiotic production, competition, enzymes generation, and plants defense induction).

3.1. Biotrophic Mycoparasitism

Generally, biotrophic mycoparasites acquire nutrients from living host cells through haustoria (which allow intimate interactions with host cells). There are two types of biotrophic parasites: *destructive* and *balanced* parasites (Boosalis 1964). Balanced parasites are those that impose a low degree or no apparent harm to the host, whereas destructive parasites are named after those that may kill or inflict partial damage to the host (during development or growth of parasites) (Barnett 1963; Barnett 1964; Jeffries and Young 1994; Kranz 1981). Examples of biotrophic mycoparasites include *Cosmospora, Melanospora, Nitschkia,* and *Anthostomella.* These mycoparasites confer control of plant pathogenic

Fusarium causing FHB and root rots, *Hypoxylon*-bark canker, *Nectria*-bark canker, *Cryptodiaporthe*-bark canker, and *Naemacyclus*-coniferous needle diseases (Vujanovic et al. 1999; Vujanovic 2002; Vujanovic et al. 2003; Vujanovic and Labrecque 2007).

In order to secure nutrients from the living host cells, there are several contact modes for biotrophic mycoparasites to parasitize the host. Table 2 illustrates a few common contact modes for the biotrophic mycoparasites to absorb nutrients from the host or launch host-parasite interactions (Barnett, 1964; Boosalis, 1964; Jeffries and Young, 1994).

3.1.1. Effects of Biotrophic Mycoparasites on the Host

According to Jeffries and Young (1994), effects caused by necrotrophic mycoparasites are more harmful compared to biotrophic mycoparasites. One of the effects caused by biotrophs is a reduction of the growth rate and sporulation capacity of the host. For instance, growth rate and sporulation were reduced in *Pilaira anomala* by *Piptocephalis* spp. in culture (Jeffries and Young 1994; Wood and Cooke 1986). Berry (1959) reported suppression in mycelial growth of the host *Mycotypha microspora* when challenged with *P. virginiana.* In some studies, experimental results show that hyphal apical tips/regions of the host are more susceptible to mycoparasite attack (England 1969; Jeffries and Young 1994). England (1969) and Wood and Cooke (1986) described the formation of unusual morphology in host hyphae, such as hyphal swelling and abnormal branching due to a mycoparasite. In some cases, the life cycle of the host was affected by mycoparasites (Jeffries and Young 1994).

Table 2. Typical structures in biotrophic mycoparasitic interactions, which are used to improve host cells attachment, hosts colonization and nutrients acquisition from host cells

Mycoparasite	Mycoparasitic contact structures	Reference
Piptocephalis virginiana	- Formation of appressorium-like swelling - Haustorium production - Germ tubes tropism towards host hyphae	Berry and Barnett 1957
Calcarisporium parasiticum	- Host hyphae tropism towards germinated spores - Small buffer cell formation	Barnett and Lilly 1958
Gonatobotrys simplex	- Formation of special contact branches - Globose to finger-like shape - Partial gripping the host hyphae	Whaley and Barnett 1963
Gonatobotrys fuscum	- Two-celled, thin, and with flattened apex structure formation	Shigo 1960
Sphaerodes retispora	- Formation of hook-shape cells	Harveson and Kimbrough 2001
Melanospora zamiae	- Formation of hook-shape cells	Jordan and Barnett 1978
Sphaerodes quadrangularis	- Formation of hook- and clamp-like cells	Goh and Vujanovic 2010a
Sphaerodes mycoparasitica	- Formation of hook- and clamp-like cells	Goh and Vujanovic 2010b

3.1.2. Fusarium-Associated Mycoparasites

Several *Fusarium* species, such as *F. oxysporum* and *F. culmorum,* were reported as the hosts for biotrophic and necrotrophic mycoparasites. Table 3 shows the phytopathogenic diseases caused by *Fusarium* spp. as well as *Fusarium* hosts for mycoparasites (references included). *Pythium* spp. were described as both biotrophic and necrotrophic mycoparasites (Hoch and Fuller 1977).

3.1.3. Sphaerodes-Fusarium Mycoparasitism

Sphaerodes retispora was the first melanosporaceous fungus being described as biotrophic mycoparasite of *Fusarium* species, especially *F. oxysporum* (Harveson and Kimbrough 2001). Later, *S. quadrangularis* was identified as mycoparasite of *F. avenaceum* recently (Goh and Vujanovic 2010a) and *S. mycoparasitica* was described as mycoparasite of *F. avenaceum, F. oxysporum* and *F. graminearum* (Vujanovic and Goh 2009; Goh and Vujanovic 2010b). Furthermore, *S. mycoparasitica* was proposed to be *polyspecific* on *F. avenaceum, F. oxysporum, F. graminearum, F. culmorum,* and *F. equiseti* (Vujanovic and Goh 2010). However, *S. retispora* and *S. quadrangularis* were tested and identified as *monospecific* on either *F. oxysporum* or *F. avenaceum,* respectively (Vujanovic and Goh 2010).

Table 2.3. Some Fusarium-associated mycoparasites

Mycoparasite	Host	*Fusarium* disease	Reference
Biotrophic			
Nitschkia parasitans	*F. aquaeductum*	Fusarium canker	Vujanovic 2002
Persiciospora moreaui	*F. oxysporum*	Fusarium wilt	Harveson and Kimbrough 2000
Gliocephalis hyalina	*Fusarium* spp.	Fusarium diseases	Jacobs et al. 2005
Stephanoma phaeospora	*Fusarium* sp.	Fusarium diseases	Hoch 1978
Sphaerodes retispora	*F.oxysporum*	Watermelon wilt	Harveson and Kimbrough 2001
S. mycoparasitica	*Fusarium* spp.	Fusarium Head Bligh, root rot	Vujanovic and Goh 2009
S. quadrangularis	*F. avenaceum*	Fusarium Head Bligh	Goh and Vujanovic 2010
Persiciospora moreaui	*F.oxysporum*	Watermelon wilt	Harveson et al. 2002
Melanospora zamie	*F.oxysporum*	Watermelon wilt	
Necrotrophic			
Trichoderma harzianum	*Fusarium* spp.	Crown and root rot of tomatoes	Paulitz and Bélanger, 2001
Biotrophic/ Necrotrophic			
Pythium spp.	*F.culmorum*	Brown root rot of barley	Davanlou et. al. 1999

Different levels of relationship or diversity or evolution can be studied by phylogenetic analyses through employing different rRNA genes. Phylogenetic relationship analyses over a broad range of fungal taxonomic levels could be achieved by using comparative studies on the nucleotides sequences acquired from ribosomal RNA (rRNA) genes (White et al. 1990). For instance, nuclear small-subunit rDNA sequences are useful for studying distantly related microorganisms. Internal transcribed spacer (ITS) region was examined before; however this region was less informative for Melanosporales group due to lack of number of sequences available for *Melanospora, Sphaerodes, Persiciospora* and other genera within the Melanosporales order (unpublished data). Based on phylogenetic study of large subunit rDNA sequences for seven *Sphaerodes* species, *Sphaerodes* strains were proposed to evolve from being a plant-associated fungus into *Fusarium*-associated fungal species and later from being a *monospecific* (on either *F. oxysporum* – white pigment or *F. avenaceum* – red pigment) (eg. *S. retispora* and *S. quadrangularis*) into *polyspecific* (on both white and red pigment *fusaria*) (eg. *S. mycoparasitica*) (Vujanovic and Goh 2010). Furthermore, ability of *S. mycoparasitica* sexual spores to germinate was improved drastically when the spores were treated with host-*Fusarium* filtrates (eg. *F. avenaceum* and *F. oxysporum*) (Goh and Vujanovic 2010c). There was no or less spore germination reported in the treatments with non-host-*Fusarium* filtrates (eg. *F. proliferatum* and *F. sporotrichioides*). However, more researches have to be done on this domain to verify the evolutionary process and specialization of *Sphaerodes* species.

3.2. Bitrophic (Fungal-Fungal) Interactions

Fungal-host or fungal-fungal interactions can be studied by examining the dual culture or *in vitro* antagonism test (Carisse et al. 2001; Li et al. 2003). Parasite and the host fungi were inoculated together (a few cm apart from each other) on potato dextrose agar (PDA)-coated microscopic slides (Hoch 1974; Hoch and Fuller 1977), or sterile glass cover slips coated with molten water agar (Laing and Deacon 1991), so, it would be easier to observe the host-parasite interactions under the microscope. Moreover, Chet and associates (1981) proposed that agar-coated slides with a cellophane membrane layer could provide better microscopic resolution. Microscopic techniques (light microscope, scanning electron microscope (SEM), transmission electron microscope (TEM), video microscopic methods, and fluorescent microscope) were employed to study host-parasite interactions in detail (Bélanger and Deacon 1996; Chet et al. 1981; Hoch and Fuller 1977; Jeffries and Young 1994; Laing and Deacon 1991; Jacobs et al. 2005; Lorang et al. 2001). More recently, Green fluorescent protein (GFP) was proposed to be useful for studying the interactions between parasite and host (in bitrophic system) and parasite-host-plant (in tritrophic system) (Neveu et al. 2007; Lorang et al. 2001).

3.3. Tritrophic (Parasite-Host-Plant) Interactions

Other than bitrophic interactions, tritrophic interactions are also very interesting topics to explore. Tritrophic interaction approach allows assessing plant response and determines whether mycoparasites are directly or indirectly beneficial to the plants. In the paper by Neveu et al. (2007), GFP technology was utilized to study tritrophic interactions (between plant-powdery mildew pathogen-*Pseudozyma flocculosa*). However, other kinds of tritrophic interactions, such as parasite-host-insect, were reported by Posada et al. (2004) on the topic of tritrophic interaction between *Syspastospora parasitica* (mycoparasite), *Beauveria bassiana* (host), and *Leptinotarsa decemlineata* (insect). Tritrophic interactions can also be analyzed through quantitative real-time PCR approaches. For instance, Filion and associates (2003) proposed the real-time PCR quantification method to quantify amounts of *Fusarium* pathogen and *Glomus* mycorrhizae DNAs in bean plants as well as surrounding mycorrhizosphere. Quantitative real-time PCR technique was also used to detect and quantify colonization of pathogenic *Sclerotinia* sp. sclerotia resting structures by potential biocontrol *Trichoderma* strains in soils (Kim and Knudsen 2008). This method was also employed to analyze colonization of arbuscular mycorrhizal (AM) fungus in plant roots (Isayenkov et al. 2004). In order to study tritrophic interaction between spruce, rust fungus, and opportunistic ascomycetes, quantitative real-time PCR was applied to study the relationship among these three organisms under natural conditions (Hietala et al. 2008). Furthermore, real-time quantitative PCR techniques were used to study phytopathogenic and antagonistic fungi in association with plants (Schena et al. 2004).

4. Future Perspectives

Highthroughput technology, genome sequencing of fungi, fungal proteomics, plants and associated fungi, transcriptomic analyses, availability of mutant collections, RNA interference lines and plants transformed with fluorescent tags have all led to new perspectives on fungus-plant, fungus-fungus and fungus-fungus-plant interactions (Bonfante and Genre 2010; Omann and Zeilinger 2010), and on mycoparasitism in particular (Morissette et al. 2008). In this review we focus on the mechanisms that govern the interaction of biotrophic mycoparasites with fungal pathogens and plants.

4.1. Bitrophic (Fungal-Fungal) and Tritrophic (Parasite-Host-Plant) Interactions

4.1.1. Fluorescent Microscopy

Green fluorescent protein (GFP) and other associated fluorescent proteins are playing important roles in the domain of mycology (Lorang et al. 2001). Furthermore, β-glucuronidase (GUS) reporter gene and GFP were transformed into necrotrophic mycoparasitic *Clonostachys rosea* to study colonization of barley roots by this fungus and ecological studies (Lübeck et al. 2002). GFP transformed necrotrophic *Trichoderma virens* mycoparasite was also used to examine parasitic interactions between *T. harzianum* and phytopathogenic fungi (Sarrocco et al. 2006). Moreover, GFP was also used in tritrophic interaction studies, such as *Trichoderma*-pathogen-plant interaction under *in vivo* conditions (Lu et al. 2004). Even though, green fluorescent protein is well-known for studying fungal-fungal interactions, however there is very little information or no researches are found on transformation of Melanosporaceous biotrophic mycoparasites with GFP. Therefore, it is very important to obtain GFP transformed *Fusarium*-associated biotrophic mycoparasite (eg.

Sphaerodes mycoparasite) for more in-depth study on the aspects of *Sphaerodes-Fusarium, Sphaerodes*-plant root and *Sphaerodes-Fusarium*-plant root interactions.

4.1.2. Electron Microscopy

Scanning electron microscopy (SEM) and transmission electron microscopy (TEM) were also considered as useful tools for bi- and tri-trophic interactions studies in biotrophic mycoparasites. *Fusarium-Stephanoma phaeospora* biotrophic mycoparasite was studied in great details with SEM and TEM (Hoch 1978). Interactions between mycoparasite and host in other biotrophic mycoparasites, such as *Gliocephalis hyaline, Piptocephalis virginiana* and many other biotrophs were also carried out using SEM and TEM (Jacobs et al. 2005; Jeffries 1995; Jeffries and Young 1994; Manocha and Zhonghua 1997). *Sphaerodes-Fusarium* interactions can also be examined by using higher magnification microscopic techniques in future studies, such as TEM and SEM to reveal the cytological information related to this unique relationship.

4.1.3 Gene Expressions

Most of the studies carried out previously were focusing on different genes expression by necrotrophic mycoparasitic *Trichoderma* species during the parasitic interactions (Flores et al. 1007; Kullnig et al. 2000; Rocha-Ramírez et al. 2002) there was not so much information on biotrophic mycoparasite gene expression especially *Fusarium*-associated mycoparasitic fungi. Furthermore, expression of *Fusarium* gene during the interactions with nectrophic mycoparasite was also reported (Lutz et al. 2003).

Therefore, gene expression studies into *Fusarium*-associated biotrophic mycoparasite or *Fusarium* species during the parasitic interactions by real-time PCR, fluorescent reporter or pyrosequencing analysis could provide enormous details as well as new insight on whether the biotrophic mycoparasites will repress the important genes expression in *Fusarium* pathogens or being suppressed by *Fusarium* species and other information. These details could be useful for studying into biological control of *Fusarium* pathogens on wheat with potential biotrophic mycoparasitic agents.

4.1.4. Proteomic Studies

The term 'Proteome' was coined in 1996 by Wilkins and associates to define the study of proteins expressed by genome of cell, tissue or organism (Anderson and Anderson 1998). Proteome analyses are carried out using two-dimensional gel electrophoresis (2-D), chemical analysis, mass-spectrometry and few more other approaches (James 1997). Necrotrophic mycoparasitic *Trichoderma* species are the most famous mycoparasitic candidates for proteomic studies.

For instance, proteome studies to isolate variety of proteins from the *Trichoderma* mycoparasite during the interactions (Ambrosino et al. 2003) and study of tritrophic interactions using proteomic approach (Marra et al. 2006). Proteomic analytical studies could be also applicable to examine *Fusarium-Sphaerodes* (phytopathogen-biotrophic mycoparasite) and *Fusarium-Sphaerodes*-wheat (phytopathogen-biotrophic mycoparasite-plant) interactions.

4.2. Biotechnology of Specific Mycoparasites against *Fusarium* Pathogens

Enormous amount of researches were reported on the capability of necrotrophic mycoparasitic *Trichoderma* species to control different plant pathogens (Fravel 2005; Kubicek et al. 2001; Papavizas 1985; Samuels 1996). However, there is not so much information related to biocontrol of *Fusarium* fungal pathogens using Melanosporaceous biotrophic mycoparasites. Genes from necrotrophic mycoparasitic *Trichoderma* species were reported to confer disease resistance in transgenic plants against fungal pathogens, such as *Alternaria alternata, A. solani, Botrytis cinerea, Rhizoctonia solani* and other pathogens (Bolar et al. 2000; Lorito et al. 1998). Harveson et al. (2002) was the first team working on using melanosporaceous *Sphaerodes retispora* biotrophic mycoparasite to control Fusarium wilt of watermelon caused by *Fusarium oxysporum*. Later, *Sphaerodes quadrangularis* was proposed to be a potential biocontrol agent for managing *F. avenaceum* under *in vitro* conditions (Goh and Vujanovic 2010a). Most recently, *Sphaerodes mycoparasitica* biotrophic mycoparasite isolated from wheat and asparagus plantations was found to parasitize and cause significant negative effects on a wide varieties of *Fusarium* pathogens, such as *F. oxysporum, F. avenaceum, F. graminearum, F. culmorum,* and *F. equiseti* (Goh and Vujanovic 2010b; Vujanovic and Goh 2009; Vujanovic and Goh 2010). Furthermore, this *S. mycoparasitica* biotrophic mycoparasite was found only parasitize mycelia of and germinate in filtrates of *Fusarium* hosts (Goh and Vujanovic 2010b; Goh and Vujanovic 2010c; Vujanovic and Goh 2009), but not on non-host *Fusarium* species. Therefore, we proposed that *Sphaerodes* biotrophic mycoparasites as an eco-friendly alternative for biocontrol of *Fusarium* species on wheat. More future studies could be carried out to assess ability of *Sphaerodes* or melanosporaceous biotrophic mycoparasites as potential biocontrol agents on *Fusarium* pathogens of wheat and barley crops.

Researches into looking for novel genes from *Fusarium*-associated biotrophic mycoparasites might led to identification of new genes that could confer *Fusarium* diseases resistance in plants or crops, such as wheat and barley. Therefore, more researches have to be done in "omics" for providing more information and details to the fields of biotechnology in crop improvement as well as protection.

REFERENCES

Agrios GN. 2005. Plant Pathology. 5th ed. Elsevier Academic Press, Amsterdam, Netherlands.

Almeida FBR, Cerqueira FM, Silva RN, Ulhoa CJ, and Lima AL. 2007. Mycoparasitism studies of *Trichoderma harzianum* strains against *Rhizoctonia solani*: evaluation of coiling and hydrolytic enzyme production. *Biotechnol Lett.* 29: 1189-1193.

Ambrosino P, Scala V, Ruocco M, Marra R, Woo S, Scala F, and Lorito M. 2003. Proteomic studies of antagonist *Trichoderma* strains. *J Plant Pathol.* 85: 301.

Anderson NL and Anderson NG. 1998. Proteome and proteomics: New technologies, new concepts, and new words. *Electrophoresis*, 19: 1853-1861.

Anuratha CS and Gnanamanickam SS. 1990. Biological control of bacterial wilt caused by *Pseudomonas solanacearum* in India with antagonistic bacteria. *Plant and Soil*, 124: 109-116.

Assante G, Maffi D, Saracchi M, Farina G, Moricca S, and Ragazzi A. 2004. Histological studies on the mycoparasitism of *Cladosporium tenuissimum* on urediniospores of *Uromyces appendiculatus. Mycol Res.* 108: 170-182.

Bai G and Shaner G. 2004. Management and resistance in wheat and barley to Fusarium head blight. *Annu Rev Phytopathol.* 42: 135-161.

Bai GH, Plattner R, Desjardins A, and Kolb FL. 2001. Resistance to Fusarium head blight and deoxynivalenol accumulation in wheat. *Plant Breeding,* 120: 1–6.

Barnett HL. 1963. The nature of mycoparasitism by Fungi. Annu Rev Microbiol 17: 1-14.

Barnett HL. 1964. Mycoparasitism. *Mycologia*, 56: 1-19.

Barnett HL and Lilly VG. 1958. Parasitism of *Calcarisporium parasiticum* on species of *Physalospora* and related fungi. West Va Agr Expt Sta Bull 420T.

Bélanger RR and Deacon JW. 1996. Interaction specificity of the biocontrol agent *Sporothrix flocculosa*: a video microscopy study. *Phytopathol.* 86: 1317–1323.

Benhamou N, Rey P, Chérif M, Hockenhull J, and Tirilly Y. 1997. Treatment with the mycoparasite *Pythium oligandrum* triggers induction of defense-related reactions in tomatio roots when challenged with *Fusarium oxysporum* f. sp. *Radicis-lycopersici. Phytopathol.* 87: 108-122.

Bennett JW and Klich M. 2003. Mycotoxins. Clin Microbiol Rev 16: 497-516.

Berry CR. 1959. Factors affecting parasitism of *Piptocephalis virginiana* on other Mucorales. *Mycologia*, 51: 824-832.

Berry CR and Barnett HL. 1957. Mode of parasitism and host range of *Piptocephalis virginiana. Mycologia*, 49: 374-386.

Bolar JP, Norelli JL, Wong K-W, Hayes CK, Harman GE, and Aldwinckle HS. 2000. Expression of endochitinase from *Trichoderma harzianum* in transgenic apple increases resistance to apple scab and reduces vigor. *Phytopathol.* 90: 72-77.

Bonfante P., Genre A. 2010. Mechanisms underlying beneficial plant – fungus interactions in mycorrhizal symbiosis. Nature Communications DOI: 10.1038/ncomms1046

Boosalis MG. 1964. Hyperparasitism. *Annu Rev Phytopathol.* 2: 363-376.

Bridges JR. 1995. Exotic pests: major threats to forest health. In: L. G. Eskew, comp. Forest health through silviculture: *Proceedings of the 1995 National Silviculture Workshop*, Mescalero, New Mexico, May 8-11, 1995. Gen. Tech. Rep. RM-GTR-267. Fort Collins, CO: U.S. Department of Agriculture, Forest Service, Rocky Mountain Forest and Range Experiment Station: 105-113.

Butler EE. 1957. *Rhizoctonia solani* as a parasite of fungi. Mycologia 49: 354-373.

Calman AL, Tewari JP and Mugala M. 1986. *Fusarium avenceum* as one of the causal agents of seedling blight of Canola in Alberta. *Plant Dis.* 70: 694.

Carisse O, El Bassam S, and Benhamou N. 2001. Effect of *Microsphaeropsis* sp. Strain P130A on germination and production of sclerotia of *Rhizoctonia solani* and interaction between the antagonist and the pathogen. *Phytopathol.* 91: 782-791.

Chelkowski J. 1998. Distribution of *Fusarium* species and their mycotoxins in cereal grains. In: Sinha, K.K., and Bhatnagar, D. Mycotoxins in agriculture and food safety. CRC Press, Florida, USA, pp. 44-64.

Chen S, Liu XZ and Chen FJ. 2000. Hirsutella minnesotensis sp. nov., a New Pathogen of the Soybean Cyst Nematode. *Mycologia*, 92: 819-824.

Chet I, Harman GE, and Baker R. 1981. *Trichoderma hamatum*: Its hyphal interactions with *Rhizoctonia solani* and *Pythium* spp. Microbial Ecol 7: 29-38.

Cortes-Penagos C, Olmedo-Monfil V, and Herrera-Estrella A. 2007. The nature of fungal mycoparasitic biocontrol agents. In: Chincholkar, S.B. and Mukerji, K.G., eds. *Biological control of plant diseases.* New York: The Haworth Press, Inc., pp. 327-353.

Dal Bello GM, Mónaco CI, and Simón MR. 2002. Biological control of seedling blight of wheat caused by *Fusarium graminearum* with beneficial rhizosphere microorganisms. *World J Microbiol Biotech.* 18: 627-636.

Davanlou M, Madsen AM, Madsen CH, and Hockenhull J. 1999. Parasitism of macroconidia, chlamydospores and hyphaer of *Fusarium culmorum* by mycoparasitic *Pythium* species. *Plant Pathol.* 48: 352-359.

Desjardins AE and Proctor RH. 2007. Molecular biology of *Fusarium* mycotoxins. *Int J Food Microbiol.* 119: 47-50.

England WH. 1969. Relation of age of two host fungi to development of the mycoparasite *Piptocephalis virginiana. Mycologia*, 61: 586-592

Fernando WGD, Miller JD, Seaman WL, Seifert K, and Paulitz TC. 2000. Daily and seasonal dynamics of airborne spores of *Fusarium graminearum* and other *Fusarium* speices sampled over wheat plots. *Can J Bot.* 78: 497-505.

Fernandez MR and Jefferson PG. 2004. Fungal populations in roots and crowns of common and durum wheat in Saskatchewan. *Can J Plant Pathol.* 26:325-334.

Fernandez MR, Basnyat P and Zenter RR. 2007a. Response of wheat root pathogens to crop management in eastern Saskatchewan. *Can J Plant Sci.* 87:953-963.

Filion M, St-Arnaud M, and Jabaji-Hare SH. 2003. Quantification of *Fusarium solani* f. sp. *phaseoli* in mycorrhizal bean plants and surrounding mycorrhizosphere soil using real-time polymerase chain reaction and direct isolations on selective media. *Phytopathol.* 93: 229-235.

Flores A, Chet I and Herrera-Estrella A. 1997. Improved biocontrol activity of *Trichoderma harzianum* by over-expression of the proteinase-encoding gene *prb1. Curr Genet.* 31: 30-37.

Food and Agricultural Organization of the United Nations. 2009. Food and Agricultural commodities production: Countries by commodity. Available at: http://www.fao.org/es/ess/top/commodity.html?lang=en&item=15&year=2005 (verified 2009-18-03).

Foroud NA and Eudes F. 2009. Trichothecenes in cereal grains. Int J Mol Sci 10: 147-173.

Fravel DR. 2005. Commercialization and implementation of biocontrol. *Ann Rev Phytopathol.* 43: 337-359.

Fravel D, Olivain C, and Alabouvette C. 2003. *Fusarium oxysporum* and its biocontrol. *New Phytol.* 157: 493-502.

Gilbert J, Abramson D, McCallum B, and Clear R. 2001. Comparison of Canadian *Fusarium graminearum* isolates for aggressiveness, vegetative compatibility, and production of ergosterol and mycotoxins. *Mycopathol.* 153: 209-215.

Gnanamanickam SS, Vasudevan P, Reddy MS, Kloepper JW, and Défago G. 2002. Principles of biological control. In: Gnanamanickam, S.S., eds. *Biological control of crop diseases.* New York: Marcel Dekker Inc., pp. 1-9.

Goh YK, Daida P and Vujanovic V. 2009. Effects of abiotic and biotic stress factors on chlamydospore formation in *Fusarium graminearum* and *Fusarium sporotrichioides. Biocontrol Sci Tech.* 19: 151-167.

Goh YK and Vujanovic V. 2010a. *Sphaerodes quadrangularis* biotrophic mycoparasitism on *Fusarium avenaceum. Mycologia*, 102: 757-762.

Goh YK and Vujanovic V. 2010b. Biotrophic mycoparasitic interactions between *Sphaerodes mycoparasitica* and phytopathogenic *Fusarium* species. *Biocontrol Science and Technology*, 20: 891-902.

Goh YK and Vujanovic V. 2010c. Ascospore germination patterns revealed Ascomycetous biotrophic mycoparasite specificity to *Fusarium* hosts. Botany (in press).

Harris LJ, Desjardins AE, Plattner RD, Nicholson P, Butler G, Young JC, Weston G, Proctor RH, and Hohn TM. 1999. Possible role of trichothecene mycotoxins in virulence of *Fusarium graminearum* on maize. *Plant Dis*. 83: 954–960.

Harveson RM and Kimbrough JW. 2000. First report of *Persiciospora moreaui*, a parasite of *Fusarium oxysporum*, in the western hemisphere. *Mycotaxon,* 76: 361-365.

Harveson RM, Kimbrough JW. 2001. Parasitism and measurement of damage to *Fusarium oxysporum* by species of *Melanospora, Sphaerodes,* and *Persiciospora.* Mycologia 93: 249-257.

Harveson RM, Kimbrough JW and Hopkins DL. 2002. Novel use of a pyrenomycetous mycoparasite for management of Fusarium wilt of watermelon. Plant Dis 86: 1025-1030.

Hietala AM, Solheim H, and Gunnar C. 2008. Real-time PCR-based monitoring of DNA pools in the tri-trophic interaction between Norway spruce, the rust *Thekopsora areolata,* and an opportunistic ascomycetous *Phomopsis* sp. *Phytopathol*. 98: 51-58.

Hoch HC. 1978. Mycoparasitic relationships. IV. *Stephanoma phaeospora* parasitic on a species of *Fusarium*. Mycologia 70: 370-379.

Hoch HC and Fuller MS. 1977. Mycoparasitic relationships. I. Morphological features of interactions between *Pythium acanthicum* and several fungal hosts. *Arch Microbiol.* 111: 207-224.

Howard R. 2003. Diseases of wheat – Fungal by Alberta Agriculture and Rural Development. Available at: http://www1.agric.gov.ab.ca/$department/ deptdocs.nsf/all/prm7802. (verified on: 2009-01-02).

Howell CR. 2003. Mechanisms employed by *Trichoderma* species in the biological control of plant diseases: The history and evolution of current concepts. *Plant Dis.* 87: 4-10.

Iavicoli A, boutet E, Buchala A, and Métraux JP. 2003. Induced systematic resistance in *Arabidopsis thaliana* in response to root inoculation with *Pseudomonas fluorescens* CHA0. *Mol Plant-Microbe Interact.* 16: 851-858.

Isayenkov S, Fester T, and Hause B. 2004. Rapid determination of fungal colonization and arbucule formation in roots of *Medicago truncatula* using real-time (RT) PCR. *J Plant Pathol.* 161: 1379-1383.

Jacobs K, Holtzman K, and Seifert KA. 2005. Morphology, phylogeny and biology of *Gliocephalis hyaline,* a biotrophic contact mycoparasite of *Fusarium* species. *Mycologia*, 97: 111-120.

James P. 1997. Protein identification in the post-genome era: the rapid rise of proteomics. *Quarterly Review of Biophysics*, 30: 279-331.

Jeffries P. 1995. Biology and ecology of mycoparasitism. Can J Bot 73 (Suppl. 1): S1284-1290.

Jeffries P, Young TWK. 1994. Interfungal parasitic relationships. CAB International, Wallingford.

Jordan EG and Barnett HL. 1978. Nutrition and parasitism of *Melanospora zamiae. Mycologia*, 70: 300-312.

Kim TG and Knudsen GR. 2008. Quantitative real-time PCR effectively detects and quantifies colonization of sclerotia of *Sclerotinia sclerotiorum* by *Trichoderma* spp. *Appl Soil Ecol.* 40: 100-108.

Kloepper JW, Rodríguez-Kábana R, McInroy JA, and Young RW. 1992. Rhizosphere bacteria antagonistic to soybean cyst (*Heterodera glycines*) and root-knot (*Meloidogyne incognita*) nematodes-Identification by fatty acid analysis and frequency of biological control activity. *Plant and Soil*, 139: 75-84.

Knudsen IMB, Hockenhull J, and Jensen DF. 1995. Biocontrol of seedling diseases of barley and wheat caused by Fusarium culmorum and Bipolaris sorokiniana: effects of selected fungal antagonists on growth and yield components. Plant Pathol 44: 467-477.

Kranz J. 1981. Hyperparasitism of biotrophic fungi. In: Blakeman, J.P., ed. *Micobial Ecology of the phylloplane*. London: Academic Press, pp. 327-352.

Kubicek CP, Mach RL, Peterbauer CK, Lorito M. 2001. *Trichoderma:* From genes to biocontrol. *J. Plant Pathol.* 83: 11-23.

Kullnig C, Mach RL, Lorito M, and Kubicek CP. 2000. Enzyme diffusion from *Trichoderma atroviride* (= *T. harzianum* P1) to *Rhizoctonia solani* is a prerequisite for triggering of *Trichoderma ech42* gene expression before mycoparasitic contact. *Appl Environ Microbiol.* 66: 2232-2234.

Kurze S, Bahl H, Dahl R, and Berg G. 2001. Biological control of fungal strawberry diseases by *Serratia plymuthica* HRO-C48. *Plant Dis.* 85: 529-534.

Laing SAK and Deacon JW. 1991. Video micrscopical comparison of mycoparasitism by *Pythium oligandrum, P. nunn* and an unnamed *Pythium* species. Mycol Res 95: 469-479.

Li GQ, Huang HC, and Acharya SN. 2003. Antagonism and biocontrol potential of *Ulocladium atrum* on *Sclerotinia sclerotiorum. Biol Control.* 28: 11-18.

Liddell CM. 2003. Systematics of *Fusarium* species and allies associated with fusarium head blight. In: *Fusarium Head Blight of Wheat and Barley*; Leonard, K.J., Bushnell, W.R., Eds.; The American Phytopathological Society: St. Paul, Minnesota, USA; Chapter 2, pp. 35-43.

Lorang JM, Tuori RP, Martinez JP, Sawyer TL, Redman RS, Rollins JA, Wolpert TJ, Johnson KB, Rodriguez RJ, Dickman MB, and Ciuffetti LM. 2001. Green fluorescent protein is lighting up fungal biology. *Appl Environ Microbiol.* 67: 1987-1994.

Lorito M, Woo SL, Fernandez IG, Colucci G, Harman GE, et al. 1998. Genes from mycoparasitic fungi as a source for improving plant resistance to fungal pathogens. *PNAS*, 95: 7860-7865.

Lübeck M, Knudsen IMB, Jensen B, Thrane U, Janvier C, and Funck Jensen D. 2002. GUS and GFP transformation of the biocontrol strain *Clnostachys rosea* IK726 and the use of these marker genes in ecological studies. *Myco Res.* 106: 815-826.

Lu Z, Tombolini R, Woo S, Zeilinger S, Lorito M, and Jansson JK. 2004. *In vivo* study of *Trichoderma*-pathogen-plant interactions, using constitutive and inducible green fluorescent protein reporter systems. *Appl Environ Microbiol.* 70: 3073-3081.

Lutz MP, Feichtinger G, Défago G, and Duffy B. 2003. Mycotoxigenic *Fusarium* and deoxynivalenol production repress chitinase gene expression in the biocontrol agent *Trichoderma atroviride* P1. *Appl Environ Microbiol.* 69: 3077-3084.

Magan N, Hope R, Cairns V, and Aldred D. 2003. Post-harvest fungal ecology: Impact of fungal growth and mycotoxin accumulation in stored grain. Eur J Plant Pathol 109: 723-730.

Manocha MS and Zhonghua Z. 1997. Immunocytochemical and cytochemical localization of chitinase and chitin in the infected hosts of a biotrophic mycoparasite, *Piptocephalis virginiana. Mycologia*, 89: 185-194.

Marra R, Ambrosino P, Carbone V, Vinale F, Woo SL, Ruocco M, et al. 2006. Study of the three-way interaction between *Trichoderma atroviride*, plant and fungal pathogens by using a proteomic approach. *Curr Genet.* 50: 307-321.

Mavragani DC. 2008. Molecular and ecological studies of fungal biodiversity on durum wheat grown in rotation with Pulses and Canola, [M Sc Thesis] Saskatoon, Saskatchewan: University of Saskatchewan, Canada. p.104.

Miller SS, Reid LM, and Harris LJ. 2007. Colonization of maize silks by *Fusarium* graminearum, the causative organism of gibberella ear rot. *Can J Bot.* 85: 369-376.

Mills JT. 1990. Mycotoxins and toxigenic fungi on cereal grains in western Canada. *Can J Physiol Pharmacol.* 68: 982-986.

Mirocha CJ, Xie W and Filho ER. 2003. Chemistry and detection of *Fusarium* mycotoxins. In: *Fusarium Head Blight of Wheat and Barley*; Leonard, K.J., Bushnell, W.R., Eds.; The American Phytopathological Society: St. Paul, Minnesota, USA; Chapter 6, pp. 144-164.

Morissette, D., Dauch A.L., Beech R., Masson L., Brousseau R. and Jabaji-Hare S.H. 2008. Isolation of mycoparasitic-related transcripts by SSH during interaction of the mycoparasite Stachybotrys elegans with its host Rhizoctonia solani. *Current Genetics*, 53: 67-80.

Neveu B, Labbé C, and Bélanger R. 2007. GFP technology for the study of biocontrol agents in tritrophic interactions: A case study with *Pseudozyma flocculosa. J Microbiol Meth.* 68: 275-281.

Omann M, and Zeilinger S. 2010. How a Mycoparasite Employs G-Protein Signaling: Using the Example of Trichoderma. Journal of Signal Transduction. DOI:10.1155/2010/123126.

Papavizas GC. 1985. *Trichoderma* and *Gliocladium:* Biology, ecology, and potential for biocontrol. *Ann Rev Phytopathol.* 23: 23-54.

Paulitz TC and Bélanger RR. 2001. Biological control in greenhouse systems. *Annu Rev Phytopathol.* 39: 103-133.

Rocha-Ramírez V, Omero C, Chet I, Horwitz BA, and Herrera-Estrella A. 2002. *Trichoderma atroviride* G-protein α-subunit gene *tga1* is involved in mycoparasitic coiling and conidiation. *Eukaryotic Cell*, 1: 594-605.

Pal KK and McSpadden Gardener B. 2006. Biological control of plant pathogens. The Plant Health Instructor DOI: 10.1094/PHI-A-2006-1117-02.

Peraica M, Radic B, Lucic A, and Pavlovic M. 1999. Toxic effects of mycotoxins in humans. *Bull World Health Organ.* 77: 754-766.

Posada F, Vega FE, Rehner SA, Blackwell M, Weber D, Suh S-O, and Humber RA. 2004. *Syspastospora parasitica,* a mycoparasite of the fungus *Beauveria bassiana* attacking the Colorado potato beetle *Leptinotarsa decemlineata*: A tritrophic association. *J Insect Sci.* 4: 24.

Prescott JM, Burnett PA, Saari EE, Ransom J, Bowman J, de Milano W, Singh RP, and Bekele G. 1986 Wheat diseases and pests: A guide of field identification. CIMMYT, mexico, D.F.

Samuels GJ. 1996. *Trichoderma:* a review of biology and systematics of the genus. *Mycol Res.* 100: 923-935.

Sarrocco S, Mikkelsen L, Vergara M, Funck Jensen D, Lübeck M, and Vannacci G. 2006. Hisopathological studies of sclerotia of phytopathogenic fungi parasitized by a GFP transformed *Trichoderma virens* antagonistic strain. *Mycol Res*. 110: 179-187.

Schena L, Nigro F, Ippolito A, and Gallitelli D. 2004. Real-time quantitative PCR: a new technology to detect and study phytopathogenic and antagonistic fungi. *Eur J Plant Pathol*. 110: 893-908.

Shigo AL. 1960. Parasitism of *Gonatobotryum fuscum* on species of *Ceratocystis*. *Mycologia*, 52: 584-598.

Smiley RW and Patterson LM. 1996. Pathogenic fungi associated with Fusarium foot rot of winter wheat in the semiarid Pacific Northwest. Plant Dis 80: 944-949.

Sudakin DL. 2003. Trichothecenes in the environment: Relevance to human health. *Toxicol Lett*. 143: 97-107.

Sutton JC. 1982. Epidemiology of wheat head blight and maize ear rot caused by *Fusarium graminearum*. Can J Plant Pathol 4: 195-209.

Ueno Y. 1983. Trichothecenes: Chemical, Biological and Toxicological Aspects. Elsevier Scientific Publishers: Amsterdam.

Vujanovic V. 2002. A new variety of *Nitschkia parasitans* hyperparasitic on *Nectria cinnabarina* from American beech. Mycologia 82:121–126.

Vujanovic V. 2008. Novel biotrophic mycoparasites for biological control of *Fusarium* pathogens. *J Plant Pathol*. 90 (3, Supplement). S3.43.

Vujanovic V, Goh YK. 2009. *Sphaerodes mycoparasitica* sp. nov., a new biotrophic mycoparasite on *Fusarium avenaceum, F. graminearum* and *F. oxysporum*. Mycol Res DOI: 10.1016/j.mycres.2009.07.018.

Vujanovic V and Goh YK. 2010. *Sphaerodes* biotrophic mycoparasitism on two new *Fusarium* hosts. *Mycotaxon*, 114: (in press)

Vujanovic V, Hamel C, Jabaji-Hare S, and St-Arnaud M. 2002. Development of a selective mycobutanil agar (MBA) medium for the isolation of *Fusarium* species from asparagus fields. *Can J Microbiol*. 48: 841-847.

Vujanovic V and Labrecque M. 2007. Potential pathogenic and biocontrol ascomycota associated with green wall structures of basket willow (*Salix viminalis*) revealed by phenotypic characters and ITS phylogeny. *Biocontrol*. 53: 413-426.

Vujanovic V, Neumann P, and Laflamme G. 2003. A new record of *Cosmospora epishaeria* hyperparasitic on *Hypoxylon* species from American beech. Mycotaxon 87: 249-253.

Weindling R. 1932. *Trichoderma lignorum* as a parasite of other soil fungi. *Phytopathol*. 22: 837-845.

Whaley JW and Barnett HL. 1963. Parasitism and nutrition of *Gonatobotrys simplex*. *Mycologia*, 55: 195-210.

White TJ, Bruns TD, Lee SB, and Taylor JW. 1990. Analysis of phylogenetic relationships by amplification and direct sequencing of ribosomal DNA genes. In: Innis, M.A., Gelfand, D.H., Sninsky, J.J., and White, T.J., eds., *PCR Protocols: A guide to methods and applications*. New York: Academic Press, pp. 315-322.

Wiese MV. 1977. *Compendium of Wheat Diseases*. ed. The American Phytopathological Society, St. Paul, MN.

Wiese MV. 1987. *Compendium of Wheat Diseases*. 2nd ed. The American Phytopathological Society, St. Paul, MN.

Wiese MV, Murray TD, and Forster RL. 2000. Common names of plant diseases: Diseases of wheat (*Triticum* spp. L.). Available at: http://www.apsnet.org/online/common/names/wheat.asp (verified on 2009-01-01).

William KJ, Dennis JI, Smyl C, and Wallwork H. 2002. The application of specific-specific assays based on the polymerase chain reaction to analyze Fusarium crown rot of durum wheat. *Aust Plant Pathol.* 31:119-127.

Wilson M, Campbell HL, Ji P, Jones JB, and Cuppels DA. 2002. Biological control of bacterial speck of tomato under field conditions at several locations in North America. *Phytopathol.* 92:1284-1292.

Wood SN and Cooke RC. 1986. Effect of *Piptocephalis* species on growth and sporulation of *Pilaira anomala*. *Trans Br Mycol Soc.* 86: 672-674.

Yergeau E, Filion M, Vujanovic V, and St-Arnaud M. 2005. A PCR-denaturing gradient gel electrophoresis approach to assess *Fusarium* diversity in asparagus. *J Microbiol Methods*, 60: 143-154.

In: Wheat: Genetics, Crops and Food Production
Editor: Maria T. Almeida, pp. 315-331
ISBN 978-1-61209-307-9

Chapter 11

WHEAT IN BANGLADESH: PROFITABILITY, PRODUCTION PRACTICES AND ENVIRONMENTAL AND MANAGERIAL FACTORS AFFECTING YIELD

Sanzidur Rahman[1] and M. Kamrul Hasan[2]
[1]School of Geography, Earth and Environmental Sciences,
University of Plymouth, Drake Circus, Plymouth, PL4 8AA, United Kingdom
[2]Planning and Evaluation Division,
Bangladesh Agricultural Research Institute (BARI),
Gazipur - 1701, Bangladesh

ABSTRACT

Wheat is the second most important cereal crop in Bangladesh and its production is expanding rapidly. However, wheat production is very sensitive to environmental production conditions and managerial practices. The present chapter first provides a detailed account of the profitability and production practices of wheat cultivation in Bangladesh using a sample survey of 293 households from three regions. The chapter then examines the effect of environmental production conditions and managerial factors on wheat yield using a reduced form regression analysis. Results reveal that wheat cultivation is profitable as indicated by Benefit-Cost Ratio of 1.41 with significant variation among regions. There are significant variations in production practices of wheat cultivation across regions as well. A host of environmental and managerial factors adversely affect wheat yield explaining a substantial 77% of variation in its yield performance. Wheat yield is significantly lower in low lying areas and poor soils. Also, yield is significantly affected by pest infestation and weather variations. Among the managerial factors, farmers' education and access to various sources of agricultural information significantly improves wheat yield, whereas a delay in sowing and fertilization significantly reduces yield. Farm size category as well as geography does

---

[1] Address for correspondence: Dr. Sanzidur Rahman; Associate Professor in Rural Development. School of Geography, Earth and Environmental Sciences; University of Plymouth; Drake Circus; Plymouth, PL4 8AA, UK; Phone: +44-1752-585911; Fax: +44-1752-585998; E-mail: srahman@plymouth.ac.uk;
[2] E-mail: khasan41@yahoo.com

matter. Productivity of wheat is significantly higher for large farmers whereas it is significantly lower in Jamalpur region. Policy implications include soil fertility improvement; development of wheat varieties that are suitable for low lying areas and are resistant to insect/pest attacks as well as can withstand weather variations; improvement in managerial practices through extension services; and investment in farmers' education.

Keywords: Wheat profitability, production practices, environmental and managerial factors, Bangladesh.

1. Introduction

Wheat (*Triticum aestivum* L.) is one of the most important cereal crops with respect to both acreage and production in Bangladesh. The annual production of wheat in Bangladesh is about 0.74 million tons which is cultivated in 0.40 million hectares of land with an average yield of 1.84 t/ha (BBS, 2008). This level of wheat yield is less than two-thirds of the world average of yield 2.92 t/ha (Allen, 2008). One unique feature of wheat in Bangladesh at present is 100% adoption of high yielding varieties as opposed to rice. Also, the use rate of modern inputs in wheat production is very high. Nevertheless, On-Farm Research Division (OFRD) of Bangladesh Agricultural Research Institute (BARI) reports that there is still a yield gap of 41% to 61% between farmers' practice and recommended package of the research station. Wheat yield with recommended package is 3.2 t/ha whereas actual production at farm level varies between 1.3 to 1.9 t/ha (OFRD, 2001). However, the best practice farmers can produce 2.8 t/ha as compared to 1.9 t/ha by the average farmers, thereby, revealing a 29% yield gap (Hasan, 2005). Such a yield gap between best practice farmers and average farmers amounts to a loss of 25% of gross margin (Tk. 9,875/ha or US$169/ha).

Production performances of wheat vary significantly from one region to another within Bangladesh. Also wide variation in yields is observed among the farmers within a particular location mainly due to improper use of resources and variation in environmental production conditions. Also, management practices and input use are likely to be influenced by various socio-agro-economic factors. The variability in yields may also be influenced by the differences in farm resources used in production. Differences in production practices and fluctuations in input costs may affect the mix and level of inputs used. Furthermore, farm operator's characteristics, although influence costs to some extent, but influence management decision to a large extent, which in turn affects crop yield.

Farmers' performance is believed to be lower than their potential capacity due to underutilization of the recently developed production technologies available at the research stations. A host of socioeconomic constraints may explain why farmers are not utilizing the practices and inputs identified as capable of raising yield. Hence, planners and policy makers need up-to-date information on production practices adopted by the farmers, level and extent of resource use and types of environmental and managerial factors affecting production performance, so that policies and programmes can be directed to address those problems.

Given this backdrop, the aims of this chapter are to: (a) examine the production practices and resource use in wheat farming; (b) estimate costs, returns and profitability of wheat farming; and (c) identify the environmental and managerial factors affecting wheat yield using a survey data of 293 wheat farmers from three regions of Bangladesh.

The chapter proceeds as follows. The next section briefly describes the analytical framework, study areas and the data. Section 3 presents the results. The final section concludes and draws policy implications.

2. METHODOLOGY

2.1. Profitability Analysis

The primary and most common goal of a farm is profit maximization. As most farms try to receive maximum profit in a perfectly competitive market situation, conditions for maximum profit were given emphasis in the present study. Profit or net return is the difference between total value product (TVP) and the total factor cost (TFC). Total factor costs included all kinds of variable and fixed costs incurred in the production process. A farm will not know its maximum profit unless the TVP is compared with TFC. Farmers' profit was also shown by gross margin (GM) analysis, where only variable costs were deducted from total revenue.

The TVP was the value of output and is given by

$$TVP = py = p * TPP = g(y) * f(x_i) = g[f(x_i)] * f(x_i) \quad (1)$$

where, p is the unit price of output; y is the quantity of output and x_i stands for the ith input.

On the other hand, TFC of a product includes all kinds of variable and fixed cost items involved in the production process; and is given by

$$\text{Total factor cost, } TFC = rx_i + b = h(x_i) * x_i + b \quad (2)$$

where, r is the factor price, which in general is a function of the quantity of the factor used $[i.e., r = h(x_i)]$ and b is the fixed costs. Given the definition of TVP in equation (1) and TFC in equation (2), we can define net profit equation as follows:

Net profit, π = TVP – TFC

or,

$$\pi = g[f(x_i)] * f(x_i) - [h(x_i) * x_i + b] \quad (3)$$

2.2. Modelling Influence of Environmental and Managerial Factors on Wheat Yield

A reduced form regression model is used to examine the influence of environmental and managerial factors on wheat yield. This is because Rahman and Hasan (2008; 2009) reported on the impacts of key production inputs including environmental factors on wheat production and managerial factors on production efficiency by using a stochastic production frontier and inefficiency effects model. In this chapter, we are particularly interested in examining solely the relative importance of the environmental and managerial factors explaining variations in observed wheat yield and measure their marginal effects. Therefore, a reduced form regression is sufficient for this purpose. The general model is written as:

$$Y_i = f(E_i, M_i) + \varepsilon_i, \qquad (4)$$

where Y_i is the output, E_i is the vector of relevant environmental variables that control production conditions and M_i is the vector of managerial factors that are largely under the control of the farmer; and ε is the error term. The model is estimated with Ordinary Least Squares (OLS) procedure as this has the BLUE (Best Linear Unbiased Estimator) properties.

2.3. Selection of the Study Area and Sample Farmers[3]

Wheat is cultivated almost all over the country, though the intensity of planted area and land suitability are not equal in all regions. Selection of the study area was conducted by first constructing a wheat area index for each greater district[4]. The wheat area index for the *j*th district is expressed as:

$$WI_j = (Area_j / GCA_j) * 100, \qquad (5)$$

where *WI* is the wheat area index, *Area* is the wheat area and *GCA* is the gross cropped area. Based on this index, wheat growing regions were classified into three levels of intensity: high intensity (WI>8.0), medium intensity (4.01<WI<8.0), and low intensity areas (WI<4.0).

A multistage sampling procedure was adopted to select the sample farmers. First, three wheat growing regions (two from high intensity areas – Dinajpur and Rajshahi, and one from medium intensity areas – Jamalpur) were selected purposively[5]. The selected three districts/regions[6] together cover 31% of the total wheat area of the country (Table 1). Also, each selected district belonged to different agro-ecological zones (AEZ) of Bangladesh, namely, AEZ-3, AEZ 11 and AEZ-9, respectively[7]. Dinajpur is located in the north-west, Rajshahi in the mid-west and Jamalpur in the mid-north of Bangladesh. In the second stage, one *upazila* (sub-district) from each district and one union from each *upazila* were selected at

[3] This section is adapted from Rahman and Hasan (2008).

[4] Although there are 64 districts in Bangladesh, most of the secondary data are still reported at the level of these 21 former greater districts.

[5] The low intensity area is excluded because it is assumed that wheat production has limited potential in these districts.

[6] In this study the term district and region are used interchangeably to emphasize the large spatial variation between our study areas.

[7] There are a total of 29 agro-ecological zones which cut across many of the 21 greater districts/regions.

random. Next, three *mouzas* (one from each union) were selected at random for primary data collection from farm households. However, due to an insufficient number of households in one *mouza*, a fourth *mouza* was also selected at random to fulfil the required sample size. In the third stage, a number of steps were followed to select the households to ensure a high level of representation. At first, a sampling frame of wheat growing holdings was constructed with the assistance of village leaders, record book at the union council office and other key informants. The list included the names of household heads and their land holdings in the selected *mouzas*. These farm holdings were then stratified into three standard farm-size categories commonly adopted in Bangladesh (e.g., Hossain, 1989). Then, a total of 293 wheat producing households were selected following a stratified random sampling procedure. Two sets of structured questionnaires were administered. These questionnaires were pre-tested prior to finalization. The survey covered wheat growing period from November 2003 to April 2004.

Table 1. Selection of the study area and sample size

Study area	Area selection criteria			Farm size categories			
	Wheat area index (WI)	Intensity Rank (Out of 21 greater districts)	% of total wheat area	Large farms (2.0 ha and above)	Medium farms (1.01 to <2.0 ha)	Small farms (up to 1.0 ha)	All categories
Dinajpur	16.94	2	16	33 (92)	29 (86)	39 (122)	101 (300)
Rajshahi	9.12	4	11	19 (32)	32 (49)	52 (228)	103 (309)
Jamalpur	7.55	6	4	8 (11)	25 (46)	56 (178)	89 (235)
All area	-	-	31	60 (135)	86 (181)	147 (528)	293 (844)

Notes: Figures in parentheses indicate sampling frame.
Source: Adapted from Rahman and Hasan (2008).

2.4. The Empirical Model

The reduced form regression of a mixed structure is written as:

$$\ln Y_i = \alpha_0 + \sum_{j=1}^{6} \alpha_j \ln E_{ij} + \sum_{k=1}^{11} \beta_k \ln M_{ik} + \sum_{m=1}^{2} \gamma_m R_{im} + \varepsilon_i \quad (6)$$

where Y_i is the wheat yield (kg/hectare); E_{ij} is jth environmental factors for the ith farmer; M_{ik} is the kth managerial factor for the ith farmer; R_{im} is the dummy variable for districts; ε is the error term; ln is the natural logarithm; α, β, and γ are the parameters to be estimated. We have specified the model with a mixture of actual values as well as logged values of the variables (for those that are non-zero across observations) thereby accommodating non-linearity in the modelling structure. The dependent variable is specified in natural logarithm. A total of six environmental production condition variables (E), 11 variables representing managerial and socio-economic characteristics of the farmer (M) and two regional dummies (R) were used in the full specification. Table 2 presents the definitions, units of measurement, and summary statistics for all the variables used in the regression.

Table 2. Definition, measurement and summary statistics of variables

Variables	Measure	Mean	Standard deviation	Minimum	Maximum
Dependent variable					
Wheat yield	Kg per hectare	2395.13	393.88	974	3163
Environmental factors					
Land suitability	Indexed (1 = Medium high land – most suitable; 2 = High land – suitable; 3 = Low land – not suitable)	1.49	0.74	1.00	3.00
Soil type	Indexed (1 = loamy – good; 2 = sandy loam – average; 3 = clay loam – poor)	1.86	0.71	1.00	3.00
Weed infestation[a]	Indexed (1 = 1 – 10%, 2 = 11 – 20% of crop yield)	0.73	0.53	0.00	2.00
Pest infestation[a]	Indexed (1 = 1 – 10%, 2 = 11 – 20% crop yield)	0.80	0.49	0.00	2.00
Weather[a]	Indexed (1 = 1 – 10%, 2 = 11 – 20%, 3 = 21 – 30%, 4 = 31 – 40% of crop yield)	0.69	0.68	0.00	4.00
Late sowing[a]	Indexed (1 = 1 – 10%, 2 = 11 – 20% of crop yield)	0.68	0.80	0.00	2.00
Soil fertility[a]	Indexed (1 = 1 – 10%, 2 = 11 – 20%, 3 = 21 – 30% of crop yield)	0.50	0.81	0.00	3.00
Regional dummies					
Jamalpur region	Dummy (1 = Yes, 0 = No)	0.35	0.48	0.00	1.00
Rajshahi regions	Dummy (1 = Yes, 0 = No)	0.30	0.46	0.00	1.00
Managerial variables					
Age of the farmer	Years	46.99	12.08	19.00	80.00
Education of the farmer	Completed years of schooling	4.88	4.12	0.00	14.00
Sowing date[b]	Indexed (1 = if sown during optimum time, i.e., November 15 – 30th, 2 = slightly late, 1 – 7th December, 3 = moderately late, 8 – 14th December, 4 = extreme late, 15 – 30th December)	2.11	0.94	1.00	4.00
Timing of first top-dressing of urea fertilizer[c]	Indexed (1 = 10 – 21 DAS [days after sowing], 2 = 22 – 28 DAS, 3 = 29 till maximum DAS)	1.22	0.44	1.00	3.00
Timing of first weeding[c]	Indexed (1 = 16 – 21 DAS, 2 = 22 – 28 DAS, 3 = 29 – 35 DAS, 4 = no weeding at all)	3.26	1.17	1.00	4.00

Variables	Measure	Mean	Standard deviation	Minimum	Maximum
Timing of first irrigation[c]	Indexed (1 = 15 – 21 DAS, 2 = 22 – 28 DAS, 3 = 29 – 35 DAS, 4 = 36 – 42 DAS, 5 = 43 – 60 DAS or no irrigation)	1.57	1.08	1.00	5.00
Mechanical power	Dummy (1 = Used, 0 = No)	0.61	0.49	0.00	1.00
Source of procuring seed	Indexed (1 = Research centre or BADC – good source, 2 = Own processed – average, 3 = Local market or neighbour – poor)	2.17	0.63	1.00	3.00
Link with extension services[d]	Frequency of contact in the past year (number)	13.84	16.50	0.00	42.00
Sources of agricultural information[e]	Number of sources	2.11	0.97	0.00	5.00
Training	Dummy (1 = received training on wheat production in the past 5 years, 0 = No)	0.14	0.35	0.00	1.00
Farm size category	Indexed (Small farms = land owned upto <1.00 ha; medium farms = land owned between 1.00 – 1.99 ha; .and large farms = land owned from 2.00 ha and above)	1.70	0.79	1	3
Total number of observations		293			

Note: [a] = figures are based on farmer's own account of his/her crop loss due to each specific factors.

[b] = wheat cultivation is very sensitive to sowing date. The optimum time of planting is November 15 – 30th. Failure to sow wheat by November 30th reduces crop yield by an estimated 1.3% per day.

[c] = to ensure optimum production of wheat, three operations need to be done simultaneously in the first spell. These are, applying first irrigation, first weeding and first top-dressing of urea fertilizer, all within 17 – 21 days after sowing.

[d] = information was collected on the nature of extension link: 0 = no link, 1 = weekly contact, 2 = fortnightly, 3 = monthly, 4 = quarterly. The information was then converted into frequencies in one year using: quarterly = 4 times, monthly = 12 times, fortnightly = 24 times (deducted two weeks to account for official holidays), and weekly = 42 times (deducted 10 weeks to account for official holidays and other missing days).

[e] = figures are number of sources that the farmers reported as his/her sources of information. A total of six sources were recorded. These are: electronic media (radio/television), block supervisor (lowest unit of extension service), NGO, neighbours, printed media (e.g., leaflets), and fertilizer/pesticide dealer.

Source: Adapted and expanded from Rahman and Hasan (2008).

3. Results

3.1. Production Practices and Input Use in Wheat Farming

The input use rates in wheat farming for each region are presented in Table 3. It is clear from Table 2 that significant regional differences exist with respect to resource use in wheat farming confirmed by formal F-tests for testing differences among regions.

Human labour is an important resource in farming especially in a country where large number of population (about 62%) is engaged in agriculture. Two types of human labour namely: family supplied and hired labour is used in agricultural production. Cost of hired labour was calculated by including direct cash and kind expenses on hiring casual labour. On the other hand, the cost of family labour was calculated on the basis of opportunity cost approach (i.e., the cost of hiring casual labour was imputed in the calculation of family supplied human labour costs). Human labour use was measured in person-days which is equivalent of 8 hours of work. Human labour use is 25% higher in Jamalpur region as compared to Rajshahi region. However, when sources of supply of labour use are examined, we see that Rajshahi uses highest proportion of hired labour (41% of total labour) whereas use of hired labour in Jamalpur is half that level (at 22% of total), implying that most of the labour in Jamalpur is supplied from the family. The lowest level of labour use in Rajshahi might be due to the expansion of mechanization in some of the activities regarding land preparation and threshing.

Animal power is another important resource in the production process. Like human labour, two types of animal power: namely family supplied and hired animal power services are used in wheat production. Cost of animal power was calculated considering direct cash and kind expenses on hiring animal power. On the other hand, the cost of farm supplied animal power was calculated on the basis of the opportunity cost approach. Animal power was measured as pair-days which constituted approximately 6 hours of working time of two animals. In the case of cost of animal power, the human labour engaged with the pair was calculated separately. We see from Table 3 that the use of animal power services is also significantly lower in Rajshahi as compared to Jamalpur and Dinajpur, which are similar. Traditionally, animal power is used for ploughing land several times followed by laddering to level the land for planting. However, due to severe shortage of livestock resources, use of power tiller has been on the rise. Our result shows that 34% of the farmers in Rajshahi used only power tiller for land preparation as compared with other two regions where animal power is the dominant input or at least a combination of animal power and power tiller is used. In fact, there is a significant relationship between the number of times a land is ploughed and wheat yield. Farmers who ploughed land for 6 times (10% of farmers did) received wheat yield of 2,488 kg/ha as compared to those who ploughed only 2 times (9% of farmers did) received 2,363 kg/ha.

The quantity of seed used by farmers for wheat cultivation is presented Table 3. Although significant differences exist in seed use rates, farmers in Jamalpur tend to use family supplied seed more than the other two regions. On an average 164 kg/ha seed were used by the farmers, of which 58% percent were own preserved or home supplied.

Table 3. Input use in wheat production

Variables	Input use rates per hectare				
	Jamalpur	**Rajshahi**	**Dinajpur**	**Overall**	**F-test[a] for regional differences**
Inputs					
Human labour (person days)	114.48	91.88	109.64	104.87	42.86***
Family supplied	88.89	54.55	77.02	72.73	21.64***
Hired	25.55	37.27	32.58	32.10	4.96***
Animal power (animal pair days)	24.73	15.22	25.21	21.55	9.88***
Family supplied	22.47	10.20	22.31	18.10	15.32***
Hired	2.27	5.10	2.90	3.45	1.88
Seed (kg)	153.78	165.43	171.60	164.02	22.54***
Family supplied	97.11	94.80	92.90	94.85	0.06
Purchased	56.67	70.63	78.70	69.17	1.74
Farm yard manure (kg)	770.64	232.07	2664.18	1234.03	78.64***
Fertilizer (nutrients) (kg)	136.93	158.43	146.31	147.72	10.16***
Output					
Wheat yield (kg)	2189.40	2477.24	2492.68	2395.13	19.72***

Note: [a] = One-way ANOVA using the Generalised Linear Model (GLM).

*** significant at 1 percent level ($p<0.01$)

** significant at 5 percent level ($p<0.05$)

* significant at 10 percent level ($p<0.10$)

The seed rate used by the farmers was much higher than the recommended rate (120 kg/ha) in all the locations. This may be to compensate for the loss incurred by bird attack in the wheat field and low germination ability of the seeds used. About 94% of the sampled farmers used only one variety called *Kanchan* (released in 1983) although a relatively older variety *Sonalika* (released in 1974) and a newer variety *Shatabdi* (released in 2000) were also used by a very limited number of farmers, i.e., 4% and 2%, respectively.

Organic and inorganic fertilizers play an important role in the crop production system. Plant nutrients from inorganic sources (i.e., fertilizers) are applied to cope with declining soil fertility but cannot play a major role in improving the soil quality, which organic fertilizers can do. Inorganic fertilizers today hold the key to the success of the crop production systems of Bangladesh agriculture, being responsible for about 50 percent of the total production (BARC, 1997 cited in Hasan, 2005). Farmers in the study areas applied both farm yard manure (FYM) and inorganic fertilizers in the form of urea, triple super phosphate (TSP), muriate of potash (MP), di-ammonium phosphate (DAP) and gypsum in the wheat field. FYM is a combination of cowdung, cattle feed wastes, animal urine etc. applied to the crop field as organic fertilizer. The quantities of FYM were measured in kilograms. The nutrient content of fertilizers used was measured in kg. The cost of FYM and fertilizers was calculated at the prevailing market price. The cost of family supplied FYM was measured on the basis of the opportunity cost approach. The most dramatic difference is in the use of FYM among regions. Use of FYM in Dinajpur is 10 times more than in Rajshahi and 2.5 times more than in Jamalpur. This may be due to the fact that cowdung, the principal component of FYM was used as cooking fuel. Nevertheless, even the highest level of use in Dinajpur (2.6 t/ha) is still well below the recommended dose of 7-10 t/ha. Although farmers in Rajshahi uses highest level of inorganic fertilizers which is closer to the recommended dose of 180 – 220 kg/ha, the differences among regions is not that dramatic as compared with the use of FYM. In other words, very low level of FYM use in Rajshahi is not compensated by corresponding high use rate of inorganic fertilizers.

Significant yield differences also exist among regions with lowest yield rate in Jamalpur estimated at 2.2 t/ha as compared to almost similar yield levels in Rajshahi and Dinajpur estimated at 2.5 t/ha. It is well known that Rajshahi and Dinajpur are traditional wheat growing areas whereas Jamalpur is an intensive agricultural region, implying that geography does matter in productivity performance of wheat.

3.3. Costs, Returns and Profitability in Wheat Production

Having observed significant differences in input use rates of most of the inputs and corresponding wheat yield, we now examine the costs, returns and profitability in wheat production among regions (Table 4). It is clear from Table 4 that significant differences among region exist in all elements of costs under consideration. Although human labour use was highest in Jamalpur (Table 3), the cost per hectare was highest in Dinajpur largely due to wage differences (see mid-panel of Table 4). As expected, the use of mechanical power is significantly higher in Rajshahi since we have seen lowest use of animal power services in this region. Similarly, the cost of FYM was significantly higher in Dinajpur and almost negligible in Rajshahi. The cost of irrigation is lowest in Jamalpur whereas in Dinajpur the cost is almost 2.5 times higher.

Table 4. Cost, return and profitability in wheat production

Variables	Taka[a] per hectare				
	Jamalpur	Rajshahi	Dinajpur	Overall	F-test[b] for regional differences
Costs					
Human labour	6782.42	5796.14	7058.85	6530.99	36.06***
Animal power	1275.08	874.14	1409.81	1180.58	7.84***
Mechanical power	417.10	2304.60	556.30	1128.61	281.56***
Seed	2043.72	2284.64	2372.88	2241.88	40.21***
Fertilizer cost	2609.39	3210.51	2910.28	2924.42	18.62***
Manures cost	317.27	103.56	1130.48	522.46	77.52***
Irrigation cost	345.64	781.54	869.30	679.39	121.03***
Interest on capital	265.13	333.87	317.03	307.19	76.51***
Land rent	3258.34	3507.43	3541.57	3443.54	15.35***
Total variable cost (TVC)	14544.92	16571.39	17396.88	16240.40	36.78***
Fixed cost (FC)	3523.49	3841.23	3858.43	3750.63	20.76***
Total cost (TC = TVC+FC)	18068.37	20412.62	21255.31	19991.03	40.31***
Input and output prices					
Wheat price (Tk/kg)	11.35	11.78	11.90	11.69	47.68***
Fertilizer price (Tk/kg)	9.11	9.83	9.64	9.55	149.65***
Labour wage (Tk/person-day)	59.29	63.20	64.58	62.49	61.99***
Animal power price (Tk/animal pair-day)	52.12	56.67	56.16	55.11	22.85***
Seed price (Tk/kg)	13.29	13.82	13.83	13.66	34.68***
Returns and profitability					
Gross return (GR)	24784.88	29175.20	29658.92	28008.37	48.03***
Wheat grain	23617.28	27839.29	28155.08	26665.69	47.88***
Straw	1167.66	1336.02	1503.88	1342.74	46.53***
Gross margin (GR – TVC)	10239.96	12603.82	12262.04	11767.97	16.89***
Net return (GR – TC)	6716.48	8762.51	8403.57	8017.29	13.95***
Benefit-cost ratio (BCR = GR/TC)	1.37	1.43	1.41	1.41	4.23**

Note: [a] = Exchange rate of USD 1.00 = Taka 63.76 in 2004-05 (BB, 2010).
[b] = One-way ANOVA using the Generalised Linear Model (GLM).
*** significant at 1 percent level ($p<0.01$).
** significant at 5 percent level ($p<0.05$)

Significant differences also exist among regions with respect to all input and output prices as well (Table 4). In our sample, wheat price varied from Tk 11.00–12.25 per kg; fertilizer price (average price of 5 types of fertilizers used) varied from 7.32 to 9.69 per kg; labour wage varied from Tk 52–65 per person day; animal power price varied from Tk 42–65 per animal pair-days; and seed price varied from Tk 12–14 per kg, respectively. Such significant differences of input and output prices alone are bound to have significant impacts on profitability derived from wheat among regions.

Looking at all the cost elements, it is clear that the main source of variation in costs of wheat production is the total variable cost (TVC). This is because the margin of difference in fixed cost (FC) is much lower among regions although these are also significantly different. For example, TVC is 20% higher in Dinajpur as compared to Jamalpur whereas FC is only 9.5% higher. As a result, gross return and net return is also 20% and 25% higher in Dinajpur as compared to Jamalpur. The differences in costs and returns between Rajshahi and Dinajpur is however very low, ranging from 2% and 5% only.

In terms of the final assessment of profitability, i.e., the Benefit-Cost Ratio (BCR), we see that BCR is highest in Rajshahi estimated at 1.43 and lowest in Jamalpur estimated at 1.37. Nevertheless, despite significant differences in input use and production practices of wheat among regions, it remains profitable in all the regions as evidenced from BCR. This level of BCR is also comparable to other Asian economies. For example, the BCR of wheat is 1.06 and 1.23 in Indian Haryana and Pakistan Punjab, respectively although their yield levels were substantially higher than observed in Bangladesh, estimated at 4.22 t/ha and 3.26 t/ha (Erenstein, 2010). On the other hand, the BCR of wheat in irrigated and dryland areas of Iran is much higher, estimated at 1.97 and 2.56 with a comparable grain yield under irrigated condition (2.9 t/ha) but substantially lower yield under dryland condition (only 1.0 t/ha), respectively (Ghorbani et al., 2010). The implication is that although yield, costs and return structure differ substantially amongst countries, wheat production remains profitable in places where it is grown.

3.3. Effects of Environmental and Managerial Factors on Wheat Yield

Table 5 presents the parameter estimates of the reduced form regression of the impacts of environmental and managerial factors on wheat yield. Twelve of the total 19 parameters are significantly different from zero at 10% level at least in the regression with an adjusted R-squared value of 0.77 implying a very good fit (Table 5). The results reveal that a substantial 77% of the variation in wheat yield is explained by these environmental and managerial factors, which are much higher than expected. Among the six environmental factors included in the analyses, four of them significantly affect wheat yield. “Poor soil fertility status” alone explained 59% of the total variation in wheat yield (see last column of Table 5) followed by unsuitable land type, pest infestation and weather variation. All these four environmental factors together explain 62% of the variation in wheat yield thereby, highlighting vulnerability of the wheat farming system to environmental production conditions.

Table 5. Regression results of environmental and managerial factors affecting wheat yield

Variables	Coefficients	t-ratio	Adjusted R-squared contribution
Constant	8.0668***	86.50	
Environmental factors			
Land suitability	-0.0739***	-4.41	0.024
Soil type	0.0051	0.34	
Weed infestation	-0.0083	-0.59	
Pest infestation	-0.0247*	-1.94	0.002
Weather variation	-0.0171*	-1.71	0.001
Soil fertility status	-0.0989***	-10.41	0.589
Regional variables			
Rajshahi	-0.0003	-0.02	
Jamalpur	-0.0828**	-2.06	0.047
Managerial factors			
Age	-0.0388*	-1.78	0.002
Education	0.0024*	1.65	0.009
Sowing date	-0.1325***	-8.84	0.075
Timing of first top-dressing of urea fertilizer	-0.0547*	-1.90	0.006
Timing of first weeding	-0.0176	-0.57	
Timing of first irrigation	0.0141***	2.61	0.003
Source of procuring seed	0.0153	0.87	
Link with extension services	0.0010	0.20	
Sources of agricultural information	0.0154**	1.99	0.003
Training	-0.0169	-0.96	
Farm size category	0.0317**	2.34	0.002
Model diagnostics			
Adjusted R-squared	0.77		
$F_{(19, 273)}$	52.63***		
Number of observations	293		

Note: *** significant at 1 percent level (p<0.01)
** significant at 5 percent level (p<0.05)
* significant at 10 percent level (p<0.10)

The managerial factors are also important in explaining variation in wheat yield. A total of seven managerial factors out of 11 factors included in the analysis significantly affect wheat yield. Diverse sources of agricultural information and education of the farmers significantly improve wheat yield whereas a delay in sowing date and timing of fertilization significantly reduce wheat yield. Also, older farmers produce significantly less wheat per hectare. Wheat yield is significantly higher for large farms. All these seven managerial factors taken together explain about 10% of variation in wheat yield. Regional factor alone explains 4.7% of variation in wheat yield. For example, wheat yield is significantly lower in Jamalpur region (also seen in Table 2).

Since a mixed structure is used in the regression, coefficients on the variables cannot directly reveal the magnitude of influence. Table 6 presents the marginal effects (in terms of elasticities) of the environmental and managerial factors on wheat yield[1]. A 1% improvement in land suitability (i.e., moving out from low lying areas to medium highland) will increase wheat yield by 0.07%. Similarly, a 1% improvement in the index of soil fertility status will improve wheat yield by 0.05%. It is difficult to quantify what will be the actual level of soil nutrients, since data on exact measure of available soil nutrients could not be collected in this survey (e.g., as seen in Rahman and Parkinson, 2007). However, the importance of soil fertility status to raise wheat productivity is clear. Again, a 1% reduction in pest attack and weather variation will significantly improve wheat yield by 0.02% and 0.01% respectively. Among the managerial factors, the most prominent effect is the delay in sowing. A 1% percent delay in sowing date from its optimum range (i.e., within 15th–30th of November) will reduce wheat yield by a substantial 0.13%. Similarly, a delay in the application of first dose of fertilizer from its optimal time (i.e., within 10-21 days after sowing) will reduce wheat yield by 0.05%. Access to information from diverse sources significantly improves wheat yield by 0.03%. A 1% improvement in educational level of the farmer will increase wheat yield by 0.01%.

Table 6. Marginal effects (elasticities) of environmental and managerial factors on wheat yield (excluding dummy variables)

Variables	Coefficients	t-ratio
Environmental factors		
Land suitability	-0.0739***	-4.41
Soil type	0.0051	0.34
Weed infestation	-0.0061	-0.59
Pest infestation	-0.0176*	-1.94
Weather variation	-0.0118*	-1.71
Soil fertility stats	-0.0493***	-10.41
Managerial factors		
Age	-0.0388*	-1.78
Education	0.0118*	1.65
Sowing date	-0.1325***	-8.84
Timing of first top-dressing of urea fertilizer	-0.0547*	-1.90
Timing of first weeding	-0.0176	-0.57
Timing of first irrigation	0.0221***	2.61
Source of procuring seed	0.0153	0.87
Link with extension services	0.0014	0.20
Sources of agricultural information	0.0325**	1.99
Farm size category	0.0317**	2.34

Note: *** significant at 1 percent level ($p<0.01$)
** significant at 5 percent level ($p<0.05$)
* significant at 10 percent level ($p<0.10$)

[1] The marginal effects of dummy variables are excluded because of their discrete nature.

4. Conclusions and Policy Implications

The present chapter provides a detailed account of production practices, costs, returns and profitability of wheat farming in Bangladesh as well as examines the impacts of environmental production conditions and managerial factors on wheat yield. Our results demonstrate that production practices in terms of resource use differ significantly among regions. As a result the cost structure and costs (particularly the total variable cost) as well as gross and net returns also differ significantly among regions. Total cost, returns and profits are lowest in Jamalpur as compared to largely similar results for Rajshahi and Dinajpur, the traditional wheat growing areas.

The environmental production conditions, within which the farmer operates, are considered vital but are often arbitrarily omitted in standard economic analysis. Our results demonstrate the vulnerability of Bangladeshi wheat farmers to environmental production conditions as well as managerial factors. A host of environmental and managerial factors alone accounts for 77% of the variability in wheat yield which is rather high. Among the environmental factors, the most prominent factor affecting wheat yield is "poor soil fertility status" followed by "unsuitable land type". Pest attacks and weather variation also significantly affects wheat yield. Among the managerial factors, delay in sowing and timing of fertilization significantly reduce wheat yield. On the other hand, farmers' education and access to diverse sources of information significantly improves yield, which is encouraging.

A number of specific policy implications can be drawn from our results. First, soil fertility improvement seems essential to raise wheat productivity. This may be addressed through adopting soil conservation practices and/or improving crop rotation practices (e.g., including soil health enhancing crops, such as pulses and oilseeds, in the system). Of the nine total cropping patterns observed among the sample farmers, most followed rice-based cropping. Only two patterns included jute in the system and none included any pulse or oilseed crops, which is potentially highly detrimental to soil health in the long run. Second our study shows that poor land type significantly reduces productivity. Therefore, research effort should be geared towards developing varieties that are suitable for marginal areas. Evidence suggests that wheat production in marginal lands accounts for 25% of global production and that research innovation has led to significant improvement in yield growth in these areas, particularly in drought and high temperature environments (Lantican et al, 2003). Third, pest infestation and weather variation also affects wheat productivity. Therefore, research effort should also focus on developing wheat varieties that are insect/pest resistant and can withstand weather variations. Marasas et al., (2003) highlighted the economic benefit of long-term crop breeding research undertaken by CIMMYT on leaf-rust resistance of modern wheat for the developing economies. They estimated a NPV of US$ 36.5 billion in 1990 dollar yielding a BCR of 27:1, thereby highlighting the significant benefit that can be accrued from investing in developing varieties that are pest/disease resistant. Fourth, is the improvement in managerial practices (e.g., timely sowing and fertilizer application). These can be addressed through strengthening agricultural extension services and improvements in rural infrastructure. Agricultural extension can be singled out as one of the important sources of information dissemination directly relevant to agricultural production practices, particularly in nations like Bangladesh where farmers has very limited access to information. This is reinforced by the fact that many studies found a significant influence of extension education on adoption of land-improving technologies and/or conservation (e.g. Neupane et al., 2002; Baidu-Forson, 1999; Adesina and Baidu-Forson, 1995). Fifth, investment in education for the farming population seems crucial. Asadullah and Rahman (2009) reported significant influence of education on farm productivity of rice in Bangladesh.

The challenges to realize all of these policy options are formidable. However, given the clear influence of these factors on wheat productivity, every effort should be made to implement these policy options which could significantly curb dependence on rice as the main staple in Bangladeshi diet.

REFERENCES

Adesina, A.A., Baidu-Forson, J. 1995. Farmers' perception and adoption of new agricultural technology: evidence from analysis in Burkina Faso and Guinea, West Africa. *Agricultural Economics*. 13: 1 – 9.

Allen, E.W. 2008. International wheat outlook for 2008/09. Economic Research Service Report #WHS-2008-01. *United States Department for Agriculture (USDA)*, Washington, D.C.

Asadullah, N., Rahman, S. 2009. Farm productivity and efficiency in rural Bangladesh: the role of education revisited. *Applied Economics*, 41: 17-33.

Baidu-Forson, J. 1999. Factors influencing adoption of land-enhancing technology in the Sahel: lessons from a case study in Niger. *Agricultural Economics*, 20: 231 – 239.

BBS, 2008. *Statistical Yearbook of Bangladesh 2006.* Bangladesh Bureau of Statistics, Dhaka, Bangladesh.

BB, 2010. Economic Trends (monthly publications) April 2010. Bangladesh Bank, Dhaka.

Erenstein, O. 2010. A comparative analysis of rice–wheat systems in Indian Haryana and Pakistan Punjab. *Land Use Policy*, 27: 869-879.

Ghorbani, R., Mondani, F., Amirmoradi, S., Feizi, H., Khorramdel, S., Teimouri, M., Sanjani, S., Anvarkhah, S., Aghel, H. 2010. A case study of energy use and economical analysis of irrigated and dryland wheat production systems. *Applied Energy* (in press).

Hasan, M.K. 2005. Yield and benefit gaps in wheat production: comparison between two farmer management practices. *Seminar on Higher Agricultural Education and Research in Bangladesh: Prospects and Challenges.* Bangladesh Agricultural University, Mymensingh, Bangladesh.

Hossain, M. 1989. *Green Revolution in Bangladesh: Impact on Growth and Distribution of Income.* University Press Limited, Dhaka, Bangladesh.

Lantican, M.A., Pingali, P.L., Rajaram, S. 2003. Is research on marginal lands catching up? The case of unfavourable wheat growing environments. *Agricultural Economics,* 29: 353 – 361.

Marasas, C.N., Smale, M., Singh, R.P. 2003. The economic impact of productivity maintenance research: breeding for leaf rust resistance in modern wheat. *Agricultural Economics,* 29: 253 – 263.

Morris, M., Chowdhury, N., Meisner, C. 1996. Economics of wheat production in Bangladesh. *Food Policy*, 21: 541 – 560.

Neupane, R.P., Sharma, K.R. and Thapa, G.B. 2002. Adoption of agroforestry in the hills of Nepal: a logistic regression analysis. *Agricultural Systems*, 72: 177 – 196.

OFRD, 2001. Annual Report, 1999-2000. *On-Farm Research Division.* Bangladesh Agricultural Research Institute (BARI), Gazipur, Bangladesh.

Rahman, S., Parkinson, R.J. 2007. Soil fertility and productivity relationships in rice production system, Bangladesh. *Agricultural Systems.* 92: 318-333.

Rahman, S., Hasan, M.K. 2008. Impact of environmental production conditions on productivity and efficiency: the case of wheat producers in Bangladesh. *Journal of Environmental Management*, 88: 1495-1504.

Rahman, S., Hasan, M.K. 2009. Wheat in Bangladesh: Yield growth, production performance and determinants. In Davis, S. and Evan, G. (eds.) Soybean and Wheat Crops: Growth, Fertilization and Yield. Nova Science Publishers, New York.

In: Wheat: Genetics, Crops and Food Production
Editor: Maria T. Almeida, pp. 333-345
ISBN 978-1-61209-307-9

Chapter 12

AGRICULTURAL POTENTIAL OF FUNGAL ENDOPHYTES OF GRASSES, CEREALS AND WHEAT

M. Hubbard*, J. Germida[2†] and V. Vujanovic[1,‡]

[1]Department of Food and Bioproduct Science, College of Agriculture, University of Saskatchewan, Saskatchewan, Canada

[2]Department of Soil Science, College of Agriculture, University of Saskatchewan, Saskatchewan, Canada

ABSTRACT

Fungal endophytes spend at least part of their lifecycle living within healthy plant tissues and form associations with the majority of terrestrial plant species. Endophytic fungi of grasses include arbuscular mycorrhizal fungi (AMF) and non-mycorrhizal fungal endophytes. While these two distinct groups of mycoendosymbionts can co-occur in plants, the functions and diversity of non-AMF endophytes is less well understood than that of AMF. Although the diversity of fungal endophytes of wheat is incompletely characterized, its potential can be appreciated based on the diversity of these organisms in other grass and cereal species. These mycobionts can benefit host plants in a variety of ways, including enhanced plant biomass formation and tolerance for biotic and/or abiotic stress. These benefits can be conferred during seed germination (mycovitality) or later in the lifecycle of the host (mycoheterotrophy). The mechanisms by which fungal endophytes and host plants interact are poorly understood, but may include fungal metabolite production, scavenging of reactive oxygen species by the fungal partner, phytohormone production or changes in plant gene expression. A more complete appreciation of the biology of endophytic fungi of grasses in general, and wheat in particular, is likely to be invaluable in developing strategies to utilise these organisms in sustainable agricultural production. In light of the agricultural importance of wheat, a

* E-mail: michelle.hubbard@usask.ca;
† E-mail: jim.germida@usask.ca;
‡ Corresponding author: Dr. Vladimir Vujanovic, Department of Food and Bioproduct Science, College of Agriculture, University of Saskatchewan, 51 Campus Dr.Saskatoon, SK, S7N 5A8Tel+1 (306) 966-5048E-mail: vladimir.vujanovic@usask.ca

further study on the use of endophytic fungi in biocontrol and biofertilizer applications is merited.

Introduction

Grasses and cereals in general and wheat in particular are found on all continents except Antarctica and play a major role in human food production. Thus, understanding the factors that impact cereal growth is of current scientific interest and practical importance. Endosymbiotic relationships with micro-fungi are a key area of cereal and grass biology that is incompletely understood but has great potential. The most well known form of fungal endosymbionts are arbusuclar mycorrhizal fungi (AMF). However, there are non-mycorrhizal mycoendosymbionts, generally referred to as fungal endophytes, which differ from AMF in terms of taxonomy (Peterson et al. 2004), morphology (Jumpponen 2001; Jun et al. 2002; Peterson et al. 2004) and requirements for growth (Declerck et al. 2005).

Fungal endophytes are defined as fungi that live, for at least part of their life cycle, within healthy plant tissues (Wilson 1993). These organisms can confer a diverse range of benefits to their plant hosts, including biocontrol of pathogens and pests (e.g. Latch 1993; Prestidge and Gallagher 1988), abiotic stress tolerance (Marquez et al. 2007; Rodriguez et al. 2008), herbivory deterrence (Roberts et al. 2005) and increased biomass accumulation and/or yield (Deshmukh and Kogel 2007). Although endophytic fungi are found in almost all terrestrial plants (Trappe 1987), little is known about the diversity of mycoendophytes of wheat.

This chapter has three main objectives: 1) to discuss the diversity of fungal endophytes in grasses and cereals and, by extrapolation, in wheat in terms of the differences between AMF and non-mycorrhizal mycoendosymbionts and ways in the latter can be classified. 2) This chapter also aims to examine the interactions between endophytic fungi and their grass hosts and 3) to address the potential practical applications of fungal endophytes of wheat in agricultural production.

Diversity of Fungal Endophytes

A diverse array of fungal organisms has been found in association with grasses and cereals. This diversity includes fungal arbuscular mycorrhizal fungi (AMF) and non-mycorrhizal fungal endophytes.

Despite the fact that non-AMF endophytic fungi co-occur with and may be more abundant than AMF in grass roots (Mandyam and Jumpponen 2008; Perez-Naranjo 2009; Santos-Gonzalez et al. 2007; Weishampel and Bedford 2006), the diversity of the former has been studied much less comprehensively. While a fairly large body of work is available on AMF-wheat associations (e.g. Almaca and Ortas 2010; Gao et al. 2010; Reddy 2009; reviewed in Hata et al. 2010), studies on non-mycorrhizal mycoendosymbionts associated with wheat is relatively sparse, with Abdellatif et al. (2009) being one of the few examples. Fungal organisms associated with wheat, which may not be endophytic in growth habit, include Ascomycota, Basidiomycota and Zygomycota (Mavragani 2008) while some of the few confirmed fungal endophytes of wheat include Ascomycota mitosporic strains belonging to class Dothideomycetes and class Eurotiomycetes (Abdellatif et al. 2009; Hibbet et al. 2007; Kiffer and Morelet 2000). Because of this scarcity of data specific to wheat, it is necessary to speak in terms of the potential of fungal endophytes of wheat and to hypothesise on this potential based on what is currently known about endophytic fungi in grass and cereal species. In addition, because the biology the AMF and their associations with wheat is relatively well known, compared to that of non-mycorrhizal fungal endophytes, it is instructive to explore the similarities and differences between AMF and fungal endophytes.

Fungal endophytes and AMF differ in terms of phylogenetics, morphology, colonization structures, plant tissues colonization, mode of plant-to-plant transmission, culturability, environmental preference and interactions with host plants (Table 1). Unlike AMF, which belong to the phylum Glomeromycota (Peterson et al. 2004), fungal endophytes belong to the sub-kingdom dikarya and to the phyla Ascomycota and Basidiomycota (Jumpponen 2001). Endophytic fungi are multicellular organisms composed of hypha partitioned by cross walls or septa (e.g. Jun et al. 2002; Peterson et al. 2004).

The colonization structures formed by fungal endophytes are more diverse than those formed by AMF. While AMF are known to produce intracellular arbuscules, intercellular hyphae and, in some cases, vesicles (e.g. Kobae and Hata 2010; Urcelay and Battistella 2007), endophytic fungi can produce intracellular hyphal coils and intercellular hypha which can vary in terms of cell length-to-width ratio and orientation relative to the long axis of the plant organ (Abdellatif et al. 2009). One similarity between AMF and fungal endophytes is the capacity to form either intercellular or intracellular colonization structures (Abdellatif et al. 2009).

While AMF colonize only plant roots, fungal endophytes can colonize roots, shoots, or both (reviewed in Rodriguez et al. 2009). In addition, while AMF are only transmitted horizontally, with fungi passing from plant to plant by way of vegetative structures (Cavalli-Sforza and Feldman 1981), fungal endophytes can be transmitted either horizontally or vertically with maternal plants passing fungi on to offspring via seed infections (Cavalli-Sforza and Feldman 1981; reviewed in Rodriguez et al. 2009; Saikkonen et al. 2002).

AMF appear to be able to grow under a more limited set of environmental conditions than non-AMF endophytic fungi. While many non-mycorrhizal mycoendosymbionts can be cultured *in vitro* (e.g. Abdellatif et al. 2009; Marquez et al. 2007), AMF are obligate biotrophs (Declerk et al. 2005). In addition, in more external environments, such as deeper layers of soil or high altitudes, non-AMF fungal endosymbionts become more dominant, while AMF become less dominant (Medina-Roldán et al. 2008; Perez-Naranjo 2009; Schimdt et al. 2008).

Interactions between plant hosts and fungal endophytes also differ from those between plants and AMF in that, while AMF are generally beneficial to their plant host, and occasionally commensal (reviewed in Auge 2001), the interaction between fungal endophytes and grass hosts can range from mutualism to commensalism to parasitism (Declerck et al. 2005; Jumpponen 2001; Redman et al. 2001).

A wide range of classification systems has been proposed for making sense of the enormous diversity of fungal endophytes. However, none of the classification systems proposed to date fully reflects the nature of endophytic fungi and their functions in symbiosis with plants (reviewed in Hyde and Soytong 2008 and Saunders et al. 2010). This state of affairs is to be expected because scientific study of fungal endophytes is a relatively nascent field. It is likely that, as the field progresses, classification systems of greater clarity and utility will be developed.

Table 1. Comparision of fungal endophytes and Arbuscular Mycorrhizal Fungi (AMF)

	Fungal Endophytes	**Arbuscular Mycorrhizal Fungi**	**References**
Taxanomic Classification	Asco and Basidiomycota	Glomeromycota	1) Abellatiff, et al. 2009; Walker, et al. 2005 2) Peterson, et al. 2004
Hyphal Compartmentalization	Mulitcellular hyphae	Unicellular, coenocytic, hyphae	3) Jun et al. 2002 4) Peterson, et al. 2004
Colonization Structures	Diverse, complex and varied. E.g. hyphal coils and knots	Less diverse. E.g. Arbusucules and vesicles	5) Abellatiff, et al. 2009 6) Kobae and Hata 2010; Urcelay and Battistella 2007
Inter or Intracellular Colonization	Both	Both	7) Abellatiff, et al. 2009 8) Peterson, et al. 2004
Plant Organs Colonized	Roots and/or shoots and/or seeds	Roots only	9) Rodriguez, et al. 2009 10) Kirk, et al. 2001
Vertical or Horizontal Transmission	Both	Horizontal	11) Rodriguez, et al. 2009; Saikkonen, et al. 2002 12) Peterson, et al. 2004
Culturable	Yes/No	No	13) e.g. Marquez, et al. 2007 14) Declerk, et al. 2005
Dominance in Extreme Environments	Increases	Decreases	15) & 16) Perez-Naranjo 2009; Schmidt, et al. 2008, Medina-Roldan, et al. 2008
Interactions with Host Plant	Continum from parasite to mutalist. Tolerance of biotic and/or abiotic stress	Mutualist or commensal. Nutrient absorption, tolerance or abiotic and/or biotic stress.	17) Auge 2001 18) Declerk, et al. 2005; Jumpponen 2001; Redman, et al. 2001

Both fungal endophytes and mycorrhizal fungi have been classified based on the host plant (Harley and Smith 1983; Brundett 1991). However, this method of classification is most useful for fungi with a very limited host range and is not particularly informative regarding the endosymbiotic fungal partner. In addition, in the context of this chapter, which focuses on the wheat host, this classification system is unhelpful.

Endophytic fungi have also been classified based on the phylogenetic identity of the fungal partner (reviewed in Roderiguez et al. 2009). While this classification system may be helpful when applied to well-characterized fungal endophytes, such as clavicipitaceous fungal endophytes, it is less helpful when applied to fungi whose taxonomic identity is less well understood. Fungal endophytes have also been classified based on qualitative and sometimes incomplete descriptions of the colonization structures formed. The designation "dark septate endophytes" (DSE), which has been in use since 1922 (Melin 1922), is an example of this type of classification. However, the term DSE can lead to confusion because some endophytes not normally considered DSE can appear to have dark septa (reviewed in Roderguez et al. 2009) and many non-AMF endophytic fungi have hyline hyphae (e.g. Haselwandter and Read 1982).

Because some fungal endophytes colonize only above ground (reviewed in Schadl et al. 2004), or only below ground (e.g. *Piriformospora indica*, reviewed in Singh et al. 2000), plant tissue, endophytic fungi can also be classified based on the plant tissues or organs colonized. In addition, fungal endophytes can be classified based on the types of mechanisms, discussed in more detail in the next section, involved in plant-fungal interactions.

Finally, attempts have been made at comprehensive classification systems, which take into account a range of the traits of fungal endophytes and their plant hosts mentioned above. For example, Rodriguez et al. (2009) divided fungal endophytes into four groups based on host range, tissue(s) colonized, *in planta* level of colonization, *in planta* biodiversity, horizontal or vertical transmission and type of benefits conferred. This classification system also incorporated the mechanisms by which fungal endophytes and their plant hosts interact.

PLANT-ENDOPHYTE INTERACTIONS

Although a great diversity of mechanisms of plant-endophyte interactions has been documented, no comprehensive classification of endophytic fungi has been proposed (Hyde and Soytong 2008; Saunders et al. 2010) and relatively few plant-non-AMF mycoendosymbiont interactions have been studied in wheat. Thus, this chapter will focus on potential wheat-endophyte interactions based on interactions documented in grasses and cereals closely related to wheat. Mutualistic mycosymbiont-host associations can occur at various stages of the plant life cycle. A beneficial relationship between fungi and seeds in which the fungal partner helps to maintain seed viability and promote seed germination can be referred to as mycovitality (Vujanovic and Vujanovic 2007). In contrast, fungal-derived enhancement of plant survival and growth at post-germination life stages can be described as mycoheterotrophy (Peterson et al. 1998). The mechanisms involved in mycovitality and mycoheterotrophy include morphological, physiological, epigenetic, metabolic, hormonal or enzymatic interactions and can result in abiotic stress tolerance, biocontrol, herbivory deterrence, increased biomass, yield and/or nutrient absorption (Table 2).

Table 2. Mechanisms by whoch fungal endophytes confer benefits to host plants

	Benefits						
	Myocovitality*	Mychoheterotrophy					
Potential Mechanism(s)	Seed Germination/ Breaking Dormancy	Abiotic Stress Tolerance	Biocontrol	Herbivory Deterence	Increased Biomass and/or Yield	Nutrient Absorption	References
Altered plant morphology/physiology	X_1	X_2					1) Vujanovic, et al. 2000 2) Bayat, et al. 2009; Richardson, et al. 1993
Osmolite accumulation		X_3					3) Hill, et al. 1990; Richardson, et al. 1992
Altered cell wall elasticity		X_4					4) Malinowski 1995; Malinowski and Belesky 2000; White, et al.1992
Epigenetic changes	$?_5$	X_6	X_7			X_8	5) Vujanovic, et al. 2000 6) Walker, et al. 2005 7) Vujanovic, et al. 2009 8) Smith, et al. 2009
ROS scavenging	$?_9$	X_{10}					9) Vujanovic, et al. 2000 10) Rodrigues, and Redman 2005; Tanaka, et al. 2006
Production of fungal metabolites				X_{11}			11) Shiba and Sugawara 2008
Systemic Acquired Resitance (SAR)	$?_{12}$		X_{13}				12) Vujanovic, et al. 2000 13) Rodriguez and Redman 2008; Vallad and Goodman 2004
Phytohormone production by fungal endophyte	X_{14}						14) Tudzynski 2005; Vujanovic, et al. 2000; Vujanovic and Vujanovic 2007
Nutrient solubilizing enxyme production	X_{15}				X_{16}	X_{17}	15) Vujanovic, et al. 2000 16) & 17) Auge, et al. 2001

* There is currently only one sub-category for mycovitality because research into other potential benefits associated with this phenomenon is still ongoing.

Fungal endophytes may enhance wheat tolerance for abiotic stress by altering plant physiology. For example, based on findings in *Neotyphodium* colonized grasses, endophyte colonized wheat may have increased root surface area for water and ion uptake (Bayat et al. 2009) and/or altered in stomatal conductance (Richardson et al. 1993) to minimize moisture lose.

Because fungal endophytes of the cool season grass, tall fescue, have been shown to promote drought tolerance by increasing the accumulation of non-structural, osmotically active, carbohydrates (Hill et al. 1990; Richardson et al. 1992) and altering cell wall elasticity (Malinowski 1995; White et al. 1992; reviewed in Malinowski and Belesky 2000) it is reasonable to hypothesise that similar mechanisms may be at work in fungal endophyte colonized wheat.

Epigenetic changes, defined as changes in gene expression that do not arise from changes in the underlying DNA sequence (Bird 2007), in host plants could be induced by the presence of endophytic fungi. Epigenetic 'reprogramming' could be maintained through cycles of cell division and may be inherited by subsequent generations, even in the absence of fungal endophyte. Such reprogramming has been observed in barley as a result of colonization by the endophytic fungus *Piriformospora indica* in the presence of abiotic stress (Walker et al. 2005). In addition, arbuscular mycorrhizae can lead to altered plant protein expression, such as increased phosphate transporter synthesis by the host plant (Burleigh and Harrison, 1997, 1999; Harrison and Van Buuren, 1995; reviewed in Auge et al. 2001 and Smith et al. 2009). Epigenetic reprogramming of wheat as a result of colonization by fungal endophytes could hypothetically lead to increased production of proteins associated with wheat response to abiotic stress. For example, increased production for dehydrins (Brini et al. 2007) and/or elongation factors which function as heat shock proteins (Bukovnik et al. 2009) might occur in the presence of heat and drought stress, respectively. Vujanovic et al. (2009) found differential protein expression in endophyte colonized wheat in the presence of a pathogenic *Fusarium* species; this altered protein expression coincided with dramatically enhance *Fusarium* pathogen tolerance.

Rodriguez and Redman (2005) hypothesized that fungal endophytes may confer tolerance to abiotic stresses such as heat or drought by scavenging reactive oxygen species (ROS) produced by the plant as a stress response. This hypothesis is supported by the fact that ROS are thought to play a role in the communication between *Neotyphodium* endophytes and their hosts, which are closely related to wheat (Tanaka et al. 2006).

Fungal endophytes may function as biocontrol agents by competing with potential pathogens for infection sites and/or nutrients, producing metabolites that are toxic to pathogens and/or by re-programming host gene expression via systemic acquired resistance (SAR) and/or Induced Systemic Resistance (ISR) (reviewed in Vallad and Goodman 2004). However, as fungal endophytes have been shown to display a continuum of mutualistic to commensal to parasitic interactions with their plant hosts (reviewed in Rodriguez and Redman 2008), careful laboratory, greenhouse and field trials of a specific endophyte under a range of conditions should be carried out to ensure that a potential biocontrol agent does not itself act as a plant pathogen.

Given that Clavicipitaceous fungal endophytes, such as *Neotyphodium* sp. are known to deter herbivores and insect pests on ryegrass via the production of loline alkaloids (Shiba and Sugawara 2008), fungal endophytes of wheat may confer similar benefits. However, as loline alkaloids can have detrimental effects of livestock health (reviewed in Porter 1995), there presence in wheat could conceivably have a negative impact on human health.

Fungal endophytes may contribute positively to wheat germination (mycovitality) and growth by increasing nutrient solublization and absorption or by producing as growth-promoting phytohormones (Vujanovic et al. 2000), such as gibberellins which certain fungi are known to produce (reviewed in Tudzynski 2005). Enhanced nutrient solublization and absorption could potentially occur via the production of nutrient solubilizing enzymes, similar to mycorrhizae (reviewed in Auge et al. 2001).

Discussion and Potential Outcomes

The study of fungal endophytes of wheat has many potential benefits, including both an increase in scientific understanding and in the development of practical applications. Understanding the diversity of mycoendosymbionts and the ways in which endophytic microfungi differ from AMF may highlight the greater ease with which fungal endophytes could be manipulated for agricultural applications. In addition, insight into the mechanisms of wheat-endophyte interactions could lead to more effective agricultural applications.

Endophytic fungi show potential as biocontrol (Latch 1993; Prestidge and Gallagher 1988; Shiba and Sugawara 2008) and biofertilizer (Deshmukh and Kogel 2007; Marquez et al. 2007; Rodriguez et al. 2008) agents of wheat. The fact that these Asco- and Basidiomycetes can by cultured under laboratory conditions permits the development of easily handled application, including peat (Hynes et al. 1995) or liquid based seed treatments containing fungal spores and/or hyphae and soil-based inoculation. Although peat based seed treatments have been frequently used, this method has proved to be problematic, both in terms of plant responses in research trials and from the point of view of agricultural equipment and producer convenience (e.g. Hynes et al. 1995). Liquid base inoculation methods are a promising method of applying beneficial fungal endophytes to wheat because they not only permit the quantification of the dose of colony forming units applied per seed, but could also be uniformly applied to a large number of seeds prior to, or during, planting. In addition, producers sometimes prefer liquid based seed inoculation methods because of convenient application methods that permit a single worker to apply the inoculant and plant the treated seeds (Walley et al. 2004). However, trials on the effectiveness of liquid inoculants and requirements for agents to bind the fungal organism to the seeds are needed. Soil-based inoculation has been shown to be the most effective method of delivering nitrogen-fixing biofertilizers (Clayton et al. 2004a; Clayton et al. 2004b).

In addition, the dikaryotic, multicellular nature of non-mycorrhizal mycoendosymbionts, coupled with the diverse plant-to-plant transmission methods (reviewed in Roderguez et al. 2009), opens the door to more innovative agricultural applications. Insight into the most advantageous applications will likely be gained by a more completely understanding of the molecular and cellular mechanisms behind cereal-mycoendophyte interactions.

Fungal endophytes of wheat could also be important in sustainable agricultural production. This could be achieved by substituting mycoendophytic biocontrol agents for pesticides, growth and/or yield promoting endophytes for fertilizers and/or drought tolerance promoting mycoendosymbionts for irrigation. The use of fungal endophytes to mitigate crop loses on marginal land may prove especially important given that stresses such as drought, heat and soil salinity and likely to increase in frequency and severity as climate change progresses (IPCC 2007). The use of the fungal endophyte seed and/or soil treatments could function as a form of "insurance" which would not decrease wheat performance under stress-free conditions, both which could reduce loses should abiotic stress be encountered during the growing season.

Many fungal endophytes are naturally occurring in the environment and are studied in research laboratories without genetic modification (e.g. Abdellatif et al. 2009; Marquez et al. 2007). Because of this, combined with the fact that inoculation of wheat which certain fungal endophytes can confer benefits similar to those achieved by the introduction of exogenous genes, fungal endophytes show potential as an alternative to genetically modified organisms (GMOs).

CONCLUSION

Fungal endophytes of grasses and cereals are highly diverse Asco- and Basidiomycetes which, unlike AMF, can be cultured *in vitro*. This trait could facilitate the efficient manipulation of mycoendophytes in biocontrol and biofertilizer applications. A greater understand of the molecular and physiological mechanisms behind the interactions between fungal endophytes and their plant hosts will likely facilitate beneficial use of mycoendosymbionts of wheat, while avoiding deleterious impacts.

REFERENCES

Abdellatif L, Bouzid S, Kaminskyj S and Vujanovic V (2009) Endophytic hyphal compartmentalization is required for successful symbiotic Ascomycota association with root cells. *Mycological Research*. 113: 782 – 791.

Almaca A and Ortas I (2010) Growth response of maize plants (Zea mays L.) to wheat and lentil pre-cropping and to indigenous mycorrhizae in field soil. *Spanish Journal of Agricultural Research*. 8: S131 – S136.

Auge RM (2001) Water relations, drought and vesicular-arbuscular mycorrhizal symbiosis. *Mycorrhiza*. 11: 3 – 42.

Bayat F, Mirlohi A and Khodambashi M (2009) Effects of endophytic fungi on some drought tolerance mechanisms of tall fescue in a hydroponics culture. *Russian Journal of Plant Physiology*. 56: 510 – 516.

Bird A (2007) Perceptions of epigenetics. *Nature*. 447: 396 – 398.

Brini F, Hanin M, Lumbreras V, Irar S, Pagès M and Khaled MK (2007) Functional characterization of DHN-5, a dehydrin showing a differential phosphorylation pattern in two Tunisian durum wheat (*Triticum durum* Desf.) varieties with marked differences in salt and drought tolerance. *Plant Science*. 172: 20 – 28.

Brundett M (1991) Mycorrhizas in natural ecosystems. *Advances in Ecological Research*. 21: 171 – 313.

Bukovnik U, Fu J, Bennet M, Prasa PVV and Ristic Z (2009) Heat tolerance and expression of protein synthesis elongation factors, EF-Tu and EF-1α, in spring wheat. *Functional Plant Biology*. 36: 234 – 241.

Burleigh SH and Harrison MJ (1997) A novel gene whose expression in *Medicago truncatula* roots is suppressed in response to colonization by vesicular-arbuscular mycorrhizal (VAM) fungi and to phosphate nutrition. *Plant Molecular Biology*. 34: 199 –208.

Burleigh SH and Harrison MJ (1999) The down-regulation of *Mt4*-like genes by phosphate fertilization occurs systemically and involves phosphate translocation to the shoots. *Plant Physiology*. 119: 241 – 248.

Cavalli-Sforza LL and Feldman M (1981) *Cultural Transmission and Evolution: A Quantitative Approach*. Princeton, New Jersey: Princeton University Press.

Clayton GW, Rice WA, Lupwayi NZ, Johnston AM, Lafond GP, Grant CA and Walley F (2004a) Inoculant formulation and fertilizer nitrogen effects on field pea: Nodulation, N_2 fixation and nitrogen partitioning. *Canadian Journal of Plant Science*. 84: 79 – 88.

Clayton GW, Rice WA, Lupwayi NZ, Johnston AM, Lafond GP, Grant CA and Walley F (2004b) Inoculant formulation and fertilizer nitrogen effects on field pea: Crop yield and seed quality. *Canadian Journal of Plant Science*. 84: 89 – 96.

Declerk S, Strulla DG and Fortin JA (eds) (2005) *In vitro* culture of mycorrhizas. In: Soil Biology, Ed: Varma A. Springer-Verlag. 4: 388.

Deshmukh SD and Kogel KH (2007) *Piriformospora indica* protects barley from root rot caused by *Fusarium graminearum*. *Journal of Plant Diseases and Protection*. 114: 263 – 268.

Gao X, Akhter F, Tenuta M, Flaten DN, Gawalko EJ and Grant CA (2010) Mycorrhizal colonization and grain Cd concentration of field-grown durum wheat in response to tillage, preceding crop and phosphorus fertilization. *Journal of the Science of Food and Agriculture*. 90: 750 – 758.

Harley JL and Smith SE (1983) *Mycorrhizal Symbiosis*. Academic Press, London.

Harrison M J and Van Buuren M L (1995) A phosphate transporter from the mycorrhizal fungus *Glomus versiforme*. *Nature*, 378: 626 – 629.

Hatta S, Kobae Y and Banba M (2010) Interactions between plants and arbuscular mycorrhizal fungi. *International Review of Cell and Molecular Biology*. 281: 1 – 48.

Hibbet DS, Binder M, Bischoff JF, Blackwell M, Cannon PF, Eriksson OE, Huhndorf S, James T and Kirk PM, Luecking R, Lumbsch HT, Lutzoni F, Matheny PB, Mclaughlin DJ, Powell MJ, Redhead S, Schoch CL, Spatafora JW, Stalpers JA, Vilgalys R, Aime MC, Aptroot A, Bauer R, Begerow D, Benny GL, Castlebury LS,Crous PW, Dai Y-C, Gams W, Geiser DM, Griffith GW, Gueidan C, Hawksworth DL, Hestmark G, Hosaka K, Humber RA, Hyde KD, Ironside JF, Koljalg, Urmas K; Cletus P, Larsson K-H, Lichtwardt R, Longcore J, Miadlikowska J, Miller A, Moncalvo J-M, Mozley-Standridge S, Oberwinkler F, Parmasto E, Reeb V, Rogers JD, Roux C, Ryvarden L, Sampaio JP, Schuessler A, Sugiyama J, Thorn RG, Tibell L, Untereiner WA, Walker C, Wang Z, Weir A, Weiss M, White MM, Winka K, Yao Y-J and Zhang N (2007) A higher-level phylogenetic classification of the fungi. *Mycological Research*. **111**: 509 – 547.

Haselwandter K and Read DJ (1982) The significance of root-fungus association in two *Carex* species of high-alpine plant communities. *Oecologia*. 53: 352 – 354.

Hill NS, String WC, Rottinghaus GE, Belesky DP, Parrott WA and Pope DD (1990) Growth, morphological, and chemical component responses of tall fescue to *Acremonium coenophialum*. *Crop science*. 30: 156 – 161.

Hyde KD and Soytong K (2008) The fungal endophyte dilemma. *Fungal Diversity*. 33: 163 – 173.

Hynes R.K, Craig KA, Covert D, Smith RS and Rennie RJ (1995) Liquid rhizobial inoculants for lentil and field pea. *Journal of Production Agriculture*. 8: 547 – 552.

Jumpponen A (2001) Dark septate endophytes: Are they mycorrhizal? *Mycorrhiza*. 11: 207 – 211.

Jun J, Abubaker J, Rehrer C, Pfeffer PE, Shachar-HillY and Lammers PJ (2002) Expression in an arbuscular mycorrhizal fungus of genes putatively involved in metabolism, transport, the cytoskeleton and the cell cycle. *Plant and Soil*. 244: 141–148.

Kiffer E and Morelet M (2000) The Deuteromycetes Mitosporic Fungi: Classification and Generic Keys, Science Publisher Inc., Enfield, NH, Plymouth.

Kirk PM, Cannon PF, David JC and Stalpers J (2001) *Ainsworth and Bisby's Dictionary of the Fungi*. 9th ed. CAB International, Wallingford, UK.

Kobae Y and Hata S (2010) Dynamics of periarbuscular membranes visualised with a fluorescent phosphate transporter in arbuscular mycorrhizal roots of rice. *Plant and Cell Physiology*. 51: 341 – 353.

Latch GCM (1993) Physiological interactions of endophytic fungi and their hosts. Biotic stress tolerance imparted to grasses by endophytes. *Agriculture, Ecosystems and Environment*. 44: 143 – 156.

Malinowski D (1995) Rhizomatous ecotypes and symbiosis with endophytes as new possibilities of improvement in competitive ability of meadow fescue (*Festuca pratensis* Huds.). Ph.D. Thesis. Diss ETH No. 11397. Swiss Fed. Inst. Technology, Zurich.

Malinowski DP and Belesky DP (2000) Adaptations of endophyte-infected cool-season grasses to environmental stresses: Mechanisms of drought and mineral stress tolerance. *Crop Science*. 40: 923 – 940.

Mandyam K and Jumpponen A (2008) Seasonal and temporal dynamics of arbuscular mycorrhizal and dark septate endophytic fungi in a tallgrass prairie ecosystem are minimally affected by nitrogen enrichment. *Mycorrhiza*. 18: 145 – 155.

Marquez LM, Redman RS, Rodriguez RJ and Roossinck MJ (2007) A virus in a fungus in a plant-three way symbiosis required for thermal tolerance. *Science*. 315: 513 – 515.

Mavragani D (2008) Molecular and ecological studies of fungal biodiversity on durum wheat grown in rotation with pulses and canola. University of Saskatchewan Thesis. etd-07212008-143535. http://library2.usask.ca/theses/available/etd-07212008-143535/

Medina-Roldán E, Arredondo J, Huber-Sannwald E, Chapa-Vargas L and Olalde-Portugal V (2008) Grazing effects on fungal root symbionts and carbon and nitrogen storage in a shortgrass steppe in Central Mexico. *Journal of Arid Environments*. 72: 546 – 556.

Melin E (1922) On the mycorrhizas of *Pinus sylvestris* L. and *Picea abies* Karst. A preliminary note. *Journal of Ecology*. 9: 254 – 257.

Perez-Naranjo JC (2009) Dark septate and arbuscular mycorrhizal fungal endophytes in roots of prairie grasses. University of Saskatchewan Thesis. etd-12222009-170513. http://library2.usask.ca/theses/available/etd-12222009-170513/

Peterson, R.L., H.B. Massicotte and L.H. Melville (2004) *Mycorrhizas: anatomy and cell biology*. National Research Council Research Press. ISBN 978-0-660-19087-7.

Peterson RL, Uetake Y and Zelmer C (1998) Fungal symbiosis with orchid protocorms. Symbiosis. 25: 29 – 55.

Porter JK (1995) Analysis of endophyte toxins: fescue and other grasses toxic to livestock. *Journal of Animal Science*. 73: 871 – 880.

Prestidge RA and Gallagher RT (1988) Endophyte fungus confers resistance to ryegrass: Argentine stem weevil larval studies. *Ecological Entomology*. 13: 429 – 435.

Redman RS, Dunigan DD and Rodriguez RJ (2001) Fungal symbiosis from mutualism to parasitism: who controls the outcome, host or invader? *New Phytologist*. 151: 705 – 716.

Richardson MD, Chapman GWJr., Hoveland CS and Bacon CW (1992) Sugar alkohols in endophyte-infected tall fescue. Crop Science. 32: 1060 – 1061.

Richardson MD, Hoveland CS, Carl S and Bacon CW (1993) Photosynthesis and stomatal conductance of symbiotic and nonsymbiotic tall fescue. *Crop Science*. 33: 145 – 149.

Roberts CA, West CP, Spiers DE, eds (2005). *Neotyphodium in Cool-Season Grasses*. Blackwell. ISBN 978-0813801896.

Rodriguez RJ, Henson J, Volkenburgh EV, Hoy H, Wright L, Beckwith F, Kim Y-O and Redman RS (2008) Stress tolerance in plants via habitat adapted symbiosis. *International Society for Microbial Ecology Journal*. 2: 404 – 416.

Rodriguez R and Redman R (2005) Balancing the generation and elimination of reactive oxygen species. *Proceedings National Academy of Science USA*. 102: 3175 – 3176.

Rodriguez R and Redman R (2008) More than 400 million years of evolution and some plants still can't make it on their own: plant stress tolerance via fungal symbiosis. *Journal of Experimental Botany*. 59: 1109 – 1114.

Rodriguez RJ, White JF, Arnold AE and Redman RS (2009) Fungal endophytes: diversity and functional roles. New Phytologist. 182: 314 – 330.

Saikkonen K, Ion D and Gyllenberg M (2002) The persistence of vertically transmitted fungi in grass metapopulations. Proceedings of the Royal Society: *Biological Sciences*. 269: 1397 – 1403.

Santos-Gonzalez JC, Finlay, RD and Tehler A (2007) Seasonal dynamics of arbuscular mycorrhizal fungal communities in roots in a seminatural grassland. *Applied and Environmental Microbiology*. 73: 5613 – 5623.

Saunders M, Glenn AE and Kohn LM (2010) Exploring the evolutionary ecology of fungal endophytes in agricultural systems: using functional traits to reveal mechanisms in community processes. *Evolutionary Applications*. 3: 525 – 537.

Schadl CL, Leuchtmann A and Spiering MJ (2004) Symbiosis of grasses with seedborne fungal endophytes. *Annual Review of Plant Biology*. 55: 315 – 340.

Schmidt SK, Sobieniak-Wiseman LC, Kageyama SA, Halloy SRP and Schadt CW (2008) Mycorrhizal and dark-septate fungi in plant roots above 4270 meters elevation in the Andes and Rocky Mountains Arctic Antarctic and Alpine Research. 40: 576 – 583.

Shiba T and Sugawara K (2008) Fungal loline alkaloids in grass endophyte associations confer resistance to the rice leaf bug, *Trigonotylus caelestialium*. *Entomologia Experimentalis et Applicata*. 130: 55 – 62.

Singh A, Sharma J, Rexer K-H and Varma A (2000) Plant productivity determinants beyond minerals, water and light: *Piriformospora indica* – A revolutionary plant growth promoting fungus. *Current Science*. 79: 1548 – 1554.

Smith SE, Facelli E, Pope S and Smith FA (2009) Plant performance in stressful environments: interpreting new and established knowledge of the roles of arbuscular mycorrhizas. *Plant Soil*. 326: 3 – 20.

Tanaka A, Christensen MJ, Takemoto D, Park P and Scott B (2006) Reactive oxygen species play a role in regulating a fungus–perennial ryegrass mutualistic interaction. *The Plant Cell*. 18: 1052 – 1066.

Trappe JM (1987) *Phylogenetic and ecologic aspects of mycotrophy in the angiosperms from an evolutionary standpoint*. Ecophysiology of VA Mycorrhizal Plants, G.R. Safir (EDS), CRC Press, Florida.

Tudzynski B (2005) Gibberellin biosynthesis in fungi: genes, enzymes, evolution, and impact on biotechnology. *Applied Microbiology and Biotechnology*. 66: 597 – 611.

Urcelay C and Battistella R (2007) Mycorrhizal colonization among different herbaceous plant functional types from central Argentina. *Ecologia Austral*. 17: 179 – 188.

Vallad GE and Goodman RM (2004) Systemic acquired resistance and induced systemic resistance in conventional agriculture. *Crop Science*. 44: 1920 – 1934.

Vujanovic V, Abdellatif L, Daida P, Bouzid S and Ferandez M (2009) Comparisons of *Fusarium avenaceum* isolates originated from Saskatchewan Prairie Ecozones by vegetative compatibility, rDNA analyses, heat tolerance and pathogenicity on wheat. *Canadian Journal of Plant Pathology* (submitted).

Vujanovic V, St-Arnaud M, Barabe D and Thibeault G (2000) Viability testing of orchid seed and the promotion of colouration and germination. *Annuals of Botany*. 86: 79 – 86.

Vujanovic V and Vujanovic J (2007) Mycovitality vs. mycoheterotrophy in plants: Where lies the dormancy in terrestrial pants with minute seed? *Symbiosis*, 44: 93 – 100.

Walker F, Achatz B, Baltruschat H, Fodor J, Becker K, Fischer M, Heier T, Hückelhoven R, Neumann C, von Wettstein D, Franken P and Kogel K-H (2005) The endophytic fungus *Piriformospora indica* reprograms barley to salt-stress tolerance, disease resistance, and higher yield. *Proceedings National Academy of Science USA*. (102): 13386 – 13391.

Walley F, Clayton G, Gan Y and Lafond G (2004) Performance of rhizobial inoculant formulations in the field. Online. *Crop Management*, doi:10.1094/CM-2004-0301-03-RV.

Weishampel PA and Bedford BL (2006) Wetland dicots and monocots differ in colonization by arbuscular mycorrhizal fungi and dark septate endophytes. *Mycorrhiza*. 16: 495 – 502.

Wilson D (1993) Fungal endophytes: out of sight but should not be out of mind. *Oikos*. 68: 379 – 384.

White RH, Engelke MC, Morton SJ, Johnson-Cicalese JM and Ruemmele BA (1992) *Acremonium* endophyte effects on tall fescue drought tolerance. *Crop Science*. 32: 1392 – 1396.

In: Wheat: Genetics, Crops and Food Production
Editor: Maria T. Almeida, pp. 347-362
ISBN 978-1-61209-307-9

Chapter 13

FUSARIUM GRAMINEARUM SENSU STRICTO CHEMOTYPE DISTRIBUTION IN THE ARGENTINEAN WHEAT CROPPING AREA

Cora L. Alvarez* and Virginia E. Fernández Pinto
PRHIDEB-PROPLAME CONICET, Biodiversity and Experimental Biology Department, School of Sciences, University of Buenos Aires (UBA). Ciudad Universitaria, Buenos Aires, Argentina

ABSTRACT

Fusarium species are worldwide causal agents of *Fusarium* Head Blight (FHB) in cereals. Head Blight can not only reduce yield but also produce fungal toxins that seriously damage grain quality for human and animal consumption. Several studies have shown that *F. graminerarum sensu stricto* and deoxynivalenol (DON) are the main contaminants detected in wheat in Argentina. Chemotype characterization has been extensively used to characterize *F. graminearum* for its toxigenic potential. Many reports have shown differences over space and time of its chemotype distribution, suggesting that this approach is valuable to map a population and identify population changes in the field. The analysis of the trichothecene chemotype distribution of the *F. graminearum sensu stricto* strains across the Argentinean wheat cropping area has revealed that the 3-DON chemotype is always present. Differences in chemotype distribution may result in the shift in toxin accumulation in grains. However, no definitive proof on the correlation between chemotype detection and grain contamination has been found to date. This is mainly due to the variability of toxin production found among strains and to the fact that the samples analyzed have been taken from different origins and environmental conditions. Thus, the combination of genetic chemotyping with chemical analysis of toxin production is desirable. Many authors who have studied this important pathogen on wheat suggest the presence of only one population with high genetic variation in Argentina. Whether or not this variation implies differences in chemotype is an issue that needs further studies. Another point of remarkable interest is the increase in the presence

* PRHIDEB-PROPLAME CONICET, Biodiversity and Experimental Biology Department, School of Sciences, University of Buenos Aires (UBA). Ciudad Universitaria, Pabellón II, 4° piso, C1428 EHA, Buenos Aires, Argentina. cora@bg.fcen.uba.ar

of other casual agents of FHB, such as *F. poae*, a relatively weak pathogen compared with *F. graminearum*, but capable of producing a large number of mycotoxins, including type A and type B trichothecenes, beauvericin and enniatins. All these aspects show the need for a better knowledge of the population dynamics and distribution of the pathogens associated to wheat in the large production area of Argentina.

The objectives of this study were to identify the chemotype distribution of *F. graminearum* isolates from wheat obtained in different regions of the main wheat production area in Argentina and to determine whether there are variations in this distribution among crop seasons or geographical locations.

WHEAT PRODUCTION AREA

Wheat (*Triticum aestivum* L.) is one of the most important cereal crops in Argentina, both as a product for local consumption and as an export commodity (9,645,491 tons in 2007 http://faostat.fao.org/site/342/default.aspx). The production area, which is of about 6,000,000 ha, is divided into five sub-regions (I to V, with sub-region II and V divided into North (N) and South (S)), according to agrometeorological conditions (Figure 1). The main production area is located in the Buenos Aires province (Zones IIS, IV and VS), East of La Pampa province (Zone VS) and South of Santa Fe province (Zone IIN). This extended area presents very different conditions of temperature and humidity.

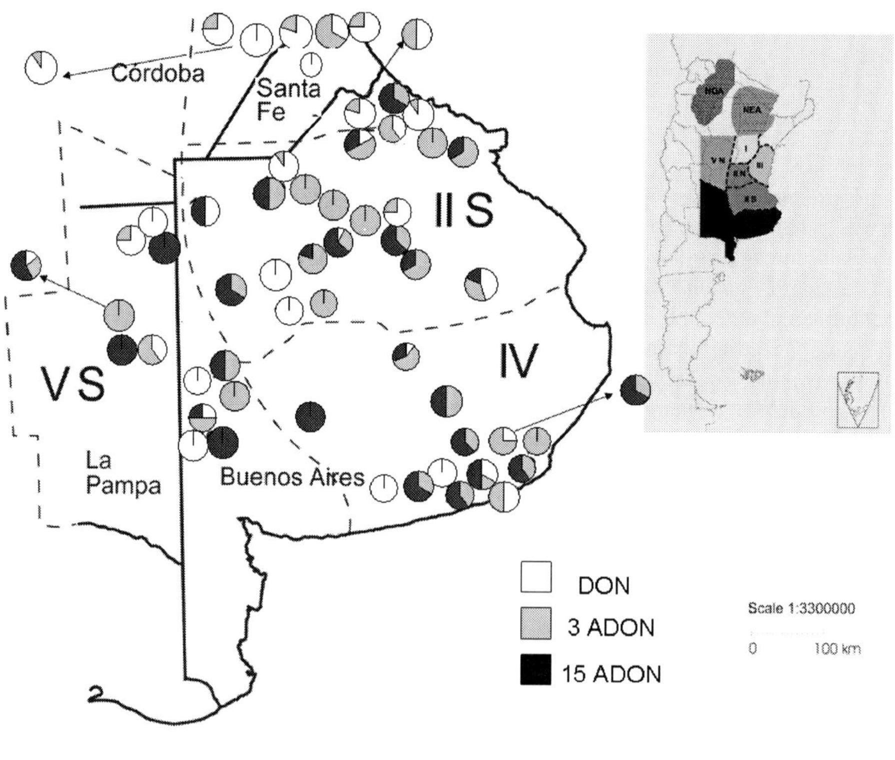

Figure 1. Map of the wheat cultivation area in Argentina (right). Five subregions are considered according to agrometeorological conditions [I to V, with II and V divided into North (N) and South (S)]. The origin of the isolates studied here and their chemotype are indicated on the left. Arrows show a different chemotype found in the same location.

FUSARIUM HEAD BLIGHT

Fusarium Head Blight (FHB) is an important disease affecting the production of wheat worldwide. *Fusarium graminearum* (Schwabe) (teleomorph *Gibberella zeae* (Schwein.)) is considered the main cause of FHB worldwide (Xu and Berrie 2005). In Argentina, the main pathogen isolated from blighted heads is also *F. graminearum* (Galich, 1996). FHB severity depends on climatic, agronomic and genetic factors. Frequent rainfall, high humidity and prolonged dew coinciding with the flowering and early kernel-fill periods of the crop favor disease infection and development. In Argentina, infection levels are variable, showing the highest FHB incidence in the most humid regions of the cropping area (region II in Figure 1) (Moschini *et al.*, 2001).

ECONOMIC IMPORTANCE

The worst outbreaks occurred in 1945-46, 1978, 1985, and 1993. Although yield losses varied among zones, they were estimated to average between 20-30% (de Galich 1989; de Galich and Galich 1994; Kohli et al. 1996). In 1993, the highest estimated losses reached 50% in areas of no-till over maize stubble. The extent of the damage was magnified by a considerable loss in the trading value of the grain resulting from low weight per hectoliter, presence of scabby grains, and mycotoxin contamination. In the most affected areas, nearly all production traded was below the established standards. To assess this problem, a wide sampling over the entire wheat growing region was carried out. Analysis results of whole samples from primary elevators, port terminal elevators, and flour mills showed that 57% of the grain was marketed within established standards. Among these samples, 11% had mycotoxin deoxynivalenol (DON) concentrations above 2 ppm, while 53% of the samples below the standard exceeded this level (Luqui and Gomez 1995). In the southeast wheat growing area, the main durum wheat region, severe epidemics occurred in 1963, 1976, 1978, and 1985, with crop losses as high as 70%. In 1976, durum wheat accounted for 20% of the wheat production of the area.

The epidemic was so severe that 60% of samples from farms fell below the established standards. Consequently, and due to other contributing factors, the planting area was heavily reduced. Eight years later, the proportion of durum wheat was reduced to 3.5% and was displaced by bread wheat, which is more resistant to FHB. As a result, durum wheat production in some years was less than the home market demand and had to be imported (de Galich 1989; Garcia 1989; Jenson 1995). Recently, there has been a trend to increase the area sown to bread wheat owing to the new possibilities offered by the MERCOSUR (common market of the Southern Cone countries of South America).

The disease is of significance because it reduces the yield and quality of the grain and results in the contamination of wheat kernels with mycotoxins, primarily trichothecenes (Desjardins 2006). The most common trichothecenes produced by *F. graminearum* are DON, its acetyl derivatives 3-acetyldeoxynivalenol (3-ADON) and 15-acetyldeoxynivalenol (15-ADON), nivalenol (NIV), and its acetylated derivative 4-acetyl-nivalenol (4-ANIV or fusarenone X, FUS-X) (Desjardins 2006).

Fusarium Graminearum

Until recently, the name '*Fusarium graminearum*' was used to describe what we now know to be a polyphyletic grouping of distinct fungal species. Before 1999, homothallic fungi causing FHB of wheat and barley were known as *F. graminearum* 'group 2' to distinguish them from the heterothallic fungus causing crown and foot rot on grains, which was called *F. graminearum* 'group 1' (Francis and Burgess, 1977). It is now known that these fungi are separate biological and phylogenetic species, each producing their own characteristic sexual state. The former 'group 1' is now known as *Fusarium pseudograminearum* and has a sexual stage called *Gibberella coronicola* (Aoki and O'Donnell, 1999). Recognition by morphological characters sometimes is not enough for accurate identification of fungal isolates at the species level. *Fusarium graminearum* 'group 2' is now known to be a monophyletic species complex. A genealogical concordance phylogenetic species recognition approach was used to identify species limits between strains morphologically identified as *F. graminearum*. DNA sequence comparisons have been made to determine whether concordant monophyletic genealogies of orthologous regions occur at independent loci (O'Donnell et al., 2004). Such concordance is the basis for recognition of phylogenetic species limits (Taylor et al., 2000). In the first analysis of DNA sequence, 99 strains collected from a variety of substrates from around the world were identified as seven phylogenetically distinct and biogeographically structured lineages (O'Donnell et al., 2000). Further studies based on DNA sequence of eleven nuclear genes and three intergenic regions led O'Donnell et al. (2004) to describe nine lineages within *F. graminearum* and to elevate these lineages to the rank of species. The nine species were collectively referred to as the *F. graminearum* species complex (FGSC). More recently, Starkey et al. (2007) described two novel species within the FGSC by phylogenetic analyses of DNA sequences of 13 genes. These species now have been formally named, with *F. graminearum sensu stricto* retained for the species most commonly associated with FHB worldwide.

As identification of *Fusarium* species is critical to predict the potential mycotoxigenic risk of the isolates, appropriate DNA sequences and species-specific PCR assays should be used in the characterization of the isolates. In *Fusarium,* translation elongation factor 1-α (TEF1-α) and β-tubulin genes appear to occur consistently as single copy genes, and show a high level of sequence polymorphism among closely related species (Rahjoo et al., 2008). TEF1-α has become the marker of choice as a single-locus identification tool in *Fusarium* (Geiser et al., 2004).

The most geographically widespread species of the FGSC is *F. graminearum sensu stricto* (syn. *F. graminearum* Lineage 7), which predominates in wheat in North and South America and in Europe (Làday et al. 2004; Zeller et al. 2003; 2004). *Fusarium asiaticum* (syn. *F. graminearum* Lineage 6) predominates in wheat in Korea and eastern China (Gale et al. 2002).

In Argentina, where *F. graminearum sensu stricto* has also been reported to be the main cause of FBH (Alvarez et al., 2006; Ramirez et al., 2007; Sampietro et al., 2010), FBH epidemics occur often due mainly to favorable weather conditions (Moschini et al. 2001). Other *Fusarium* species less pathogenic in wheat, including *F. equiseti*, *F. semitectum*, *F. poae, F. acuminatum*, *F. moniliforme*, *F. subglutinans*, *F. moniliforme* var. *anihophilum*, *F.*

oxysporum, and *F. Solani,* have been isolated from infected grains (Carrera 1954; Dalcero et al. 1996; de Galich 1989; Lori et al. 1992).

There has been a recent report of 67 isolates of *Fusarium* spp. obtained from samples of wheat grains from the Pampas region from 15 locations distributed in Buenos Aires, Entre Ríos, Santa Fe and Córboba provinces during the 2006 and 2007 wheat-growing seasons (Kikot et al., 2010). In that work, the identification of species from monosporic isolates was carried out by morphological characterization and use of species-specific PCR-based assays. Both identification criteria were necessary and complementary for the species determination, since in some cases the molecular identification was not specific. Scanty presence of *F. graminearum* was observed in the 2006 wheat-growing season coinciding with the lack of favorable meteorological conditions for producing FHB infection events. In contrast, high presence of *F. graminearum* isolates was observed in the 2007 wheat-growing season, in accordance with a moderate incidence of the disease according to spatial distribution of FHB incidence values. Toxins were not measured (Kikot et al., 2010).

In another recent work, different species of the *Fusarium graminearum* complex were detected by TEF 1-α gene sequences. Results of this work showed that 3 out of 10 samples obtained from the 2007-2008 wheat crop season were *F. graminearum* and that 7 out of 10 samples from maize were *F. graminearum* (3), *F. meridionale* (1) and *F. boothii* (3). Other *Fusarium* species detected were *F. subglutinans* (1), *F. andiyazi* (1), *F. proliferatum* (2), *F. verticillioides* (11) and *F. thapsinum* (1) (Sampietro et al., 2010).

Toxin Production

Several toxins have been identified during wheat production in years of epidemic FHB development. All reports note the preponderance of the trichothecene deoxynivalenol (DON). This is expected since *F. graminearum* is the most important *Fusarium* species known to produce this toxin (Marasas et al. 1984).

The first toxin survey, undertaken after the epidemic of 1985, was based on 84 injured grain samples non-randomly collected . Several *Fusarium* species were isolated and all samples were found to be infected by *F. graminearum.* In addition, all samples were contaminated with DON, with levels ranging from 0.1-25 ppm; however, almost half of the samples had levels ranging between 0.3-0.5 ppm. In 8% of samples, the toxin T2 was present at levels of 0.3-0.5 ppm. No zearalenone (ZEA) was found (Banchero et al. 1986).

In another work, a total of 24 *F. graminearum* isolates were obtained from 82 wheat samples of the 1985/1986 crop season, collected at several locations of Buenos Aires and Santa Fe provinces. Their production of mycotoxins was studied by Thin-Layer Chromatography (TLC). Eight out of the 24 isolates produced toxins. Six out of these eight isolates were able to produce DON together with either 3- or 15-ADON, whereas two of them also produced ZEA. No evidence for the presence of NIV or FUS-X was found (Faifer et al., 1990).

During the severe FHB epidemic of 1993, DON was the only toxin reported. In Córdoba province, one of the most affected areas, the mean DON level found in a milling sample was 1.06 ppm, i.e. lower than the internationally accepted level for grain. Furthermore, there was no strong correlation between the level of DON and the percentage of *Fusarium* infection (Dalcero 1996; Luqui and Gomez 1995).

In 2003, Lori et al. studied the distribution and contamination of DON and its relationship with *F. graminearum* kernel invasion during two consecutive harvests over a total of 147 samples (cultivars and lines) of durum wheat from five locations of the Southern of Buenos Aires Province, and analyzed fusarotoxins by TLC. They found that none of the varieties and lines was free of *F. graminearum* infection. In the first harvest, fungal invasion was very low. A total of 55% out of the 40 samples showed DON contamination, but only 10% showed levels higher than 2 ppm. In the second harvest, a crop year conducive to scab development, the highest level of *F. graminearum* kernel invasion observed was 42% on a sample from the humid area (eastern Buenos Aires Province). DON was detected in 47 (78.2%) of the 60 samples analyzed and 19 (31.6%) showed levels of DON higher than those established in the guidelines in Canada and the USA for food and feedstuff. In the two years studied, all locations situated in the humid area showed levels ranging from 0 to > 8 ppm. Within the durum wheat area differences among locations were found (Lori et al., 2003).

During the 1993 harvest period there was a high incidence of FHB in Argentina, and *F. graminearum* was the main *Fusarium* species isolated. About 80% of the 40 samples analyzed showed DON contamination. The levels of DON found ranged between 300 and 4500 μg/kg. Only five samples showed levels of DON higher than those established in the guidelines in Canada and the USA for food and feedstuffs (Dalcero et al., 1997).

In another work, DON and contaminant mycoflora in freshly harvested wheat in 1993 was carried out on wheat heads from the main production area of Argentina. *F. graminearum* was the predominant *Fusarium* species. An analysis of DON natural contamination was performed on a limited number (44) of samples. DON contamination levels in positive samples ranged from 0.2 to 30 ppm (González et al., 1996).

Regarding durum wheat (*Triticum durum*), a mycological survey carried out on a limited number of samples (60) obtained during the 1996 crop season from the main production area of Argentina showed that *Alternaria alternata* and *F. graminearum* were the predominant fungal species. DON contamination levels in positive samples ranged from 26 to 6400 μg/kg (González et al., 1999).

A total of 120 freshly harvested wheat samples from the 2004 season in nine locations from Northern Buenos Aires Province, Argentina, were analyzed for trichothecene natural occurrence and associated mycoflora. The trichothecenes T-2 tetraol, T-2 triol, HT-2 and T-2 toxins (HT-2, T-2), diacetoxyscirpenol (DAS), NIV, DON, 3-ADON and 15-ADON were analyzed by gas chromatography and electron capture detection. *Alternaria alternata, F. graminearum, F. poae* and *F. semitectum* were the predominant fungal species identified as endogenous mycoflora. The type A trichothecenes detected were HT-2 and T-2 triol toxins whereas the type B trichothecenes detected were DON, NIV and 3-ADON. Based on 120 samples, the incidences were 21.7% for 3-ADON, 22.5% for HT-2, 27.5% for T-2 triol and 85% for DON. NIV was confirmed in one sample. Mean levels of trichothecene positive samples were between 7 and 2788 μg/kg (González et al., 2008).

A mycological survey was carried out in the Southeastern part of Buenos Aires Province, in order to identify the mycobiota and the main mycotoxins present in fermented feed, wheat grain and corn grain samples from the 2005-2006 crop season. Fifty-six wheat samples were cultured for fungal quantification, isolation and identification, and analyzed for DON, ZEA, T-2 toxin and aflatoxins (AFLA). *Penicillium* (42%), *Fusarium* (27%) and *Alternaria* (25%)

were the most frequently genera recovered from wheat. DON and ZEA were detected in 22 out of 45 wheat samples, while T-2 toxin was detected in one sample. Wheat samples were negative for AFLA (Roigé et al., 2009).

CHEMOTYPE

The chemotype of a microbial or plant species has traditionally been defined as its chemical phenotype, including its profile of natural products. Since the 19th century, chemotyping has been useful for morphologically distinguishing similar organisms. Nowadays, natural product chemistry and other classical morphological and nonmorphological methods retain a critical place in an integrated approach to biological systematics (Desjardins, 2008).

With the discovery of genes for the biosynthesis of natural products, chemistry and genetics have entered a new interrelationship. Gene sequences can potentially be used to predict natural product profiles of plants or cultured microbes or perhaps even in food or agricultural samples. It is clear that PCR-based genotyping is a rapidly growing field of diagnostic testing that will contribute to, although not replace, the analysis of natural products of agriculturally important fungi, plants, and plant products. It is critical and important that there remains a differentiation between genotype data obtained by DNA analysis and chemotype data obtained by chemical analysis (Desjardins, 2008). The term "chemotype" is a chemical phenotype (or natural product profile), i.e. the "chemical expression of a genotype" (Desjardins, 2008).

Three chemotypes, which are strain-specific profiles of trichothecene metabolites, have been found in *F. graminearum* (Miller et al., 1991). They produce Nivalenol (NIV chemotype), which is a C-4 oxygenated derivative of DON, DON and an acetyl ester derivative of DON at 15-position oxygen (15-ADON chemotype), or DON and an acetyl ester derivative of DON at 3-position oxygen (3-ADON chemotype). The biosynthesis pathway of trichothecene toxin has been well characterized in *F. graminearum* (Desjardins, 2006). The 15-ADON chemotype is predominant in North America, while the 3-ADON chemotype is predominant in some areas of Asia, including China, Australia and New Zealand (Mirocha et al., 1989). Recent molecular surveillance has shown that the 3-ADON chemotype is replacing 15-ADON from eastern to western Canada (Ward et al., 2008), where 3-ADON chemotype isolates accumulate more trichothecenes than 15-ADON chemotype isolates. The expansion of 3-ADON populations suggests the need to re-evaluate chemotype diversity (Starkey et al., 2007). The NIV and both DON chemotypes are common in China, Japan, Korea, and Nepal (Desjardins, 2006).

In a study carried out in 1987 (Lori et al., 1992), when FHB incidence was low, 406 samples were analyzed from 17 sites over the traditional wheat growing region. Results showed that more than 90% of the *Fusarium* spp. isolated were *F. graminearum,* and that among them, 89.4% were toxin producers (type B trichothecenes and ZEA producers; no type A trichothecenes were found). Ninety-one *Fusarium* spp strains belonging to *F. graminearum*, *F. equiseti*, *F. moniliforme*, *F. m* var *anthophilum*, *F. acuminatum*, *F. solani* and *F. oxysporum* were isolated and identified. The production of trichothecenes and ZEA was evaluated after growing them on peeled and polished rice. From the strains studied,

82.4% produced trichothecenes, and could be divided into two groups: a) strains producing DON and its precursor 3-ADON; and b) strains producing NIV and its precursor 4-acetylnivalenol or fusarenone-x (FUS-X). The results showed that 48.3% of the strains were DON producers, 7.7% NIV producers and 26.4% produced both DON and NIV. A total of 89,4% of *F. graminearum* strains were toxigenic; among them 55.9% were DON chemotype, 10.3% NIV chemotype and 33.8% produced DON and NIV. Besides, 63,2% of the strains also produced ZEA, there being no relation between the DON and NIV groups with regard to ZEA production. The geographic distribution of different toxigenic strains led to the conclusion that there are regional differences (Lori et al. 1992).

In another work, by using Gas Chromatography and High Pressure Liquid Chromatography (HPLC), Fernández Pinto et al., (2008) determined the ability of 33 isolates of *F. graminearum,* obtained from 19 wheat samples of the 2001/2002 crop, to produce trichothecenes on autoclaved rice. Ten out of the 33 isolates were found to produce DON (0.1- 29 mg kg^{-1}), 13 to produce both DON (1.0- 708 mg kg^{-1}) and NIV (0.1- 6.2mg kg^{-1}), 12 to produce 3-ADON (0.1- 14 mg kg^{-1}), 13 to produce 15-ADON (0.1- 1.9 mg kg^{-1}), 10 to produce FUS-X (0.1- 2.4 mg kg^{-1}), and 7 to produce ZEA (0.1- 0.6 mg kg^{-1}). Natural occurrence of mycotoxins showed that 13 out of 19 samples were contaminated with DON in a range of 0.3 to 70 mg kg^{-1} and that 2 samples were contaminated with DON (7.5 and 6.7 mg kg^{-1}) and NIV (0.05 and 0.1 mg kg^{-1}), respectively (Fernández Pinto et al., 2008).

The mycotoxin production of DON, NIV and 3-ADON was also assayed on isolates from three localities in Argentina: San Antonio de Areco, Alberti and Marcos Juarez during the 2002 crop season (Ramírez et al., 2006). Toxin production on irradiated wheat revealed that 90% of the 70 strains of *F. graminearum* tested produced only DON and that 10% were able to produce DON and very low amounts of 3-ADON, which was occasionally detected as a minor co-metabolite. DON production varied greatly among the *F. graminearum* isolates, ranging from 1.0 to 7.6 μg g^{-1}). No isolate produced NIV. However, the authors did not specify the chemotype for each locality (Ramírez et al., 2006).

A recent report of the mycotoxin production of 144 *F. graminearum sensu stricto* isolates from 33 locations representing fields located in subregions IIS, IV and VS from the 2001, 2003 and 2004 crop seasons has reported that DON was produced in almost all isolates (138 out of 144). Isolates producing 15-ADON were predominant over the isolates producing 3-ADON (114 and 75, respectively). Nivalenol was produced by 34 of the isolates. This toxin was found simultaneously with DON, but DON was always produced in higher quantities. Only 2% of the isolates produced FUS-X and ZEA (Alvarez et al., 2009).

Another study focused on the mycotoxigenic potential and DON production of 10 *F. graminearum* species complex from the 2007-2008 crop season showed that 3 out of 10 isolates obtained from wheat samples were DON producers, whereas 7 out of 10 isolates from maize samples were identified as *F. graminearum* (3), *F. meridionale* (1) and *F. boothii* (3). The last two species belonged to the NIV chemotype, while *F. graminearum sensu stricto* was DON producer (Sampietro et al., 2010).

All the information available in *Fusarium graminearum* chemotypes found in wheat in Argentina is summarized in Table 1. A total of 250 isolates were analyzed to trace chemotype changes from different subregions and years. Data are organized according to the profile's production of DON and their acetylated derivatives 3- and 15-ADON to produce a map of distribution of chemotypes (Figure 1).

Table 1. Number of isolates producing Deoxynivalenol (DON), 3-acetyldeoxynivalenol (3-ADON), 15-acetyldeoxynivalenol (15-ADON), Nivalenol (NIV) and Fusarenone X (FUS X) in different locations of the main wheat production area in Argentina (with the exception of Ramírez et al. (2006) in which the authors did not specify the chemotype for each locality)

	DON	3-ADON	15-ADON	NIV	FUS X	Total	Reference
Buenos Aires province							
25 de Mayo	11	4	7	2		11	Alvarez et al., 2009
Alberdi							Ramírez et al., 2006
Areco	1	1				1	Alvarez et al., 2009
Arrecifes	2	3		2	1	5	Fernández Pinto et al., 2008
Azul	10	6	3			10	Alvarez et al., 2009
Balcarce	4	3			1	5	Lori et al., 1992
Balcarce	3	1	2	2		3	Alvarez et al., 2009
Barrow	5					5	Lori et al., 1992
Belgrano	11	4	2	1		11	Alvarez et al., 2009
Bolivar	2	2				2	Alvarez et al., 2009
Bordenave	2					2	Lori et al., 1992
Bragado	1	1				1	Alvarez et al., 2009
Cardales	3	2	1			3	Alvarez et al., 2009
Cardenau	8	3	5	3		8	Alvarez et al., 2009
Carhue	2					2	Lori et al., 1992
Carlos Casares	5	4	1	2	2	5	Alvarez et al., 2009
Coronel Suárez	2	2		2	1	5	Lori et al., 1992
Coronel Suárez	2	2		1	1	2	Alvarez et al., 2009
General Alvarado	10	4	6	5		10	Alvarez et al., 2009
General Pinto	2	1	1	1		2	Alvarez et al., 2009
General Pueyrredón		1		1	1	2	Fernández Pinto et al., 2008
General Pueyrredón	1	1		1		1	Alvarez et al., 2009
General Viamonte	1	1				1	Alvarez et al., 2009
General Villegas	1		1	1	1	2	Fernández Pinto et al., 2008
Guamini	2	1	1	1		2	Alvarez et al., 2009
H. Yrigoyen	1					1	Alvarez et al., 2009
La Dulce	3				1	5	Lori et al., 1992
Laprida	1		1	1		1	Alvarez et al., 2009
Lincoln	2	2		1		2	Alvarez et al., 2009
Loberia	2	1	3			6	Alvarez et al., 2009
Miramar	6	3				6	Lori et al., 1992
Necochea	10	4	6			10	Alvarez et al., 2009
Nueve de Julio	14	4	9	3		14	Alvarez et al., 2009
Pehuajo	2					2	Alvarez et al., 2009
Pergamino	5	1				5	Lori et al., 1992
Pergamino	2	1		1	1	4	Fernández Pinto et al., 2008
Pigue	4	2	1	3		4	Alvarez et al., 2009
Pla	4			3	3	5	Lori et al., 1992
Saavedra	2		2	2		2	Alvarez et al., 2009
Saladillo		2	1	2	2	3	Fernández Pinto et al., 2008
Salto	1	3	2	5	3	8	Fernández Pinto et al., 2008
San Antonio de Areco							Ramírez et al., 2006
San Cayetano	3	1	2			3	Alvarez et al., 2009
San Pedro		1	2	1		7	Fernández Pinto et al., 2008
Tandil	10	5	5	1		10	Alvarez et al., 2009
Trenque Lauquen	3	1	2			3	Alvarez et al., 2009
Entre Ríos province							
Paraná	4	1		3	2	5	Lori et al., 1992
La Pampa province							

Table 1. (continued)

	DON	3-ADON	15-ADON	NIV	FUS X	Total	Reference
Agustoni	1		1			1	Alvarez et al., 2009
Anguil	5	3		4	3	5	Lori et al., 1992
Conhello		1			1	1	Fernández Pinto et al., 2008
Conhello	7	2	4	3		7	Alvarez et al., 2009
Conhelo	1	1			1	1	Lori et al., 1992
General Pico	4	1			1	4	Lori et al., 1992
Intendente Alvear	1					1	Alvarez et al., 2009
La Pampa, capital	1		1			1	Alvarez et al., 2009
Córdoba province							
Marcos Juárez	3			2	3	5	Lori et al., 1992
Marcos Juárez							Ramírez et al., 2006
Villa del Rosario	3	1		2	4	4	Lori et al., 1992
Santa Fe province							
Oliveros	3	2		1		3	Lori et al., 1992
Rafaela	3				3	5	Lori et al., 1992
Reconquista	4	1		2	1	5	Lori et al., 1992
Total	206	91	72	65	37	250	

An analysis of trichothecene chemotype distribution across the wheat cropping area revealed that the 3-ADON chemotype is the most common chemotype (cita?). Results of this work showed that 91 out of 250 isolates produced DON and 3-ADON, while 72 produced DON and 15-ADON (Table 1).

The map of distribution of chemotypes was carried out considering only the DON chemotype because the NIV chemotype was detected only by Lori et al. (1992) in 2 of the 76 isolates of *F. graminearum* (Lori et al., 1992). In the other isolates, NIV was co-produced with DON, but there are no data about the quantities of each toxin. In other reports, this toxin was found always simultaneously with DON, but since DON was produced in higher quantities than NIV, isolates that co-produce DON and NIV were considered as belonging to the DON chemotype (Fernández Pinto et al., 2008; Alvarez et al., 2009).

Regarding the distribution along the main wheat production area in Argentina, there seems to be a tendency for the presence of the 3-ADON chemotype in the North and of the 15-ADON chemotype in the South (Figure 1). These results indicate the need to increase vigilance over crop seasons and geographical areas to establish a clear tendency in chemotype distribution. These results contrast with those obtained in North America, where the 15-ADON chemotype is predominant, but show coincidence with Canada, where the 3-ADON chemotype frequency increases with a pronounced longitudinal cline in which the frequency of 3-ADON producers is highest in the East and decreases from East to West (Ward et al., 2008). This expansion of 3-ADON populations, along with evidence of chemotype diversity in the United States (Starkey et al., 2007), suggests the need to re-evaluate chemotype diversity.

DIVERSITY

The diversity of *F. graminearum* has been extensively studied worldwide. Several methods have been used to characterize *F. graminearum* isolates, including Vegetative Compatibility Grouping (VCG) (Bowden & Leslie, 1992; Bowden & Leslie, 1994; Mc Callum *et al.*, 2001), morphological and molecular methods (Cumagun *et al.*, 2004; Desjardins *et al.*, 2004; Toth *et al.*, 2004), analyses of ergosterol, and mycotoxin profile (Gilbert *et al.*, 2001). Mycelial compatibility grouping (MCG) has been applied to study the diversity of *F. graminearum* strains by Mc Callum *et al.* (2003). Mycelial compatibility (MC) is based on the ability of two isolates to fuse and form one colony. Kohn *et al.* (1991) suggested that MCGs represent genetically different individuals and that each MCG is genotypically unique. MC allows the observation of barrage zones of thick, raised mycelium at the junctions of vegetative incompatible *F. graminearum* isolates. The *F. graminearum* population analyzed in Argentina by Alvarez et al. (2007) showed a high level of diversity, being greater for VCG than for MCG. Ramírez et al. (2006) obtained similar results on *F. graminearum* genotypic variation using VCG. Amplified Restriction Polymorphism Fragments (AFLPs) have also demonstrated high levels of genotypic variability. The similarity coefficient for the two most closely related isolates was found to be 92%, thus suggesting high heterogeneity in the population (Alvarez et al., 2009). Diversity could be explained by sexual recombination because perithecia can be observed frequently in nature in the residue of wheat, maize or small grains. Migration of genotypes among fields could be important in maintaining the high level of diversity (Schmale et al., 2006). Besides sexual recombination, there are other mechanisms, such as parasexuality or mitotic recombination, to generate genetic variability (Leslie, 1993).

Diversity in the *F. graminearum* population from Argentinean wheat has been scarcely studied. Considering the importance to know the casual agent of FHB, it is of special concern to know the diversity of this pathogenic fungus. Further studies will be necessary to characterize its genetic diversity including molecular aspects and aggressiveness of *F. graminearum* populations from Argentina in order to design better control strategies for disease management and suitable methods to analyze the toxin contents in food and feed.

FUSARIUM POAE

In different cereal surveys from Argentina, *F. poae* has been among the most frequently isolated FHB pathogen. Regarding wheat cultivation, *F. graminearum* and *F. poae* were the most frequent species found in wheat samples harvested in nine locations of the northern Buenos Aires province in 2004 (González et al., 2008). In a study based on 70 isolates, also from Buenos Aires province, 71.43% were assigned to *F. graminearum* and 7.14% to *F. poae* (Alvarez et al., 2007). Although the available information and research on FHB are focused on *F. graminearum*, studies of *F. poae* pathogenicity are needed to know about the pathways involved in plant resistance/susceptibility and mycotoxin accumulation (Stenglein, 2009). The main type B trichothecenes produced by *F. poae* are NIV, DON and FUS-X (Desjardins, 2006).

In a work carried out by Somma et al., 81 strains of *F. poae* isolated from durum wheat kernels and identified by species-specific PCR and TEF-1α gene sequence analysis were genetically characterized by the AFLP technique and analyzed by HPLC for their ability to produce beauvericin (BEA) and trichothecene mycotoxins. A high level of variability was observed by AFLP, with low level of genetic similarity among the strains, approximately 61%. Most of the strains (95%) produced BEA at < 2,655 μg/g, 88% produced NIV at < 865 μg/g and 76% produced FUS-X at < 167 μg/g. These data show that *F. poae* can produce high amounts of BEA together with trichothecenes, and can represent a high potential mycotoxin risk for wheat colonized by this species (Somma et al., 2009).

Differences in the distribution of chemotypes could contribute to the characterization of the contamination in a regional scale (Bottalico, 2002). This review presents a preliminary map of distribution of chemotypes. However, the different toxin production profiles found in the main wheat production area analyzed here confirm the need for further analyses considering a higher number of isolates in order to do a better characterization of *F. graminearum* population. These results also indicate the need of continuous mycotoxin monitoring, because mycotoxins that either have a low incidence or are unknown may emerge or be introduced into the food chain as a result of effects arising from climate change, global trade and technological changes in the processing industries (Magan and Aldred, 2007).

BIBLIOGRAPHY

Alvarez CL, Azcarate MP, Terminiello L, Cabral D, Vaamonde G and Fernández Pinto V, 2007. Genotypic diversity and mycotoxin production of *Fusarium graminearum* isolates from wheat cultivated in Argentina. *Research and Reviews in Biosciences*, 1: 172-180.

Alvarez CL, Azcarate PM and Fernández Pinto V, 2009. Toxigenic potential of *Fusarium graminearum sensu stricto* isolates from wheat in Argentina. *International Journal of Food Microbiology* 135:131-135.

Aoki T and O'Donnell K, 1999. Morphological and molecular characterization of *Fusarium pseudograminearum* sp. nov., formerly recognized as the Group 1 population of *F. graminearum*. *Mycologia* 91, 597-609.

Banchero ES, de Miguel MS, de Godoy H, Di Giulio AM, Dubois M, de Souto G, Gómez M, Resnik FZ and Torres ME, 1986. Evaluación de la presencia de fusarium y micotoxinas en muestras de trigo de la campaña 1985/86 y su relación con la calidad comercial. DITE M 929. Secretaría de Agricultura, Ganadería y Pesca. Junta Nacional de Granos. Instituto Nacional de Tecnología Agropecuaria. Buenos Aires, Argentina.

Bottalico A and Perrone G, 2002. Toxicogenic *Fusarium* species and mycotoxins associated with head blight in small-grain cereals in Europe. *European Journal of Plant Pathology* 108, 611-624.

Bowden RL and Leslie JF, 1992. Nitrate-nonutilizing mutants of *Gibberella zeae* (*Fusarium graminearum*) and their use in determining vegetative compatibility. *Experimental Mycology* 16: 307-315.

Bowden RL and Leslie JF, 1994. Diversity of *Gibberella zeae* at small spatial scales. *Phytopathology* 84: 1140-1146.

Cumagun CJR, Bowden RL, Jurgenson JE, Leslie JF and Miedaner T, 2004. Genetic mapping of pathogenicity and aggressiveness of *Gibberella zeae* (*Fusarium graminearum*) toward wheat. *Phytopathology* 94: 520-526.

Dalcero A, Torres A, Etcheverry M, Chulze S and Varsavsky E, 1997. Occurence of deoxynivalenol and *Fusarium graminearum* in Argentinian wheat. *Food Additives and Contaminants* 14, 11-14.

Desjardins AE, Jarosz AM, Plattner RD, Alexander NJ, Brown DW and Jurgenson JE, 2004. Patterns of trichothecene production, genetic variability, and virulence to wheat of *Fusarium graminearum* from smallholder farms in Nepal. *Journal of Agricultural and Food Chemistry*. 52: 6341-6346.

Desjardins AE, 2006. *Fusarium* Mycotoxins: Chemistry, Genetics and Biology. The American Phytopathological Society St Paul, Minnesota USA.

Desjardins AE, 2008. Natural Product Chemistry Meets Genetics: When Is a Genotype a Chemotype? *Journal of Agriculture and Food Chemistry*, 56, 7587–7592.

Faifer G, De Miguel M and Godoy H, 1990. Patterns of mycotoxin production by *Fusarium graminearum* isolated from Argentine wheat. *Mycopathologia 1*09: 165-170.

Fernández Pinto V, Terminiello L, Basilico JC and Rittieni A, 2008. Natural occurrence of Nivalenol and mycotoxigenic potential of *Fusarium graminearum* strains in wheat affected by Head Blight in Argentina. *Brazilian Journal of Microbiology*, 39: 157-162.

Francis RG and Burgess LW, 1977. Characteristic of two populations of *Fusarium roseum "graminaerum*" in eastern Australia. *Transactions of the British Mycological Society* 68, 421-427.

Gale L, Ward T, Balmas V and Kistler H, 2003. Detection of distinct subpopulations of *Fusarium graminearum*, lineage 7. (Abstract) National *Fusarium Head Blight Forum Proceedings*, page 139.

Galich MTV de, 1989. Avances en el Control de la Fusariosis del Trigo en Argentina. In Taller Sobre la Fusariosis de la Espiga en América del Sur. M. M. Kohli(editor) Mexico, D. F. CIMMYT. Page 27-31.

Galich AN and de Galich MTV de, 1996. Enfermedades del trigo en el área central norte de la región cerealera argentina. Informe Técnico 121. E.E.A Marcos Juárez, INTA Córdoba, Argentina.

Garcia CB, 1989. Evaluación de resistencia a fusariosis en campo mediante inoculación artificial. In: M.M. Kohli (ed.), Taller sobre fusariosis de la espiga en America del Sur, pp. 77-84. Mexico, D.P.: CIMMYT.

Geiser DM, Giménez-Gasco MM, Kang S, Makalowska I, Veeraraghava N, Ward TJ, Zhang N, Kuldau GA and O'Donnell K, 2004. FUSARIUM-ID v. 1.0: A DNA Sequence Database for Identifying *Fusarium*. European Journal of Plant Pathology 110: 473-479.

Gilbert J, Abramson D, McCallum B and Clear R, 2001. Comparison of Canadian *Fusarium graminearum* isolates for aggressiveness, vegetative compatibility, and production of ergosterol and mycotoxins. *Mycopathology* 153: 209-215.

González HHL, Pacin A, Resnik SL and Martinez EJ, 1996. Deoxynivalenol and contaminant mycoflora in freshly harvested Argentinian wheat in 1993. *Mycopathologia* 135, 129-134.

González HHL, Martínez EJ, Pacin A y Resnik SL, 1999. Relationship between *Fusarium graminearum* and *Alternaria alternata* contamination and deoxynivalenol occurrence on Argentinian durum wheat. *Mycopathologia* 144, 97 - 102.

González HHL, Moltó GA, Pacin EA, Resnik SL, Zelaya MJ, Masana M and Martínez EJ, 2008. Trichothecenes and Mycoflora in Wheat Harvested in Nine Locations in Buenos Aires Province, Argentina. *Mycopathologia* 165:105–114.

Jensen CA, 1995. Mejoramiento de trigo candeal en la Chacra Experimental Integrada, Barrow. In: M.M. Kohli, J. Nisi, and S. Rajaram (eds.), EI mejoramiento de trigo en Argentina. Treinta años de investigación cooperativa con el CIMMYT, pp. 37-40. Balcarce, Argentina. Uruguay: CIMMYT.

Kikot GE, Moschini R, Consolo VF, Rojo R, Salerno G, Hours RA, Gasoni L, Arambarri AM and Alconada TM, 2010. Occurrence of Different Species of Fusarium from Wheat in Relation to Disease Levels Predicted by a Weather-Based Model in Argentina Pampas Region, *Mycopathologia.* DOI 10.1007/s11046-010-9335-0.

Kohli MM, 1995. Conceptos básicos en el manejo de las enfermedades de cultivo. In "Curso de manejo de las enfermedades de trigo" Centro Internacional de capacitación INTA-CIMMYT. 20 Pages.

Kohn LM, Stasovski E, Carbone I, Royer J and Anderson JB, 1991. Mycelial incompatibility and molecular markers identify genetic-variability in field populations of *Sclerotinia sclerotiorum*. *Phytopathology* 81: 480-485.

Láday M, Juhász A, Mulè G, Moretti A, Sezcsi A and Logrieco A, 2004. Mitochondrial DNA diversity and lineage determination of European ioslates of *Fusarium gramineraum* (*Gibberella zeae*). *European Journal of Plant Pathology110*: 545-550.

Leslie JF, 1993. Fungal vegetative compatibility. *Annual Review of Phytopathology* 31: 127-15.

Lori GA, Carranza MR, Violante A, Rizzo I and Alippi HE, 1992. *Fusarium spp* en trigo, capacidad toxicogénica y quimiotaxonomía de las cepas aisladas en la Argentina. *Agronomie* 12: 459-467.

Lori GA and Sisterna MN, 2001. Occurrence and distribution of *Fusarium* spp. associated with durum wheat seed from Argentina. *Journal of Plant Pathology* 83: 63-67.

Lori GA, Sisterna MN, Haidukowski M, Rizzo I, 2003. *Fusarium graminearum* and deoxinivalenol in the durum wheat area of Argentina. *Microbiological Research* 158: 29-35.

Luqui AB and Gómez M, 1995. Evaluación de la incidencia del porcentaje de granos dañados por *Fusarium* en el nivel de toxinas Don y T 2 en trigo durante la campaña 1993/1994. Secretaría de Agricultura, Ganadería y Pesca, Buenos Aires (Argentina). Instituto Argentino de Sanidad y Calidad Vegetal. Buenos Aires (Argentina). Ene 1995. 21 p..

Magan N and Aldred D, 2007. Post-harvest control strategies: minimizing mycotoxins in the food chain. International Journal of Food Microbiology 119: 131-139.

Marasas WFO, Nelson PE y Toussoun TA, 1984. Toxigenic *Fusarium* species: Identity and Mycotoxicology. Pennsylvania State University Press, University Park PA.

Mc Callum BD, Tekauz A and Gilbert J, 2001. Vegetative compatibility among *Fusarium graminearum* (*Gibberella zeae*) isolates from barley spikes in southern Manitoba. *Canadian Journal of Plant Pathology* 23: 83-87.

Mc Callum BD, Tekauz A and Gilbert J, 2004. Barrage zone formation between vegetatively incompatible *Fusarium graminearum* (*Gibberella zeae*) isolates. *Phytopathology* 94: 432-437.

Miller JD, Greenhalgh R, Wang YZ and Lu M, 1991. Trichothecene chemotype of three *Fusarium* species. *Mycologia* 83: 121-130.

Mirocha CJ, Abbas HK, Windels CE and Xie W, 1989. Variation in deoxynivalenol, 15-acetyldeoxynivalenol, 3-acetyldeoxynivalenol, and zearalenone production by *Fusarium graminearum* strains. *Applied and Environmental Microbiology* 55:1315-1316.

Moschini RC, Pioli R, Carmona M and Sacchi O, 2001. Empirical predictions of wheat head blight in the Northern Argentinean Pampas Region. *Crop Science* 41: 1541-1545.

O'Donnell K, Kistler HC, Tacke BK and Caspar HH, 2000. Gene genealogies reveal global phylogeographic structure and reproductive isolation among lineages of *Fusarium graminearum,* the fungus causing wheat scab. *Proceedings of the National Academy of Sciences USA* 97: 7905-7910.

O'Donnell K, Ward TJ, Geiser DM, Kistler HC and Aoki T. 2004. Genealogical concordance between the mating type locus and seven other nuclear genes supports formal recognition of nine phylogenetically distinct species within the *Fusarium graminearum* clade. *Fungal Genetics and Biology* 41: 600-623.

Rahjoo V, Zad J, Javan-Nikkhah M, Mirzadi Gohari A, Okhovvat SM, Bihamta MR, Razzaghian J and Klemsdal SS, 2008. Morphological and molecular identification of *Fusarium* isolated from maize ears in Iran. *Journal of Plant Pathology* 90: 463–468.

Ramírez ML, Reynoso MM, Farnochi MC and Chulze S, 2006. Vegetative compatibility and mycotoxin chemotypes among *Fusarium graminearum* (*Gibberella zeae*) isolates from wheat in Argentina. *European Journal of Plant Pathology* 115: 139-148.

Roigé MB, Aranguren SM, Riccio MB, Pereyra S, Soraci AL and Tapia MO, 2009. Mycobiota and mycotoxins in fermented feed, wheat grains and corn grains in Southeastern Buenos Aires Province, Argentina. Revista Iberoamericana de Micología 26(4):233–237.

Sampietro DA, Marín P, Iglesias J, Presello DA, Vattuone MA, Catalan CAN and Gonzalez Jaen MT, 2010. A molecular based strategy for rapid diagnosis of toxigenic *Fusarium* species associated to cereal grains from Argentina. *Fungal Biology* 114:74-81.

Schmale DG, Leslie J F, Zeller K A, Saleh A A, Shields E J and Bergstrom GC, 2006. Genetic structure of atmospheric populations of *Gibberella zeae*. *Phytopathology* 96: 1021-1026.

Somma S, Alvarez CL; Ricci V, Ferracane L, Ritieni A, Logrieco A, Moretti A, 2010. Trichothecene and beauvericin mycotoxin production and genetic variability in *Fusarium poae* isolated from wheat kernels from northern Italy. *Food additives and Contaminants*, 27 (5): 729-737.

Starkey DE, Ward TJ, Aoki T, Gale LR, Kistler HC, Geiser DM, Suga H, Toth B, Varga J and O'Donnell K, 2007. Global molecular surveillance reveals novel *Fusarium* head blight species and trichothecene toxin diversity. *Fungal Genetics and Biology* 12: 1745-1776.

Stenglein SA, Dinolfo MI, Moreno MV, Rogers WJ and Salerno GL, 2008. Avances en el estudio de la ocurrencia de *Fusarium poae* en la Argentina. In: Mirassón H., Faraldo M.L., Baudino E.M. (eds). Proceedings of the VII Congreso Nacional de Trigo, V Simposio Nacional de Cereales de Siembra Otoño-Invernal, I Encuentro del MERCOSUR, Mar del Plata, Argentina.

Taylor JW, Jacobson DJ, Kroken S, Kasuga T, Geiser DM, Hibbett DS and Fisher MC, 2000. Phylogenetic species *Fungal Genetics and Biology* 31: 21-32.

Tóth B, Mesterházy A, Nicholson P, Téren J and Varga J, 2004. Mycotoxin production and molecular variability of European and American *Fusarium culmorum* isolates. *European Journal of Plant Pathology*, 110: 587–599.

Ward TJ, Clear RM, Rooney AP, O'Donnell K, Gaba D, Patrick S, Starkey DE, Gilbert J, Geiser DM and Nowicki TW, 2008. An adaptive evolutionary shift in *Fusarium* head blight pathogen populations is driving the rapid spread of more toxigenic *Fusarium graminearum* in North America. *Fungal Genetics and Biology* 45, 473–484.

Xu XM and Berrie AM, 2005. Epidemiology and mycotoxigenic fungi associated with *Fusarium* ear Blight and apple blue mold: a review. *Food additives and Contaminants* 22: 290-301.

Yu J, Chang PK, Ehrlich KC, Cary JW, Bhatnagar D, Cleveland TE, Payne GA, Linz JE, Woloshuk CP and Bennett JW, 2004. Clustered pathway genes in aflatoxin biosynthesis. *Applied Environment Microbiology*, 70: 1253–1262.

Zeller KA, Vargas JI, Valdovino Ponce G, Leslie JF and Bowden RL, 2003. Population genetic and lineage composition among *Gibberella zeae* in North and South America. *XXII Fungal Genetics Newsletter* 50 Supplement. 446 pages.

Zeller KA, Bowden RL and Leslie JF, 2004. Population differentiation and recombination in wheat scab populations of *Gibberella zeae* in the United States. *Molecular Ecology* 13: 563-571.

In: Wheat: Genetics, Crops and Food Production
Editor: Maria T. Almeida, pp. 363-391

ISBN 978-1-61209-307-9

Chapter 14

THE BIODIVERSITY OF THE MICROBIOTA OF TRADITIONAL ITALIAN SOURDOUGHS

Cristiana Garofalo, Lucia Aquilanti* and Francesca Clementi
Sezione di Microbiologia Alimentare, Ambientale ed Industriale, Dipartimento di Scienze Alimentari, Agro-Ingegneristiche, Fisiche, Economico-Agrarie e del Territorio (SAIFET), Università Politecnica delle Marche, Ancona, Italy

ABSTRACT

Sourdough is a mixture of cereal flour and water which is fermented by a heterogeneous population of lactic acid bacteria (LAB) and yeasts and is traditionally used as inoculum for the production of sourdough bread and other baked products.

Yeasts are primarily responsible for leavening, while LAB metabolism is the main cause of acidification and flavour formation, being only partially responsible for leavening. Sourdough fermentation enhances dough properties, improves the volume, texture, flavour, and nutritional value of the final baked products, retards the staling process, and protects the baked goods from mould and bacterial spoilage.

The sourdough microbiota usually originate from flour, dough ingredients, the environment, and a stable microbial consortium is reached in a short time. The stability of a ripe sourdough depends on several factors including the type and the quality of the flour and the technological process parameters applied to the fermentation. As a consequence of the heterogeneity of these ecological determinants and because of the artisan- and region-dependent handling of sourdoughs, mature sourdough microbiota differ greatly in the number and type of species found. The prevalence of certain species and strains confers typical characteristics to the final products resulting in a wide regional variety of baked products.

In Italy, sourdough technology is used in more than 30% of bakeries for the production of more than 200 different types of sourdough breads, snacks, pizza and various sweet baked products such as *Panettone*, *Pandoro* and *Colomba*.

* Corresponding author: Lucia Aquilanti, Ph.D., Tel: +39-071-2204985, Fax: +39-071-2204988, E-mail: l.aquilanti@univpm.it, Sezione di Microbiologia Alimentare, Ambientale ed Industriale, Dipartimento di Scienze Alimentari, Agro-Ingegneristiche, Fisiche, Economico-Agrarie e del Territorio (SAIFET), Università Politecnica delle Marche, via Brecce Bianche, 60131, Ancona, Italy

Some Italian sourdoughs have been extensively characterized for their microflora. This chapter deals with the microbial ecology of Italian sourdoughs examined both with traditional methods and using novel molecular techniques; the importance of using a combined analytical approach to explore the microbial communities of sourdoughs has often been demonstrated.

INTRODUCTION

Sourdough is a mixture of water and flour from different cereal grains (wheat, rye, rice, etc.), fermented by a heterogeneous population of lactic acid bacteria (LAB) and yeasts (De Vuyst and Vancanneyt, 2007). It is traditionally used as an inoculum for the manufacture of sourdough bread and other baked goods (Robert et al., 2009). According to their production technologies, sourdoughs can be divided into three types (Böcker et al., 1995). Type I sourdoughs are continuously propagated through daily additions of flour and water at room temperature to keep the microorganisms in an active state (this process is commonly referred to as "refreshment"). A defined pool of microbial species provides the well-adapted microflora of type I sourdoughs; among LAB, *Lactobacillus sanfranciscensis* predominates, being characterized by a high souring activity and a high resistance to microbial contamination. Other LAB species also occur in this cereal ecosystem, namely *L. pontis*, *L. brevis*, *L. fermentum*, *L. paralimentarius*, and *L. plantarum.* Among yeasts, maltose-negative species, like *Candida humilis*, *C. milleri*, and *Saccharomyces exiguus* predominate together with *S. cerevisiae.*

Type II sourdoughs are semi-fluid silo preparations incubated at high temperatures, sometimes greater than 30°C, to speed up the process, and are characterized by long fermentation periods, mostly ranging from 2 to 5 days. These sourdoughs are generally used as dough-souring supplements in bread-making processes and are dominated by acid-resistant LAB species, like *L. brevis*, *L. panis*, *L. fermentum*, *L. reuteri*, *L. frumenti*, and *L. pontis*.

Finally, type III sourdoughs are dried preparations of industrial doughs containing LAB resistant to the drying process. Unlike type I sourdoughs, doughs of types II and III require the addition of baker's yeast (*S. cerevisiae*) as a leavening agent (Corsetti and Settanni, 2007; Scheirlinck et al., 2007a).

In modern bakery technology, type I sourdoughs have the largest application as natural leavening agents for the manufacture of a variety of leavened baked goods due to their numerous advantages with respect to baker's yeast. Indeed, sourdough fermentation enhances dough properties, improves the volume, texture, flavour, and nutritional value of baked goods, retards staling, and protects the end products against mould and bacterial spoilage (De Vuyst and Vancanneyt, 2007). These properties largely depend on the metabolic activities of yeasts and LAB during fermentation. The first microbial group is primarily responsible for the leavening of the dough, while the latter is mainly responsible for acidification, proteolysis, and the production of aroma compounds, ethanol, exopolysaccharides, as well as anti-mould and anti-ropiness substances (Corsetti and Settanni, 2007; De Vuyst et al., 2009).

The typical microbiota of type I sourdoughs mainly originate from flour; indeed, cereal grains are naturally contaminated by heterogeneous communities of moulds, yeasts and bacteria, whose species composition is greatly affected by climatic conditions, physical damage due to insects or mould attacks, and the use of preservatives (Corsetti et al., 2007a).

The vegetable matter (e.g. grape must, raisins, apples, figs, lemon or orange peel, bran, hay and even horse dung) traditionally added to the initial mixture of water and flour (Foschino et al., 2004) and the bakery environment (Scheirlinck et al., 2007a; Scheirlinck et al., 2009) have been indicated as further sources of yeasts and LAB in type I sourdoughs. Recently, the addition of selected starter cultures has also been documented (Siragusa et al., 2009; De Vuyst et al., 2009).

During the continuous propagation of type I sourdough, a microbial association with a remarkably high stability is established within a few days (Scheirlinck et al., 2008; De Vuyst et al., 2009). The persistence of such an association depends on several ecological determinants, such as the chemical and enzymatic composition of flours, the addition of peculiar ingredients, as well as technological process parameters, namely temperature of leavening and maintenance, dough hydration and yield, number of refreshments, fermentation time, use of starter cultures, final pH, and redox potential. As a consequence of the selective pressure exerted by these determinants, mature sourdoughs differ greatly in both the number and type of microbial species (De Vuyst et al., 2009). To date, just under 50 LAB species, have been isolated from type I sourdoughs (Hammes et al., 2005). Lactobacilli undoubtedly dominate [*L. sanfranciscensis*, *L. plantarum* and *L. brevis* (Gobbetti, 1998)], but strains ascribed to *Leuconostoc*, *Weissella*, *Pediococcus*, *Lactococcus*, *Enterococcus*, and *Streptococcus* have also been occasionally isolated (Corsetti and Settanni, 2007).

In recent years, strains assigned to novel *Lactobacillus* species have been isolated from sourdoughs in different countries: *L. paralimentarius* and *L. siliginis* from Japanese (Cai et al., 1999) and Southern Korean sourdoughs (Aslam et al., 2006), respectively; *L. hammesii* and *L. nantensis* from French sourdoughs (Valcheva et al., 2005; Valcheva et al., 2006); *L. mindensis* and *L. spicheri* from German wheat (Ehrmann et al., 2003) and rice sourdoughs (Meroth et al., 2004) respectively; *L. acidifarinae*, *L. zymae* and *L. namurensis* from Belgian sourdoughs (Vancanneyt et al., 2005; Scheirlinck et al., 2007b); and finally, *L. rossiae* from Central Italy wheat sourdoughs (Corsetti et al., 2005).

During sourdough propagation one or two *Lactobacillus* species have often been found, with cell numbers three or four orders of magnitude higher than those of the adventitious microbial flora (Corsetti and Settanni, 2007; De Angelis et al., 2007).

Regarding yeasts, *S. cerevisiae* is undoubtedly the species most frequently found in type I sourdoughs. In addition, other yeast species have been found associated in this peculiar sourdough ecosystem, namely *S. exiguus*, *C. milleri*, *C. krusei*, *Pichia norvegensis*, and *Hansenula anomala* but with cell numbers significantly lower than LAB (Gobbetti, 1998).

The microbial ecology of type I sourdoughs greatly affects the typical features of baked goods produced with this leavening agent. Accordingly, hundreds of different traditional sourdough breads exist in Europe (De Vuyst and Vancanneyt, 2007). In Italy, type I sourdoughs are used for the manufacture of more than 200 different leavened baked products, including bread, snacks, pizza, and cakes (Corsetti and Settanni, 2007; Minervini et al., 2010; Zannini et al., 2010). The latter include both regional specialities and sweet leavened goods, that are commercialized and consumed all over the Italian territory during the two main religious festivities, Christmas (*Panettone* and *Pandoro*) and Easter (*Colomba*). These products are particularly appreciated by consumers thanks to their unique flavour, high nutritional value, and enhanced shelf-life.

Due to the many differences in both the recipe (type of flour and additional ingredients) and the process parameters used in traditional Italian bread-making, a great variability exists in the distribution of LAB and yeast taxa among sourdoughs from different geographical areas. As a general rule, *Triticum aestivum* wheat is the main crop used for the production of bread, due to the superior performance of this flour for fermentation and kneading, while *T. durum* wheat is mainly used for the manufacture of pasta. However, mostly in the regions of Central and Southern Italy, such as Abruzzo, Campania, Basilicata, Calabria, Apulia, Sicily, and Sardinia, *T. durum* wheat flour, alone or in association with *T. aestivum* wheat flour, is also used for bread-making (Corsetti et al., 2007a).

To date, the microbial ecology and population dynamics of some Italian type I sourdoughs have been extensively investigated. Most of these studies were aimed at finding a correlation between the composition of the sourdough population and the geographical location; a further aim was often the preservation and exploitation of the immense patrimony of autochthonous yeasts and LAB occurring in these cereal ecosystems.

In this chapter, the main results of the investigations performed with either traditional microbiological methods or novel molecular approaches on Italian type I sourdoughs destined for the manufacture of "designation of origin" and local breads, snacks, pizza or sweet leavened goods in different Italian geographical areas are summarized and comparatively discussed.

1. Bread

The great diversification among European agricultural products has been recognized at Community level leading to the definition of two different kinds of origin labels for agricultural products and their derivatives: the Protected Designation of Origin (PDO) and the Protected Geographical Indication (PGI) (EC Regulation 2081/1992). The PDO is assigned to those products which have a strong bond with the area of origin. Two conditions have to be satisfied in order to assign the PDO label: the raw materials must be produced and processed in the indicated geographical region and the features of the product must be strongly dependent on the geographical environment of origin (soil, climate, microbiota, human factors, etc.). Similar to PDO, the PGI denomination defines a strong bond with the territory but it does not require all the production phases to occur in the same geographical area. In 2006, the Europen Community also introduced the Traditional Speciality Guaranteed (TSG) label, which was aimed at protecting productions characterized by either traditional ingredients or production processes (EC Regulation 509/2006). Up to now only a few breads have obtained one of these Community denominations: *Pane di Altamura* PDO (Altamura bread), *Pagnotta del Dittaino* PDO, *Coppia ferrarese* PGI, *Pane casereccio di Genzano* PGI, and *Cornetto di Matera* PGI (Cornetto of Matera) (Minervini et al., 2010).

Altamura bread, which is the most popular bread in Apulia (Southern Italy) and nowadays throughout Italy, was the first bread to obtain the PDO (EC Regulation 1291/2003). This bread is exclusively produced in small Altamura bakeries using a blend of 80% *T. durum* wheat flour and 20% *T. aestivum* wheat flour, water and salt. The sourdough has to be refreshed at least three times before its addition to the final dough. The microbial traits of sourdoughs destined for the manufacture of Altamura bread are discussed in section 1.2.2. The bread is produced in the shape of a large loaf weighing 0.5, 1 or 2 kg. It is characterized by a 3 mm thick crust, a typical yellow crumb, whose colour is due to the semolina pigments, with homogeneous bubbles and a maximum 33% moisture. It is greatly appreciated by consumers for its long shelf life (at least one week) and its typical flavour (Huy et al., 2006; Minervini et al., 2010).

Pagnotta del Dittaino is the other bread which has obtained the PDO denomination (EC Regulation 516/2009). It is typically produced in Enna, in the centre of Sicily (Southern Italy), with 70% *T. durum* wheat flour from *Simeto*, *Duilio*, *Arcangelo*, *Mongibello*, *Ciccio*, *Colosseo* cultivar and 30% from 16 other cultivar varieties from Valle del Dittaino. Final doughs are moulded into round portions of about 0.5-1 kg. The crust is 3-4 mm thick; the crumb is light yellow and very elastic with small bubbles and a uniform texture, with moisture up to 38% (Minervini et al., 2010).

Coppia ferrarese is produced in Ferrara (Emilia Romagna, Northern Italy) with *T. aestivum* wheat flour, water, salt, olive oil, and baker's yeast. It obtained the European PGI denomination in 2001 (EC Regulation 2036/2001). The microbial community of type I sourdoughs destined for the manufacture of this typical Italian bread are described in section 1.1.3. *Coppia ferrarese* has a particular shape since two ribbons of dough are knotted together in the centre with their ends twisted to form four spokes, called "*crostini*". Each loaf has a weight, which varies from 80 to 250 g; loaves are quite crispy, friable, and characterized by a hard, golden and smooth crust and a crumb with small pores in the four spokes but larger cells in the central part (Huy et al., 2006; Minervini et al., 2010).

Pane casereccio di Genzano is prepared in the area around Genzano, a town 30 Km from Rome (Lazio, Central Italy). It obtained PGI denomination from the EC in 1997 (EC regulation 2325/1997) and it was the first cereal product to obtain this label. It is made with *T. aestivum* wheat flour, water salt and sourdough. This bread has a 3 mm thick crust while the crumb maintains its moisture and elasticity. The crumb is characterized by large pores. It has a particular flavour and long shelf life, until 7-8 days (Huy et al., 2006; Minervini et al., 2010).

Cornetto of Matera is typically manufactured in Matera (Basilicata, Southern Italy) with *T. durum* wheat flour accounting for at least 20% from local and old cultivar such as *Senatore Cappelli* and *Duro Lucano*. Other ingredients are water, salt, and sourdough. The microbial ecology of sourdough destined for the manufacture of this typical Italian bread is described in section 1.2.3. This bread has recently obtained the European PGI denomination (EC Regulation 160/2008). The peculiar manufacturing process starts by mixing 1 kg of flour with 250 g of pulp of mature fresh fruits previously macerated in 250-300 cl of water. The dough is left to ferment for 10-12 h at 26-30°C. This first fermentation is followed by several refreshments. At the end of the production process, loaves weigh 1 or 2 kg; they have a crescent shape with a 3 mm thick crust, yellow crumb with heterogeneous bubbles (from 2 to 60 mm diameter), and a unique taste and flavour (Minervini et al., 2010).

1.1. *Triticum aestivum* Wheat Breads

1.1.1. Lombardy

Galli and colleagues (1988) characterized the microbial ecology of 20 sourdoughs from Lombardy (Northern Italy); among these, two (a type I and a dry sourdough) were destined for the manufacture of wheat bread. An isolation campaign of yeasts and LAB was carried out from all the sourdoughs; yeasts were identified with the API 50 CH system plus fermentation tests, whereas the taxonomic affiliation of LAB was determined on the basis of the utilisation of different sugars and other physiological properties. Type I sourdoughs for wheat bread were found to be characterized by the close association between the heterofermentative species *L. sanfranciscensis* and the maltose-negative yeast *C. milleri*. The dry sourdough harboured a different and complex microbial population, including heterofermentative species, dominated by *L. brevis* and *Leuconostoc mesenteroides*, homofermentative species, namely *L. plantarum* and *Pediococcus acidilactici*, and yeasts ascribed to *S. cerevisiae*.

1.1.2. Veneto

Lombardi and colleagues (2007) analysed the microbial ecology of six wheat sourdoughs (four destined for bread production and two destined for pastry products) and one wheat bran sourdough for bread production collected from artisan bakeries located in Veneto (Northern Italy), specifically around Vicenza. A total of 136 LAB and 61 yeasts were isolated; these underwent phenotypical characterization coupled to molecular analysis such as Randomly Amplified Polymorphic DNA (RAPD) analysis and rRNA gene sequencing. Among LAB, *L. sanfranciscensis* was the only species detected in wheat bran sourdough and in sourdoughs for sweet goods. Wheat sourdoughs for bread production were characterized by the presence of *L. sanfranciscensis* in association with heterofermentative species, such as *Leuc. citreum* and *L. rossiae,* and homofermentative species, such as *L. plantarum*, *Lactococcus lactis*, *Pd. parvulis*. A great intra-specific genetic variability was seen among the isolates ascribed to *L. sanfranciscensis*. Concerning yeasts, a high homogeneity was found in terms of species diversity, with a prevalence of the species *S. cerevisiae* and *C. humilis*. However, a high strain diversity was seen within these taxa.

1.1.3. Emilia Romagna

Vernocchi and colleagues (2004a) investigated the physico-chemical traits and the yeast dynamics of a wheat sourdough, which had been continuously propagated for 8 years in an artisan bakery in Ferrara (Emilia Romagna) for the manufacture of typical Ferrara bread (*Coppia ferrarese*). LAB and yeasts were isolated from sourdough and dough samples collected at five sampling points throughout the entire bread-making process; all the isolates were subjected to microscopic examination and conventional physiological tests, while the yeasts alone underwent RAPD analysis for the assessment of the intra-specific genetic variability. The identification of LAB showed the dominance of mesophilic heterofermentative bacilli at all stages of Ferrara bread production. For yeasts, a heterogeneous *C. milleri* population, accounting for 65.8% of the isolates and consisting of different biotypes, was found to dominate.

1.1.4. Marche

The microbial ecology of nine type I wheat sourdoughs collected from one semi-industrial and eight artisan bakeries in the Marche region (Central Italy) was investigated using a culture-dependent approach (Osimani et al., 2009). LAB and yeasts were counted on suitable solid media and selected colonies were streaked to purity. The 36 yeast and 118 LAB isolates were preliminary identified by Restriction Fragment Length Polymorphism (RFLP) analysis and Amplified Ribosomal DNA Restriction Analysis (ARDRA), respectively, whereas the diagnosis of species was finally verified by sequencing a portion of the 16S and

5.8S-ITS rDNA from one or more isolates of each RFLP and ARDRA group, respectively. An extremely high LAB diversity was seen across the sourdough ecosystems investigated. *L. plantarum* occurred in all the sourdough samples analysed, followed by *L. paralimentarius*, *L. curvatus*, *L. sanfranciscensis*, and *W. confusa*. Other LAB species were found with lower frequencies, namely *L. sakei*, *L. rhamnosus*, *L. helveticus*, *L. gallinarum*, *L. brevis*, *L. casei* spp. *carnosus*, *L. paracasei* spp. *paracasei*, *L. fermentum*, *L. casei sensu latu*, *L. salivarius*, *Leuc. citreum*, *Leuc. pseudomesenteroides*, *Weissella cibaria*, and *Enterococcus faecalis*. By contrast, a lower species diversity was seen among yeasts, with *S. cerevisiae* uniquely recovered in three out of four sourdoughs used for the isolation of this microbial group and found in coexistence with *C. humilis*. This latter species was dominant in only one sourdough. The main technological traits of a selected pool of yeast and LAB isolates were further investigated as a basis for a preliminary selection of promising strains to be exploited by the local baking industry in sourdough fermentation processes. The overall results of this study demonstrated the high biodiversity of the sourdough ecosystems investigated and the occurrence of wild yeast and LAB strains suitable for bread-making.

1.1.5. Umbria

In 1994, Gobbetti and colleagues published an ecological study on 24 wheat sourdoughs collected in some semi-industrial bakeries located in Umbria (Central Italy). Among these, 22 sourdoughs were destined for bread production whereas the remaining two were destined for Panettone production. The specific aim of the study was the identification and characterization of sourdough LAB and yeasts as a basis for the selection of starter cultures to be used in the manufacture of typical Italian breads. The 493 presumptive LAB were identified using traditional tests (determination of Gram reaction, catalase activity, growth at 15°C and 45°C, production of CO_2 from glucose and NH_3 from arginine, and esculin hydrolysis) and miniaturized assays (API 50 CH system). Three hundred and eighty-four randomly isolated yeasts were screened on the basis of their colony and cell morphology and their assimilation patterns using miniaturized assays (ATB 32 systems). A microbial map of the 24 sourdoughs was built up. Among LAB, *L. brevis* subsp. *lindneri*, subsequently classified as *L. sanfranciscensis* (Vogel et al., 1994), prevailed, accounting for 49% of the bacterial isolates, followed by *L. plantarum*, *L. farciminis*, *L. acidophilus*, *L. fermentum*, *L. fructivorans*, *L. alimentarius*, and *L. brevis*, whereas *S. cerevisiae* dominated the yeast community, accounting for 66% of the isolates, followed by *C. krusei*, *S. exiguus* and *H. anomala*. Interestingly, *S. cerevisiae* was found to be mostly associated with *L. brevis* subsp. *lindneri* and/or *L. fructivorans* and with *L. plantarum* and/or *L. alimentarius*. The dominance of *S. cerevisiae* in the same sourdoughs was later confirmed by Rossi (1996), who found the species *C. krusei* and *S. exiguus* together with members of the species *P. norvegensis*.

In the year 2005, a further study was carried out by Corsetti and colleagues on wheat sourdoughs in the same Italian region with the purpose of identifying sourdough LAB able to produce antimicrobial substances. In this study, a bacteriocinogenic strain ($CS1^T$) not assignable to known species was isolated and classified as *Lactobacillus rossiae* sp. nov., in honour of Professor Jone Rossi of the University of Perugia (Umbria, Italy) for her valuable contribution to dairy and sourdough microbiology.

1.1.6. Abruzzo

Valmorri and colleagues (2006) conducted a study aimed at detecting, isolating, and characterizing lactobacilli from 20 sourdoughs produced without the addition of baker's yeast in artisan bakeries throughout the Abruzzo region (Central Italy). All the sourdoughs were made from *T. aestivum* wheat flour, fermented at room temperature (23-28°C) for about 6 h, and stored at refrigeration temperature. The isolation campaign allowed the recovery of 343 Gram-positive, catalase-negative, non-motile, and rod-shaped cultures. On the basis of growth temperature, CO_2 production, arginine and esculin hydrolysis, and carbohydrate fermentation pattern, the 343 presumptive lactobacilli were divided into seven clusters. A representative portion of each cluster, forming a total of 127 isolates, was genetically identified by means of the multiplex PCR assay developed by Settanni and colleagues (2005) and sequence analysis of the 16S rRNA gene. The association between *L. plantarum* and *L. sanfranciscensis* was found in most of the sourdoughs analysed. In two sourdoughs *L. plantarum* and *L. paralimentarius* were the unique LAB species; the latter facultatively heterofermentative microorganism also occurred in further six sourdoughs. Three isolates belonging to *L. rossiae* and three isolates belonging to *L. alimentarius* were found in further three and one sourdoughs, respectively. Among the obligately heterofermentative species, *L. brevis* was largely present throughout Abruzzo, while *L. fermentum* was poorly detected, being isolated from four sourdoughs in association with the sole species *L. plantarum*. Finally, 12 isolates belonging to the *L. graminis/L. sakei/L. curvatus* group, and two rod-shaped isolates ascribed to *W. cibaria* were also collected.

In the same year Settanni and colleagues (2006) developed a single-step identification assay for the differential detection of 16 *Lactobacillus* species, that are generally associated with the sourdough ecosystem, by combining a previously described multiplex PCR assay (Settanni et al., 2005) with PCR-Denaturing Gradient Gel Electrophoresis (PCR-DGGE). This combined approach was used to analyse four artisan wheat sourdoughs collected in Abruzzo at the end of the fermentation period, namely 6 h at about 25°C. Analyses showed a prevalence of *L. plantarum* and *L. sanfranciscensis*.

Corsetti and colleagues (2007b) continued the characterization of the microflora of the Abruzzo sourdoughs previously collected by Valmorri and colleagues (2006) with the aim to count, isolate, and identify coccal-shaped LAB, and to follow the fate of some *E. faecium* and *Pd. pentosaceus* strains, which had been isolated from wheat grains in laboratory-made doughs with sterile wheat flour. Twenty-one Gram-positive and catalase-negative cocci were collected from six sourdoughs. Coccal-shaped LAB were found at concentrations ranging approximately between 10^4 and 10^6 CFU g^{-1}. The isolates were physiologically divided into two main groups on the basis of different phenotypic characteristics, including morphology, growth temperature, CO_2 production from glucose, NH_3 production from arginine, esculin hydrolysis, and growth on kanamycin aesculin azide (KAA) agar. Genotypic characterization by RAPD analysis was fully in agreement with physiological grouping. Eleven strains were further subjected to 16S rRNA gene sequencing and multiplex PCR, and so identified as *E. faecium* (n = 8 strains) and *Pd. pentosaceus* (n = 3 strains). Laboratory-made doughs were continuously propagated in order to follow the evolution of these subdominant LAB in combination with the key sourdough species *L. sanfranciscensis*. The co-fermentations of *E. faecium/L. sanfranciscensis* and *Pd. pentosaceus/L. sanfranciscensis* were carried out under the conditions typically used for propagation of type I sourdough, namely continuous daily refreshments at 30°C. *E. faecium* and *Pd. pentosaceus* were found to be stronger acidifiers than *L. sanfranciscensis* at the beginning of the sourdough fermentation process. These species, inhibiting indigenous microorganisms other than LAB by lowering the pH, might prepare the environment for the establishment of the typical species (e.g. *L. sanfranciscensis*) of mature sourdoughs. Nevertheless, once *L. sanfranciscensis* had colonized doughs (ca. 10^9 CFU g^{-1}), it determined a rapid decrement in coccal-shaped LAB, which remained at subdominant concentrations (ca. 10^7 CFU g^{-1}).

Recently Valmorri and colleagues (2010) have also investigated the yeast community of the 20 sourdoughs collected in the year 2006 (Valmorri et al., 201), with a combination of phenotype- and genome-based methods. A polyphasic analytical approach, consisting of culture-dependent and -independent methods was applied and a complete picture of the yeast populations depicted. *S. cerevisiae* was detected in 18 sourdoughs, and in 16 of them it was the only species found. In one sourdough, it was associated with *C. milleri*; this latter was the only species found in another sourdough, whereas a unique association between *C. krusei* and *Torulaspora delbrueckii* was seen in one further sourdough. The high variability of the *S. cerevisiae* strains, as revealed by the RAPD analysis, suggested that they were not derived from the baker's yeast.

1.1.7. Molise and *Campania*

In 1995, Iorizzo and colleagues investigated seven wheat sourdoughs collected from artisan bakeries in Molise (Southern Italy). Fifty-eight LAB were isolated and identified with phenotype-based assays. More than half (33) of these isolates showed an obligately heterofermentative metabolism, most of them being ascribed to *L. brevis* (60% of the isolates) and the rest to *L. fructivorans* (18%), *L. buchneri* (15%), and *L. cellobiosus* (6%). Among facultatively heterofermentative species, *L. sake* (64%), nowadays classified as *L. sakei*, was predominant followed by *L. plantarum* (36%). Concerning the 32 yeast isolates, 78% were ascribable to *S. cerevisiae*, while the remaining 22% to *C. humilis*.

A few years later Succi and colleagues (2003) analysed the occurence of yeasts isolated from 13 samples of sourdough made from *T. aestivum* wheat flour collected from artisan bakeries located in Molise and Campania. All the samples were produced without the use of baker's yeast. Seventy-five isolates were observed for colony shape, cell morphology, and spore formation. The isolates were then tested for growth at 15 and 37°C on yeast peptone dextrose (YPD) medium, ability to produce H_2S, and resistance to NaCl, whereas the identification system API 20 C AUX was used to validate the diagnosis of species. The analyses carried out on 75 isolates allowed the assignment of 58 strains to *S. cerevisiae*, five to *C. colliculosa*, four to *C. lambica*, three to *C. krusei* and to *C. valida* and two to *C. glabrata*. RAPD mostly confirmed the phenotypic identification. Moreover this technique made it possible to discriminate different strains of *S. cerevisiae*, the main species present in all the sourdough samples. Such a heterogeneity among the isolates was ascribed by the authors to different factors, such as the type of flour and the ingredients utilised, the production technology and the particular environment where the various bakeries were located.

Gatto and Torriani (2004) monitored the changes in the species composition of both yeast and LAB populations during a sourdough fermentation process relying on the use of a homemade sourdough starter continuously propagated for more than 20 years in Molise to produce local bread. This traditional wheat sourdough was manufactured without the addition of baker's yeast and fermentation was conducted at 28°C for 24 h. For each sourdough sample pH, total titrable acidity (TTA), and microbial composition were assessed by using an integrated approach, including conventional culturing procedures and PCR-DGGE analyses. The overall results indicated changes in the LAB population structure and the presence of a complex association of LAB species, including several obligate or facultative heterofermentative lactobacilli. In particular *L. sanfranciscensis*, *L. brevis*, *L. arizonensis-L. plantarum* group and *L. kimchii-L. paralimentarius* prevailed throughout the fermentation process. In contrast with the variability found in LAB, the yeast population was uniform and represented only by the species *S. cerevisiae*.

Reale and colleagues (2005) explored the LAB ecosystem of 16 samples of artisan sourdoughs from *T. aestivum* wheat flour produced in different areas of Molise. The identification of LAB was carried out using traditional methods coupled with a genetic RAPD analysis. Initially, 98 isolates were characterized by their morphological aspects, Gram stain and catalase reaction. Subsequently, 84 rod-shaped Gram-positive and catalase-negative bacteria were tested for growth at 15°C and 45°C, ammonia production from arginine, and gas production from glucose. In addition to the preliminary tests, API 50 CHL was used to univocally identify LAB. The comparative use of these techniques permitted the identification of *L. sanfranciscensis* (63.1% of the isolates) as the predominant species followed by *L. brevis* (15.5%), *L. brevis* ssp. *lindneri* (13.1%), *L. plantarum* (7.1%) and *L. fermentum* (1.2%).

1.2. *Triticum durum* Wheat Breads

1.2.1. Friuli Venezia Giulia

The microbial ecology of three type I (samples Q, C, B) and one type III (sample L) sourdoughs produced with durum wheat four without the addition of starter cultures in four artisan bakeries located in Friuli Venezia Giulia (Northern Italy) was investigated using culture-dependent and -independent methods (Iacumin et al., 2009). A total of 320 LAB and 80 yeast isolates were identified by PCR-DGGE coupled with sequencing and nested-PCR coupled with DGGE analysis, respectively. *L. plantarum* and *L. brevis* were dominant in type I sourdoughs, while *W. cibaria, L. plantarum* and *Pd. pentosaceus* prevailed in the type III sourdough. Among yeasts, *S. cerevisiae* was the sole species found in all but one sourdough (sample B), where *C. humilis* and *C. milleri* were exclusively detected. The PCR-DGGE analysis of the microbial DNA and RNA extracted directly from the sourdough samples allowed several bacterial species not detected with the culture-dependent approach to be revealed, namely *L. brevis* from sample L, *L. fructosus* and *Lc. lactis* from sample Q, *L. durionis* and *L. paralimentarius* from sample C, and *L. sanfranciscensis* from all the sourdough samples. Two hypotheses were formulated by the authors to explain such a discrepancy, namely the use of selective media not favourable for the growth of all the LAB species occurring in the sourdough samples considered and the occurrence of LAB cells in a viable not-cultivable (VNC) status. The outcome of this study allowed further interesting considerations to be made. On the one hand, the non-competitive association between *L. sanfranciscensis* and *L. plantarum*, detected in all the sourdough samples analysed, confirmed the high adaptation of these two species to the cereal environment, which has previously been ascribed to the ability of these microorganisms to utilise different flour carbohydrates. On the other hand, the non-competitive association between *S. cerevisiae* and *L. plantarum* (sourdoughs C, L and Q) was confirmatory of the stimulation of this yeast by the occurrence of this particular bacterial counterpart, as even suggested by Gobbetti (1998). Finally, the inability to isolate *L. sanfranciscensis* from sourdoughs C, L and Q, where *S. cerevisiae* dominated (sourdough C, L and Q), was in line with the negative effect of maltose-positive yeasts such as *S. cerevisiae* on the metabolism of maltose-positive LAB like *L. sanfranciscensis*.

1.2.2. Apulia

Ottogalli and colleagues (1996) reported microbiological data concerning sourdoughs for *Panettone* cake and Altamura bread (at that moment referred to as *Pugliese* bread, literally Apulia bread), collected in 26 years of research in collaboration with small and big industries in Lombardy. Concerning the microbiological analyses on sourdoughs for Altamura bread the lactic microflora was composed of *L. plantarum, L. brevis, L. fermentum, L. fructivorans* at 10^7-10^9 CFU g^{-1} levels. In addition, different coccal-shaped LAB belonging to *Streptococcus, Pediococcus* and *Leuconostoc* genera were present at 10^5-10^9 CFU g^{-1}. The yeasts were normally represented by *S. cerevisiae* either as a natural contaminant or, more frequently, purposely added.

A few years later, Corsetti and colleagues (2001) carried out a study on phenotypical and molecular identification of LAB and yeasts from 25 sourdoughs produced in semi-industrial bakeries in Apulia as a basis for the selection of starter cultures for the production of local typical breads. The bakeries were located all over the regional territory: six sourdough samples were collected from bakeries located in Lecce, eight from Bari, three from Brindisi, and eight from Foggia. Each bakery produced several typical breads. Seventeen of the 25 sourdoughs were produced using the wheat flour species *T. durum* alone or in association with the wheat flour species *T. aestivum*. In all except for six bakeries, baker's yeast was added to the sourdough. The time of fermentation to produce the sourdoughs varied considerably from 3 to 24 h, while the temperature was generally within the 20-28°C range. Three hundred and seventeen presumptive LAB were isolated and clustered in eight groups according to their morphological and physiological characteristics. Seventy-five randomly isolated yeast cultures were first clustered according to their colony and cell morphology. About 38% of the presumptive LAB and yeast isolates belonging to each cluster were phenotypically identified with API 50 CHL and ATB 32 systems, respectively. Sequence analysis of the 16S rRNA gene and 16S/23S rRNA spacer region confirmed the phenotype-based identification. The microbial distribution was very heterogeneous and among the obligately heterofermentative species *L. sanfranciscensis* was the most widespread, being isolated from 17 sourdoughs, while among the facultatively heterofermentative species *L. alimentarius* was the most commonly found, being isolated from nine sourdoughs, seven of which were prepared with *T. durum* wheat flour. *L. alimentarius* was found largely associated with *L. sanfranciscensis* apparently in substitution of *L. plantarum*, the latter normally associated with *L. sanfranciscensis. L. plantarum* strains were only isolated from sourdoughs prepared with the *T. aestivum* wheat flour, and therefore corresponded to only 7% of the isolates. In a few sourdoughs (four samples) only one species was isolated, while the other sourdoughs showed mixed populations of LAB. Interestingly, 14 of the 25 sourdoughs did not contain associations of homo- and heterofermentative LAB and some of them were characterized by a very complex heterofermentative microflora (e.g. *L. sanfranciscensis, L. brevis, Leuc. citreum* and *W. confusa*). The association of *L. alimentarius* and *L.*

sanfranciscensis seemed to be characteristic of sourdoughs incubated for the longest time (24 h) whereas *Leuc. citreum* was often isolated in sourdoughs not added with baker's yeast. Furthermore, most of the sourdoughs produced in the province of Lecce contained the typical association of *L. alimentarius* and *L. sanfranciscensis* while in the province of Foggia the association between *L. sanfranciscensis* and *Leuc. citreum* was characteristic. Other species isolated in lower percentages were: *L. fermentum*, *Lc. lactis*, *W. confusa*, *L. acidophilus*, *L. delbrueckii* ssp. *delbrueckii*. All the sourdoughs, including those without the addition of baker's yeast, contained *S. cerevisiae*. *S. exiguus* and *C. krusei* were only isolated from two sourdoughs. Based on overall results, differences in either the recipes or the technological processes have been hypothesized to greatly influenced the microbial composition of the sourdoughs analysed.

Ricciardi and colleagues (2002) performed the microbial characterization of durum wheat sourdoughs destined for the manufacture of Cornetto of Matera and Altamura bread. One hundred and sixty-three LAB and 80 yeasts were isolated from eight sourdoughs for Altamura bread and further identified by phenotypic tests. *L. plantarum* was the main bacterial species detected, followed by *L. casei* and *L. paracasei*, *L. brevis* and *Leuc. mesenteroides*. Among yeasts, *S. cerevisiae* was the only species found; interestingly, a high heterogeneity was seen in the leavening ability of the isolates ascribed to this species, thus suggesting the occurence of different phenotypes and hence strains.

A few years later, Ricciardi and colleagues (2005) again focussed their attention on the lactic microflora of Altamura bread. Seven sourdough samples were collected from different artisan producers. A total of 111 LAB strains was randomly isolated and characterized using a set of 29 phenotypic tests. The phenotype-based identification of these isolates was further confirmed by SDS-PAGE of whole cell proteins. Overall, 88% of the isolates were identified as *L. plantarum*, *L. casei* and *paracasei*, 12% as *L. brevis*, and *Leuc. mesenteroides*. Intriguingly *L. plantarum* was the species most frequently isolated, representing 49% of the isolates, whereas *L. sanfranciscensis* and *L. alimentarius* were never detected. The results of this study confirmed those previously obtained by Ricciardi and colleagues (2002) but were quite unexpected with respect to those of Corsetti and colleagues (2001), who found that *L. plantarum* acted as a minority species and *L. alimentarius* was frequently associated with *L. sanfranciscensis*. Another peculiarity of the sourdoughs destined for the manufacture of Altamura bread was the high percentage (88%) of homofermentative isolates, which was feasibly attributed to the addition of baker's yeast during the manufacture of this baked product. Furthermore, Altamura bread sourdoughs were found to be completely dominated by *S. cerevisiae*.

1.2.3. Basilicata

The microbial characterization of durum wheat sourdoughs destined for the manufacture of Cornetto of Matera was carried out by Ricciardi and colleagues in the year 2002. Twelve sourdoughs added with baker's yeast were sampled from six bakeries located in Matera (Basilicata). A total of 407 LAB and 173 yeasts were isolated and identified by phenotype-based tests. *L. plantarum*, *L. pentosus*, *L. curvatus*, *L. fermentum*, and *Weissella* spp. were the predominant LAB while *S. cerevisiae* was the sole yeast species detected.

A few years later, Zotta and colleagues (2007, 2008) characterized 41 LAB previously isolated by Ricciardi and colleagues (2002) from Cornetto of Matera sourdoughs by means of enzymatic tests, SDS-PAGE of whole cell proteins, and phenotype-based assays aimed at assessing the ability of the isolates to produce acid, antimicrobial compounds, and exopolysaccharides. The cultures were identified as *L. plantarum* (49%), *Leuc. mesenteroides* (17%), *L. curvatus* (15%), *L. paraplantarum* (12%), *W. cibaria* (5%) and *L. pentosus* (2%). Finally, three strains ascribed to *W. cibaria*, *L. plantarum* and *S. cerevisiae* were selected on the basis of their technological properties and used for the formulation of a defined starter culture, which was further tested in model sourdough fermentations.

1.2.4. Sicily

Pulvirenti and colleagues (2001) carried out a research study aimed at identifying the yeast species in 14 samples of sourdough taken from different bakeries throughout Sicily. A total of 50 strains were isolated and identified by conventional and molecular procedures such as PCR/RFLP analysis of the Non-Transcribed Spacer 2 (NTS2), PCR of inter-δ regions, PCR/RFLP analysis of the Internal Transcribed Spacer (ITS) region, Pulsed Field Gel Electrophoresis (PFGE) analysis, and partial sequencing of the 26S rRNA gene. *S. cerevisiae* was the prevailing species, followed by *Issatchenkia orientalis*, *C. milleri* and *C. humilis*. The occurence of *S. cerevisiae* old type *anamensis* was also highlighted, thus suggesting a strong relationship between this species and the particular environment of Sicilian sourdoughs.

Two years later, Gullo and colleagues (2003) monitored the yeast population dynamics in a sourdough used for the production of durum wheat bran flour bread. Sourdough samples were taken from an industrial bakery, at different intervals of time during the sourdough rebuilding phase and 138 yeast isolates were obtained from the isolation campaign. Assimilation tests coupled to a molecular method (PCR/RFLP of the ITS region) were employed for yeast identification and *C. humilis* was found to be the dominant species. Furthermore, the yeast isolation and analyses were repeated after 4 and 8 weeks, at 84 and 168 rebuildings, and the occurence of *C. humilis* was steady over time, thus indicating that the conditions in which the sourdough was kept exerted a strong selective pressure in favour of this yeast species.

Soon after, Pulvirenti and colleagues (2004) performed a study aimed at identifying the dominant yeast species in 35 homemade sourdoughs, prepared as previously described by Romano and colleagues (2001) and collected from artisan bakeries in four different locations in Sicily. A combination of phenotype- and genome-based (PCR-RFLP of ITS regions) techniques was used for the identification at species level of the 185 isolates, whereas for the sole *S. cerevisiae* isolates, strain differentiation was achieved through the analysis of inter-δ regions. *S. cerevisiae* was the only species detected in 91% of the sourdoughs analysed, while *C. humilis*, *I. orientalis* and *S. exiguus* were each dominant in 3% of the samples. Finally, the analysis of the inter-δ regions of *S. cerevisiae* strains allowed a high genetic polymorphism to be highlighted, even among isolates collected from the same sourdough sample.

In the year 2005, Randazzo and colleagues studied bacterial and *Lactobacillus* communities of nine artisan durum wheat sourdoughs from different areas of Sicily (Modica and Ispica) using a combination of culture-dependent and -independent techniques. The latter included a protocol for DNA extraction directly from sourdough and PCR-DGGE of rRNA genes, using bacterial universal primers as well as *Lactobacillus*-specific primers. The sourdough samples were produced according to the most common locally used traditional procedure: grape juice or honey was added to durum wheat flour and kneaded with warm water and the dough was incubated for 16 h at room temperature. The first sourdough was again mixed with flour and tepid water and incubated for a further 16 h at the same temperature. The sourdough was then daily propagated by back slopping for several days and the ripe sourdough was used as inoculum for bread-making. After isolation and phenotypical characterization, molecular identification of the isolates was based on RFLP analysis and 16S rRNA gene sequencing. Forty-five sourdough LAB were isolated and 17 were identified as *L. sanfranciscensis*, 14 as *L. pentosus*, five as *L. plantarum*, four as *L. kimchii/L. alimentarius*, and three as *L. casei*. As phenotypic characterization was ambiguous, giving the growth of few isolates belonging to *L. sanfranciscensis* at 45°C unable to ferment maltose, the presence of the species was confirmed by sequence analysis. The PCR-DGGE approach showed the dominance of *L. sanfranciscensis*, although in two sourdoughs, the species *L. sanfranciscensis*, *L. fermentum*, and *L. alimentarius/L. kimchii* were also found using *Lactobacillus*-specific PCR-DGGE systems.

1.2.5. Sardinia

Catzeddu and colleagues (2006) studied the LAB composition of 25 durum wheat sourdoughs used for the manufacture of four Sardinian typical breads, namely the thin crisp *Carasau* and *Zichi*, the thin soft *Spianata* and the crumb *Moddizzosu*. The main traits of these breads are briefly summarized. *Carasau*, also known as "*carta da musica*" (music paper), due to the particular sound it produces when chewed, was originally produced in the internal areas of Sardinia (Southern Italy), but nowadays baked all over the regional territory. It is a round-shaped bread of about 40 mm in diameter and it is 1-2 mm thick, with no crumb. It is characterized by a long shelf-life due to the double baking process which reduces the moisture of the product to about 6% (Huy et al., 2006; Minervini et al., 2010). Since *Carasau* is prepared only occasionally, its sourdough is normally kept at -20°C and then propagated by adding water and flour and fermented for 6-8 h. *Zichi* is produced mostly in the Sassari district, in two different types: it can be soft or dry. It has a round shape with 30 cm diameter and is 1 cm thick, without crumb, and a light golden colour (Minervini et al., 2010). *Spianata,* also known as *Pane di Ozieri*, is manufactured in the North-Central area of Sardinia (Logudoro) and it is destined for daily consumption but even used for special occasions or ceremonies. This bread is 2-5 mm thick and soft; it is a white-golden colour and it has a circular shape with variable diameter (from 30-40 cm or from 18-23 cm), without crumb. In recent years the producers of *Spianata* have applied for PGI recognition for this product (Minervini et al., 2010). *Moddizzosu* is prepared especially in the South of the island and is characterized by a weight of 0.5-1 kg, a hard golden-brown 2-4 mm thick crust, and a white soft crumb with bubbles of medium size. It has a long shelf-life (about 1 week). Recently even the producers of *Moddizzosu* have applied for the designation of the PGI label (Minervini et al., 2010). *Zichi*, *Spianata*, and *Moddizzosu* are produced daily and their manufacture involves the addition of baker's yeast to the dough during the baking process.

The sourdoughs used for the manufacture of these four typical Sardinian breads were sampled from artisan bakeries located in different areas by Catzeddu and colleagues (2006). These sourdoughs were prepared with durum wheat flour and fermented at room temperature. All the sourdough samples were analysed with both molecular (sequencing of the 16S rRNA gene and RAPD analysis) and multivariate statistical methods. In total, 210 presumptive LAB were isolated and all of them were confirmed to be Gram- positive, catalase-negative, able to grow at 15°C, and unable to grow at 45°C. The 16S rDNA partial sequencing and similarity search allowed the identification of 12 LAB species, namely *L. pentosus*, *L. farcimins*, *L. brevis*, *L. alimentarius*, *L. plantarum*, *Pd. pentosaceus*, *Leuc. citreum*, *L. sanfranciscensis*, *W.*

confusa, *L. sakei*, *L. casei*, and *L. zeae*. Overall, *L. pentosus* and *L. alimentarius* were found in all of the sourdoughs. *L. pentosus* was the prevailing species, representing 89% of the isolates from *Moddizzosu* sourdoughs and 36% of the isolates from *Carasau* sourdoughs. The other species were found only in some sourdoughs: *L. sanfranciscensis* was detected in *Carasau*, *Spianata* and *Zichi* sourdoughs; *L. brevis* in *Carasau* and *Zichi* sourdoughs; *L. plantarum* in *Carasau* and *Spianata* sourdoughs; and *W. confusa* in *Zichi* sourdoughs only. Intriguingly *L. sanfranciscensis,* which is well known as a key sourdough species, was isolated with low frequency. In *Zichi* and *Carasau* sourdoughs the species were quite homogeneously disseminated among samples, while *Moddizzosu* was characterized by a lower species diversity, with *L. sakei*, *L. pentosus*, and *L. alimentarius* as the sole lactobacilli found. The RAPD analysis of the isolates revealed a clear intra-specific strain diversity, especially within *L. sanfranciscensis* from *Zichi*, *W. confusa* from *Spianata*, and *L. pentosus* from *Carasau*.

1.3. Rye Bread

Rye is a cereal with an ancient origin (Bronze age) and widespread cultivation, mainly explainable by its great adaptability to different environmental conditions. It has been used for bread-making since the IV century B.C. in North-Central and North-East Europe. In Italy, it is mainly produced in the Alpine territory, which encompasses the Italian regions of Valle d'Aosta, Trentino Alto Adige, and Lombardy. To our knowledge, only one specific study has been performed on rye sourdoughs. Indeed, in the year 1988, Galli and colleagues characterized the microbial ecology of three sourdoughs destined for the manufacture of rye bread. Only heterofermentative bacteria ascribed to *L. sanfranciscensis*, *Leuc. mesenteroides*, and *L. confusus/fermentum* were recovered together with *S. cerevisiae* and *C. milleri.*

2. PIZZA

Pizza is one of the most widespread meals in the world, but its origin is in Naples. It is a flat leavened bread traditionally topped with tomatoes and mozzarella cheese, although a variety of vegetables and even animal source foods (e.g. eggs, anchovies, tuna, etc.) may also be added. From September 2010 *Pizza* is officially included in the list of Italian TSG products.

Coppola and colleagues (1996) first characterized the microbial population of ten leavened doughs for *Pizza* sampled from restaurants in Naples (Campania, Southern Italy). The authors also focussed their attention on the chemical and rheological properties of these doughs. A total of 144 isolates, including 22 yeasts and 122 bacteria, was collected. Five different species of yeasts and 20 species of bacteria were consistently identified with traditional microbiological methods. Among yeasts, *C. versatilis* was predominant (nine isolates), followed by *S. cerevisiae* (eight isolates), *C. krusei* (two isolates), *Dekkera intermedia* (two isolates), and *Debaryomyces hansenii* (one isolate). Of the bacteria, 70 isolates were homofermentative and 52 heterofermentative. Within the first group, *L. sake*, nowadays classified as *L. sakei*, and *L. plantarum* showed the main occurrence, whereas

Leuc. gelidum and *Leuc. mesenteroides* were dominant among heterofermentative bacteria. Minority species such as *L. paracasei*, *L. pentosus*, *L. agilis*, *L. confusus*, *L. viridescens*, *L. brevis* var. *lindneri*, *L. kefir*, *Leuc. paramesenteoides*, *Leuc. dextranicum*, *Leuc. amelibiosum*, *Leuc. argentinum*, *W. paramesenteoides*, *E. faecium*, *E. faecalis*, *E. hirae*, and *E. raffinosus*, were also found. One homofermentative strain ascribed to the genus *Lactococcus* was also isolated, but it could not be referred to the known species. Finally, 19 homofermentative cocci could not be identified either at the genus or at the species level.

This study was the basis for subsequent research carried out by the same group (Coppola et al., 1998) to determine the influence of various starter cultures on the microbiological, chemical, and physical characterization of doughs for *Pizza*. Fourteen different starter cultures containing one strain of *S. cerevisiae* with a combination of *L. plantarum*, *L. sanfranciscensis*, *E. faecium*, and *Leuc. mesenteroides* were employed to that end. Interestingly the most favourable association, allowing the best growth of bacteria and yeasts, included *L. plantarum* and *Leuc. mesenteroides.* Significant bacterial growth could also be detected in the doughs fermented by *S. cerevisiae* in association with *L. sanfranciscensis* and *Leuc. mesenteroides*. The doughs with the lowest bacterial content were those fermented by *S. cerevisiae* and *L. sanfranciscensis.* However, despite these microbial characteristics, the starter in question did not fail in leavening performance.

3. Sweet Leavened Goods

The sourdoughs for the production of sweet leavened goods share common features, namely long time of ripening (34 h on average), low incubation temperature (<20°C), low dough yield (150%), anaerobic state (by binding the sourdough with canvas), and high acidity (final pH<4.0). Baker's yeast is never added during most production processes, such as those used for *Panettone* and *Colomba*. Furthermore, most manufacturers use the same sourdough for the preparation of different sweet baked goods throughout the year (Foschino et al., 2004).

3.1. Panettone

Among sweet leavened goods, *Panettone* is undoubtedly one of the most appreciated and worldwide renowned Italian bakery specialities. It is a typical Christmas cake originally manufactured in Lombardy, specifically in Milan, but currently produced and consumed all over the Italian territory and even distributed at international level (Garofalo et al., 2008). It has a cylindrical shape and a soft texture, with elongated cellular structure and a unique aroma. In general, the production process requires three sourdough refreshments through the addition of water and flour to gradually increase the leavening mass. The sourdough is usually added as 5-15% of the final dough. After the refreshments, the dough is first added with sugar and fats in order to obtain the so-called "white dough", and later with eggs ("yellow dough"), raisins, and candied fruits (Huy et al., 2006; Zannini et al., 2010). In September 2003, the artisan Panettone producers located in Milan formed the Committee of Master Confectioners and made an application to obtain PGI denomination from the EC.

According to the specifications of the production disciplinary for this bakery speciality, sourdough is the sole leavening agent allowed.

In the early '70s, Galli and Ottogalli (1973) carried out the first microbial characterization of sourdoughs destined for the manufacture of *Panettone* using culture-dependent techniques. Sourdough and dough samples were collected during a *Panettone* production cycle from three factories located in Milan; these were subjected to both viable counting and isolation of LAB and yeasts. The isolates were identified by microscopic examination, observation of the colony morphology and sierological analyses. Only slight differences were detected among the three productions considered. Yeast and LAB viable counts were around 10^6-10^7 CFU g^{-1} and 10^7-10^8 CFU g^{-1}, respectively. Yeasts were identified as asporogenous strains of *S. exiguus* (*Torulopsis holmii*) whereas LAB were mainly ascribed to heterofermentative lactobacilli with intermediate characteristics between *L. brevis* and *L. cellobiosus*, and to homofermentative lactobacilli identified as *L. plantarum*. Sub-dominant species ascribed to *Citrobacter* and *Enterobacter* were also found.

Later, Galli and colleagues (1988) characterized the microbial ecology of two sourdoughs and a final dough destined for the manufacture of *Panettone*. No microbial differences were detected at the different stages of production, where heterofermentative bacteria such as *L. brevis/fermentum* and *L. sanfranciscensis* were found to dominate. Among yeasts, *T. holmii*, *S. cerevisiae* and *C. stellata* were found.

In the year 1994, Gobbetti and colleagues published an ecological study on 24 wheat sourdoughs collected in semi-industrial bakeries located in Umbria. The two sourdoughs destined for the production of *Panettone* were characterized by the typical association of *L. brevis* subsp. *lindneri*, subsequently classified as *L. sanfranciscensis* (Vogel et al., 1994) and *S. exiguus*.

A few years later Ottogalli and colleagues (1996) reported microbiological data concerning sourdoughs for the production of this traditional Italian Christmas cake, where *L. sanfranciscensis* was found to prevail with viable counts ranging from 10^8 CFU g^{-1}, at the starting point, to 10^9 CFU g^{-1}, at the end of the ripening period. *C. holmii* was also found, with viable counts going from 10^6 CFU g^{-1} to 10^7 CFU g^{-1}.

Foschino and colleagues (1999) analysed the microbial and chemical compositions of ten sourdoughs destined for the manufacture of *Panettone*, *Cornetto*, and *Brioche* and collected from small and industrial bakeries throughout Lombardy. Overall, 64 LAB and 28 yeast strains were isolated and identified with conventional methods and species-specific PCR assays. A very homogeneous species distribution was highlighted in the majority of the sourdoughs analysed and in particular in those for *Panettone*, with *L. sanfranciscensis* as the prevailing species among lactobacilli and *C. holmii* as the unique yeast identified. In a few sourdoughs *L. brevis* flanked or substituted *L. sanfranciscensis*.

Some years later, Foschino and colleagues (2004) investigated the composition of the yeast communities in sourdoughs destined for the manufacture of *Panettone*, *Pandoro*, *Cornetto*, and *Brioche*. Sourdoughs were collected from eight bakeries located in Northern Italy, mainly in Lombardy. Yeast cells were isolated and their morphological aspect was observed microscopically. The physiological traits of each isolate were investigated as well as the assimilation capabilities on different carbon sources using different miniaturized assays. Furthermore, different molecular methods (RAPD, mitochondrial-DNA restriction enzyme analysis, and sequencing of the ITS between the 18S and 26S rRNA genes) were comparatively evaluated for the univoque identification of the isolates. *C. humilis* was found to dominate in the three sourdoughs destined for the manufacture of *Panettone*, whereas a high degree of intra-specific homogeneity was seen among isolates ascribed to *C. humilis,* as shown by the RAPD fingerprinting.

More recently, Picozzi and colleagues (2006) developed a rapid analytical tool to trace the occurence of *L. sanfranciscensis* throughout the entire production process for *Panettone*, including the baked product. Different bacterial DNA extraction methods were comparatively evaluated and a species-specific PCR assay, relying on the use of a primer set targeting the 16S rRNA gene, was further developed.

Two years later, Garofalo and colleagues (2008) investigated the yeast and LAB dynamics throughout the production process for the following three *Panettone* varieties in a semi-industrial bakery located in the Marche region: i) a *Panettone* manufactured according to the traditional recipe ii) a "sugar-free" *Panettone* manufactured with malt instead of sucrose and added with baker's yeast to speed up the last stage of leavening; and iii) a round-shaped *Panettone*-like cake available on the market under the name of *Ciambellona*. The PCR-DGGE technique was applied to both the microbial DNA extracted directly from the sourdough and dough samples, and to the DNA recovered from the bulk of LAB and yeast colonies harvested from suitable media. Overall, a low species diversity was detected in all the sourdough and dough ecosystems investigated. The analysis of the DNA recovered directly from the sourdough and dough samples showed that *L. sanfranciscensis* and *L. brevis* were the dominant species in all three fermentation processes considered. However, by analysing the DNA recovered from the bulk cells, a different species association was found in the final dough of *Panettone* and *Ciambellona*, where a shift was seen in the bacterial pattern towards *L. plantarum* in association with *L. brevis*. Furthermore, *Leuc. mesenteroides* was also detected in the final dough of *Panettone*. Concerning yeast dynamics, a high population stability was highlighted in *Panettone* and *Ciambellona* production processes, where *C. humilis* was found to dominate the two productions using both culture-dependent and independent approaches. The sourdough for *Ciambellona* production was also characterized by the occurence of *C. milleri.* As expected, in the production of "sugar-free" *Panettone* a substitution of *C. humilis* with *S. cerevisiae* was seen after the addition of the compressed yeast. Intriguingly, despite the introduction of high loads of *S. cerevisiae* during the fermentation process, *C. humilis* was able to survive, as emerged from the analysis of the viable yeast community.

In the year 2009 Venturi and colleagues investigated the microbial populations throughout the production process of a *Panettone* cake and a typical sweet leavened product, known as *Biscotto del Lagaccio* (Lagaccio biscuit), from Liguria (Northern Italy). The sourdough and doughs for the preparation of *Panettone* were sampled at a factory located near Turin (Piemonte, Northern Italy). The LAB and yeasts isolates were identified with molecular methods, namely ARDRA and RFLP of the ITS region, respectively. The microbial metabolites were also determined by High Performance Liquid Chromatography (HPLC) analysis. The sourdough was dominated by *L. sanfranciscensis* in coexistence with *C. milleri* and *S. cerevisiae.* In the final dough, the LAB population, and specifically *L. sanfranciscensis* which is very sensitive to a changing environment, showed a drastic reduction probably due to the addition of other ingredients besides to water and flour.

3.2. Colomba

Colomba is a dove-shaped sweet leavened cake, which originated in Milan (Lombardy), and is traditionally manufactured for Easter. It has a light texture and an elongated porous structure with icing of egg white, sugar, shelled almonds, with granulated sugar on top and candied fruit in the dough. The technological steps for obtaining *Colomba* are the same as for the production of *Panettone* expect for the unique final shape (Huy et al., 2006; Zannini et al., 2010).

Vernocchi and colleagues (2004b) published a study aimed at investigating the yeast population dynamics and the evolution of the main physicochemical parameters (pH, carbohydrates, and organic acid content) in sourdoughs and doughs collected during the production cycle of *Colomba* at a factory located in Ravenna (Emilia Romagna). The preparation of *Colomba* followed its specific procedure, starting with a continuously reproduced sourdough fermented at 24°C for 19 h. Seven sampling points along the production cycle were chosen and LAB and yeasts were counted and isolated from the corresponding sourdough and dough samples. A total of 86 yeasts and 48 LAB were isolated. LAB isolates were studied by microscopic examination and by conventional physiological tests (API 50 CHL). The Gram-positive and catalase-negative isolates were further examined for cell morphology, growth in MRS broth at 15 and 45°C, and production of CO_2 from glucose. All the yeast isolates were subjected to a preliminary morphological (morphology and spore formation) and physiological (ability to utilise carbon compounds and to ferment the following carbohydrates: glucose, galactose, sucrose, threalose, maltose, lactose, and raffinose) characterization. RAPD analyses were then used in order to precisely assess the yeast species and strains. A dominance of mesophilic heterofermentative bacilli was found in all stages of *Colomba* production. *S. cerevisiae* was the main yeast species detected and strains isolated after the addition of compressed yeast showed a very high level of similarity. On the contrary, among *S. cerevisiae* isolates from the sourdough and doughs collected before the addition of baker's yeast a great biodiversity was seen. The fact that some of those strains derived from sourdough make the finding even more interesting since *S. cerevisiae* is usually sensitive to the low pH and high level of acetic acid content typical of that environment. Therefore the *S. cerevisiae* strains isolated in this study represent a unique potential starter culture for the manufacture of baked goods. *C. milleri* was also found to be a dominant species with notable strain polymorphism among isolates.

3.3. Pandoro

Pandoro is a traditional Christmas cake originally manufactured in Verona; its name "*pan de oro*" is feasibly derived from its golden colour, which is mainly due to the great amount of egg and fat used for its manufacture. *Pandoro* is a conical trunk, shaped into an octagonal star; it has a very soft dough with no external crust and is usually covered with icing sugar. It has a uniform texture, which is due to the addition of up to 1% baker's yeast to the final dough. The use of this leavening agent and the high fat percentage (22-27%) are the main traits which differentiate *Pandoro* from *Panettone* and *Colomba* (Huy et al., 2006; Zannini et al., 2010).

Based on the results obtained by Foschino and colleagues (2004), the *Pandoro* sourdough was found to be dominated by *C. humilis* and *S. cerevisiae.* RAPD analyses applied to *S. cerevisiae* isolates obtained from different bakeries highlighted a great similarity between them and a certain distance from the reference strains originated from compressed yeast.

3.4. Cornetto and Brioche

Cornetto and *Brioche* are snacks, rich in butter and sometimes filled with jam, cream or chocolate, which are consumed for breakfast or at coffee-time.

In the year 1988, Galli and colleagues characterized the microbial ecology of four sourdoughs for *Brioche* production. Among the heterofermentative species *L. sanfranciscensis* predominate, followed by *Leuc. mesenteroides*, *Leuc. pseudomesenteroides*, and *Leuc. dextranicum*, whereas among the homofermentative bacteria *L. plantarum* was dominant. Concerning yeasts, *C. stellata* was mainly detected, followed by *S. exiguus* and *S. cerevisiae*.

A few years later, three and four sourdoughs destined for the preparation of *Cornetto* and *Brioche*, respectively were used for the isolation of LAB and yeasts by Foschino and colleagues (1999). Two *Cornetto* sourdoughs were characterized by the occurrence of *L. sanfranciscensis* in association with *C. holmii,* while the third was characterized by *L. sanfranciscensis* together with *S. cerevisiae.* As far as the *Brioche* sourdoughs are concerned, three were characterized by the presence of *L. sanfranciscensis* and *L. brevis*, in coexistence with *C. holmii* or *S. cerevisiae*, whereas the remaing one, showing the highest pH and the lowest acidity, unusually consisted of *L. delbruechii* subsp. *delbruechii* and *S. cerevisiae.*

In 2004, Foschino and colleagues carried on the investigation of the composition of yeast communities in sourdoughs destined for the manufacture of *Cornetto.* Two of the four sourdoughs under study were dominated by *S. cerevisiae*, whereas the other two were dominated by *C. humilis* and *S. pastorianus*, respectively.

3.5. Lagaccio Biscuit

Venturi and colleagues (2009) investigated the microbial populations throughout the production process for Lagaccio biscuit, a dry biscuit which originated in Genoa (Liguria). This product has an oval-shape, is 2 cm wide and is prepared with wheat flour, milk, water, sugar, soft butter, salt, and anise. Its preparation involves four sourdough refreshments with water and flour (4 h fermentation for each refreshening step) and two subsequent dough fermentation steps 8-10 h each. The final dough is moulded into a long shape; after baking, it is left to cool for one day and is finally cut into oblique slices and each slice is toasted on both sides (Zannini et al., 2010). In the sourdough and doughs sampled from a factory located near Genoa by Venturi and colleagues (2009) and used for the production of this sweet baked specialty, the species *L. sanfranciscensis* was found in coexistence with *C. milleri* and *S. cerevisiae*.

CONCLUSION

Italian sourdoughs are characterized by a wide species heterogeneity (Table 1 and Table 2) as a result of different traditional regional procedures as well as environmental conditions and raw materials.

L. sanfranciscensis was found to be the main LAB species in Italian sourdoughs and the association of *L. sanfranciscensis* and *L. plantarum* was detected in most of the wheat sourdoughs for bread and sweet baked goods production, as well. *L. brevis* and *L. fermentum* were also quite diffused followed by *Leuc. mesenteroides, L. alimentarius*, *L. sakei*, *Leuc. citreum*, *L. casei*, *L. paralimentarius*, *L. pentosus*, *Lc. lactis*, and *W. cibaria*. Exceptions were found in *T. durum* wheat sourdoughs from Apulia, which were mainly characterized by *L. sanfranciscensis* and *L. alimentarius*, the latter apparently in substitution of *L. plantarum*. Intriguingly, in sourdoughs for Altamura bread, *L. plantarum* was the dominant species while *L. sanfranciscensis* and *L. alimentarius* were not detected at all.

In Southern Italy and specifically in Sicily and Sardinia, *L. pentosus* was found to be dominant in *T. durum* bread sourdoughs. This species was also identified, with low frequencies, in *T. durum* wheat sourdoughs destined for the manufacture of Cornetto of Matera in Basilicata and *Pizza* in Campania thus indicating that the technological processes rather than the geographical area are the cause for the selection of this species.

Intriguingly, bread sourdoughs collected in the Marche region harboured the higher LAB species diversity among the Italian sourdoughs analysed, feasibly due to the different technological parameters applied, raw materials used and bakery environments. A similar conclusion can be drawn from the ecological study of *Pizza* sourdoughs that highlighted a great LAB species biodiversity.

Concerning yeasts, *S. cerevisiae* appeared to dominate almost all the Italian sourdoughs investigated, often in coexistence with *C. milleri*, *C. humilis*, *C. krusei*, and *T. holmii/C. holmii/S. exiguus*.

As far as sweet leavened goods are concerned, a common feature of type I sourdoughs employed in Italian productions is the stable occurrence of *L. sanfranciscensis* associated with *S. cerevisiae* and/or maltose-negative yeasts, namely *C. milleri*, *C. humilis* and *T. holmii/C. holmii/S. exiguus*, thus confirming the impact of the technological parameters applied to these sourdoughs independently of the geographical areas.

Notably, until a short time ago, recovery of *S. cerevisiae* from sourdoughs was merely ascribed to adventitious contamination with compressed baker's yeast. The recent molecular analyses carried out on sourdoughs clearly highlighted the high intraspecific genetic variability of *S. cerevisiae* strains, thus suggesting the occurrence of heterogeneous populations not originating from baker's yeast.

In conclusion, in-depth studies of sourdough biodiversity and population dynamics are fundamental for basic ecological knowledge and for practical applications, such as starter formulation, the control of the leavening process and the manufacture of standardized high-quality products.

Table 1. Sourdough lactic acid bacteria distribution. AB = *Triticum aestivum* Wheat Bread; DB = *Triticum durum* Wheat Bread; RB = Rye Bread; P = Pizza; S = Sweet leavened goods.

Species	Lombardy	Friuli Venezia Giulia	Veneto	Liguria	Emilia Romagna	Marche	Umbria	Abruzzo	Molise	Campania	Apulia	Basilicata	Sicily	Sardinia
Enterococcus faecalis						AB				P				
Enterococcus faecium								AB		P				
Enterococcus hirae										P				
Enterococcus raffinosus										P				
Lactobacillus acidophilus							AB				DB/AB			
Lactobacillus agilis										P				
Lactobacillus alimentarius							AB	AB			DB/AB		DB	DB
Lactobacillus arizonensis									AB					
Lactobacillus brevis	AB/S	DB				AB/S	AB	AB	AB		DB/AB			DB
Lactobacillus buchneri									AB					
Lactobacillus casei						AB					DB		DB	DB
Lactobacillus casei spp. *carnosus*						AB								
Lactobacillus cellobiosus	S								AB					
Lactobacillus confusus	RB									P				
Lactobacillus curvatus						AB		AB				DB		
Lactobacillus delbruechii subsp. *delbruechii*	S										DB			
Lactobacillus durionis		DB												
Lactobacillus farciminis							AB							DB
Lactobacillus fermentum	RB/S					AB	AB	AB	AB		DB/AB	DB	DB	
Lactobacillus fructivorans							AB		AB		DB			
Lactobacillus fructosus		DB												
Lactobacillus gallinarum						AB								
Lactobacillus graminis								AB						
Lactobacillus kefir										P				
Lactobacillus helveticus						AB								
Lactobacillus kimchii									AB				DB	
Lactobacillus paracasei										P	DB			
Lactobacillus paracasei spp. *paracasei*						AB								
Lactobacillus paralimentarius		DB				AB		AB	AB					
Lactobacillus paraplantarum												DB		
Lactobacillus pentosus										P		DB	DB	DB
Lactobacillus plantarum	AB/S	DB	AB			AB/S	AB	AB	AB	P	AB	DB	DB	DB
Lactobacillus rhamnosus						AB								
Lactobacillus rossiae			AB				AB	AB						
Lactobacillus sakei						AB		AB	AB	P				DB
Lactobacillus salivarius						AB								
Lactobacillus sanfranciscensis	AB/RB/S	DB	AB/S	S		AB/S	AB/S	AB	AB	P	DB/AB		DB	DB
Lactobacillus viridescens										P				
Lactobacillus zeae														DB
Lactococcus lactis		DB	AB								DB/AB			

Leuconostoc amelibiosum										P				
Leuconostoc argentinum										P				
Leuconostoc citreum			AB			AB					DB/AB			DB
Leuconostoc dextranicum	S									P				
Leuconostoc gelidum										P				
Leuconostoc mesenteroides	AB/RB/S					S				P	DB	DB		
Leuconostoc paramesenteroides										P				
Leuconostoc pseudomesenteroides	S					AB								
Pediococcus acidilactici	AB													
Pediococcus parvulis			AB											
Pediococcus pentosaceus		DB						AB						DB
Weissella cibaria		DB				AB		AB				DB		
Weissella confusa						AB					DB			DB
Weissella paramesenteoides										P				

Table 2. Sourdough yeasts distribution. AB = *Triticum aestivum* Wheat Bread; DB = *Triticum durum* Wheat Bread; RB = Rye Bread; P = Pizza; S = Sweet leavened goods.

Species	Lombardy	Friuli Venezia Giulia	Veneto	Liguria	Emilia Romagna	Marche	Umbria	Abruzzo	Molise	Campania	Apulia	Basilicata	Sicily	Sardinia
Candida colliculosa									AB	AB				
Candida glabrata									AB	AB				
Candida humilis	S	DB	AB			AB/S			AB				DB	
Candida krusei							AB	AB	AB	AB/P	DB			
Candida lambica									AB	AB				
Candida milleri	AB/RB	DB		S	AB/S	S		AB					DB	
Candida stellata	S													
Candida valida									AB	AB				
Candida versatilis										P				
Debaryomyces hansenii										P				
Dekkera intermedia										P				
Hansenula anomala							AB							
Issatchenkia orientalis													DB	
Saccaromyces cerevisiae	AB/RB/S	DB	AB	S	S	AB/S	AB	AB	AB	AB/P	DB	DB	DB	
Saccaromyces cerevisiae anamensis													DB	
Saccaromyces pastorianus	S													
Torulopsis holmii/Candida holmii/Saccaromyces exiguus	S						AB/S				DB		DB	
Torulaspora delbrueckii								AB						
Pichia norvegensis							AB							

REFERENCES

Aslam, Z., Im, W. T., Ten, L. N., Lee, M. J., Kim, K. H. & Lee, S. T. (2006). *Lactobacillus siliginis* sp. nov., isolated from wheat sourdough in South Korea. *International Journal of Systematic and Evolutionary Microbiology*, 56, 2209-2213.

Böcker, G., Stolz, P. & Hammes, W. P. (1995). Neue Erkenntnisse zum Ökosystem Sauerteig und zur Physiologie des sauerteig-typischen Stämme *Lactobacillus sanfrancisco* und *Lactobacillus pontis*. *Getreide Mehl und Brot*, 49, 370-374.

Cai, Y., Okada, H., Mori, H., Benno, Y. & Nakase, T. (1999). *Lactobacillus paralimentarius* sp. nov., isolated from sourdough. *International Journal of Systematic and Evolutionary Microbiology*, 49, 1451-1455.

Catzeddu, P., Mura, E., Parente E., Sanna, M. & Farris, G. A. (2006). Molecular characterization of lactic acid bacteria from sourdough breads produced in Sardinia (Italy) and multivariate statistical analyses of results. *Systematic and Applied Microbiology*, 29, 138-144.

Coppola, S., Pepe, O., Masi, P. & Sepe, M. (1996). Characterization of leavened doughs for Pizza in Naples. *Advances in Food Sciences*, 18, 160-162.

Coppola, S., Pepe, O. & Mauriello, G. (1998). Effect of leavening microflora on pizza dough properties. *Journal of Applied Microbiology*, 85, 891-897.

Corsetti, A. & Settanni, L. (2007). Lactobacilli in sourdough fermentation. *Food Research International*, 40, 539-558.

Corsetti, A., Lavermicocca, P., Morea, M., Baruzzi, F., Tosti, N. & Gobbetti, M. (2001). Phenotypic and molecular identification and clustering of lactic acid bacteria and yeasts from wheat (species *Triticum durum* and *Triticum aestivum*) sourdoughs of Southern Italy. *International Journal of Food Microbiology*, 64, 95-104.

Corsetti, A., Settanni, L., Van Sinderen, D., Felis, G. E., Dellaglio, F. & Gobbetti, M. (2005). *Lactobacillus rossii* sp. nov. isolated from wheat sourdough. *International Journal of Systematic and Evolutionary Microbiology*, 55, 35-40.

Corsetti, A., Settanni, L., Chaves López, C., Felis, G. E., Mastrangelo, M. & Suzzi, G. (2007a). A taxonomic survey of lactic acid bacteria isolated from wheat (*Triticum durum*) kernels and non-conventional flours. *Systematic and Applied Microbiology*, 30, 561-571.

Corsetti, A., Settanni, L., Valmorri, S., Mastrangelo, M. & Suzzi, G. (2007b). Identification of subdominant sourdough lactic acid bacteria and their evolution during laboratory-scale fermentations. *Food Microbiology*, 24, 592-600.

De Angelis, M., Di Cagno, R., Gallo, G., Curci, M., Siragusa, S., Crecchio, C., Parente, E. & Gobbetti, M. (2007). Molecular and functional characterization of *Lactobacillus sanfranciscensis* strains isolated from sourdoughs. *International Journal of Food Microbiology*, 114, 69-82.

De Vuyst, L. & Vancanneyt, M. (2007). Biodiversity and identification of sourdough lactic acid bacteria. *Food Microbiology*, 24, 120-127.

De Vuyst, L., Vrancken, G., Ravyts, F., Rimaux, T. & Weckx, S. (2009). Biodiversity, ecological determinants, and metabolic exploitation of sourdough microbiota. *Food Microbiology*, 26, 666-675.

Ehrmann, M. A., Müller, M. R. & Vogel, R. F. (2003). Molecular analysis of sourdough reveals *Lactobacillus mindensis* sp. nov. *International Journal of Systematic and Evolutionary Microbiology,* 53, 7-13.

Foschino, R., Terranno, R., Mora, D. & Galli, A. (1999). Microbial characterization of sourdoughs for sweet baked products. *Italian Journal of Food Science*, 11, 19-28.

Foschino, R., Gallina, S., Andrighetto, C., Rossetti, L. & Galli, A. (2004). Comparison of cultural methods for the identification and molecular investigation of yeasts from sourdoughs for Italian sweet baked products. *FEMS Yeast Research*, 4, 609-618.

Galli, A. & Ottogalli, G. (1973). Aspetti della microflora degli impasti per panettone. *Annals of Microbiology* 23, 39-47.

Galli, A., Franzetti, L. & Fortina, M. G. (1988). Isolation and identification of sour dough microflora. *Microbiologie-Aliments-Nutrition*, 6, 345-351.

Garofalo, C., Silvestri, G., Aquilanti L. & Clementi, F. (2008). PCR-DGGE analysis of lactic acid bacteria and yeast dynamics during the production processes of three varieties of Panettone. *Journal of Applied Microbiology*, 105, 243-254.

Gatto, V. & Torriani, S. (2004). Microbial population changes during sourdough fermentation monitored by DGGE analysis of 16S and 26S rRNA gene fragments. *Annals of Microbiology*, 54, 31-42.

Gobbetti, M. (1998). The sourdough microflora: interactions of lactic acid bacteria and yeasts. *Trends in Food Science and Technology*, 9, 267-274.

Gobbetti, M., Corsetti, A., Rossi, J., La Rosa, F. & De Vincenzi, S. (1994). Identification and clustering of lactic acid bacteria and yeasts from wheat sourdoughs of central Italy. *Italian Journal of Food Science*, 6, 85-94.

Gullo, M., Romano, A. D., Pulvirenti, A. & Giudici, P. (2003). *Candida humilis*-dominant species in sourdoughs for the production of durum wheat bran flour bread. *International Journal of Food Microbiology*, 80, 55-59.

Hammes, W. P., Brandt, M. J., Francis, K. L., Rosenheim, M., Seitter, F. H. & Vogelmann, S. (2005). Microbial ecology of cereal fermentations. *Trends in Food Science and Technology,* 16, 4-11.

Huy, Y. H., Corke, H., De Leyn, I., Nip, W. K. & Cross, N. A. (2006). Bakery products: Science and Technology, *Wiley-Blackwell* (Eds).

Iacumin, L., Cecchini, F., Manzano, M., Osualdini M., Boscolo D., Orlic S. & Comi, G. (2009). Description of the microflora of sourdoughs by culture-dependent and culture-independent methods. *Food Microbiology*, 26, 128-135.

Iorizzo, M., Coppola, R., Sorrentino, E. & Grazia, L. (1995). Microbiological characterization of sourdough from Molise. Caratterizzazione microbiologica di paste acide molisane. *Industrie Alimentari*, 34, 1290-1294.

Lombardi, A., Zilio, F., Andrighetto, C., Zampese, L. & Loddo, A. (2007). Microbiological characterization of sourdoughs of Veneto region. Caratterizzazione microbiologica di impasti acidi veneti. *Industrie alimentari*, 46, 147-151.

Meroth, C. B., Hammes, W. P. & Hertel, C. (2004). Characterisation of the microbiota of rice sourdoughs and description of *Lactobacillus spicheri* sp. nov. *Systematic and Applied Microbiology*, 27, 151-159.

Minervini, F., Rizzello, C. G. & Sanna, M. (2010). I pani tipici. In "Biotecnologia dei prodotti lievitati da forno". *Casa Editrice Ambrosiana* (Eds.), Chapter 14, 263-284.

Osimani, A., Zannini, E., Aquilanti, L., Mannazzu, I., Comitini, F. & Clementi, F. (2009). Lactic acid bacteria and yeasts from wheat sourdoughs of the Marche region. Batteri lattici e lieviti da madri acide di farina di grano tenero della regione Marche. *Italian Journal of Food Science*, 21, 269-286.

Ottogalli, G., Galli, A. & Foschino, R. (1996). Italian bakery products obtained with sour dough: characterization of the typical microflora. *Advances in Food Sciences*, 18, 131-144.

Picozzi, C., D'Anchise, F. & Foschino, R. (2006). PCR detection of *Lactobacillus sanfranciscensis* in sourdough and Panettone baked product. *European Food Research and Technology*, 222, 330-335.

Pulvirenti, A., Caggia, C., Restuccia, C., Gullo, M. & Giudici, P. (2001). DNA fingerprinting methods used for identification of yeasts isolated from Sicilian sourdoughs. *Annals of Microbiology*, 51, 107-120.

Pulvirenti, A., Solieri, L., Gullo, M., De Vero, L. & Giudici, P. (2004). Occurrence and dominance of yeast species in sourdough. *Letters in Applied Microbiology*, 38, 113-117.

Randazzo, C. L., Heilig, H., Restuccia, C., Giudici, P. & Caggia, C. (2005). Bacterial population in traditional sourdough evaluated by molecular methods. *Journal of Applied Microbiology*, 99, 251-258.

Reale, A., Tremonte, P., Succi, M., Sorrentino, E. & Coppola, R. (2005). Exploration of lactic acid bacteria ecosystem of sourdoughs from the Molise region. *Annals of Microbiology*, 55, 17-22.

Ricciardi, A., Paraggio, M., Salzano, G., Andreotti, G., De Fina, M. & Romano, P. (2002). Microflora of the sourdoughs used to produce two typical Southern Italy breads: Cornetto of Matera and Altamura bread. La microflora degli impasti utilizzati per la produzione del Cornetto di Matera e del Pane di Altamura. *Tecnica Molitoria*, 8, 780-787.

Ricciardi, A., Parente, E., Piraino, P., Paraggio, M. & Romano, P. (2005). Phenotypic characterization of the lactic acid bacteria from sourdoughs for Altamura bread produced in Apulia (Southern Italy). *International Journal of Food Microbiology*, 98, 63-72.

Robert, H., Gabriel, V. & Fontagné-Faucher, C. (2009). Biodiversity of lactic acid bacteria in French wheat sourdough as determined by molecular characterization using species-specific PCR. *International Journal of Food Microbiology*, 135, 53-59.

Romano, A. D., Gullo M. & Caggia, C. (2001). Le tecnologie di produzione del pane in Sicilia. *Tecnica Agricola*, 53, n.1-2 gennaio-giugno, 2001.

Rossi, J. (1996). The yeasts in sourdough. *Advances in Food Sciences*, 18, 201-211.

Scheirlinck, I., Van der Meulen, R., Van Schoor, A., Vancanneyt, M., De Vuyst, L., Vandamme, P. & Huys, G. (2007a). Influence of geographical origin and flour type on diversity of lactic acid bacteria in traditional belgian sourdoughs. *Applied and Environmental Microbiology*, 73, 6262-6269.

Scheirlinck, I., Van der Meulen, R., Van Schoor, A., Cleenwerck, I., Huys, G., Vandamme, P., De Vuyst, L. & Vancanneyt, M. (2007b). *Lactobacillus namurensis* sp. nov., isolated from a traditional Belgian sourdough. *International Journal of Systematic and Evolutionary Microbiology*, 57, 223-227.

Scheirlinck, I., Van der Meulen, R., Van Schoor, A., Vancanneyt, M., De Vuyst, L., Vandamme, P. & Huys, G. (2008). Taxonomic structure and stability of the bacterial community in belgian sourdough ecosystems as assessed by culture and population fingerprinting. *Applied and Environmental Microbiology*, 74, 2414-2423.

Scheirlinck, I., Van der Meulen, R., De Vuyst, L., Vandamme, P. & Huys, G. (2009). Molecular source tracking of predominant lactic acid bacteria in traditional Belgian sourdoughs and their production environments. *Journal of Applied Microbiology*, 106, 1081-1092.

Settanni, L., Van Sinderen, D., Rossi, J. & Corsetti, A. (2005). Rapid differentiation and in situ detection of 16 sourdough *Lactobacillus* species by Multiplex PCR. *Applied and Environmental Microbiology*, 71, 3049-3059.

Settanni, L., Valmorri, S., van Sinderen, D., Suzzi G., Paparella A. & Corsetti, A. (2006). Combination of Multiplex PCR and PCR-Denaturing Gradient Gel Electrophoresis for monitoring common sourdough-associated *Lactobacillus* species. *Applied and Environmental Microbiology*, 72, 3793-3796.

Siragusa, S., Di Cagno, R., Ercolini, D., Minervini, F., Gobbetti, M., & De Angelis, M. (2009). Taxonomic structure and monitoring of the dominant population of lactic acid bacteria during wheat flour sourdough Type I propagation using *Lactobacillus sanfranciscensis* starters. *Applied and Environmental Microbiology*, 75, 1099-1109.

Succi, M., Reale, A., Andrighetto, C., Lombardi, A., Sorrentino, E. & Coppola, R. (2003). Presence of yeasts in southern Italian sourdoughs from *Triticum aestivum* flour. *FEMS Microbiology Letters*, 225, 143-148.

Valcheva, R., Korakli, M., Onno, B., Prévost, H., Ivanova, I., Ehrmann, M. A., Dousset, X., Gänzle, M. G. & Vogel, R. F. (2005). *Lactobacillus hammesii* sp. nov., isolated from French sourdough. *International Journal of Systematic and Evolutionary Microbiology*, 55, 763-767.

Valcheva, R., Ferchichi, M. F., Korakli, M., Ivanova, I., Gänzle, M. G., Vogel, R. F., Prévost, H., Onno, B. & Dousset, X. (2006). *Lactobacillus nantensis* sp. nov., isolated from French wheat sourdough. *International Journal of Systematic and Evolutionary Microbiology*, 56, 587-591.

Valmorri, S., Settanni, L., Suzzi, G., Gardini, F., Vernocchi, P. & Corsetti, A. (2006). Application of a novel polyphasic approach to study the lactobacilli composition of sourdoughs from the Abruzzo region (central Italy). *Letters in Applied Microbiology*, 43, 343-349.

Valmorri, S., Tofalo, R., Settanni, L., Corsetti, A. & Suzzi, G. (2010). Yeast microbiota associated with spontaneous sourdough fermentations in the production of traditional wheat sourdough breads of the Abruzzo region (Italy). *Antonie van Leeuwenhoek*, 97, 119-129.

Vancanneyt, M., Neysens, P., Dewachter, M., Engelbeen, K., Snauwaert, C., Cleenwerck, I., Van der Meulen, R., Hoste, B., Tsakalidou, E., De Vuyst, L. & Swings, J. (2005). *Lactobacillus acidifarinae* sp. nov. and *Lactobacillus zymae* sp. nov. from wheat sourdough. *International Journal of Systematic and Evolutionary Microbiology*, 55, 615-620.

Venturi, M. Guerrini, S. & Vincenzini, M., (2009). Microbiology of two traditional sourdough production processes in Italy: Lagaccio biscuit and Panettone cake. *Annals of Microbiology*, 59, 82.

Vernocchi, P., Valmorri, S., Dalai, I., Torriani, S., Gianotti, A., Suzzi, G., Guerzoni, M. E., Mastrocola, D. & Gardini, F. (2004a). Characterization of the yeast population involved in the production of a typical italian bread. *Journal of Food Science*, 69, 182-186.

Vernocchi, P., Valmorri, S., Gatto, V., Torriani, S., Gianotti, A., Suzzi, G., Guerzoni, M. E. & Gardini, F. (2004b). A survey on yeast microbiota associated with an Italian traditional sweet-leavened baked good fermentation. *Food Research International*, 37, 469-476.

Vogel, R.F., Böcker, G., Stolz, P., Ehrmann, M., Fanta, D., Ludwig, W., Pot, B., Kersters, K., Schleifer, K.H. & Hammes, W.P. (1994). Identification of Lactobacilli from sourdough and description of *Lactobacillus pontis* sp. nov. *International Journal Systematic Bacteriology*, 44, 223-229.

Zannini, E., Garofalo, C. & Clementi, F. (2010). I prodotti dolciari lievitati italiani. In "Biotecnologia dei prodotti lievitati da forno". *Casa Editrice Ambrosiana* (Eds.), Chapter 13, 243-262.

Zotta, T., Ricciardi, A. & Parente, E. (2007). Enzymatic activities of lactic acid bacteria isolated from Cornetto di Matera sourdoughs. *International Journal of Food Microbiology*, 115, 165-172.

Zotta, T., Piraino, P., Parente, E., Salzano, G. & Ricciardi, A. (2008). Characterization of lactic acid bacteria isolated from sourdoughs for Cornetto, a traditional bread produced in Basilicata (Southern Italy). *World Journal of Microbiology and Biotechnology*, 24, 1785-1795.

In: Wheat: Genetics, Crops and Food Production
Editor: Maria T. Almeida, pp. 393-410
ISBN 978-1-61209-307-9

Chapter 15

PRIMARY BIOREFINERY OF WHEAT STRAW

Hongzhang Chen* andYanjun Li
State Key Laboratory of Biochemical Engineering, Institute of Process Engineering, Chinese Academy of Sciences, Beijing 100190, PR China

ABSTRACT

Agricultural residues, mainly cereal straws, have gradually been recognized as very important renewable biomass. Wheat straw is a relatively less heterogeneity material compared to maize and other crop straws, facilitates the subsequent harvesting, baling, transport, and the final fractionation and degradation. However, at present wheat straw hasn't been effectively utilized, which is otherwise an important and abundant biorefinery feedstock for production of bioenergy, biochemicals and biomaterials. The utilization of wheat straw has great advantages such as reducing dependence on fossil resources, alleviating food shortage caused by producing corn-based ethanol and avoiding air pollution from straw combustion. Therefore, to study and develop an array of integrated utilization methods of wheat straw is of significant importance. Our group has devoted more than two decades' work in primary biorefinery of wheat straw; formed several key platform techniques and preliminarily constructed an ecological industry chain system. Steam explosion pretreatment of wheat straw is an ideal method to cope with the complex and recalcitrant structure with the merits of high efficiency, simple operation and clean processing. After steam explosion, wheat straw can be processed to various products generally by three pathways viz. mechanical pathway, biochemical pathway and physicochemical pathway. The present chapter summarizes these achievements in detail.

1. INTRODUCTION

The petrochemical industry has tremendously boosted the world's development and promoted human beings' living standard. However, as the exhaustion of fossil resources such as coal, oil and natural gas, more and more attention has been paid to exploiting alternative

* Corresponding author: Chen Hongzhang. Tel: +86-10-82627067, Fax: +86-10-82627071. E-mail address: hzchen@home.ipe.ac.cn

renewable resources. Lignocellulose, accounting for c. 90% of the energy fixed by plant photosynthesis, is the main focus due to its abundance and low cost. Crop straws possess a considerable proportion of lignocellulosic feedstocks. At present, these abundant agricultural residues generally being burned or plowed back into the soil directly, not only causes severe air pollution or sowing problems, but also is a great waste of resources. This situation seems inapt, given the high fuel prices and the great demand for reducing greenhouse gas emission as well as air pollution[1].

A biorefinery was initially viewed as a facility that integrated biomass conversion processes and equipment to produce fuels, power, and chemicals from biomass by NREL[2]. Biomass utilization, especially for agricultural residues, the biorifinery strategy is perfectly required because of the material heterogeneity. The traditional single utilization technology not only wastes much of the biomass resource but also causes pollution to the environment from discarding non-targeted components. Therefore, Chen et al. (2009) brought out the concept of fractionated conversion which was defined as "a process in which preparation of biomass products depends on the compositions and characteristics of the raw materials"[3].This component separation and solitary utilization methods of cereal straws can not only make use of each component's value, but also facilitate other component processing; for example, the removal of ash in rice straw promoted cellulose enzymatic hydrolysis[4] and could obtain nano-silicon dioxide simultaneously[3]. Meanwhile, fractionated conversion of complicated biomass is inevitably associated with multiple products and process integration, thus "material-oriented, process-oriented and product-oriented" was stressed as a criterion for cereal straws processing[5]. In sum, by producing multiple products and integrating processes, biorefineries can maximize the values derived from cereal straws and turn its processing into real opportunities.

2. Resources and Characteristics of Wheat Straw

Wheat is the second largest crop grown in the world after rice and the third largest in China after rice and maize. According to the grain market report by International Grains Council on 23rd September, 2010, the world total grain production in 2010/11 is forecast at 1,741 million tons, and the world wheat production forecast is maintained at 644 million tons[6], among which China wheat production is estimated to exceed 115 million tons by the Ministry of Agriculture of China. Assuming residue/crop ratio of 1.3, there are about 850 million tons of wheat residues annually produced globally[7], and the wheat residues accounted for 25% of the world total agricultural residues[8]. This large quality of wheat straw is a great resource if properly exploided otherwise may cause environmental and social problems.

Wheat straw is a complex mixture of cellulose (35.1%), hemicellulose (27.1%) and lignin (5.3% Klason lignin), as three main components, and a small amount of soluble substrates (also known as extractives) and ash (6.0%)[9]. Although with the similar composition, wheat straw is more homogeneity than maize straw on the tissue and organ levels, which significantly facilitates its machinery collection procedures including harvesting, baling and transport. The somewhat mature baling techniques of miscanthus can be easily transferred to wheat straw collection. Furthermore, the compression molding of wheat straw needs less energy input due to the relatively less heterogeneity feature and the straw bales can reach high density therefore enhances the calorific value. The density of natural wheat straw is 50-120 kg/m^3 and more than 1000 kg/m^3 after compression molding, meanwhile the efficiency of heat utilization increases from 10%-30% to more than 40%[10].

However, the overall chemical composition of wheat straw could differ depending on wheat species, growing environments, even on different morphological fractions. Different chemical components of wheat straw, cellulose, hemicellulose and lignin, and different morphological fractions of wheat straw have different characteristics and functions[3]. Hence the fractionation conversion strategy of wheat straw is of great necessity. Except for the components difference, the recalcitrance of wheat straw is another characteristic to be concerned. The cellulose strains of wheat straw are bundled together and tightly packed in such a way that neither water nor enzyme can penetrate through the structure. Hemicellulose serves as a connection between lignin and cellulose fibers and it is readily hydrolyzed. Lignin is covalently linked to cellulose and xylan (predominant hemicellulose carbohydrate polymer in wheat straw) such that lignin-cellulose-xylan interactions exert a great influence on digestibility of wheat straw. Due to this structural complexity, efficient pretreatment of wheat straw is indispensable. A number of pretreatment methods have been developed and applied for wheat straw biomass, which are roughly categorized into physical, physicochemical, chemical and biological processes. Steam explosion which was invented by Mason[11] and initially applied to the pulping process, is a promising physicochemical pretreatment means of wheat straw. We developed the low-pressure and non-pollution (without any chemical additive) steam explosion technology based on the knowledge of chemical composition and structural differences between wood and straw, and carried on intensive study on structural change and fractionation mechanism of the components of cereal straws[12-14]. After the steam explosion pretreatment, a series of primary biorefinery schemes of wheat straw have been developed.

3. Overall Description of Wheat Straw Primary Biorefinery

The primary biorefinery of steam exploded wheat straw (SEWS) often deploys three pathways to accomplish, namely mechanical pathway, biochemical pathway and physicochemical pathway (Figure 1). With regarding to the mechanical pathway, there are two methods generally being used. The first method is water and ethanol extraction of SEWS, water extraction liquid contains soluble substances released from wheat straw by steam explosion such as xylooligosaccharides, furfural, acetic acid and formic acid which can be further purified, subsequently the lignin can be extracted with ethanol, while the solid fibers can be fermented to a variety of products. The second one is airstream/waterstream fractionation, separating SEWS into long fibers and short fibers, which suited for pulping and fermentation, respectively. Within the biochemical pathway, novel solid state fermentation (SSF) technologies, including the gas double-dynamic solid-state fermentation (GDSFF) and solid-state fermentation on inert support (IS-SSF), are highlighted because of the good performance and wide application.

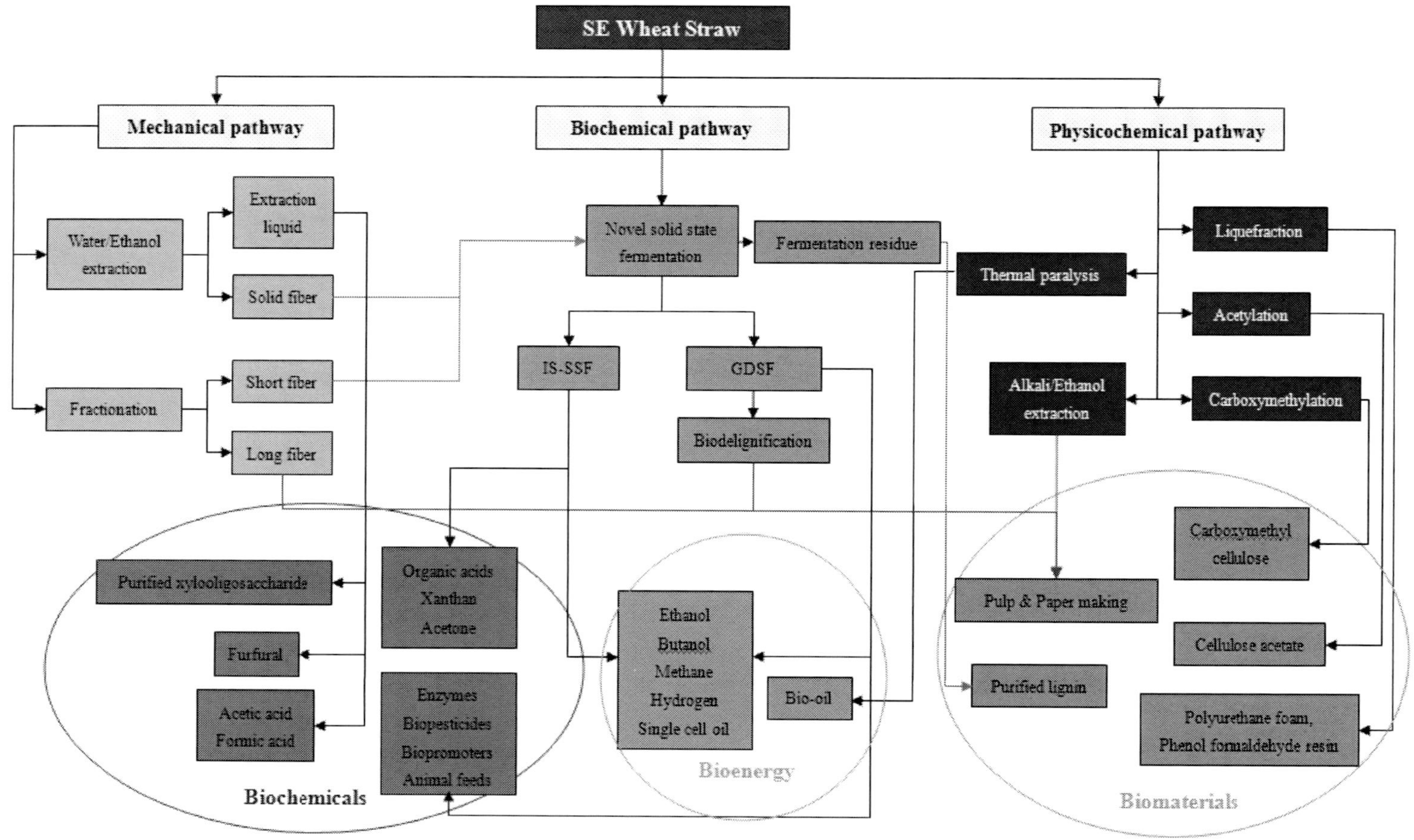

Figure 1. Overall schematic diagram of ecological industry chain of wheat straw primary biorefinery.

GDSFF is especially applicable for filamentous fungi to ferment SE wheat straw to enzymes, biopesticides, animal feeds, etc. IS-SSF technology can be extensively adopted to produce all kinds of products, for example, organic acids, xanthan, ethanol, butanol, hydrogen and single cell oil. In addition, a number of indispensable physicochemical operation units are commonly carried out in purifying and converting some constituents of wheat straw into value-added biomaterials. For instance, through acetylation and carboxymethylation reactions, SEWS can be modified to cellulose acetate and carboxymethyl cellulose, respectively; after liquefraction, SEWS can be converted to polyurethane foam and phenol formaldehyde resin.

4. PROCESS ELABORATION OF SEVERAL BIOREFINERY SECTIONS OF WHEAT STRAW

4.1. Unpolluted Fractionation of SEWS by Water and Ethanol Extraction—Mechanical Pathway I

A flow chart of the fractionation procedure was shown in Figure 2. The chipped wheat straw was initially steam exploded for 4.5 min with moisture of 34.01%, a pressure of 1.5 MPa without acid or alkali, and subsequently water washed to fractionate the hemicellulose which could be fermented by xylose-utilizing microorganisms. The remains were ethanol extracted to separate the cellulose and lignin, suitable for pulp and production of biomaterials, respectively [9].

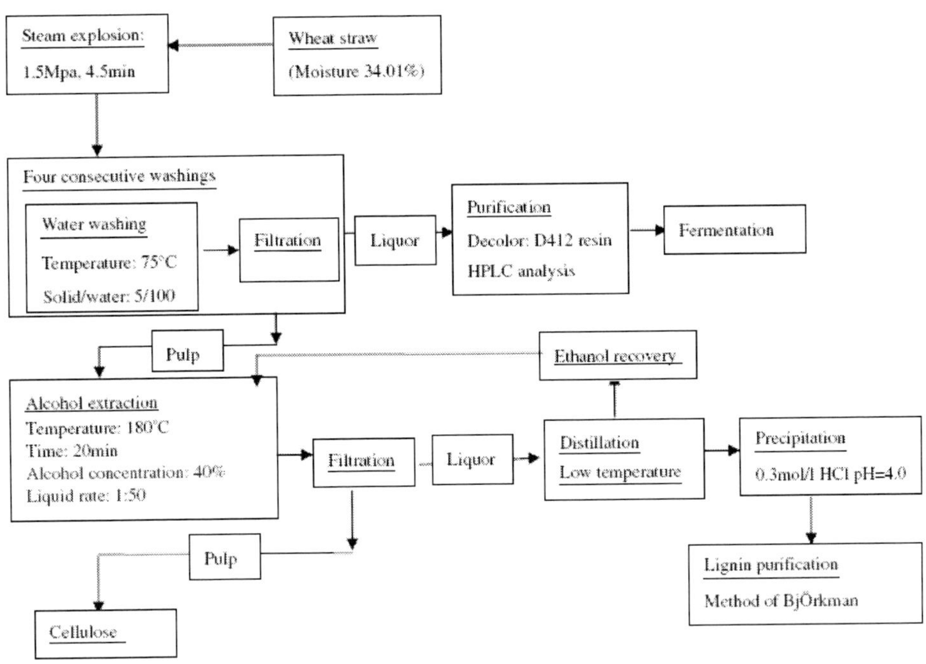

Figure 2. Flow diagram of the fractionation process and operational conditions used in each stage [9].

Hemicellulose sugars were recovered by water countercurrent extraction. The optimum extraction temperature and solid/liquid ratio were 75°C and 5/100, respectively, and the sugars concentration was 30.12 g/L and the recovery rate of hemicellulose sugars was 80%. Due to the soluble lignin and pigment in the hemicellulose extract, the washing solution showed a filemot color. Therefore, the extract was dicolored with chelating ion exchange resin D412. The results showed that D412 resin had an optimal decoloration effect. The HPLC and gas chromatography (GC) results of the discolored liquor showed that xylose (85.91%) was the main component. The ratio of monosaccharides to oligosaccharides was 1:9, which indicated that the sugar of steam exploded hemicellulose mostly existed in the soluble oligosaccharides form.

Water washed materials were subsequently extracted with ethanol. The optimum extraction conditions in this study were 40% ethanol, fiber/liquor ratio 1:50 (w/v), severity log(R) = 3.657 (180°C for 20 min), 0.1% NaOH. The lignin was subsequently purified using Björkman method. Under these processes the lignin recovery was approximately 75%, and 85% ethanol solvent was recovered. Results of lignin purification indicated that most of the residual sugars, pigments and degradation products of lignin were removed by the purification. Infrared spectrometry (IR) results showed the fine lignin belonged to GSH (guaiacyl (G) syringyl (S) and p-hydroxyphenyl (H)) lignin and its purity reached 85.3%. The cellulose recovery rate after ethanol extraction was 94%.

In conclusion, processing of wheat straw by steam explosion pretreatment/ethanol delignification allowed the fractionation of the three main polymers present in the lignocellulosic matrix. Results of this study demonstrated that steam explosion pretreatment combined with ethanol delignification was an effective process that allowed the fractionation of lignocellulosic materials into polymeric fractions with reasonable yields and purity.

4.2. Waterstream Fractionation of SEWS—Mechanical Pathway

Figure 3 shows the procedures of waterstream fractionation of SEWS[15]. At first, compression separation method was used to obtain the solid fibers, which then were separated to long fibers and miscellaneous cells by a sieving machine (Figure 4). For a batch operation, approximately 100 g SEWS was fed into this apparatus, the tank was filled with water and then a constant water speed of 2 L $\cdot$min^{-1} was kept, meanwhile, the stirring speed of agitator was kept to 500 r/m to mix the wheat straw for 30 min. After that, the treated materials were vacuum-filtrated to gain long fibers and miscellaneous cells. The long fibers could be used to produce pulp, while the miscellaneous cells were more readily to be enzymolyzed to soluble sugars which could be fermented to numerous biochemicals through IS-SSF technology.

Through waterstream fractionation as described above, three components, fibers (47.6%), miscellaneous cells (19.9%) and soluble constituents (32.5%), were derived (Figure 5). The soluble substances were probably short-chain xylooligosaccharides from hemicellulose hydrolysis. From SEM images, it was found that the fiber cells were intact with smooth surface, while miscellaneous cells were irregular and disordered with rough surface. Table 1 shows the different chemical compositions of fibers and miscellaneous cells derived from SEWS fractionation.

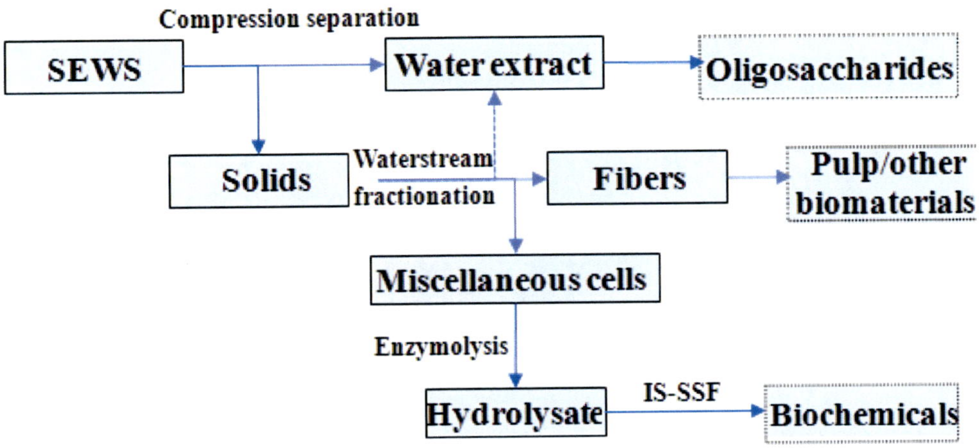

Figure 3. Flow diagram of waterstream fractionation of SEWS and potential usage of each component[15].

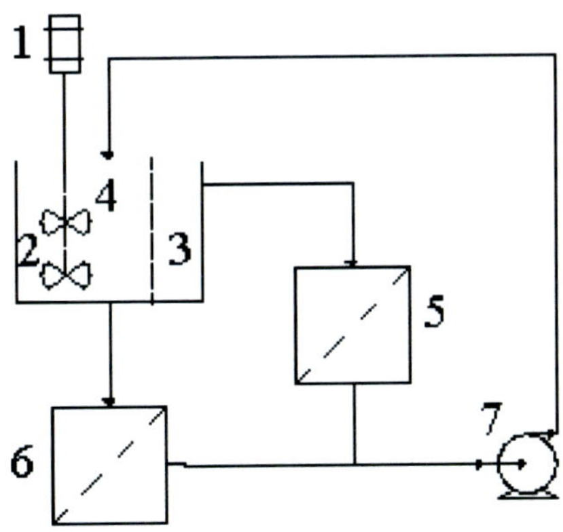

Figure 4. Schematic diagram of waterstream fractionation apparatus. 1. Electromotor; 2. Agitator, 3. Sieve, 4. Tank, 5. Filter for fiber, 6. Filter for overflow water, 7. Pump[15].

It could be seen the miscellaneous cells contained less cellulose (36.23%), relatively more lignin and ash. On the contrary, the fibers contained more cellulose (57.16) and less lignin and ash, 16.1% and 4.62%, respectively, and this chemical composition pattern was almost identical to the wood. In addition, the length, width and ratio of length to width of fibers were determined by a digital image scanning and plotting system and the results showed the average values were 1.067 mm, 13.893 μm and 76.81, respectively. All these data demonstrated the fibers derived from SEWS waterstream fractionation met the requirements for pulping.

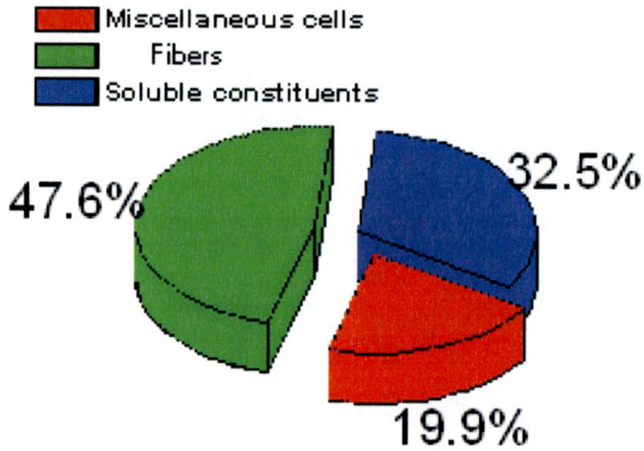

Figure 5. Proportion of three components derived from SEWS fractionation[15].

Table 1. Chemical composition of fibers, miscellaneous cells and SEWS [15]

	NDS (%)	Hemicellulose (%)	Cellulose (%)	Lignin (%)	Ash (%)
Fibers	8.66	13.36	57.16	16.21	4.62
Miscellaneous cells	1.90	18.65	36.32	27.64	15.49
SEWS	35.53	9.45	36.41	11.86	6.79

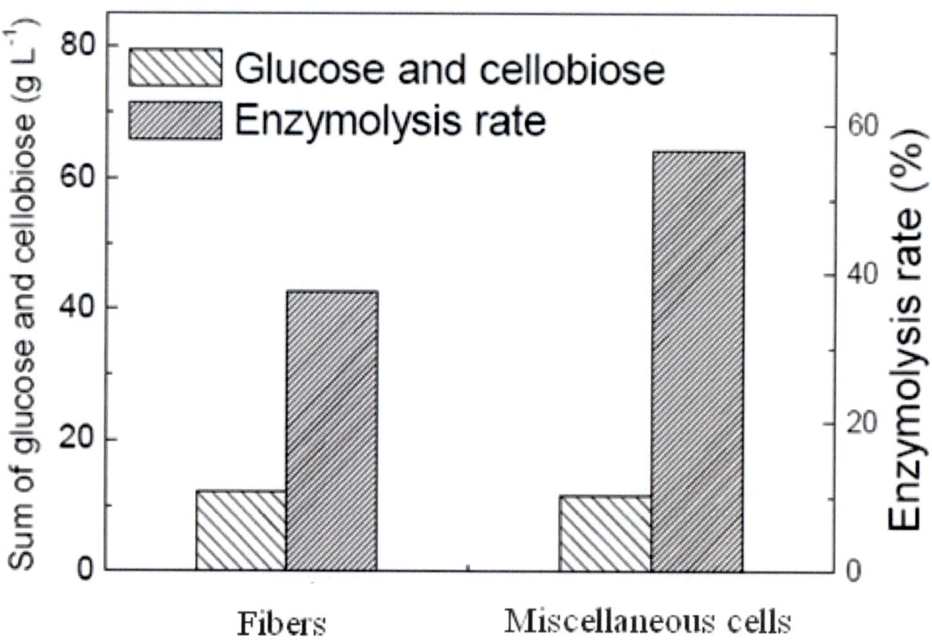

Figure 6. Comparison of enzymolysis rate of fibers and miscellaneous cells[15].

It is an important section of biorefinery to enzymatically hydrolyze wheat straw to fermentable sugars. The enzymatic hydrolysis characteristics of fibers and miscellaneous cells were detected. The results of 36 h enzymolysis with a cellulase dosage of 40 FPA/ g substrate were showed in Figure 6. The sum of glucose and cellobiose (11.66 g/L) derived from miscellaneous cells hydrolysis was slightly lower than that of 12.18 g/L from fibers. However, considering the cellulose content of miscellaneous cells was distinctly lower than that of fibers (Table 1), the actual enzymolysis rate of miscellaneous cells was significantly higher than that of fibers, 64.20% and 42.61%, respectively (Figure 6). Therefore, the miscellaneous cells were more readily to be degraded into fermentable sugars.

In conclusion, the fractionation of wheat straw by steam explosion and waterstream screening was of great necessity. Two components of the fractionation products from SEWS, fibers and miscellaneous cells, could be used for production of value-added biomaterials and for biotransfer to various biochemicals with less enzyme loading, respectively.

4.3. Cellulase Fermentation Using SEWS as Substrate by Gas Double-dynamic Solid-State Fermentation

The cellulase fermentations using SEWS by *Trichoderma reesei* YG3, *Penicillium decumbens* JUA10 in the gas double-dynamic solid-state fermentation bioreactor GDSFB[16,17]were studied. At the same time, forced aeration solid-state fermentation (FASSF) and periodic air replacement solid-state fermentation (PGRSSF), respectively simulating the functions of internal air circulation and periodic air influence without pulsation, were conducted as the control of GDSFF[18]. At the end of fermentation by *P. decumbens* JUA10, the cellulase activity (FPA) and biomass (designated as aminoglucose content) were detected as shown in Figure 7.

The highest cellulase activity and biomass both confirmed the superior effect of GDSFF, during which the heat and mass transfers were enhanced to a great extent. However, the improvements of cellulase activity and biomass were not strictly positive correlated, indicating the high cellulase activity was not only due to the high biomass, but also the better enzyme secretion ability per unit biomass.

In order to further understand the effects of GDSSF on heat and mass transfers, different layers fermentation parameters (biomass and cellulase activity) by inoculating of *T. reesei* YG3 to SEWS substrate under GDSSF were determined in comparison of those under FASSF. As shown in Figure 8, there were obvious gradients of biomass and enzyme activity on different layers, especially for biomass. The gradient differences under GDSSF were significantly less than those under FASSF, for example, the relative ratio of 3rd day biomass on upper, middle and lower layers was 1:0.87:0.67 under FASSF, compared to that of 1:0.96:0.90 under GDSSF. In a word, GDSSF effectively promoted heat and mass transfers inside the fermentor, thereby decreased the gradients of biomass and enzyme activity on different layers of SEWS substrates, and consequently exhibited good performance on cellulase solid-state fermentation.

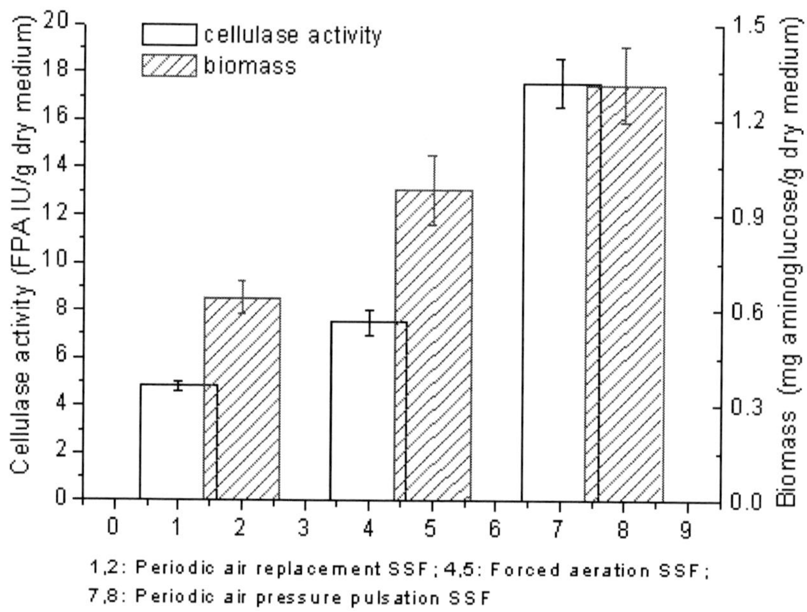

Figure 7. Comparision of cellulase activity and biomass under three different SSFs[18].

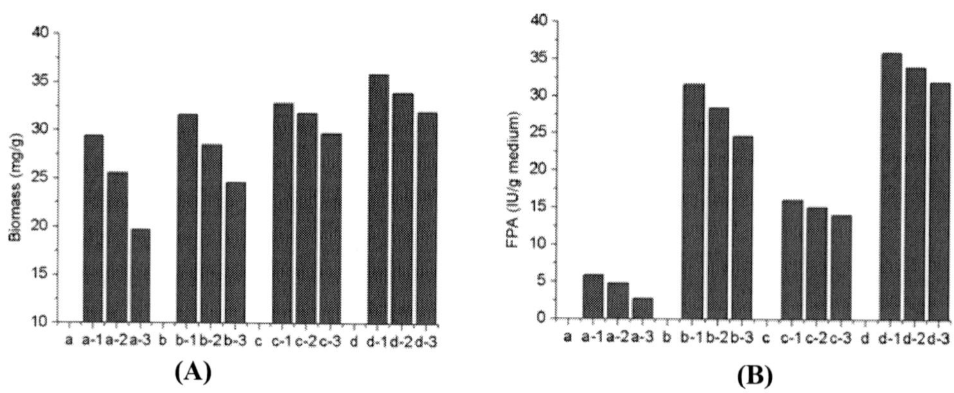

Figure 8.Biomass and cellulase activity differences on different medium layers[18]. a: The 3rd day of forced aeration SSF, b: The 7th day of forced aeration SSF, c: The 3rd day of gas double-dynamic SSF, d: The 5th day of gas double-dynamic SSF; 1. Upper level, 2. Middle level, 3. Lower level.

On the basis of achievements on GDSSF mentioned above, intensive studies about wheat straw substrate processing were conducted, such as steam explosion conditions, removal of fermentation inhibitors by water-washing or drying, additions of different ammonium salts to the medium. On the optimum conditions, the maximum cellulase activity (FPA) and medium weight loss reached 80 IU/g medium and 30%, respectively, after 120 fermentation using *T. reesei* YG3 (Figure 9). An industrial level of cellulase production employing a 100 m^3 GDSSB was accomplished by Zhesheng Bioengineering Technology Co.,Ltd, Dongping, Shandong Province. The detailed industrial production procedure was presented in Figure 10 and the average enzyme activity fluctuated within a range of 60-80 IU/g dry medium, comparable to the laboratory scale.

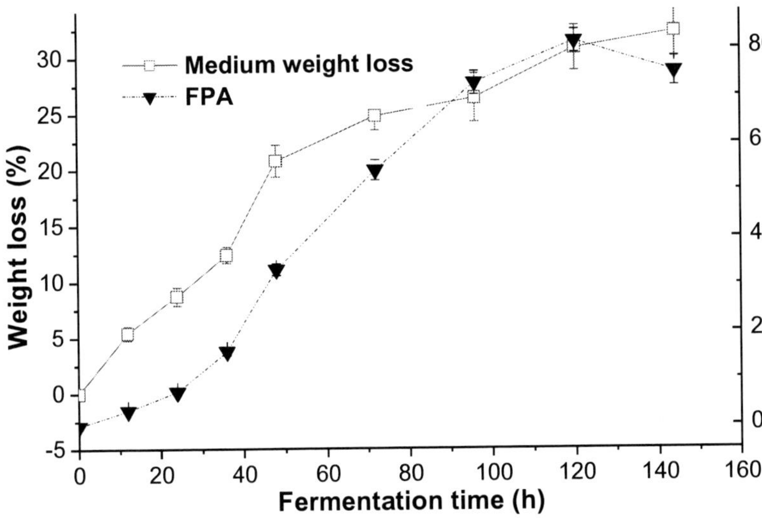

Figure 9. Cellulase production and medium weight loss curves for *T. reesei* YG3[18].

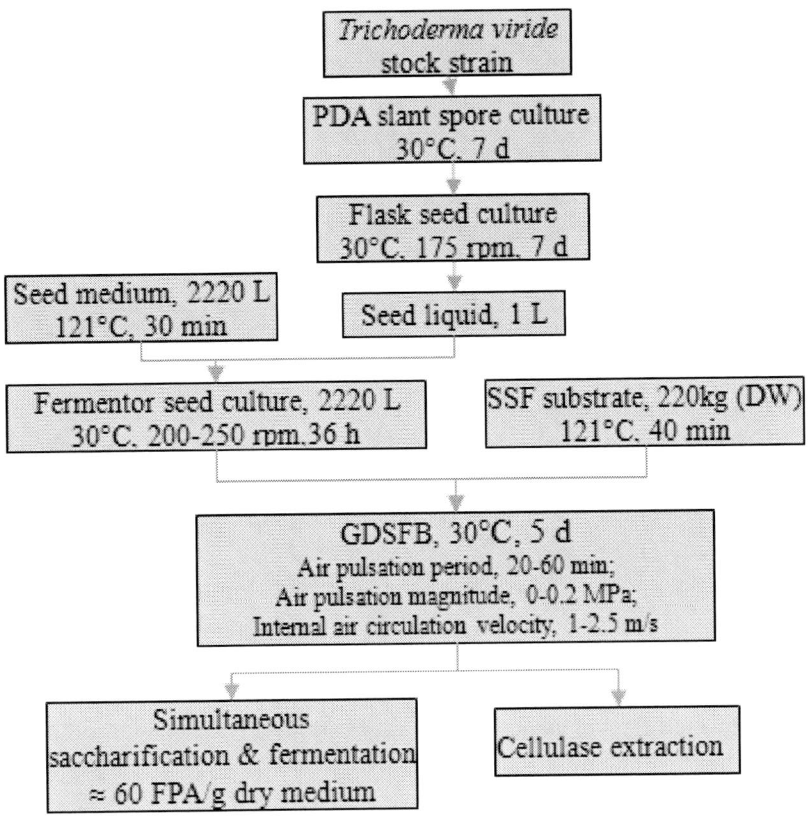

Figure 10. Flow chart of cellulase industrial production [18].

4.4. Xanthan Production with SEWS Hydrolysate on Inert Support

Xanthan gum is an extracellular heteropolysaccharide produced by *Xanthomonas campestris*. Because of its unique rheological properties, such as high viscosity and pseudoplasticity, xanthan has wide applications in industries such as food, pharmaceuticals, and oil and in other industries[19]. There are two characteristics with xanthan fermentation, one is lower sugar concentration required because fermentation will be blocked as the gum concentration is high due to its high viscosity, and the other is both glucose and xylose can be utilized by *X. campestris*. Therefore, wheat straw hydrolysate may be an ideal medium for xanthan production. Furthermore, SSF, featured as no free water needed, could be a better technology satisfying the high viscosity property of xanthan compared to SmF. Hence in combination with these considerations, a new IS-SSF technology, SEWS hydrolysate supplementing onto polyurethane foam (PUF) was developed by Zhang[15] to produce xanthan by *X. campestris*.

SEWS hydrolysate was produced by the enzyme hydrolysis method as follows: (1) Adding SEWS to tap water (pH adjusted to 4.8) with a solid-to-liquid ratio of 1:7, (2) Adding cellulase to the mixture with a dosage of 20 IU FPA/g dry SEWS, (3) Incubating in a shaking water bath at 50 °C for 48 h, (4) Centrifugation, and the supernatant was SEWS hydrolysate. The concentrations of glucose and xylose in the hydrolysate could get to 22 g/L and 12 g/L, respectively.

PUF was a porous material with the approximate pore diameter of 700 μm, and the specific surface area was calculated as around 4.5×10^{3} m^{-1} taken the pores inside as sphere. The medium was in the form of liquid films covered on the pore walls inside the PUF, and the thickness of these liquid films were estimated as 0.1 mm if the solid-to-liquid ratio was 1:10. The vast specific surface area and extreme thin liquid film would facilitate the transfers of O_2 and CO_2 between liquid-film phase and the continuous gas phase above it thus could benefit the performance of IS-SSF.

The fermentation results were shown in Figure 11, as the consumption of glucose and xylose, the biomass and xathan were increased until 60 h the xathan production reached maximum of 20.7 g/L. The initial glucose and xylose concentrations in medium were 20.4 g/L and 11.6 g/L, respectively, and at the end of fermentation (72 h) the carbohydrate (glucose and xylose) transformation rate reached 62%. Another noteworthy phenomenon was that *X. campestris* consumed glucose and xylose simultaneously (Figure 11) and there was almost no "glucose effect" which was regular in other microorganism fermentation circumstances.

Xanthan is a heteropolysaccharide with a very high molecular weight, consisting of repeating units. The sugars present in xanthan are D-glucose, D-mannose, and D-glucuronic acid. The glucoses are linked to form a β-1,4-D-glucan cellulosic backbone, and alternate glucoses have a short branch consisting of a glucuronic acid sandwiched between two mannose units. The terminal mannose moiety may carry pyruvate residues linked to the 4-and 6-positions.The internal mannose unit is acetylated at C-6. Acetyl and pyruvate substituents are linked in variable amounts to the side chains. These different contents of pyruvate and acetate account for the different characteristics of xathan. Although the average molecular weights of xathan fermented from SEWS hydrolysate and glucose were nearly equal, the pyruvate and acetate contents derived from SEWS hydrolysate were much higher than those from pure glucose (Table 2), and this might endow the xathan from SEWS hydrolysate special quality and applications.

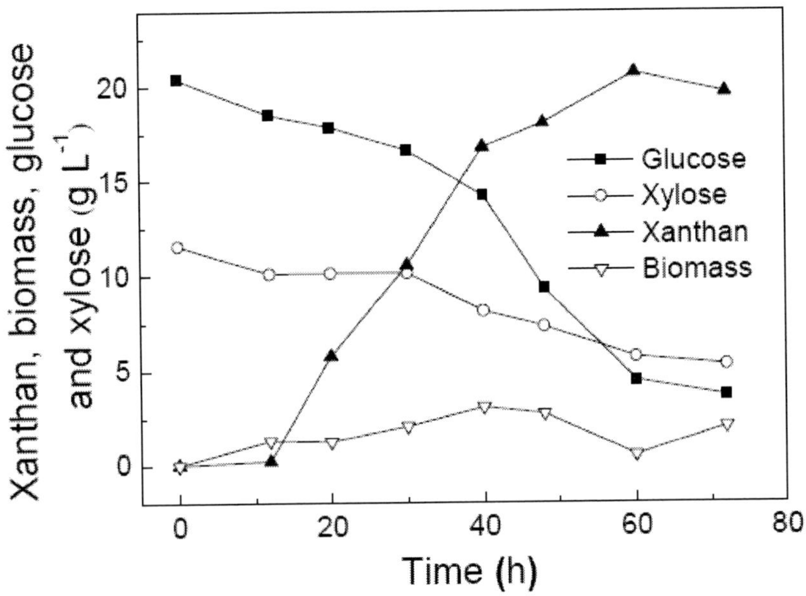

Figure 11. Biomass, xanthan, glucose and xylose curve in the fermentation with SEWS hydrolysate medium[15].

Table.2. Comparison of characteristics of xathan derived from SEWS hydrolysate and glucose [15]

	Xathan from SEWS hydrolysate	Xathan from glucose
Pyruvate content (%)	3.76	2.43
Acetate content (%)	8.46	2.55
Average molecular weight (kD)	5.63×10^5	5.98×10^5

4.5. Biomaterials Production from Wheat Straw by Physicochemical Pathways

4.5.1. Preparation of Polyurethane Foam and Phenol Formaldehyde Resin by Liquefying SEWS

The liquefaction method of SEWS for preparation of polyurethane foam was established by Wang [20] as described below. At first, mix SEWS and liquefaction reagents (mainly polyethylene glycol (PEG) and glycerol with organic acid as catalyst) evenly, subsequently pour the mixture into a 250 ml three-neck flask, then heat the flask in a oil bath apparatus equipped with continuous stirrer and reflux condenser for a certain reaction period, finally transfer the flask to water bath at room temperature.

Table 3. Formula of polyurethane foam preparation[20]

Reagent composition	Liquefied SEWS as substrate (W %)	Commercial polyol as substrate (W %)
Polyol	100	100
MDI	69.8	45.8
Water	4	4
Triethylene diamine (TEDA)	0.125	0.125
Stannous caprylate	0.125	0.125
Tween-80	2	2

According to the formula in Table 3, mix the Liquefied SEWS and other chemical reagents evenly, next add appropriate amount of diphenylmethane 4,4'-diiso-cyanate (MDI), then shake and pour into a mould of 20cm×20cm×20cm, and let it react until solid foam is formed. Figure 12 shows the polyurethane foam prepared by this method.

Figure 12. PUF based on liquefied SEWS[20].

From table 4, it could be found the characteristics of PUF based on liquefied SEWS were similar to those of PUF based on PEG, and water absorptivity of former was more superior than that of later. Due to these characteristics, PUF based on liquefied SEWS could be a promising inert support for SSF and its applicability was also verified by Wang[17] using the PUF as inert support to produce clavulanic acid (CA) and alkaline protease by solid state fermentation. It was shown that PUF not only could be used in solid state fermentation, but also made the separation of fermentation products easier.

Table 4. Characteristics of PUF based on prepared liquefied SEWS polyol and PEG[20]

	Liquefied SEWS as substrate	PEG as substrate
Apparent density (kg/m^3)	35	30~60
Tensile strength (kPa)	80	70~85
Tensile elongation (%)	160	180～300
Resilient rate (%)	30	30~35
Water absorptivity (%)	6	0.2

The liquefaction method of SEWS for preparation of phenol formaldehyde resin is similar to that mentioned before except the reagent is phenol. The liquefied SEWS production was reacted with formaldehyde in boiling water bath to synthesize phenol formaldehyde resin (Figure 13-a), and Figure 13-b shows the phenol formaldehyde resin prepared with phenol and formaldehyde. The comparison of characteristics of these two phenol formaldehyde resin were also conducted and revealed very similar properties, but the color of resin from liquefied SEWS was dark brown due to the complex chemical substances containing chromophores in SEWS.

(A) **(B)**

Figure 13. Prepared phenol formaldehyde resin[20].

a, from phenol and formaldehyde; b, from liquefaction products and formaldehyde.

4.5.2. Preparation of Carboxymethyl Cellulose from SEWS

SEWS was firstly etherified by sodium chloroacetate then alkalized by sodium hydroxide, and finally transformed to carboxymethylated product, carboxymethyl cellulose (CMC) by Yang[21]. SEM images of CMC prepared from α-cellulose and SEWS were shown in Figure 14 and the two different sourced CMC presented identical morphological features at the micron level.

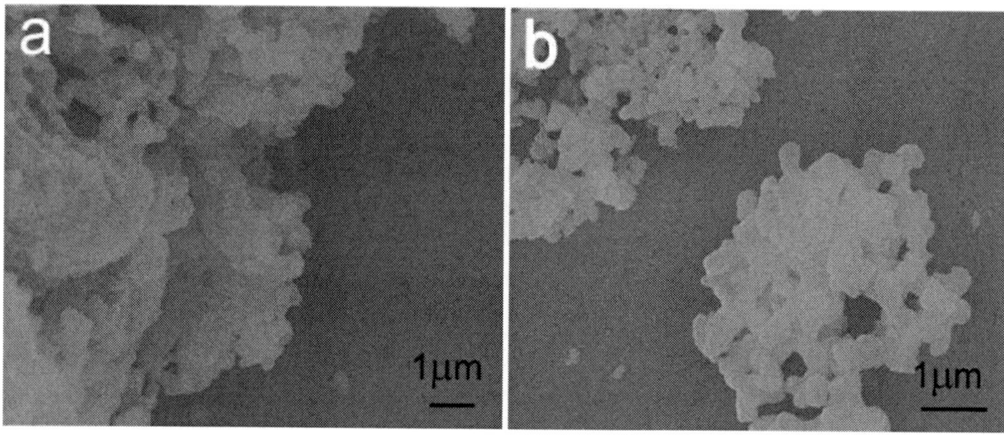

Figure 14. SEM images of CMC a, CMC prepared from α-cellulose; b, CMC prepared from SEWS[21].

4.5.3. Preparation of Cellulose Acetate from SEWS

SEWS was used as raw material for acetylation, then cellulose acetate was prepared by solvent extraction according to different solubilities of acetylated products. Compared with using α-cellulose as raw material , not only the cost was very low , but also the process was simplified greatly[22]. Figure 15 shows the cellulose acetate film prepared from SEWS as raw material.

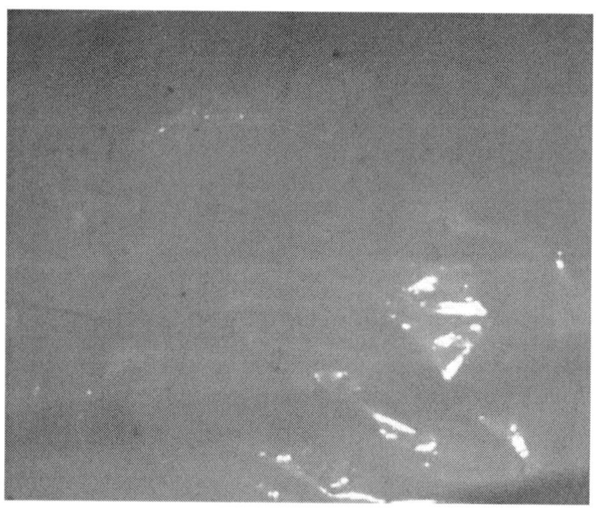

Figure 15. Cellulose acetate film prepared from SEWS[22].

In conclusion, the potential of biomaterials production from wheat straw was demonstrated by these studies mentioned above. Steam explosion seems to be an indispensable pretreatment method, which facilitates the subsequent physicochemical reactions and lowers the costs of biomaterials preparation to a large extent. The value-added biomaterials are easily degradable and production of biomaterials will account for a large proportion of wheat straw clean biorefinery.

CONCLUSION AND PERSPECTIVES

Wheat straw, accounted for 25% of the world total agricultural residues, is a relatively less heterogeneity biorefinery feedstock for production of bioenergy, biochemicals and biomaterials. After low-pressure and non-pollution steam explosion pretreatment, wheat straw can be processed to various products generally by three pathways, namely mechanical pathway, biochemical pathway and physicochemical pathway. Wheat straw is not only consisted of cellulose, hemicellulose which can be hydrolyzed to fermentable sugars, but also contains extraordinary complicated lignin, fatty substances such as cutin, wax and suberin, and relatively high content of ash. The lignin hydrolytes are a lot of phenolic compounds, which are fermentation inhibitors to some extent, but also can potentially be exploited as antimicrobial agents or medicines. The cutin, wax and suberin in the cuticle of wheat straw hampering enzymolysis or solid-state fermentation are also important resources. The ash gradients also can be separated as useful metal materials otherwise influence wheat straw combustion or hydrolysis. In conclusion, all these useful elements should play their distinct functions in the future biorefinery of wheat straw. Furthermore, on one hand as the rapid development of plant biology, the plant cell synthesis mechanism and more and more other chemical gradients synthesis and metabolic pathways are progressively elucidated. On the other hand, genetic engineering of plants is also booming. Therefore, genetic modified plants particularly for energy or material use could probably be designed, and the corresponding new biorefinery strategies are foreseeable.

ACKNOWLEDGMENT

We are grateful for the financial support by Important National Basic Research Program of China (973 Program, 2011CB707400), National key technology R&D program (2007BAD39B01) and Knowledge Innovation Program of Chinese Academy of Sciences (KSCX1-YW-11A1; KGCX2-YW-328), also grateful to those authors who provide the literature materials cited in this chapter.

REFERENCES

[1] Kim, S. and Dale, B.E. (2004). Global potential bioethanol production from wasted crops and crop residues. *Biomass Bioenergy, 26*, 361-375.

[2] NREL Biomass Research: *http://www.nrel.gov/biomass/biorefinerty.html.*

[3] Chen, H.Z., Yang, Y. and Zhang, J.X. (2009). In Singh nee' Nigam, P. and Pandey A., *Biotechnology for Agro-Industrial Residues Utilization* (pp. 327-340). Berlin: Springer.

[4] Yu, B. and Chen, H.Z. (2010). Effect of the ash on enzymatic hydrolysis of steam-exploded rice straw. *Bioresource Technology, 101*, 9114-9119.

[5] Chen, H.Z. (2009). *Process Engineering in Plant-Based Products:* Beijing: Chemical Industry Press and New York: Nova Science Publishers.

[6] International Grains Council report: http://www.igc.int/downloads/gmrsummary/gmrsumme.pdf.

[7] Talebnia, F., Karakashev, D. and Angelidaki, I. (2010). Production of bioethanol from wheat straw: An overview on pretreatment, hydrolysis and fermentation. *Bioresource Technology*, *101*, 4744-4753.

[8] Shi, Z.P. and Hua, Z.Z. (2007). *Biomass handbook*: Beijing: Chemical Industry Press.

[9] Chen, H.Z. and Liu, L.Y. (2007). Unpolluted fractionation of wheat straw by steam explosion and ethanol extraction. *Bioresource Technology*, *98*, 666-676.

[10] Li, Q.L. (2008). *New Energy and Renewable Energy*: Nanjing: Southeast University Press.

[11] Mason, W.H. (1929). *Apparatus for and process of explosion fibration of lignocellulosic material.* US Patent, No. 1655618.

[12] Chen, H.Z. and Liu, L.Y. (2007). *Technology of Steam Explosion: Principles and Applications:* Beijing: Chemical Industry Press.

[13] Chen, H.Z. (2008). *Biomass Science and Engineering:* Beijing: Chemical Industry Press.

[14] Chen, H.Z. and Qiu, W.H. (2010). Key technologies for bioethanol production from lignocellulose. *Biotechnology Advances. 28*, 556-562.

[15] Zhang, Z.G. (2010). *Straw Fractionation by Steam Explosion and Xathan Production.* Doctoral Dissertation.

[16] Chen, H.Z. and Li, Z.H. (2007) US Patent, No. 7183074 B.

[17] Chen, H.Z. and Li, H.Q. (2008). The Principle of Artificial Periodic Stimulation: A Novel Idea for Bioreactor Design. In: Columbus, F. *Biotechnology: Research, Technology and Applications*: New York: Nova Science Publishers.

[18] Li, H.Q. (2008). *Study on Cellulase Solid-state Fermentation with Gas Periodic Stimulation Intensification.* Doctoral Dissertation.

[19] Zhang, Z.G. and Chen, H.Z. (2010). Xanthan Production on Polyurethane Foam and Its Enhancement by Air Pressure Pulsation. *Applied Biochemistry and Biotechnology*, Online First.

[20] Wang, H. (2006). *Preparation of Adsorptive Material Based on Liquefying Steam-explosed Wheat Straw and Its Application in Solid State Fermentation.* Doctoral Dissertation.

[21] Yang, Y. (2009). *Preparation of Carboxymethyl Cellulose from Steam Exploded Straws and Its Fractionation.* Master Dissertation.

[22] Zhang, J.X. (2007). Acetylation of *Steam Exploded Straws and Its Fractionation.* Master Dissertation.

INDEX

B

C

D

E

F

G

H

N

O

P

Q

R

S

T

U

V